Lecture Notes in Computer Science 658

Edited by G. Goos and J. Hartmanis

Advisory Board: W. Brauer   D. Gries   J. Stoer

W0055469

R. A. Rueppel (Ed.)

Advances in Cryptology – EUROCRYPT '92

Workshop on the Theory and Application
of Cryptographic Techniques
Balatonfüred, Hungary, May 24-28, 1992
Proceedings

Springer-Verlag Berlin Heidelberg GmbH

Series Editors

Gerhard Goos
Universität Karlsruhe
Postfach 69 80
Vincenz-Priessnitz-Straße 1
W-7500 Karlsruhe, FRG

Juris Hartmanis
Cornell University
Department of Computer Science
4130 Upson Hall
Ithaca, NY 14853, USA

Volume Editor

Rainer A. Rueppel
R3 Security Engineering
Bahnhofstr. 242, CH-8623 Wetzikon, Switzerland

CR Subject Classification (1991): E.3-4, D.4.6, G.2.1

ISBN 978-3-540-56413-3   ISBN 978-3-540-47555-2 (eBook)
DOI 10.1007/978-3-540-47555-2

Originally published by Springer-Verlag Berlin Heidelberg New York in 1993.

Typesetting: Camera ready by author
45/3140-543210 - Printed on acid-free paper

Preface

A series of open workshops devoted to modern cryptology began in Santa Barbara, California in 1981 and was followed in 1982 by a European counterpart in Burg Feuerstein, Germany. The series has been maintained with summer meetings in Santa Barbara and spring meetings somewhere in Europe. At the 1983 meeting in Santa Barbara the International Association for Cryptologic Research was launched and it now sponsors all the meetings of the series.

Eurocrypt '92 in Hungary was a special meeting in many ways. For the first time, it was held in an Eastern European country. Our charming Hungarian hosts turned the conference into an unforgettable experience for all of us. Also for the first time, the General Chair and the Program Chair were based in different countries. The Program Committee was selected very internationally, which implied that joint meetings were impossible in the course of setting the program. It was encouraging to see how swiftly disputes could be resolved by electronic mail. To ease its burden, the official Program Committee of Eurocrypt '92 obtained help from many renowned researchers and scientists. Here is the final list of all those people (that I know of) who helped during the refereeing phase.

Brandt, Brickell, Charpin, Crépeau, Csirmaz, Damgård, Denes, Desmedt, Feigenbaum, Fell, Fujioka, Girault, Golic, Helleseth, Itoh, Joux, Kenyon, Koyama, Kurosawa, Landrock, Matsui, Matsumoto, McCurley, Merritt, Miyaguchi, Miyaji, Morain, Morita, Nemetz, Odlyzko, Ohta, Okamoto, Quisquater, Rueppel, Sako, Sakurai, Santha, Seberry, Shamir, Simmons, Staffelbach, Stern, Tanaka, Vajda, Valle, Yang, Yung.

The Rump Session, this time held more in the spirit of a recent results session, was chaired by Laszlo Csirmaz. Some of the presentations, after a simplified review procedure, were selected for publication in these proceedings. They can be found at the end of this volume.

For the first time, a panel discussion was organized, entitled "The Eurocrypt '92 Controversial Issue: Trapdoor Primes and Moduli". The topic was mainly motivated by the public debate on the draft standard on digital signatures proposed by NIST. The panel members produced an interesting report which is included in this volume.

Following the tradition of the series, the authors produced full papers after the meeting, in some cases with revisions. These papers form the main part of the present volume. They are placed in the same order that they took at the meeting and under the same headings, for ease of reference by those who attended.

My thanks go to the "extended" Program Committee, to the General Chair Tibor Nemetz, to the Organizing Committee, and last but not least to the authors who contributed their recent results. They all have invested their time and effort to make Eurocrypt '92 a success.

Zurich, October 1992   Rainer A. Rueppel

Contents

Stream Ciphers

Public Key I

Factoring

The Eurocrypt '92 Controversial Issue

Trapdoor Primes and Moduli

Public Key II

Pseudo-random Permutation Generators

Complexity Theory and Cryptography I

Zero-Knowledge

Digital Signatures and Electronic Cash

Complexity Theory and Cryptography II

Applications

Selected Papers from the Rump Session

Graph Decompositions
and Secret Sharing Schemes

C. Blundo[1,*], A. De Santis[1,*], D. R. Stinson[2,†], U. Vaccaro[1,*]

[1] Dipartimento di Informatica, Università di Salerno, 84081 Baronissi (SA), Italy

[2] Computer Science and Engineering Department and Center for Communication and Information Science, University of Nebraska, Lincoln, NE 68588-0115, U.S.A.

Abstract

In this paper, we continue a study of secret sharing schemes for access structures based on graphs. Given a graph G, we require that a subset of participants can compute a secret key if they contain an edge of G; otherwise, they can obtain no information regarding the key. We study the information rate of such schemes, which measures how much information is being distributed as shares as compared to the size of the secret key, and the average information rate, which is the ratio between the secret size and the arithmetic mean of the size of the shares. We give both upper and lower bounds on the optimal information rate and average information rate that can be obtained. Upper bounds arise by applying entropy arguments due to Capocelli et al [10]. Lower bounds come from constructions that are based on graph decompositions. Application of these constructions requires solving a particular linear programming problem. We prove some general results concerning the information rate and average information rate for paths, cycles and trees. Also, we study the 30 (connected) graphs on at most five vertices, obtaining exact values for the optimal information rate in 26 of the 30 cases, and for the optimal average information rate in 28 of the 30 cases.

1 Introduction

A secret sharing scheme is a method of dividing a secret S among a set $\mathcal{P}$ of participants in such a way that: if the participants in $A \subseteq \mathcal{P}$ are qualified to know the secret, then by pooling together their information, they can reconstruct the secret S; but any set $A \subseteq \mathcal{P}$, which is not qualified to know S, has absolutely no information on the secret.

Secret sharing schemes are useful in any important action that requires the concurrence of several designed people to be initiated, as launching a missile, opening a bank vault or even opening a safety deposit box. Secret sharing schemes are also used in management of cryptographic keys and multi-party secure protocols (see [12], for example).

The first secret sharing schemes considered were threshold schemes, introduced by Blakley [3] and Shamir [21]. A (k, n) threshold scheme allows a secret to be shared

---

*Partially supported by Italian Ministry of University and Research (M.U.R.S.T.) and by National Council for Research (C.N.R.) under grant 91.02326.CT12.

†Research supported by NSERC (Canada) grant A9287.

R.A. Rueppel (Ed.): Advances in Cryptology - EUROCRYPT '92, LNCS 658, pp. 1-24, 1993.
© Springer-Verlag Berlin Heidelberg 1993

among n participants in such a way that any k of them can recover the secret, but any $k-1$, or fewer, have absolutely no information on the secret (see [26] for a comprehensive bibliography on (k, n) threshold schemes).

Ito, Saito, and Nishizeki [14] described the general method of secret sharing. An access structure is a specification of all the subsets of participants who can recover the secret and it is said to be monotone if any set which contains a subset that can recover the secret can itself recover the secret. Ito, Saito, and Nishizeki gave a methodology to realize secret sharing schemes for arbitrary monotone access structures. Subsequently, Benaloh and Leichter [1] gave a simpler and more efficient way to realize secret sharing schemes for any given monotone access structure. Other general techniques handling arbitrary access structures are given by Simmons, Jackson, and Martin [27] and Martin [17].

An important issue in the implementation of secret sharing schemes is the size of shares since the security of a system degrades as the amount of the information that must be kept secret increases. If one requires that non-qualified set of participants should have no information on the secret, then the size of the shares cannot be less than the size of the secret [15]. In [1] it is proved that there exists an access structure for which any secret sharing scheme must give to some participant a share which is from a domain larger than that of the secret. This was improved by Brickell and Stinson [8], who showed that for the same access structure, the number of elements in the domain of the shares must be at least $2|S| - 1$ if the cardinality of the domain of the secret is $|S|$. Finally, Capocelli, De Santis, Gargano, and Vaccaro [10] proved, for the same access structure, that the number of elements in the domain of the shares must be at least $|S|^{1.5}$, and they showed that the bound is tight.

Ideal secret sharing schemes, that is, schemes where the shares are taken from the same domain as that of the secret, were characterized by Brickell and Davenport [6] in terms of matroids. The uniqueness of the associated matroid is established by Martin in [16]. Brickell constructed some classes of ideal schemes in [5] ,and an interesting non-existence result was proved by Seymour [20].

We also briefly mention some "extended capabilities" of secret sharing schemes that have been studied. The idea of protecting against cheating by one or more participants is addressed in [18], [28], [10], [23] and [9]. Prepositioned schemes are studied in [26]. Finally, the question of how to set up a secret sharing scheme in the absence of a trusted party is solved in [13].

Different measures are possible for the amount of secret information that must be given to participants. When we are interested in the maximum size of the shares, we can use the information rate [7], which is the ratio between the secret size and the maximum size of the shares. When we are interested in the total size of all the shares (and not just the maximum one), it is preferable to use as a measure the average information rate, which is the ratio between the secret size and the arithmetic mean of the size of all the shares [4], [16], [17].

In this paper, we study secret sharing schemes in the case where the access structure consists of the closure of a (connected) graph. We consider all 30 connected graphs on at most five vertices, and determine the exact value of the optimal information rate in all but four cases and optimal average information rate in all but two cases. For these remaining cases, we give quite good upper and lower bounds. For two infinite classes of graphs — cycles of even length (≥ 6) and paths of arbitrary length (≥ 3) — we prove that the value of optimal information rate is $2/3$. For paths and for cycles of even length

(≥ 4), we show how to realize secret sharing schemes with optimal *average* information rate. For any tree, we present a secret sharing scheme with information rate at least $1/2$, and a scheme with average information rate at least $2/3$, both of which which improve previous results.

The main tool for proving upper bounds on the information rate is the entropy approach of Capocelli, De Santis, Gargano and Vaccaro [10]. Lower bounds are obtained by construction methods based on graph decompositions. The main idea of our new method is to use different constructions for different bits of the secret and different subsets of participants. Application of these constructions requires solving a suitable linear programming problem.

The paper is organized as follows. In Section 2, we give the formal definition of secret sharing schemes and recall some basic results. In Section 3, we give our general graph decomposition construction. In Section 4, we discuss the methods for bounding information rates and prove the results mentioned above concerning cycles, paths and trees. In Section 5, we discuss the methods for bounding average information rates and prove the results concerning paths, cycles and trees. Then, in Section 6, we investigate the information rate and the average information rate for the connected graphs on at most five vertices.

2 Secret Sharing Schemes

We recall some definitions and notation from [7]. Suppose that $\mathcal{P}$ is the set of participants. Denote by Γ the set of subsets of participants which we desire to be able to determine the key; hence $\Gamma \subseteq 2^{\mathcal{P}}$. Γ is called the *access structure* of the secret sharing scheme. It seems reasonable to require that Γ be *monotone*, i.e. if $B \in \Gamma$ and $B \subseteq C \subseteq \mathcal{P}$, then $C \in \Gamma$.

For any $\Gamma_0 \subseteq 2^{\mathcal{P}}$, define the *closure* of Γ_0 to be

$$cl(\Gamma_0) = \{C : \exists B \subseteq \Gamma_0, B \subseteq C \subseteq \mathcal{P}\}.$$

Note that the closure of any set of subsets is monotone.

Let $\mathcal{K}$ be a set of q elements called *keys*. For every participant $P \in \mathcal{P}$, let $\mathcal{S}_P$ be a set of s_P elements. Elements of the sets $\mathcal{S}_P$ are called *shares*. Suppose a *dealer* D wants to a share the secret key $K \in \mathcal{K}$ among the participants in $\mathcal{P}$ (we will assume that $D \notin \mathcal{P}$). He does this by giving each participant $P \in \mathcal{P}$ a share from $\mathcal{S}_P$. We say that the scheme is a *perfect scheme* (with respect to access structure Γ) if the following two properties are satisfied:

1. if a subset $\mathcal{B}$ of participants pool their shares, where $\mathcal{B} \in \Gamma$, then they can determine the value of K,

2. if a subset $\mathcal{B}$ of participants pool their shares, where $\mathcal{B} \notin \Gamma$, then they can determine nothing about the value of K (in an information-theoretic sense), even with infinite computational resources.

Remark: In [7], Brickell and Stinson required every participant to have shares taken from the same set, say $\mathcal{S}$. This can easily be done, if desired, by taking a set $\mathcal{S}$ of cardinality $\max\{s_P : P \in \mathcal{P}\}$ and defining injections $\phi_P : \mathcal{S}_P \to \mathcal{S}$ for every $P \in \mathcal{P}$.

Throughout this paper, we confine our attention to perfect schemes, so the term "secret sharing scheme" can be taken to mean "perfect secret sharing scheme".

We will depict a secret sharing scheme as a matrix M. This matrix is not secret, but is known by all the participants. There will be $|\mathcal{P}| + 1$ columns in M. The first column of M will be indexed by D, and the remaining columns are indexed by the members of $\mathcal{P}$. In any row of M, we place the key K in the column D, and a possible list of shares corresponding to K in the remaining columns. When D wants to distribute shares corresponding to a key K, he will choose uniformly at random a row r of M having K in column D, and distribute the shares in that row to the participants (i.e. $M(r, P)$ is given to participant P, for all $P \in \mathcal{P}$).

With this matrix representation, we can present combinatorial conditions on the matrix M that will ensure that the two properties above are satisfied. These conditions are equivalent to conditions presented in [7].

1. if $\mathcal{B} \in \Gamma$ and $M(r, P) = M(r', P)$ for all $P \in \mathcal{B}$, then $M(r, D) = M(r', D)$,

2. if $\mathcal{B} \notin \Gamma$, then for every possible assignment f of shares to the participants in $\mathcal{B}$, say $f = (f_P : P \in \mathcal{B})$ (where $f_P \in S_P$ for all $P \in \mathcal{B}$), there exists a non-negative integer $\lambda(f, \mathcal{B})$ such that

$$|\{r : M(r, P) = f_P \forall P \in \mathcal{B}, M(r, D) = K\}| = \lambda(f, \mathcal{B}),$$

independent of the value of K.

An important issue in the implementation of secret sharing schemes is the size of shares, since the security of a system degrades as the amount of the information that must be kept secret increases. Define $s = \max\{s_P : P \in \mathcal{P}\}$. The *information rate* [7] of the secret sharing scheme is defined to be

$$\rho = \frac{\log q}{\log s}.$$

(We use the term "information rate" because the concept is similar to that of the information rate of an error-correcting code.) It is not difficult to see that $q \leq s$ in a perfect scheme, so the information rate satisfies $\rho \leq 1$. If a secret sharing scheme is to be practical, we do not want to have to distribute too much secret information as shares. Consequently, we want to make the information rate as close to 1 as possible. A perfect secret sharing scheme with information rate $\rho = 1$ is called *ideal*.

In many cases it is preferable to limit the sum of the size of shares over all participants. To analyze such cases we use the *average information rate* [4], [17] defined as

$$\tilde{\rho} = \frac{|\mathcal{P}| \log q}{\sum_{P \in \mathcal{P}} \log s_P}.$$

In a perfect secret sharing scheme, $q \leq s_P$ for all $P \in \mathcal{P}$, and thus $\tilde{\rho} \leq 1$. Also, $\rho = 1$ if and only if $\tilde{\rho} = 1$. It is clear that the information rate is always no greater than the average information rate; that is, $\tilde{\rho} \geq \rho$ for any scheme. Equality holds if and only if $s_P = s_{P'}$ for all $P, P' \in \mathcal{P}$.

2.1 Basic Results

We present some basic terminology from graph theory. Graphs do not have loops or multiple edges; a graph with multiple edges will be termed a *multigraph*. If G is a graph, we denote the vertex set of G by $V(G)$ and the edge set by $E(G)$. We consider undirected graphs only. In an undirected graph the pair of vertices representing any edge is unordered. Thus, the pairs (u, v) and (v, u) represent the same edge. To avoid overburdening the notation we often describe a graph G by the list of all edges $E(G)$ and each edge $(u, v) \in E(G)$ will be represented by uv. G is *connected* if any two vertices are joined by a path. The *complete graph* K_n is the graph on n vertices in which any two vertices are joined by an edge. The *complete multipartite graph* $K_{n_1,n_2,...,n_t}$ is a graph on $\sum_{i=1}^{t} n_i$ vertices, in which the vertex set is partitioned into subsets of size n_i $(1 \le i \le t)$ called *parts*, such that vw is an edge if and only if v and w are in different parts. An alternative way to characterize a complete multipartite graph is to say that the complementary graph is a vertex-disjoint union of cliques. Note that the complete graph K_n can be thought of as a complete multipartite graph with n parts of size 1.

A *stable set* or *independent set* of G is a subset of vertices $A \subseteq V(G)$ such that no two vertices in A are joined by an edge in $E(G)$. The *stability number* or *independence number* $\alpha(G)$ is defined to be the maximum cardinality of a stable set of G. A *vertex cover* of G is a subset of vertices $A \subseteq V(G)$ such that every edge in $E(G)$ is incident with at least one vertex in A. The *vertex covering number* $\beta(G)$ is defined to be the minimum cardinality of a vertex cover of G.

The *girth* of a graph G is defined to be the length of the smallest cycle in G. If G is acyclic, the girth is defined to be ∞. A *regular* graph is a graph where each vertex has degree d, for a fixed d.

We will use the notation $PS(G, \rho, q)$ to denote a perfect secret sharing scheme with access structure $cl(E(G))$ and information rate ρ for a set of q keys. Analogously, a perfect secret sharing scheme with access structure $cl(E(G))$ and average information rate $\tilde{\rho}$ for a set of q keys will be denoted by $\widetilde{PS}(G, \tilde{\rho}, q)$. Throughout this paper, we will restrict our attention to connected graphs. If a graph is not connected, it suffices to find schemes for each of its connected components. The following theorem was proved for information rate in [7]; the proof for average information rate is similar.

Theorem 2.1 *Suppose G is a graph having as its connected components G_i, $1 \le i \le t$. Suppose that there is a $PS(G_i, \rho, q)$, $1 \le i \le t$. Then there is a $PS(G, \rho, q)$. Similarly, if there is a $\widetilde{PS}(G_i, \tilde{\rho}, q)$ for $1 \le i \le t$, then there is a $\widetilde{PS}(G, \tilde{\rho}, q)$.*

Ideal schemes for connected graphs were characterized by Brickell and Davenport [6].

Theorem 2.2 *Suppose G is a connected graph. Then there exists a $PS(G, 1, q)$ (and equivalently, a $\widetilde{PS}(G, 1, q)$) for some q if and only if G is a complete multipartite graph.*

The following result from [7] specifies some values of q for which ideal schemes can be constructed.

Corollary 2.3 *Suppose $q \ge t$ is a prime power. Then there is a $PS(K_{n_1,n_2,...,n_t}, 1, q)$.*

Proof: Let $V_1, \ldots, V_t$ be the parts of the graph $K_{n_1,n_2,...,n_t}$. Let $x_1, \ldots, x_t$ be distinct elements of $GF(q)$. We will construct a matrix M having q^2 rows and $1 + \sum_{i=1}^{t} n_i$

columns. The rows of M will be indexed by $GF(q) \times GF(q)$, and the columns will be indexed by $\{D\} \cup V_1 \cup \ldots \cup V_t$. Define the entries of M by the following rule:

$$M((a,b),D) = a$$
$$M((a,b),v) = ax_i + b,$$

where $a, b \in GF(q)$ and $v \in V_i$. □

Remark: If we start with a complete graph K_t, then the matrix M, as constructed above, is a structure from combinatorial design theory known as an orthogonal array $OA(t+1, q)$.

We recall two basic results from [7]. The first result indicates that the information rate is an appropriate measure of the efficiency of a secret sharing scheme. It states that the existence of one scheme with a specified rate immediately implies the existence of a scheme with the same rate handling as many keys as desired. The result was proved for information rate in [7] and for average information rate in [17, Corollary 2.4].

Theorem 2.4 *Suppose there is a $PS(G, \rho, q)$. Then, for any positive integer n, there is a $PS(G, \rho, q^n)$. Similarly, if there is a $\widetilde{PS}(G, \tilde{\rho}, q)$, then for any positive integer n, there is a $\widetilde{PS}(G, \tilde{\rho}, q^n)$.*

If G is a graph, then G_1 is said to be a *subgraph* of G if $V(G_1) \subseteq V(G)$ and $E(G_1) \subseteq E(G)$. If $V_1 \subseteq V(G)$, then we define the graph $G[V_1]$ to have vertex set V_1 and edge set $\{uv \in E(G) : u, v \in V_1\}$. We say that $G[V_1]$ is an *induced subgraph* of G. The following theorem is obvious.

Theorem 2.5 *Suppose G is a graph and G_1 is an induced subgraph of G. If there is a $PS(G, \rho, q)$, then there exists a $PS(G_1, \rho, q)$.*

Observe that the statement of the above theorem is *not* true for average information rate.

Let $\mathcal{A}$ be an access structure such that there are four participants, A, B, C, D, such that $\{A, B\}$, $\{B, C\}$, $\{C, D\} \in \mathcal{A}$ but $\{A, C\}$, $\{A, D\} \notin \mathcal{A}$. Capocelli, De Santis, Gargano and Vaccaro [10] proved that for any secret sharing scheme for $\mathcal{A}$ the sum of the entropies of the two random variables defined by the shares given to B and C cannot be less than three times the entropy of the secret. By taking all probability distributions to be uniform, the upper bound can be stated as follows:

Theorem 2.6 *Let $\mathcal{A}$ be an access structure. If there are four participants, A, B, C, D, such that:*

$$\{A, B\}, \ \{B, C\}, \ \{C, D\} \in \mathcal{A} \text{ but } \{A, C\}, \ \{A, D\} \notin \mathcal{A},$$

then any secret sharing scheme for $\mathcal{A}$ satisfies

$$\log s_B + \log s_C \geq 3 \log q.$$

Examples of access structures that satisfy the hypotheses of the above theorem are the closure of P_3 (the path of length three), which is the graph having edge set

$$\{AB, BC, CD\};$$

and the closure of H, the graph having edge set

$$\{AB, BC, CD, BD\}.$$

Theorem 2.6 will be the main tool we use for proving upper bounds on information rate and average information rate for paths, cycles and general graphs.

3   Graph Decomposition Constructions

Suppose G is a graph and $G_1, \ldots, G_n$ are subgraphs of G, such that each edge of G occurs in at least one of the G_i's. Suppose also that each G_i is a complete multipartite graph. Then we say that $\Pi = \{G_1, \ldots, G_t\}$ is a *complete multipartite covering* (or CMC) of G. The following construction utilizing CMCs is a special case of [7, Theorem 3.5]. The extension to average information rate is straightforward.

Theorem 3.1 (CMC Construction) *Suppose G is a graph and $\Pi = \{G_1, \ldots, G_n\}$ is a complete multipartite covering of G. For $1 \leq i \leq n$, denote by t_i the number of parts in G_i, and let $t = \max\{t_i : 1 \leq i \leq n\}$. For every vertex v, define $R_v = |\{i : v \in G_i\}|$. Let $R = \max\{R_v : v \in V(G)\}$ and let $\rho = 1/R$. Finally, let $\widetilde{\rho} = |V(G)|/\sum_{v \in V(G)} R_v$. Then there is a $PS(G, \rho, q)$ and a $\widetilde{PS}(G, \widetilde{\rho}, q)$ for any prime power $q \geq t$.*

Proof: Let $q \geq t$, and for $1 \leq i \leq n$, let M_i be the matrix representing $PS(G_i, 1, q)$, which exists by Corollary 2.3. Let $\mathcal{K}$ denote a set of q keys and let $\mathcal{S}$ denote a set of q shares (which we can assume are the same for all the schemes). Then, define a matrix M as follows: for every key K, and for every $n-$tuple of rows $(r_i : 1 \leq i \leq n)$ such that r_i is a row of M_i $(1 \leq i \leq n)$ and $M_i(r_i, D) = K$ $(1 \leq i \leq n)$, define a row $(r_i : 1 \leq i \leq n)$ of M by the rule

$$M((r_1, r_2, \ldots, r_n), v) = (M_i(r_i, v) : v \in V(G_i))$$
$$M((r_1, r_2, \ldots, r_n), D) = K.$$

$\square$

Remark: It is not necessary to actually construct the matrix M of the above proof. When D wishes to share a secret K, it suffices for him to choose, for each i, $1 \leq i \leq n$, a random row r_i of M_i such that $M_i(r_i, D) = K$. Then, for $1 \leq i \leq n$ and for each $v \in G_i$, D gives $M_i(r_i, v)$ to participant v. Hence, each participant v gets a share corresponding to each G_i such that $v \in V(G_i)$.

The main result of this section is a generalization of the CMC construction. The idea is to use several decompositions, rather than just one.

Theorem 3.2 (Multiple CMC Construction) *Suppose G is a graph and for $1 \leq j \leq \ell$, suppose $\Pi_j = \{G_{j1}, \ldots, G_{jn_j}\}$ is a complete multipartite covering of G. Denote by t_{ji} the number of parts in G_{ji} $(1 \leq j \leq \ell, 1 \leq i \leq n_j)$ and define $t = \max\{t_{ji} : 1 \leq j \leq \ell, 1 \leq i \leq n_j\}$. For every vertex v and for $1 \leq j \leq \ell$, define $R_{jv} = |\{i : v \in G_{ji}\}|$. Define $R_v = \sum_{j=1}^{\ell} R_{jv}$ and let $R = \max\{R_v : v \in V(G)\}$. Finally, let $\rho = \ell/R$. Then there is a $PS(G, \rho, q^\ell)$ for any prime power $q \geq t$.*

Proof: Carry out the construction of Theorem 3.1 independently for each of ℓ keys. The details are left to the reader. $\square$

Remark: In the case $\ell = 1$, we recover the original CMC construction. Also, we observe that we cannot improve the lower bound on $\tilde{\rho}$ by taking $\ell > 1$.

Example 3.1 *Recall that P_3, the path of length three, has edges AB, BC, CD. Using one CMC, the best information rate that can be obtained for the access structure $cl(E(P_3))$ is 1/2. However, using two CMCs, we can get $\rho = 2/3$ (a result first obtained by Capocelli, De Santis, Gargano and Vaccaro [10]). The two CMCs are*

$$\{\{AB\}, \{BC, CD\}\}$$

and

$$\{\{AB, BC\}, \{CD\}\}.$$

Then $R_1 = R_4 = 2$ and $R_2 = R_3 = 3$. Hence $R = 3$ and $\rho = 2/3$. A $PS(P_3, 2/3, 4)$ can constructed. Note that if we implement the scheme, we get precisely the scheme presented in [10]. Also, either of these two CMCs yields a scheme with average information rate $\tilde{\rho} = 4/5$.

Example 3.2 *The graph H has edges AB, BC, CD, BD. From the two CMCs*

$$\{\{AB\}, \{BC, BD, CD\}\}$$

and

$$\{\{AB, BC, BD\}, \{CD\}\},$$

we can construct a $PS(H, 2/3, 9)$. Using Corollary 2.3, this scheme could be implemented as follows. Take $K = GF(3) \times GF(3)$. The dealer will choose four random elements (independently) from $GF(3)$, say $b_{11}, b_{12}, b_{21},$ and b_{22}. Given a key (K_1, K_2), the dealer distributes shares as follows: participant A receives $(b_{11} + K_1, b_{21} + K_2)$; participant B receives (b_{11}, b_{12}, b_{21}); participant C receives $(b_{12} + K_1, b_{21} + K_2, b_{22})$; and participant D receives $(b_{12} + 2K_1, b_{21} + K_2, b_{22} + K_2)$. Hence, $S_1 = GF(3) \times GF(3)$ and $S_2 = S_3 = S_4 = GF(3) \times GF(3) \times GF(3)$. Finally, observe that the first CMC yields a scheme with average information rate $\tilde{\rho} = 4/5$, while the second CMC would give $\tilde{\rho} = 2/3$.

4 Optimal Information Rates

For a positive integer q and a graph G, define

$$\rho^*(G, q) = \max\{\rho : \exists PS(G, \rho, q_0), q_0 \le q\}.$$

Then define $\rho^*(G) = \lim_{q \to \infty} \rho^*(G, q)$. Note that this limit exists and is at most 1. Also, note that the definition does *not* require that there exist a $PS(G, \rho^*(G), q)$ for any integer q. However, in all cases where we know the value of $\rho^*(G)$, we can actually construct a scheme having that information rate.

Of course, $\rho^*(G) \le 1$ for all graphs, and $\rho^*(G) = 1$ if G is a complete multipartite graph. The first non-trivial upper bounds on ρ^* were proved by Capocelli et al [10]. Using Theorem 2.6, they proved that $\rho^*(P_3) = 2/3$ and $\rho^*(H) \le 2/3$. In view of the construction given in Example 3.2, we have the following theorem.

Theorem 4.1 *Let P_3 be the graph having edges AB, BC, CD and let H be the graph having edges AB, BC, CD, BD. Then $\rho^*(P_3) = 2/3$ and $\rho^*(H) = 2/3$.*

We can also prove the following general upper bound.

Theorem 4.2 *Suppose G is a connected graph that is not a complete multipartite graph. Then $\rho^*(G) \leq 2/3$.*

Proof: We will prove that any connected graph that is not a complete multipartite graph must contain four vertices w, x, y, z such that the induced subgraph $G[w, x, y, z]$ is isomorphic to either P_3 or H (from Examples 3.1 and 3.2). The desired result then follows from Theorem 2.5.

Let G^C denote the complement of G. Since G is not a complete multipartite graph, there must exist three vertices x, y, z such that $xy, yz \in E(G^C)$ and $xz \in E(G)$. Since G is connected, there is some vertex w such that $wy \in E(G)$. Now, if $\{wx, wz\} \cap E(G) \neq \emptyset$, then $G[w, x, y, z]$ is isomorphic to P_3 or H, and we are done. So, assume that $wx, wz \in E(G^C)$. Define

$$d = \min\{d_G(y, x), d_G(y, z), d_G(w, x), d_G(w, z)\},$$

where d_G denotes the length of a shortest path (in G) between two vertices. Then $d \geq 2$. Without loss of generality, we can assume that $d = d_G(y, x)$ by symmetry. Let $y = y_0, y_1, \ldots, y_{d-1}, x$ be a path in G. Then $\{w, z\} \cap \{y_1, \ldots, y_{d-1}\} = \emptyset$. It follows that $G[y_{d-2}, y_{d-1}, x, z]$ is isomorphic to either P_3 or H, as desired. □

Hence, $\rho^*(G) = 1$ if and only if G is a complete multipartite graph; and $\rho^*(G) \leq 2/3$ if and only if G is not complete multipartite graph. Thus, there is a "gap" in the possible values for $\rho^*(G)$.

4.1 A Linear Programming Problem

We are also interested in the best possible information rate that can be obtained by applying the multiple CMC construction, Theorem 3.2. We define the quantity $\rho_C^*(G)$ which will denote this optimal rate for graph G. In view of the nature of the construction, we can construct a $PS(G, \rho_C^*(G), q)$ for all sufficiently large prime powers q. Of course, $\rho_C^*(G) \leq \rho^*(G)$.

Our main observation is that $\rho_C^*(G)$ can be computed by solving a suitable linear programming problem. We describe how this can be done in the remainder of the section.

Suppose G is a graph. We define a partial order on the CMCs of G as follows. Suppose $\Pi_j = \{G_{j1}, \ldots, G_{jn_j}\}$, $j = 1, 2$, are two CMCs of G. For every vertex v and for $j = 1, 2$, define $R_{jv} = |\{i : v \in G_{ji}\}|$. Then we define $\Pi_1 \preceq \Pi_2$ if $R_{1v} \leq R_{2v}$ for all $v \in V(G)$. Define a CMC, Π, to be *minimal* if $\Pi \preceq \Pi'$ for all CMCs Π' of G.

Now, suppose $\Pi_j = \{G_{j1}, \ldots, G_{jn_j}\}$, $1 \leq j \leq L$, comprise a complete enumeration of the minimal CMCs of G. For every vertex v and for $1 \leq j \leq L$, define $R_{jv} = |\{i : v \in G_{ji}\}|$. Consider the following optimization problem $\mathcal{O}(G)$:

$$
\begin{aligned}
\text{Minimize} \quad R_0 &= \max\{\textstyle\sum_{j=1}^L a_j R_{jv} : v \in V(G)\} \quad \text{subject to:} \\
a_j &\geq 0, \ 1 \leq j \leq L \\
\textstyle\sum_{j=1}^L a_j &= 1
\end{aligned}
$$

Theorem 4.3 *Let R^* be the optimal solution to $\mathcal{O}(G)$. Then $\rho_C^*(G) = 1/R^*$.*

Proof: Suppose $R^* = \max\{\sum_{j=1}^{L} a_j R_{jv} : v \in V(G)\}$, where a_j $(1 \leq j \leq L)$ satisfy the constraints of $\mathcal{O}(G)$. It is clear that the a_j are rational, so denote $a_j = b_j/c_j$, where $b_j, c_j \in \mathbb{Z}$, $1 \leq j \leq L$. Let C denote the least common multiple of $c_1, \ldots, c_L$. Then take Ca_j copies of Π_j for $1 \leq j \leq L$, and apply the multiple CMC construction. We get a scheme with information rate $1/R^*$; hence $\rho_C^*(G) \geq 1/R^*$.

Conversely, suppose we start with an application of the multiple CMC construction that yields the information rate $\rho_C^*(G)$. We can assume without loss of generality that only minimal CMCs are used. Suppose there are b_j copies of Π_j, $1 \leq j \leq L$. Let $B = \sum_{j=1}^{L} b_j$, and define $a_j = b_j/B$, $1 \leq j \leq L$. Then $(a_1, \ldots, a_L)$ satisfy the constraints of $\mathcal{O}(G)$, and yield $R_0 = 1/\rho_C^*(G)$. Hence, $\rho_C^*(G) \leq 1/R^*$, and we're done. $\square$

The difficulty with the problem $\mathcal{O}(G)$ is that the objective function is the maximum of several linear functions. However, we can easily obtain an "equivalent" linear programming problem $\mathcal{O}'(G)$:

$$
\begin{array}{l}
\text{Minimize } T \text{ subject to:} \\[1em]
\qquad\qquad\quad a_j \;\geq\; 0, \;\; 1 \leq j \leq L \\[0.5em]
\qquad \sum_{j=1}^{L} a_j \;=\; 1 \\[0.5em]
\qquad\qquad\quad T \;\geq\; \sum_{j=1}^{L} a_j R_{jv}, \;\; v \in V(G)
\end{array}
$$

It is easy to see that $\mathcal{O}(G)$ and $\mathcal{O}'(G)$ have the same optimal solution. Hence, we obtain the following result.

Theorem 4.4 *Let T^* be the optimal solution to $\mathcal{O}'(G)$. Then $\rho_C^*(G) = 1/T^*$.*

4.2 Information Rate for Paths and Cycles

We next establish some general results when G is a path or a cycle. P_n will denote a path of length n, that is, the graph with edges $X_1 X_2, \ldots, X_n X_{n+1}$; and C_n will denote a cycle of length n, that is, the graph with edges $X_1 X_2, \ldots, X_{n-1} X_n, X_n X_1$.

Theorem 4.5 *If $n \geq 3$, then $\rho^*(P_n) = 2/3$.*

Proof: If $n \geq 3$, $\rho^*(P_n) \leq 2/3$ by Theorem 4.2. First, suppose $n + 1$ is even. Then $\rho^*(P_n) \geq 2/3$ by using the following two CMCs:

$$\Pi_1 = \{\{X_1 X_2, \; X_2 X_3\}, \{X_3 X_4, \; X_4 X_5\}, \ldots, \{X_{n-1} X_n, \; X_n X_{n+1}\}\}$$

and

$$\Pi_2 = \{\{X_1 X_2\}, \{X_2 X_3, \; X_3 X_4\}, \ldots, \{X_{n-2} X_{n-1}, \; X_{n-1} X_n\}, \{X_n X_{n+1}\}\}.$$

If $n + 1$ is odd, then $\rho^*(P_n) \geq 2/3$ by using

$$\Pi_3 = \{\{X_1 X_2, \; X_2 X_3\}, \{X_3 X_4, \; X_4 X_5\}, \ldots, \{X_{n-2} X_{n-1}, \; X_{n-1} X_n\}, \{X_n X_{n+1}\}\}$$

and

$$\Pi_4 = \{\{X_1 X_2\}, \{X_2 X_3, \; X_3 X_4\}, \ldots, \{X_{n-1} X_n, \; X_n X_{n+1}\}\}.$$

$\square$

Theorem 4.6 *If $n \geq 3$, then $\rho^*(C_{2n}) = 2/3$.*

Proof: If $n \geq 3$, $\rho^*(C_{2n}) \leq 2/3$ by Theorem 4.2. $\rho^*(C_{2n}) \geq 2/3$ by using the following two CMCs:

$$\{\{X_1X_2, \ X_2X_3\}, \{X_3X_4, \ X_4X_5\}, \ldots, \{X_{2n-1}X_{2n}, \ X_{2n}X_1\}\}$$

and

$$\{\{X_2X_3, \ X_3X_4\}, \{X_4X_5, \ X_5X_6\}, \ldots, \{X_{2n}X_1, \ X_1X_2\}\}.$$

$\square$

Theorem 4.7 *If $n \geq 2$, then $\rho_C^*(C_{2n+1}) = (2n+1)/(3n+2)$.*

Proof: Here, we appeal to Theorem 4.4. First, we enumerate the minimal CMCs for C_{2n+1}. Take the vertices to be $X_1, X_2, \ldots, X_{2n+1}$, and perform all arithmetic operations on indices $\mathrm{mod}(2n+1)$. Define

$$\Pi_0 = \{\{(X_1X_2, \ X_2X_3\}, \ldots, \{X_{2n-1}X_{2n}, \ X_{2n}X_{2n+1}\}, \{X_{2n+1}X_1\}\}.$$

For $0 \leq j \leq 2n$, define Π_j by "adding" j to indices of Π_0 and reducing $\mathrm{mod}(2n+1)$ to the interval $1, 2, \ldots, 2n+1$. Then Π_j, $0 \leq j \leq 2n$, are the $2n+1$ minimal CMCs. We get a $(2n+1) \times (2n+1)$ matrix of values R_{jX_v}, where $R_{jX_v} = 1$ if and only if $v-j \ \mathrm{mod}\ (2n+1)$ is odd (where $v-j$ is reduced $\mathrm{mod}(2n+1)$ to the interval $1, 2, \ldots, 2n+1$). For example, in the case $2n+1 = 5$, we get the matrix

$$\begin{pmatrix} 2 & 1 & 2 & 1 & 2 \\ 2 & 2 & 1 & 2 & 1 \\ 1 & 2 & 2 & 1 & 2 \\ 2 & 1 & 2 & 2 & 1 \\ 1 & 2 & 1 & 2 & 2 \end{pmatrix}.$$

The optimal solution to $\mathcal{O}'(C_{2n+1})$ is obtained when $a_1 = \ldots = a_{2n+1} = 1/(2n+1)$; then $T = (3n+2)/(2n+1)$ and $\rho_C^*(C_{2n+1}) = (2n+1)/(3n+2)$. In applying the multiple CMC construction, we take one copy of each Π_j. $\square$

4.3 Information Rate for Trees

Brickell and Stinson proved in [7, Theorem 3.8] that for any graph G of maximum degree d, a secret sharing scheme can be realized with information rate

$$\rho \geq \frac{1}{\lceil \frac{d}{2} \rceil + 1}.$$

This was proved using the CMC construction, by decomposing G into complete bipartite graphs $K_{1,m}$ (such a decomposition is called a *star decomposition*, since $K_{1,m}$ is often called a *star*). In the case where G is regular and has girth at least 5, this result is the best that can be obtained using star decompositions [7, Theorem 3.9]. However, we can improve the lower bound whenever G is acyclic. We use star decompositions to obtain information rate equal to $1/2$ in this case.

We now describe the algorithm used to obtain this decomposition. First, we need some definitions. Let G be a connected graph and $v \in V(G)$. $Inc(v)$ denotes the set of edges incident with v:

$$Inc(v) = \{uv : uv \in E(G)\}.$$

By $Adj(v)$ we denote the set of vertices adjacent to v:

$$Adj(v) = \{u \in V(G) : uv \in E(G)\}.$$

Finally, by $degree\_one(v)$ we denote the set of vertices adjacent to v having degree one:

$$degree\_one(v) = \{u \in Adj(v) : |Inc(u)| = 1\}.$$

For any vertex $v \in V(G)$, let $G_v = G[\{v\} \cup Adj(v)]$, i.e. $V(G_v) = \{v\} \cup Adj(v)$ and $E(G_v) = Inc(v)$.

The algorithm **Covering** constructs a star decomposition of G by calling the recursive algorithm **Cover**. The algorithms are as follows:

Covering(G)
    Let $X \in V(G)$
    $\Pi \leftarrow \emptyset$
    Cover(X)
    Output the star decomposition Π

Cover(X)
    $\Pi \leftarrow \Pi \bigcup \{G_X\}$
    $B \leftarrow \{Y \in Adj(X) : |Inc(Y)| = 1\}$
    $E(G) \leftarrow E(G) - Inc(X)$
    $V(G) \leftarrow V(G) - \left(B \bigcup \{X\}\right)$
    For all $X' \in Adj(X) - B$ do **Cover**(X')

It easy to see that the algorithm **Covering** always finds a complete multipartite covering of G. If G is acyclic then each of its vertices belongs to at most two different connected subgraphs of the covering as stated by next lemma.

Lemma 4.8 *Let Π be a complete multipartite covering of a tree G obtained by applying* **Covering** *to G. Then each vertex $X \in V(G)$ belongs to at most two different subgraphs $G', G'' \in \Pi$.*

Proof: If $|Inc(X)| = 1$ and $(X, Y) \in E(G)$ with $|Inc(Y)| > 1$ then **Cover** is called on Y. The vertex Y belongs to two connected subgraphs, and the graph G' with set of edges $E(G') = E(G) - Inc(X)$ and set of vertices $V(G') = V(G) - \left(B \bigcup \{X\}\right)$ is still connected. Since G is a tree, if $|Inc(X)| > 1$ then the graph is disconnected. All connected components of the new graph are trees and **Cover** is called on each $Y \in Adj(X) - B$. Since each $Y \in Adj(X) - B$ belongs to different connected components of G, each Y belongs to at most two connected subgraphs in Π. Thus the lemma is proved. $\Box$

The following result is immediate from Lemma 4.8 and Theorem 3.1.

Corollary 4.9 *For any tree a secret sharing scheme exists with information rate $\rho \geq 1/2$.*

There is only one case in which G is connected and **Covering** gives a secret sharing scheme with information rate greater than $1/2$. This case arises when G is itself a star graph and X is chosen to be the vertex of maximum degree in G.

5   Optimal Average Information Rates

Recall that we use the notation $\widetilde{PS}(G, \tilde{\rho}, q)$ to denote a perfect secret sharing scheme with access structure $cl(E(G))$ and average information rate $\tilde{\rho}$ for a set of q keys.

For a positive integer q and a graph G, define

$$\tilde{\rho}^*(G, q) = \max\{\tilde{\rho} : \exists \widetilde{PS}(G, \tilde{\rho}, q_0), q_0 \leq q\}.$$

Then define $\tilde{\rho}^*(G) = \lim_{q \to \infty} \tilde{\rho}^*(G, q)$. Note that this limit exists and is at most 1. Also, note that the definition does *not* require that there exist a $\widetilde{PS}(G, \tilde{\rho}^*(G), q)$ for any integer q.

The following lemma is the analogue of Theorem 4.2 for the average information rate. It is a generalization of [16, Lemma 4.3.5].

Lemma 5.1 *Let G be a connected graph with n vertices. If G is a complete multipartite graph then $\tilde{\rho}^*(G) = 1$; otherwise $\tilde{\rho}^*(G) \leq n/(n+1)$.*

Proof: Assume G is a complete multipartite graph. By Theorem 2.2 an ideal scheme exists; this scheme has an average information rate equal to 1. If G is not a complete multipartite graph then, from Theorem 4.2 and Theorem 2.6, there exist two vertices in $V(G)$, X and Y with $XY \in E(G)$, such that $\log s_x + \log s_y \geq 3 \log q$. Thus $\sum_{X \in V(G)} \log s_x \geq (n+1) \log q$ so the average information rate is not greater than $n/(n+1)$.   □

5.1   A Linear Programming Problem

With respect to the information rate $\rho^*(G)$, we solved a linear programming problem to obtain a lower bound. Now, for average information rate $\tilde{\rho}^*(G)$, we will obtain an *upper* bound by solving a linear programming problem.

Let G be a graph, and define a subgraph G_1 of G as follows: $xy \in E(G_1)$ if and only if there exist vertices $w, z \in V(G)$ such that $G[w, x, y, z] = \{wx, xy, yz\}$ or $G[w, x, y, z] = \{wx, xy, yz, xz\}$. We will take $V(G_1)$ to consist of all vertices in $V(G)$ that are incident with at least one edge in $E(G_1)$ (i.e. we delete all isolated vertices from G_1). We say that G_1 is the *foundation* of G.

For example, the path P_4, having edges $\{AB\}, \{BC\}, \{CD\}, \{DE\}$, has foundation consisting of the two edges $\{BC\}, \{CD\}$.

If xy is an edge in the foundation of a graph G, then by Theorem 2.6, $\log s_x + \log s_y \geq 3 \log q$ for any secret sharing scheme with access structure $cl(E(G))$. Consider the following linear programming problem $\mathcal{A}(G)$:

$$
\begin{array}{lll}
\text{Minimize} & C = \sum_{v \in V(G)} a_v & \text{subject to:} \\
& a_v \geq 0, \ v \in V(G) & \\
& a_v + a_w \geq 1, \ vw \in E(G_1) &
\end{array}
$$

Then we have the following upper bound on the average information rate.

Theorem 5.2 *Let G be a graph with foundation G_1. Let C^* be the optimal solution to the problem $\mathcal{A}(G)$. Then*

$$
\tilde{\rho}^*(G) \leq \frac{|V(G)|}{C^* + |V(G)|}.
$$

Proof: Consider any secret sharing scheme realizing the access structure $cl(E(G))$. For every vertex $v \in V(G)$, define

$$
a_v = \frac{\log s_v}{\log q} - 1.
$$

Suppose vw is an edge of the foundation G_1. Now, from Theorem 2.6 we get $\log s_v + \log s_w \geq 3 \log q$, or $a_v + a_w \geq 1$. For any $v \in V(G)$, we have $s_v \geq q$, so $a_v \geq 0$. Hence, the a_v's, as defined above, are a feasible solution for the problem $\mathcal{A}(G)$. Hence,

$$
C^* \leq \sum_{v \in V(G)} a_v,
$$

where C^* is the optimal solution to $\mathcal{A}(G)$. It follows that

$$
C^* \leq \frac{\sum_{v \in V(G)} \log s_v}{\log q} - |V(G)|.
$$

But then we have

$$
\begin{aligned}
\tilde{\rho}(G) &= \frac{|V(G)| \log q}{\sum_{v \in V(G)} \log s_v} \\
&\leq \frac{|V(G)|}{C^* + |V(G)|},
\end{aligned}
$$

which is the bound to be proved. $\square$

Remark: Given a graph G, the foundation G_1 can be determined in polynomial time. One way to do this is to check all 4–subsets of $V(G)$. Every time we get an induced subgraph isomorphic to P_3, we can add one edge to the foundation; and every time we find an induced subgraph isomorphic to H, we can add two edges to the foundation. This algorithm requires time $O(n^4)$, where $n = |V(G)|$. Since the linear programming problem $\mathcal{A}(G)$ can be solved in polynomial time, so too can the bound of Theorem 5.2 be computed in polynomial time.

We have a couple of general observations on the linear programming problem $\mathcal{A}(G)$. Let d be a positive integer. A d–*factor* of a graph is a spanning subgraph that is regular of degree d.

Lemma 5.3 *Let G be a graph having foundation G_1. If G_1 has a d-factor for some integer $d \geq 1$, then the optimal solution to $\mathcal{A}(G)$ is $C^* = |V(G_1)|/2$.*

Proof: Let the edges in the d-factor be $x_j y_j$, $1 \leq j \leq dn/2$, where $V(G_1) = \{1, \ldots, n\}$. Then we obtain the following:

$$\frac{dn}{2} \leq \sum_{j=1}^{dn/2} (a_{x_j} + a_{y_j})$$

$$= d \sum_{i=1}^{n} a_i.$$

Hence, $C^* \geq n/2$. To obtain $C^* \leq n/2$, let $a_i = 1/2$ for $1 \leq i \leq n$. ☐

Next, note that $C^* \leq \beta(G_1)$. To see this, let W be a minimum vertex cover of G_1, and define $a_v = 1$ if $v \in W$; $a_v = 0$, otherwise. This gives a feasible solution for which $\sum_{v \in V(G)} a_v = \beta(G_1)$. In the case where G_1 is bipartite, this will in fact be the optimal solution, as follows.

Lemma 5.4 *Let G be a graph having foundation G_1. If G_1 is bipartite, then the optimal solution to $\mathcal{A}(G)$ is $C^* = \beta(G_1)$ and the optimal solution is given by $a_v = 1$ if $v \in W$, $a_v = 0$, otherwise, where W is a minimum vertex cover of G_1.*

Proof: It is well-known that the incidence matrix of a bipartite graph is a totally unimodular matrix (that is, the determinant of any square submatrix is $0, 1$ or -1). Hence, if G_1 is bipartite, the linear programming problem $\mathcal{A}(G)$ and the corresponding integer programming problem have the same optimal solution. But an optimal solution to the integer programming problem is obtained from a vertex cover, as described above. ☐

Hence, we have the following bound as an immediate consequence.

Theorem 5.5 *Let G be a graph with foundation G_1, and suppose G_1 is bipartite. Then*

$$\tilde{\rho}^*(G) \leq \frac{|V(G)|}{\beta(G_1) + |V(G)|}.$$

5.2 Vertex Covers and Secret Sharing Schemes

From Theorem 3.1, there exists a secret sharing scheme for a graph G with average information rate $\tilde{\rho} = |V(G)| / \sum_{v \in V(G)} R_v$. Suppose we construct a scheme by using a star decomposition, as in Section 4.3. Let W denote the set of centers of the stars used in the decomposition. Then W must be a vertex cover of G. Conversely, if W is a vertex cover of G, then we can use it to construct a star decomposition of G and hence a secret sharing scheme. The algorithm to do this is as follows:

Algorithm
>   Let $W = \{v_1, \ldots, v_n\}$ be a vertex cover of G
> $\Pi \leftarrow \emptyset$
>   For $i \leftarrow 1$ to n do

$$X \leftarrow v_i$$
$$\Pi \leftarrow \Pi \bigcup \{G_X\}$$
$$B \leftarrow \{Y \in Adj(X) : |Inc(Y)| = 1\}$$
$$E(G) \leftarrow E(G) - Inc(X)$$
$$V(G) \leftarrow V(G) - \left(B \bigcup \{X\}\right)$$

Output the star decomposition Π

We now show that if we construct a scheme from the star decomposition Π, then we can express $\tilde{\rho}$ as a function of $|V(G)|$, $|E(G)|$ and $|W|$. Let $\Pi = \{G_1, \ldots, G_n\}$, where $n = |W|$. Consider a star $G_i = K_{1,m}$ in the decomposition. The total number of shares in the scheme $PS(G_i, 1, q)$ is $m + 1 = |E(G_i)| + 1$. Hence, the total number of shares in the scheme for G is:

$$\sum_{v \in V(G)} R_v = \sum_{i=1}^{n} (|E(G_i)| + 1)$$
$$= |E(G)| + |W|.$$

Hence, applying Theorem 3.1, we have the following result.

Theorem 5.6 *Let G be a graph and $W \subseteq V(G)$ be a vertex covering. Then a secret sharing scheme for G exists with average information rate*

$$\tilde{\rho} = \frac{|V(G)|}{|E(G)| + |W|}.$$

Since $\tilde{\rho}$ depends only on $|W|$, finding the maximum rate among all vertex coverings is equivalent to minimizing $|W|$, i.e. determining the vertex covering number $\beta(G)$. Unfortunately, the problem of computing $\beta(G)$ is NP-hard [11]. However, for certain classes of graphs, such as bipartite graphs and chordal graphs, $\beta(G)$ can be computed in polynomial time (see [11]). We will return to this in Section 5.4.

Let us mention a couple of general bounds that can be proved by this technique. It is obvious that $W \subseteq V(G)$ is a vertex covering of G if and only if $V - W$ is a stable set of G. Hence, $\beta(G) = |V(G)| - \alpha(G)$. Using known lower bounds on the stability number of a graph, we can obtain the following corollaries to Theorem 5.6.

Corollary 5.7 *Let G be a graph with $|V(G)| = n$ and $|E(G)| = m$. Then*

$$\tilde{\rho}^*(G) \geq \frac{n(2m + n)}{m(2m + 3n)}.$$

Proof: Use Theorem 5.6 and the bound $\alpha(G) \geq n^2/(2m + n)$ ([2, Corollary 2, p. 279]).
□

Corollary 5.8 *Let G be a graph with $|V(G)| = n$ and maximum degree d. Then*

$$\tilde{\rho}^*(G) \geq \frac{1}{\frac{d+2}{2} - \frac{1}{n}\lceil \frac{n}{d+1} \rceil}.$$

Proof: Use Theorem 5.6 and the bound $\alpha(G) \geq \lceil \frac{n}{d+1} \rceil$ ([2, Corollary 2, p. 276]).   □

Note that the bound on average information rate given by Corollary 5.8 exceeds the bound on information rate proved in [7, Theorem 3.8].

5.3 Average Information Rate for Paths and Cycles

In this section we give an upper bound for average information rate for P_n, the path of length n. Then we show how to construct secret sharing schemes with optimal average information rate.

If n is equal either to 1 or to 2, then P_n is a complete multipartite graph and a secret sharing scheme with an average information rate equal to 1 exists. If n is greater than 2, then the next theorem provides the optimal average information rate.

Theorem 5.9 *The optimal average information rate of a secret sharing scheme for P_n, where $n \geq 3$, is given by*

$$\tilde{\rho}^*(P_n) = \begin{cases} \frac{2(n+1)}{3n} & \text{if } n \text{ is even} \\[2mm] \frac{2(n+1)}{3n+1} & \text{if } n \text{ is odd} \end{cases}$$

Proof: It is easy to see that the foundation of P_n consists of the edges

$$X_2 X_3, \ldots, X_{n-1} X_n,$$

so it is isomorphic to P_{n-2}. P_{n-2} is bipartite, and $\beta(P_{n-2}) = \lfloor \frac{n-1}{2} \rfloor$.
First suppose n even and $n \geq 4$. By applying Theorem 5.5 we know that $\tilde{\rho}^*(P_n) \leq 2(n+1)/3n$. We have $\tilde{\rho}^*(P_n) \geq 2(n+1)/3n$ by using the CMC II_1 from Theorem 4.5. If n is odd and $n \geq 3$, $\tilde{\rho}^*(P_n) \leq 2(n+1)/(3n+1)$ by Theorem 5.5. We obtain a secret sharing scheme with average information rate equal to $2(n+1)/(3n+1)$ by using the CMC II_3 from Theorem 4.5. □

We now consider average information rate for cycles. If n is equal either to 3 or to 4, then C_n is a complete multipartite graph and a secret sharing scheme exists with an average information rate equal to 1. If n is greater than 4, then the next theorem gives the optimal average information rate for even length cycles, while for odd length cycles it gives upper and lower bounds.

Theorem 5.10 *The optimal average information rate of a secret sharing scheme for C_n, where $n \geq 5$, satisfies*

$$\tilde{\rho}^*(C_n) = 2/3 \text{ if } n \text{ is even}$$

$$\frac{2n}{3n+1} \leq \tilde{\rho}^*(C_n) \leq \frac{2}{3} \text{ if } n \text{ is odd.}$$

Proof: It is easy to see that the foundation of C_n is again C_n. C_n is a 2-factor of itself, so $C^* = n/2$, by Lemma 5.3. Applying Theorem 5.2, we get $\tilde{\rho}^*(C_n) \leq 2/3$.
First, suppose n is even, $n \geq 6$. We have already shown in Theorem 4.6 that $\rho^*(C_n) = 2/3$. Since $\tilde{\rho}^*(C_n) \geq \rho^*(C_n)$ and since $\tilde{\rho}^*(C_n) \leq 2/3$, we obtain $\tilde{\rho}^*(C_n) = 2/3$. Next, let n be odd, $n \geq 5$. From Theorem 4.7, we have $\rho^*(C_n) \geq 2n/(3n+1)$. Since $\tilde{\rho}^*(C_n) \geq \rho^*(C_n)$ and since $\tilde{\rho}^*(C_n) \leq 2/3$, the stated bounds follow. □

5.4 Average Information Rate for Trees

In this section, we discuss upper and lower bounds on the average information rate of secret sharing schemes for trees.

For a graph G, let $degree\_one(G)$ denote the vertices in $V(G)$ having degree one. Our first observation is that the foundation of a tree T can be constructed by deleting all degree one vertices from T.

Lemma 5.11 *Let T be a tree; then the foundation of T is*

$$T_1 = T[V(T) - degree\_one(T)].$$

Proof: Let xy be an edge of T. If $\{x,y\} \cap degree\_one(T) \neq \emptyset$, then clearly, $xy \notin E(T_1)$. So assume $\{x,y\} \cap degree\_one(T) = \emptyset$. Let $wx, yz \in E(T)$, where $w \neq y, z \neq x$. Since T is a tree, $wy, wz, xz \notin E(T)$. Hence, $T[w,x,y,z] = \{wx, xy, yz\}$ and $xy \in E(T_1)$. □

Remark: It is not difficult to see that the conclusion remains true if T is any bipartite graph having girth at least six.

Here now are our upper and lower bounds on the average information rate for trees.

Theorem 5.12 *Let T be a tree and let $T_1 = T[V(T) - degree\_one(T)]$. Then we have*

$$\frac{|V(T)|}{\beta(T) + |V(T)| - 1} \leq \tilde{\rho}^*(T) \leq \frac{|V(T)|}{\beta(T_1) + |V(T)|}.$$

Proof: By Lemma 5.11, T_1 is the foundation of T. Hence, the upper bound on $\tilde{\rho}^*$ follows from Theorem 5.5. The lower bound follows from Theorem 5.6, since $|E(T)| = |V(T)| - 1$ for any tree T. □

Remarks:

1. Since T and T_1 are bipartite graphs, the vertex covering numbers can be computed in polynomial time. In fact, by Konig's Theorem, the vertex covering number of a biparitite graph equals the size of a maximum matching.

2. The reader can check that, in the special case where T is a path, the upper and lower bounds of Theorem 5.12 coincide, and they agree with Theorem 5.9.

Now, we give a general lower bound on the average information rate for trees.

Theorem 5.13 *Let T be a tree with n vertices. Then*

$$\tilde{\rho}^*(T) \geq \frac{2n}{3n - 2}.$$

Proof: In a bipartite graph G with vertex bipartition V_1, V_2, both V_1 and V_2 are vertex covers. Hence, $\beta(G) \leq \min\{|V_1|, |V_2|\} \leq (|V_1| + |V_2|)/2$. A tree is bipartite, so $\beta(T) \leq n/2$. Apply Theorem 5.12 to obtain the stated result. □

6 The Connected Graphs on at Most Five Vertices

In this section, we give upper and lower bounds on the information rate and average information rate for the connected graphs on at most five vertices. First, there are nine connected graphs on at most four vertices. Seven of these are complete multipartite graphs and admit ideal schemes: $K_2, K_3, K_{1,2}, K_4, K_{1,3}, K_{2,2}, K_{1,1,2}$. The remaining two graphs are P_3 (the path of length 3) and the graph H (from Example 3.2). We have already shown that $\rho^*(P_3) = 2/3$ (Theorem 4.5) and $\tilde{\rho}^*(P_3) = 2/3$ (Theorem 5.9). With regard to H, we have $\rho^*(H) = 2/3$ (Theorem 4.1) and $\tilde{\rho}^*(H) = 4/5$ (Example 3.2 and Theorem 5.1).

So, let's move on to the connected graphs on five vertices. There are 21 non-isomorphic connected graphs on five vertices. Of these 21 graphs, six are complete multipartite graphs and admit ideal schemes. These graphs are $K_{1,4}, K_{2,3}, K_{1,1,3}, K_{1,2,2}, K_{1,1,1,2}$ and K_5. The remaining 15 graphs are depicted in Appendix A, where we also show the minimal $CMCs$ for each graph.

The bounds on information rate and average information rate are summarized in Table 1. The lower bounds are obtained by making use of CMC constructions. Upper bounds on information rate are given by Theorem 4.2, whereas upper bounds on average information rate are given by application of Theorem 5.2.

Table 1: Information Rate and Average Information Rate

Graph	Information Rate	Average information Rate
$G_1, \ldots, G_9$	$\rho^* = 2/3$	$\tilde{\rho}^* = 5/6$
G_{10}, G_{11}	$\rho^* = 2/3$	$\tilde{\rho}^* = 5/7$
G_{12}	$5/8 \leq \rho^* \leq 2/3$	$5/8 \leq \tilde{\rho}^* \leq 2/3$
G_{13}	$3/5 \leq \rho^* \leq 2/3$	$5/7 \leq \tilde{\rho}^* \leq 10/13$
G_{14}	$3/5 \leq \rho^* \leq 2/3$	$\tilde{\rho}^* = 5/7$
G_{15}	$4/7 \leq \rho^* \leq 2/3$	$\tilde{\rho}^* = 5/7$

The first CMC for each graph in Appendix A gives rise to the scheme that attains the given lower bound for the average information rate. For the graphs $G_1, \ldots, G_{11}$, the schemes with information rate equal to $2/3$ are obtained by taking one copy of each CMC shown in Appendix A. We next consider the lower bounds on ρ^* for the remaining four graphs, $G_{12}, \ldots, G_{15}$.

- First, let $E(G_{12}) = \{AB, BC, CD, DE, AE\}$. Then G_{12} is the cycle C_5 and $\rho_C^*(G_{12}) = 5/8$ from Theorem 4.7.

- Let $E(G_{13}) = \{AB, BC, BE, EC, CD\}$. $\rho_C^*(G_{13}) = 3/5$ is realized by using the three CMCs shown in Appendix A.

- Let $E(G_{14}) = \{AB, AD, BD, BC, DE, CE\}$. $\rho_C^*(G_{14}) = 3/5$ is realized by using the three CMCs shown in Appendix A.

- Finally we consider The four minimal CMCs of G_{15} are depicted in Appendix A.

The matrix of entries R_{jv} is

$$\begin{pmatrix} 1 & 2 & 2 & 1 & 1 \\ 2 & 1 & 2 & 1 & 1 \\ 2 & 3 & 1 & 2 & 2 \\ 3 & 2 & 1 & 2 & 2 \end{pmatrix}.$$

Hence the linear programming problem to be solved is the following:

Minimize T subject to

$$a_j \geq 0, 1 \leq j \leq 4$$
$$\sum_{j=1}^{4} a_j = 1$$
$$T \geq a_1 + 2a_2 + 2a_3 + 3a_4$$
$$T \geq 2a_1 + a_2 + 3a_3 + 2a_4$$
$$T \geq 2a_1 + 2a_2 + a_3 + a_4$$
$$T \geq a_1 + a_2 + 2a_3 + 2a_4.$$

The optimal solution is

$$(a_1, a_2, a_3, a_4, T) = (1/4, 1/2, 1/4, 0, 7/4).$$

Hence, $\rho_C^*(G_{15}) = 4/7$, and this rate can be attained by taking one copy of Π_1, two copies of Π_2, and one copy of Π_3.

Now we turn to the upper bounds on average information rate. Theorem 5.1 gives the upper bound $\tilde{\rho}^* \leq 5/6$ for $G_1, \ldots, G_9$. So, there remain the six graphs $G_{10} \ldots, G_{15}$ to consider.

- Consider the graph G_{10}. The foundation of G_{10} consists of the four edges BC, BE, DC, DE. This foundation is a 2–regular graph on four vertices, so $C^* = 2$ (Theorem 5.3). Hence, by Theorem 5.2, $\tilde{\rho}^* \leq 5/7$.

- Consider the graph G_{11}. The foundation of G_{11} consists of the four edges BC, BE, DC, DE. As with G_{10}, we obtain $\tilde{\rho}^* \leq 5/7$.

- G_{12} is the cycle of length five, so $\tilde{\rho}^* \leq 2/3$ (Theorem 5.10).

- Consider the graph G_{13}. The foundation of G_{13} consists of the three edges BC, BE, CE. This foundation is a 2–regular graph on three vertices, so $C^* = 3/2$ (Theorem 5.3). Hence, by Theorem 5.2, $\tilde{\rho}^* \leq 10/13$.

- Consider the graph G_{14}. The foundation of G_{14} consists of the five edges AB, AD, BC, BD, DE. The optimal solution to the linear programming problem is $C^* = 2$. Hence, by Theorem 5.2, $\tilde{\rho}^* \leq 5/7$.

- Consider the graph G_{15}. The foundation of G_{15} consists of the five edges AB, AC, BC, CD, CE and is isomorphic to the foundation of G_{14}. As before, we get $\tilde{\rho}^* \leq 5/7$.

Acknowledgements

D. R. Stinson would like to thank Dean Hoffman, U. S. R. Murty and Qiu-rong Wu for helpful discussions. In particular, Dean Hoffman pointed out the linear programming formulation given in Theorem 4.4. Wen-ai Jackson pointed out the minimal decomposition Π_2 of graph G_{11} in Appendix A.

References

[1] J. Benaloh and J. Leichter. Generalized secret sharing and monotone functions. *Lecture Notes in Computer Science*, 403:27–35, 1990.

[2] C. Berge. *Graphs*. Second revised edition, North–Holland, 1985.

[3] G. R. Blakley. Safeguarding cryptographic keys. *AFIPS Conference Proceedings*, 48:313–317, 1979.

[4] C. Blundo. *Secret Sharing Schemes for Access Structures based on Graphs*. Tesi di Laurea, University of Salerno, Italy, 1991, (in Italian).

[5] E. F. Brickell. Some ideal secret sharing schemes. *J. Combin. Math. and Combin. Comput.*, 9:105–113, 1989.

[6] E. F. Brickell and D. M. Davenport. On the classification of ideal secret sharing schemes. *J. Cryptology*, 4:123–134, 1991.

[7] E. F. Brickell and D. R. Stinson. Some improved bounds on the information rate of perfect secret sharing schemes. *Lecture Notes in Computer Science*, 537:242–252, 1991. To appear in *J. Cryptology*.

[8] E. F. Brickell and D. R. Stinson. Some improved bounds on the information rate of perfect secret sharing schemes. Department of Computer Science and Engineering Report Series # 106, University of Nebraska, May 1990.

[9] E. F. Brickell and D. R. Stinson. The detection of cheaters in threshold schemes. *SIAM J. on Discrete Math.*, 4:502–510, 1991.

[10] R. M. Capocelli, A. De Santis, L. Gargano, and U. Vaccaro. On the size of shares for secret sharing schemes. *Lecture Notes in Computer Science*, 576:101–113, 1991. To appear in J. Cryptology.

[11] M. R. Garey and D. S. Johnson. *Computers and Intractability. A Guide to Theory of NP-Completeness*. W. H. Freeman and Company, New York, 1979.

[12] O. Goldreich, S. Micali, and A. Wigderson. How to play any mental game, *Proc. 19th ACM Symp. on Theory of Computing*, pages 218–229, 1987.

[13] I. Ingemarsson and G. J. Simmons. A protocol to set up shared secret schemes without the assistance of a mutually trusted party. *Lecture Notes in Computer Science*, 473:266–282, 1991.

[14] M. Ito, A. Saito, and T. Nishizeki. Secret sharing scheme realizing general access structure. *Proc. IEEE Globecom '87*, pages 99–102, 1987.

[15] E. D. Karnin, J. W. Greene, and M. E. Hellman. On secret sharing systems. *IEEE Transactions on Information Theory*, IT-29:35–41, 1983.

[16] K. M. Martin. *Discrete Structures in the Theory of Secret Sharing*. PhD Thesis, University of London, 1991.

[17] K. M. Martin. New secret sharing schemes from old. Submitted to *Journal of Combin. Math. and Combin. Comput.*

[18] R. J. McEliece and D. V. Sarwate. On sharing secrets and Reed-Solomon codes. *Commun. of the ACM*, 24:583–584, 1981.

[19] T. Rabin and M. Ben-Or. Verifiable secret sharing and multiparty protocols with honest majority. *Proc. 21st ACM Symp. on Theory of Computing*, pages 73–85, 1989.

[20] P. D. Seymour. On secret-sharing matroids. Preprint.

[21] A. Shamir. How to share a secret. *Commun. of the ACM*, 22:612–613, 1979.

[22] G. J. Simmons. Shared secret and/or shared control schemes. *Lecture Notes in Computer Science*, 537:216–241, 1991.

[23] G. J. Simmons. Robust shared secret schemes or 'how to be sure you have the right answer even though you don't know the question'. *Congressus Numer.*, 68:215–248, 1989.

[24] G. J. Simmons. How to (really) share a secret. *Lecture Notes in Computer Science*, 403:390–448, 1990.

[25] G. J. Simmons. An introduction to shared secret and/or shared control schemes and their application. *Contemporary Cryptology*, IEEE Press, pages 441–497, 1991.

[26] G. J. Simmons. Prepositioned shared secret and/or shared control schemes. *Lecture Notes in Computer Science*, 434:436–467, 1990.

[27] G. J. Simmons, W. Jackson, and K. Martin. The geometry of shared secret schemes. *Bulletin of the ICA*, 1:71–88, 1991.

[28] M. Tompa and H. Woll. How to share a secret with cheaters. *J. Cryptology*, 1:133–138, 1988.

Appendix A

Minimal CMCs for the Connected Graphs on Five Vertices which are not Complete Multipartite

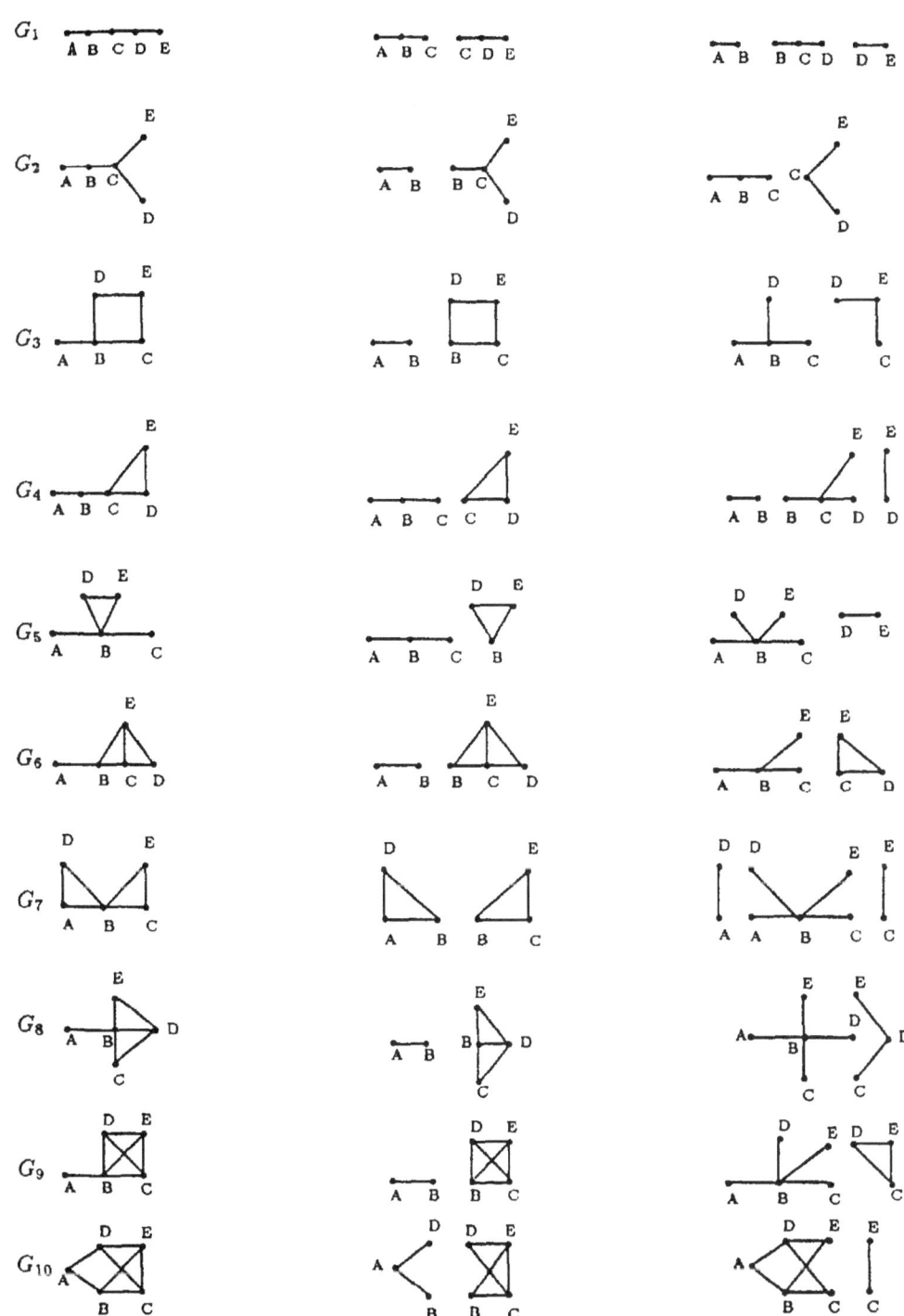

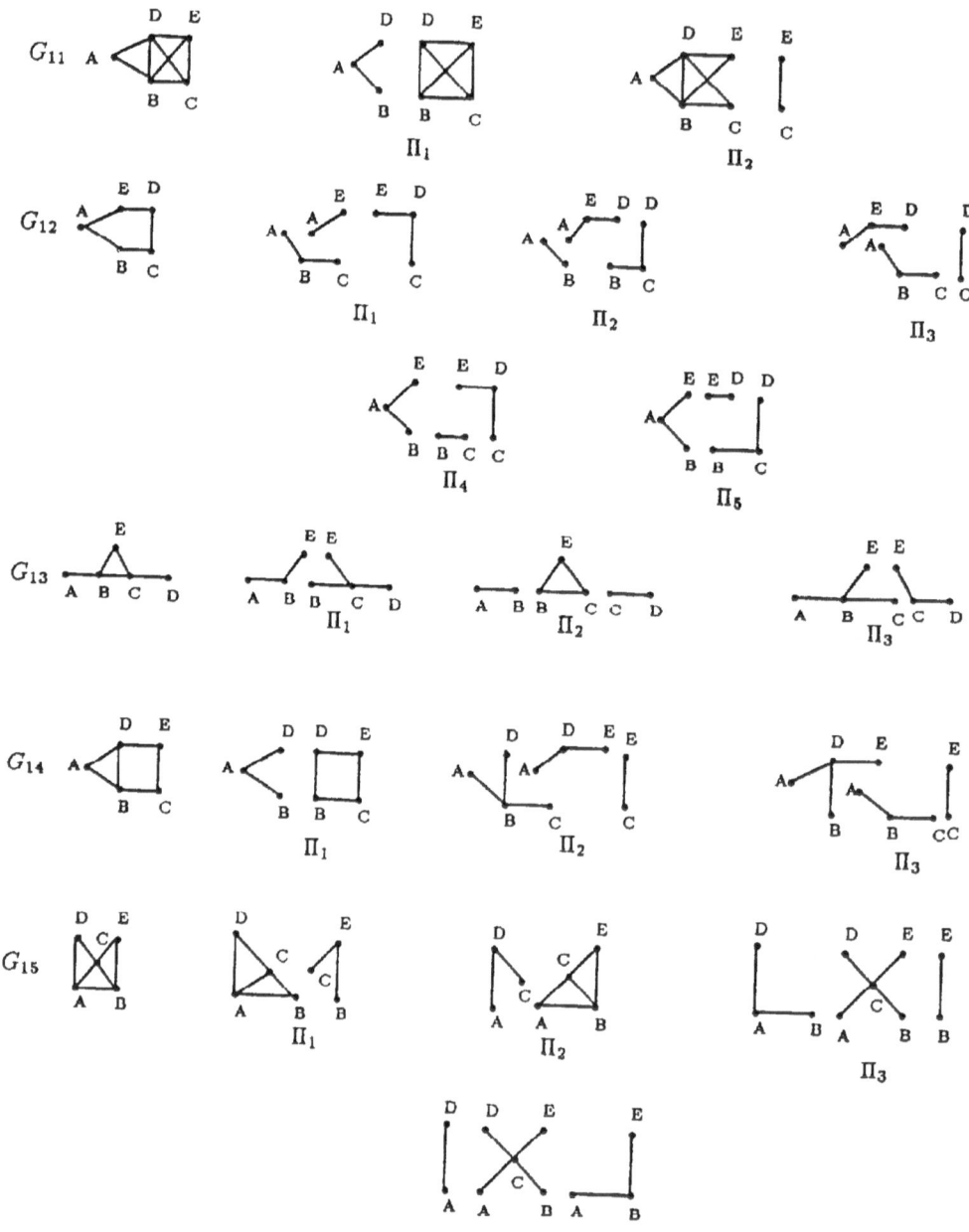

CLASSIFICATION OF IDEAL HOMOMORPHIC THRESHOLD SCHEMES OVER FINITE ABELIAN GROUPS*

(Extended Abstract)

Yair Frankel   Yvo Desmedt

Department of EE & CS
University of Wisconsin — Milwaukee
Milwaukee, WI 53201
U. S. A.

Abstract

Threshold schemes allow any t out of l individuals to recompute a secret (key). General sharing schemes are a generalization. In homomorphic sharing schemes the "product" of shares of the keys gives a share of the product of the keys. We prove that there exist infinitely many Abelian groups over which there does *not* exist an *ideal homomorphic* threshold scheme. Additionally we classify *ideal homomorphic* general sharing schemes. We discuss the potential impact of our result on the construction of general sharing schemes.

1   Introduction

General secret sharing schemes [3, 14, 9] provide a means to distribute *shares* of a secret (key) k so that any subset of individuals (shareholders) specified by an access structure can recompute the secret. Threshold schemes [3, 14] have an access structure where l out of l individuals can recompute the secret. Besides using threshold schemes to recompute a secret, they are used, for example, in fault tolerant computing [13].

---

*This work has been supported by NSF Grant NCR-9106327.

R.A. Rueppel (Ed.). Advances in Cryptology - EUROCRYPT '92, LNCS 658, pp. 25-34, 1993.
© Springer-Verlag Berlin Heidelberg 1993

Many threshold schemes such as [5, 11, 12, 14] work over a finite field. However, other structures such as the groups Z_n^*, elliptic curves and Z_n^{+1} (groups of integers modulo n with Jacobi symbol $+1$) are often used in cryptography. Giving secrets and maintaining group operations is therefore useful. Homomorphic threshold schemes over a finite Abelian group have been used in several cryptographic schemes. Indeed homomorphic threshold schemes over a finite Abelian group have been used to set up secret ballot election schemes [1]. Existing threshold authentication (and threshold signature) schemes [6] are also based on them. These have shown the usefulness of homomorphic threshold schemes over a finite Abelian group.

To guarantee that a threshold scheme is secure Stinson and Vanstone [17] speak about perfect threshold schemes. *Perfect* threshold schemes do not reveal anything about the secret k when $t - 1$ shares are used. Let S be the set of all possible shares and $\mathcal{K}$ be the set of possible secrets (keys). A threshold scheme is called *ideal* when it is perfect and when $|S|$ equals $|\mathcal{K}|$, $|S|$ is the cardinality of the set S.

Benaloh [1] defined homomorphic threshold schemes as those having the property that when $s_i \in S$ is i's share of $k \in \mathcal{K}$ and $s'_i \in S$ is i's share of $k' \in \mathcal{K}$, then $s_i \cdot s'_i$ is i's share of $k * k'$ and for such threshold schemes t shareholders can reconstruct $k * k'$ using their $s_i \cdot s'_i$.

To keep storage requirements restricted it is important to make the size of the shares in a sharing scheme as small as possible. It is well known that in perfect general sharing schemes the size of the share must be at least as large as the size of the key. Therefore ideal sharing schemes have been studied extensively. Unfortunately, ideal sharing schemes cannot be made for all access structures [2]. The maximum l for an ideal threshold scheme is dependent on t and $|\mathcal{K}|$ [11]. Other results on ideal sharing schemes encompasses a classification for ideal sharing schemes [4] and the fact that without having public information no threshold scheme can be made ideal [17]. Observe that Shamir's threshold scheme [14] and others [5, 11, 12] are homomorphic and schemes satisfying this property are becoming important.

In this paper we study *ideal homomorphic* threshold and general sharing schemes where the key space is a finite Abelian group. On the first look it seems that this study would only result in a combination of earlier obtained results. Unexpectedly we are able to exemplify a set of secrets (keys) $\mathcal{K}$ that when it forms a group and one insists that the threshold scheme is homomorphic then there does not exists an ideal threshold scheme (see Section 4). Moreover we can give infinitely many examples of this. In threshold schemes the maximum l is dependent on t and the cardinality of $\mathcal{K}$ [11]. However, in *homomorphic* threshold scheme the maximum l is a dependent on t and the algebraic structure of $\mathcal{K}$.

The results in this paper will make protocols which use homomorphic threshold schemes over a finite Abelian group (*e.g.*, see [1, 6]) more practical. For instance, being able to use an ideal scheme has a direct implication on the practicality of protocol using the homomorphic threshold scheme.

In Section 2 we overview the necessary definitions. In Section 3 we prove that $S(\cdot)$ is isomorphic to $\mathcal{K}(*)$ in any ideal homomorphic general sharing scheme over a group $\mathcal{K}(*)$. Using this property a classification of ideal homomorphic threshold schemes over a finite Abelian group is made in Section 4. A link between a geometric general sharing scheme [16] and homomorphic general sharing schemes is established in Section 5.

2 Background and notation

We now introduce formal definitions and notation used in this paper. When $\mathcal{A}$ is a set, $|\mathcal{A}|$ will denote the cardinality of the set. We now overview the definition of threshold schemes [3, 14] and homomorphic threshold schemes [1].

Definition 1 A *threshold scheme* contains two algorithms, one which creates shares of a secret key $k \in \mathcal{K}$ for l individuals so that any t individuals (t is fixed and $t \leq l$) can regenerate the secret using the second algorithm, yet less than t individuals cannot using any method. Let $\mathcal{A} = \{1, \ldots, l\}$ and S be the set of possible shares[1]. The distributor generates the tuple $S_\mathcal{A} = (s_1, \ldots, s_l)$ where $s_i \in S$ and the public directory $\mathcal{X}_\mathcal{A} = \{x'_i \mid i \in \mathcal{A}\}$.

More formally, a t-out-of-l threshold scheme satisfies:

1. $\forall \mathcal{B} \subset \mathcal{A}$ where $|\mathcal{B}| = t - 1$ holds: if $\mathsf{H}(k) \neq 0$ then $0 < \mathsf{H}(k \mid S_\mathcal{B}, \mathcal{X}_\mathcal{A}) \leq \mathsf{H}(k)$ for H the entropy function [8] and if $S_\mathcal{A} = (s_1, \ldots, s_l)$ then $S_\mathcal{B} =$

---

[1]A more general definition allows the set of shares to be different for each shareholder $i \in \mathcal{A}$ [2, 7]. All our results remain valid for the more general definition. To avoid heavy notation we assume the set of shares are identical.

$(s_{i_1}, \ldots, s_{i_{|D|}})$ where $\mathcal{B} = \{i_1, \ldots, i_{|B|}\}$.

2. $\forall \mathcal{B} \subset \mathcal{A}$ where $|\mathcal{B}| = t$, there exists a function $\eta_{\mathcal{B}, \mathcal{X}_A}$ such that $\eta_{\mathcal{B}, \mathcal{X}_A}(S_\mathcal{B}) = k$.

Schemes in which for any $\mathcal{B} \subset \mathcal{A}$ with $|\mathcal{B}| = t-1$ holds that $H(k | S_\mathcal{B}, \mathcal{X}_A) = H(k)$ are called *perfect*. When a sharing scheme is perfect and $|\mathcal{K}|/|\mathcal{S}| = 1$, it is called an *ideal* sharing scheme.

Definition 2 Let "$\cdot$" be a binary operation over $\mathcal{S}$ (so $\mathcal{S}$ is closed under "$\cdot$") and "$*$" be a binary operation over $\mathcal{K}$. If $\eta_{\mathcal{B}, \mathcal{X}_A}$ is a homomorphism from $\mathcal{S}^t(\cdot)$ to $\mathcal{K}(*)$ for each $\mathcal{B} \subset \mathcal{A}$ with $|\mathcal{B}| = t$, then the threshold scheme is called a homomorphic threshold scheme[2].

The above definitions can be easily modified to general sharing schemes allowing for any access structure.

3   Structure of shares

We first analyze the structure of $\mathcal{S}(\cdot)$ in an ideal general sharing scheme where $\mathcal{K}(*)$ is a group. We note that the definition of homomorphic threshold scheme is very general and does *not* even state whether $\mathcal{S}(\cdot)$ has any special properties such as being a group. The same statement can be made about homomorphic general sharing schemes.

Theorem 1 *If the key space $\mathcal{K}(*)$ of an ideal homomorphic t-out-of-l threshold scheme is a finite group, then the share space $\mathcal{S}(\cdot)$ is isomorphic to $\mathcal{K}(*)$.*

Proof.   In any ideal threshold scheme when $s_{i_1}, \ldots, s_{i_t} \in \mathcal{S}$ then $S_\mathcal{B} = (s_{i_1}, \ldots, s_{i_t})$ is a valid tuple of shares. Clearly, if $S'_\mathcal{B} = (s'_{i_1}, \ldots, s'_{i_t})$ is a valid tuple of shares then $S''_\mathcal{B} = (s_{i_1}, s'_{i_2}, \ldots, s'_{i_t})$ is also for the scheme to perfect. Repeating this process for $i_2, \ldots, i_t$ proves that *any* combination of t elements of $\mathcal{S}$ can be valid as t shares.

---

[2]If the threshold scheme is not perfect then the above definition must be slightly adapted.

Let $s \in S$ and $\eta_{\mathcal{B},\mathcal{X}_A}(s,\ldots,s) = k \in \mathcal{K}$. For the threshold scheme to be perfect one needs that $\eta_{\mathcal{B},\mathcal{X}_A}(s,\ldots,s,S) = \mathcal{K}$. First $s \cdot S = S$ since $\eta_{\mathcal{B},\mathcal{X}_A}(s,\ldots,s) * \eta_{\mathcal{B},\mathcal{X}_A}(s,\ldots,s,S) = \eta_{\mathcal{B},\mathcal{X}_A}(s \cdot s,\ldots,s \cdot s, s \cdot S) = k * \mathcal{K} = \mathcal{K}$. Since $s \cdot S = S$ and S is finite, there exist an element e_x for every $x \in S$ such that $x \cdot e_x = x$. From this we note that $\eta_{\mathcal{B},\mathcal{X}_A}(e_x,\ldots,e_x) = 1 \in \mathcal{K}$ since $\eta_{\mathcal{B},\mathcal{X}_A}(x,\ldots,x) = \eta_{\mathcal{B},\mathcal{X}_A}(x \cdot e_x,\ldots,x \cdot e_x)$. Now, $\eta_{\mathcal{B},\mathcal{X}_A}(x,\ldots,x,y) = \eta_{\mathcal{B},\mathcal{X}_A}(x \cdot e_x,\ldots,x \cdot e_x, y \cdot e_x) = \eta_{\mathcal{B},\mathcal{X}_A}(x,\ldots,x,y \cdot e_x)$. Since $|S| = |\mathcal{K}|$ and the scheme is perfect, $y \cdot e_x = y$. Thus e_x is a right identity element. Similarly we can prove a left identity element. So there exists an identity element $1 \in S$. Let $\psi_{i,\mathcal{B},\mathcal{X}_A}(x) = \eta_{\mathcal{B},\mathcal{X}_A}(1,\ldots,1,x)$ where x is the ith share. The mapping $\psi_{i,\mathcal{B},\mathcal{X}_A}$ is a homomorphism from S to $\mathcal{K}$. Observe that $\psi_{i,\mathcal{B},\mathcal{X}_A}$ is onto because the scheme is perfect. The fact that $|S| = |\mathcal{K}|$ implies $\psi_{i,\mathcal{B},\mathcal{X}_A}$ is bijective. Because $\psi_{i,\mathcal{B},\mathcal{X}_A}(x \cdot (y \cdot z)) = \psi_{i,\mathcal{B},\mathcal{X}_A}((x \cdot y) \cdot z)$ and because $\psi_{i,\mathcal{B},\mathcal{X}_A}$ is bijective, we must have that $S(\cdot)$ is associative and therefore $S(\cdot)$ is a group. $\square$

A *monotone access structure* satisfies the property that when a set of shareholders $\mathcal{B}$ can recompute a secret then any superset $\mathcal{B}' \supset \mathcal{B}$ can also recompute the secret. Careful examination of the above proof indicates the following.

Corollary 1 *If the key space $\mathcal{K}(*)$ of an ideal homomorphic monotone general sharing scheme is a finite group, then the share space $S(\cdot)$ is isomorphic to $\mathcal{K}(*)$.*

4 Classification

Due to [11], in any threshold scheme there is the following bound on l: $l_{\text{MAX}} \leq |\mathcal{K}| + t - 2$. Theorem 1 can be used to find a bound for l_{MAX} for ideal homomorphic threshold schemes.

Theorem 2 *Let $\mathcal{K}(*)$ be a finite Abelian group. There is an ideal homomorphic t-out-of-l threshold scheme with key space $\mathcal{K}(*)$ if and only if for each Sylow subgroup $G(*)$ of $\mathcal{K}(*)$ there is an ideal homomorphic t-out-of-l threshold scheme with key space $G(*)$.*

Proof. It is well known that each Abelian group $\mathcal{K}$ is isomorphic to $G_1 \times G_2 \times \cdots \times G_c$ where the G_i are all the different Sylow subgroups[3] in $\mathcal{K}$. Let $\psi_{i,\mathcal{B},\mathcal{X}_A}$ be as in the proof of Theorem 1. Note that to $k \in \mathcal{K}$ corresponds an (k', k'') where $k' \in G_1$ and $k'' \in G_2 \times \cdots \times G_c$, similarly due to Theorem 1, $s_i \in \mathcal{S}$ corresponds to (s_i', s_i'') $(1 \leq i \leq l)$. So, $\eta_{\mathcal{B},\mathcal{X}_A} : \mathcal{S}^t \to \mathcal{K}$ corresponds to $\eta'_{\mathcal{B},\mathcal{X}_A} : (G_1 \times (G_2 \times \cdots \times G_c))^t \to G_1 \times (G_2 \times \cdots \times G_c)$ and similarly we define $\psi'_{i,\mathcal{B},\mathcal{X}_A}$. Because $\psi'_{i,\mathcal{B},\mathcal{X}_A}$ gives a group isomorphism and because the G_j are Sylow subgroups we have $\psi'_{i_t,\mathcal{B},\mathcal{X}_A}((s_{i_t}', 1)) = (k_t', 1)$. Observe that $\eta_{\mathcal{B},\mathcal{X}_A}(s_{i_1}, \ldots, s_{i_t}) = \prod_{j \in \mathcal{B}} \psi_{j,\mathcal{B},\mathcal{X}_A}(s_j)$. One can now prove that $\eta'_{\mathcal{B},\mathcal{X}_A}((s_{i_1}', 1), \ldots, (s_{i_t}', 1)) = (k', 1)$ for some k'. Similarly $\eta'_{\mathcal{B},\mathcal{X}_A}((1, s_{i_1}''), \ldots, (1, s_{i_t}'')) = (1, k'')$. When $\eta'_{\mathcal{B},\mathcal{X}_A}((s_{i_1}', s_{i_1}''), \ldots, (s_{i_t}', s_{i_t}'')) = (a, b)$ then $a = k' \in G_1$ and $b = k'' \in G_2 \times \cdots \times G_c$, because $\eta'_{\mathcal{B},\mathcal{X}_A}$ is a function. So this induces an ideal homomorphic t-out-of-l threshold scheme with key space $\mathcal{K}' \cong G_1$. A similar argument is made for each G_i $(2 \leq i \leq c)$.

Moreover, if there exists an ideal threshold scheme for each G_i $(1 \leq i \leq c)$ then there exists one for the key space $\mathcal{K} \cong G_1 \times G_2 \times \cdots \times G_c$. $\qquad\square$

Corollary 2 *There exists an infinite number of Abelian groups $\mathcal{K}$ for which there does not exist an ideal homomorphic threshold scheme when $l > 2$, even when $l < |\mathcal{K}|$ and $t = 2$.*

Proof. Let $\mathcal{K}(*) \cong Z_2 \times Z_{q_2^{w_2}} \times \cdots \times Z_{q_c^{w_c}}(+)$ where $q_i \neq 2$ are primes and $q_i \neq q_j$ for $i \neq j$. Due to [11] the maximum l in a threshold scheme over $\mathcal{K}$ is $l_{MAX} \leq |\mathcal{K}| + t - 2$. Due to Theorem 2, when $t = 2$ in an ideal homomorphic threshold scheme then $l \leq 2$ for our $\mathcal{K}$. $\qquad\square$

Careful examination of the proof for Theorem 2 indicates that the theorem can be generalized to any monotone access structure.

Corollary 3 *Let $\mathcal{K}(*)$ be a finite Abelian group. Then there is an ideal homomorphic monotone general sharing scheme with key space $\mathcal{K}$ if and only if for each Sylow subgroup G of $\mathcal{K}$ there is an ideal homomorphic monotone general sharing scheme for key space G.*

---

[3]A Sylow p-subgroup of $\mathcal{K}$, p prime, is a subgroup whose order is the largest power of p which divides the order of $\mathcal{K}$ [10].

Corollary 4 *Let $\mathcal{K}(*)$ be a finite Abelian group. If there is an ideal homomorphic t-out-of-l threshold scheme with key space $\mathcal{K}$, then for each characteristic subgroup[4] G of $\mathcal{K}$ there is an ideal homomorphic t-out-of-l threshold scheme with key space G.*

Proof. Restrict the shares to G and use the fact that $\eta_{B,\mathcal{X}_A}(s_{i_1}, \ldots, s_{i_{|B|}}) = \prod_{j \in B} \psi_{j,B,\mathcal{X}_A}(s_j)$ (because ideal threshold schemes are erasure codes [11]). □

Corollary 4 implies that there is *no* ideal homomorphic 2-out-of-3 threshold scheme when the key space is $Z_4(+)$, but there is one when the key space is $Z_2 \times Z_2(+)$. So insisting on having a homomorphic scheme does make it not ideal.

5 A general homomorphic sharing scheme

In this section we establish a link between a geometric general sharing scheme [16] and the homomorphic property. Using finite projective geometry, a method to create sharing schemes for any monotone access structure has been developed [16]. Let us briefly review their scheme [16]. In their sharing scheme a public hyperplane V_d intersects with a secret hyperplane V_i at a point which is the secret. Points are given to each shareholder in such a way that they meet the following two conditions. First, when a set of shareholders allowed by the access structure work together, they will be able to generate the secret hyperplane V_i. When a set of shareholders not allowed by the access structure work together, they do not obtain any information about the secret point.

Lemma 1 *Let $\mathcal{K}(*)$ be any finite Abelian group. The general sharing scheme in [16] induces a perfect homomorphic sharing scheme with key space $\mathcal{K}$.*

Proof. We modify the scheme developed in [16]. When in [16] the distributor gives a point p_i to shareholders $\{j_1, \ldots, j_{h_i}\}$, the distributor here will give $s_i \in \mathcal{K}$ to shareholders $\{j_1, \ldots, j_{h_i}\}$. Let the total number of such points p_i in [16] be m, then $\prod_{1 \leq i \leq m} s_i = k$ where k is the secret and $s_1, \cdots, s_{m-1}$ have been chosen

---

[4]Characteristic subgroups are those subgroups that are mapped into themselves by all automorphisms.

as independently random elements in $\mathcal{K}$. The fact that the sharing scheme is perfect follows from the one-time-pad [15]. □

6   Conclusion

Earlier results have demonstrated that homomorphic threshold schemes are useful [1, 6]. A homomorphic *ideal* general sharing scheme where the secret domain is a group has a share domain which is isomorphic to the secret domain. A bound on the maximum l can be made for homomorphic threshold schemes over an Abelian group. This result shows that it is better not to use homomorphic threshold (or sharing) schemes when the homomorphic property is not needed.

Acknowledgment

We thank Mike Burmester (University of London) for discussions about this paper.

REFERENCES

[1] J. C. Benaloh. Secret sharing homomorphisms: Keeping shares of a secret secret. In A. Odlyzko, editor, *Advances in Cryptology, Proc. of Crypto '86 (Lecture Notes in Computer Science 263)*, pp. 251–260. Springer-Verlag, 1987. Santa Barbara, California, U.S.A., August 11–15.

[2] J. C. Benaloh and J. Leichter. Generalized secret sharing and monotone functions. In S. Goldwasser, editor, *Advances in Cryptology, Proc. of Crypto '88 (Lecture Notes in Computer Science 403)*, pp. 27–35. Springer-Verlag, 1990. Santa Barbara, California, U.S.A., August 11–15.

[3] G. R. Blakley. Safeguarding cryptographic keys. In *Proc. Nat. Computer Conf. AFIPS Conf. Proc.*, pp. 313–317, 1979. vol.48.

[4] E. F. Brickell and D. M. Davenport. On the classification of ideal secret sharing schemes. *Journal of Cryptology*, 4(2), pp. 123–134, 1991.

[5] G. I. Davida, R. DeMillo, and R. Lipton. Protecting shared cryptographic keys. In *Proceedings of the 1980 Symposium on Security and Privacy*, pp. 100–102. IEEE Computer Society, April 1980. IEEE Catalog No. 80 CH1522-2.

[6] Y. Desmedt and Y. Frankel. Shared generation of authenticators and signatures. In J. Feigenbaum, editor, *Advances in Cryptology, Proc. of Crypto '91 (Lecture Notes in Computer Science 576)*, pp. 457–469. Springer-Verlag, 1992. Santa Barbara, California, U.S.A., August 11-15.

[7] Y. Frankel, Y. Desmedt, and M. Burmester. Non-existence of homomorphic general sharing schemes for some key spaces. To be presented at Crypto '92, to appear in: Advances in Cryptology. Proc. of Crypto '92 (Lecture Notes in Computer Science), Springer-Verlag, 1992.

[8] R. G. Gallager. *Information Theory and Reliable Communications*. John Wiley and Sons, New York, 1968.

[9] M. Ito, A. Saito, and T. Nishizeki. Secret sharing schemes realizing general access structures (in English). In *Proc. IEEE Global Telecommunications Conf., Globecom'87*, pp. 99–102, Washington, DC., 1987. IEEE Communications Soc. Press. Also in "Trans. IEICE Japan" Vol. J71-A, No. 8, 1988 (in Japanese).

[10] N. Jacobson. *Basic Algebra I*. W. H. Freeman and Company, New York, 2nd edition, 1985.

[11] E. D. Karnin, J. W. Greene, and M. Hellman. On secret sharing systems. *IEEE Tr. Inform. Theory*, 29(1), pp. 35–41, January 1983.

[12] R. J. McEliece and D. V. Sarwate. On sharing secrets and Reed-Solomon codes. *Comm. ACM*, 24(9), pp. 583–584, September 1981.

[13] M. Rabin. Efficient dispersal of information for security, load balancing, and fault tolerance. *Journal of the ACM*, 36(2), pp. 335–348, April 1989.

[14] A. Shamir. How to share a secret. *Commun. ACM*, 22, pp. 612–613, November 1979.

[15] C. E. Shannon. Communication theory of secrecy systems. *Bell System Techn. Jour.*, 28, pp. 656–715, October 1949.

[16] G. J. Simmons, W. Jackson, and K. Martin. The geometry of shared secret schemes. *Bulletin of the Institute of Combinatorics and its Applications*, 1, pp. 71–88, 1991.

[17] D. R. Stinson and S. A. Vanstone. A combinatorial approach to threshold schemes. *SIAM Journal on Discrete Mathematics*, 1(2), pp. 230–236, 1988.

F.F.T. Hashing is not Collision-free

T. BARITAUD * , H. GILBERT * , M. GIRAULT **

(*) CNET PAA / TSA / SRC
38 - 40, avenue du Général Leclerc
92131 ISSY LES MOULINEAUX (France)

(**) SEPT PEM
42, rue des Coutures
BP 6243
14066 CAEN (France)

Abstract

The FFT Hashing Function proposed by C.P. Schnorr [1] hashes messages of arbitrary length into a 128-bit hash value. In this paper, we show that this function is not collision free, and we give an example of two distinct 256-bit messages with the same hash value. Finding a collision (in fact a large family of colliding messages) requires approximately 2^{23} partial computations of the hash function, and takes a few hours on a SUN3- workstation, and less than an hour on a SPARC-workstation.

A similar result discovered independently has been announced at the Asiacrypt'91 rump session by Daemen-Bosselaers-Govaerts-Vandewalle [2].

1 The FFT Hashing Function

1.1 The Hash algorithm

Let the message be given as a bit string $m_1 m_2 ... m_t$ of t bit.

The message is first padded so that its length (in bits) becomes a multiple of 128. Let the padded message $M_1 M_2 ... M_n$ consist of n blocks $M_1, ... , M_n$, each of the M_i (i=1, ... ,n) being 128-bit long.

The algorithm uses a constant initial value H_0 given in hexadecimal as

$$H_0 = 0123\ 4567\ 89ab\ cdef\ fedc\ ba98\ 7654\ 3210 \text{ in } \{0,1\}^{128}.$$

R.A. Rueppel (Ed.): Advances in Cryptology - EUROCRYPT '92, LNCS 658, pp. 35-44, 1993.
© Springer-Verlag Berlin Heidelberg 1993

Let p be the prime $65537 = 2^{16} + 1$.

We will use the Fourier transform $\quad FT_8 : \{0, \ldots, p-1\}^8 \longrightarrow \{0, \ldots, p-1\}^8$

$$(a_0, \ldots, a_7) \longrightarrow (b_0, \ldots, b_7)$$

with $\quad b_i = \sum_{j=0}^{7} 2^{4ij} a_j \mod p$, for $i = 0, \ldots, 7$.

Algorithm for the hash function h :

INPUT   : $M_1 M_2 \ldots M_n$ in $\{0,1\}^{n.128}$ (a padded message)

DO   : $H_i = g(H_{i-1}, M_i)$ for $i = 1, \ldots, n$

OUTPUT : $h(M) := H_n$

Algorithm for $g : Z_p^{16} \longrightarrow \{0,1\}^{8.16}$

INPUT $(c_0, \ldots, c_{15})$ in $\{0,1\}^{16.16}$

1. $(c_0, c_2, \ldots, c_{14}) := FT_8(c_0, c_2, \ldots, c_{14})$

2.   FOR $i = 0, \ldots, 15$ DO

$$c_i := c_i + c_{i-1}c_{i-2} + c_{c_{i-3}} + 2^i \pmod p$$

(The lower indices $i, i-1, i-2, i-3, c_{i-3}$ are taken modulo 16)

3.   REPEAT steps 1 and 2

OUTPUT $\overline{c}_i := c_i \mod 2^{16}$, for $i = 8, \ldots, 15$ (an element of $\{0,1\}^{8.16}$)

1.2 Notations

For a better clarity of our explanation, we will denote by c_i^0 $(i=0, \ldots, 15)$ the initial c_i values, and we will denote by step 3 (resp. step 4) the second pass of step 1 (resp. step2) in the algorithm for g.

When it will be necessary to avoid any kind of slip, we will denote by c_i^k $(i=0, \ldots, 15 ; k=0, \ldots, 4)$ the c_i intermediate value, after step k.

In order to simplify the expressions, we are using the following notations :

- The additions (x+y), multiplications (x.y) and exponentiations (x^y) are implicitly made modulo

p, except when the operands are lower indices.
- The = symbol denotes that the right and the left terms are congruent modulo p.
- For lower indices the additions (i+j) and substractions (i-j) are implicitly made modulo 16, and

the $\equiv$ symbol denotes that the right and the left terms are congruent modulo 16.

1.3 Preliminary remarks

The difficulty of finding collisions is related to the diffusion properties of the hashing function, i.e. the influence of a modification of an intermediate variable on the subsequent variables of the calculation.

<u>Remark 1</u> (limitation on the diffusion at steps 1 and 3)

At step 1 and 3, the input values $c_1, c_2, \dots, c_{15}$ are kept unchanged.

<u>Remark 2</u> (limitation on the diffusion at steps 2 and 4)

The diffusion introduced by the $c_{i-1} c_{i-2}$ terms in the recurrence for steps 2 and 4 can sometimes be

cancelled (if one of values e_{i-1} and c_{i-2} is 0). More precisely, let $(c_0^1, c_1^1, \dots, c_{15}^1)$ be the input to step 2 :

<u>Proposition 1</u> : If for a given value i in $\{1, \dots, 14\}$ we have $c_{i-1}^2 = c_{i+1}^2 = 0$ and if $c_{13}^1 \not\equiv i$; $c_{14}^1 \not\equiv i$;

$c_{15}^1 \not\equiv i$; $c_j^2 \not\equiv i$ for j in $\{0, \dots, 12\}$, then the impact of replacing the input value c_i^1 by a new value $e_i^1 + \Delta c_i^1$

such that $c_i^1 + \Delta c_i^1 \equiv c_i^1$, is limited to the output value c_i^2 (that means c_j^2 are not modified for $j \neq i$).

<u>Proposition 2</u> : If $e_{14}^1 = c_0^2 = 0$ and if $c_j^2 \not\equiv 15$ for j in $\{1, \dots, 11\}$ then the impact of replacing the

input value e_{15}^1 by a new value $c_{15}^1 + \Delta c_{15}^1$ such that $c_{15}^1 + \Delta c_{15}^1 \equiv c_{15}^1$, is limited to the output value c_{15}^2.

Similarly, let $(c_1^3, c_2^3, \dots, c_{15}^3)$ be the input to step 4 :

<u>Proposition 1'</u> : If for a given value i in $\{1, \dots, 14\}$ we have $e_{i-1}^4 = e_{i+1}^4 = 0$ and if $e_{13}^3 \not\equiv i$;

$c_{14}^3 \not\equiv i$; $c_{15}^3 \not\equiv i$; $c_j^4 \not\equiv i$ for j in $\{0, \dots, 12\}$, then the impact of replacing the input value c_i^3 by a new value

$e_i^3 + \Delta c_i^3$ such that $c_i^3 + \Delta c_i^3 \equiv e_i^3$, is limited to the output value c_i^4.

<u>Proposition 2'</u> : If $c_{14}^3 = c_0^4 = 0$ and if $c_j^4 \not\equiv 15$ for j in $\{1, \dots ,11\}$ then the impact of replacing the input value c_{15}^3 by a new value $c_{15}^3 + \Delta c_{15}^3$ such that $c_{15}^3 + \Delta c_{15}^3 \equiv c_{15}^3$ is limited to the output value c_{15}^4.

2 Construction of two colliding messages

2.1 Construction of a partial collision

We first find two 128-bit blocks M_1 and M'_1 which hash values $H_1 = (\overline{c}\,_8^4, \dots ,\overline{c}\,_{15}^4)$ and $H'_1 = (\overline{c'}\,_8^4, \dots ,\overline{c'}\,_{15}^4)$ differ only by their right components $\overline{c}\,_{15}^4$ and $\overline{c'}\,_{15}^4$. We will later refer to this property in saying that M_1 and M'_1 realize a <u>partial collision</u>.

Our technique for finding M_1 and M'_1 is the following : we search M_1 values such that $c_{14}^1= 0$; $c_0^2= 0$; $c_{14}^3= 0$; $c_0^4= 0$. The propositions 2 and 2' suggest that for such a message $M_1=(c_8^0, \dots , c_{14}^0, c_{15}^0)$, M_1 and the message $M'_1= (c_8^0, \dots , c_{14}^0, c_{15}^0 + 16)$ realize a partial collision with a significant probability (approximately 1/8).

There are two main steps for finding M_1.

<u>Step1</u> : Selection of $c_8^0, c_{10}^0, c_{12}^0$ and c_{14}^0

Arbitrary (e.g. random) values are taken for c_{12}^0 and c_{14}^0. The values of c_8^0 and c_{10}^0 are then deduced from these values by solving the following linear system :

$$\begin{cases} c_{14}^1 = 0 & \quad (1) \\ c_0^1 = -1 & \quad (2) \end{cases}$$

<u>Proposition 3</u> :

If $c_{13}^0 \equiv 14$ then $c_{14}^1 = 0$ and $c_0^2 = 0$ independently of the values of $c_9^0, c_{11}^0, c_{13}^0, c_{15}^0$.

<u>Proof</u> : This is a direct consequence of the definition of the g function.

<u>Step 2</u> : Selection of $c_9^0, c_{11}^0, c_{13}^0, c_{15}^0$

The values of $c_8^0, c_{10}^0, c_{12}^0, c_{14}^0$ are taken from Step 1.

We fix the values of $c_{11}^0 = 0$ and $c_{15}^0 = 0$. An arbitrary (e.g random) value is taken for c_9^0. We first

calculate the c_{12}^2 and c_{14}^3 values corresponding to the chosen value of c_9^0, c_{11}^0 and c_{15}^0 and to the temporary

value $c_{13}^0 = 14$. Based on these preliminary calculations, we "correct" the temporary value $c_{13}^0 = 14$ by a

quantity Δc_{13}^0, i.e. we replace the value $c_{13}^0 = 14$ by the value $c_{13}^0 = 14 + \Delta c_{13}^0$, and we leave the other

input values unchanged. We denote by Δc_j^i ($0 \le i \le 4$; $0 \le j \le 15$) the corresponding variations of the

intermediate variables in the H_1 calculation. We select Δc_{13}^0 in such a way that the quantity $c_{14}^3 + \Delta c_{14}^3$

(i.e the new value of c_{14}^3) is equal to zero with a good probability.

<u>Proposition 4</u> : If $c_{12}^2 \neq 0$ and $\dfrac{-c_{14}^3}{2^{4.7.7} c_{12}^2} \equiv 0$ and $c_j^2 \not\equiv 13$ for $1 \le j \le 11$ then the above values of

, c_{15}^1, c_0^2 and the value $\Delta c_{13}^0 = \dfrac{-c_{14}^3}{2^{4.7.7} c_{12}^2}$ lead to the three relations

$$\begin{cases} c_{14}^1 + \Delta c_{14}^1 = 0 & \text{(a)} \\[2mm] c_0^2 + \Delta c_0^2 = 0 & \text{(b)} \\[2mm] c_{14}^3 + \Delta c_{14}^3 = 0 & \text{(c)} \end{cases}$$

<u>Proof</u> : (a) is straightforward; (b) and (c) are direct consequences of the following relations, which result
from the definition of the g function :

$$\Delta c_{j-2}^2 = 0 \text{ for } 0 \le j \le 12 \quad ; \quad \Delta c_{13}^2 = \Delta c_{13}^0 \quad ; \quad \Delta c_{14}^2 = c_{12}^2 \cdot \Delta c_{13}^2 \quad ; \quad \Delta c_{14}^3 = 2^{4.7.7} \cdot \Delta c_{14}^2$$

We performed a large number n_1 of trials of step 1. For each trial of step 1, we made a large

number n_2 of trials of step 2. The success probability of step 2, i.e the probability that the trial of a c_9^0

value leads to a message such that (a), (b) and (c) are realized is slightly less than 1/16 (since the strongest condition in proposition 2 is : $\dfrac{-c_{14}^3}{2^{4.4.7}c_{12}^2} \cong 0$). Therefore the probability that a step 2 trial leads to a message

M_1 such that $c_{14}^1 = c_0^2 = c_{14}^3 = c_0^4 = 0$ is slightly less than $1/16 \cdot 2^{-16} = 2^{-20}$.

Moreover, the probability that such a message M_1 leads to a partial collision is basically the probability that none of the c_{i-3} mod 16 indices occurring in the calculation of c_0^2 to c_{15}^2 and c_0^4 to c_{15}^4 takes the value 15, which is close to $1/8$. So, in summary, approximatively 2^{23} partial computations of the g function were necessary to obtain a suitable message $M_1 = (c_8^0, \ldots, c_{14}^0, c_{15}^0)$, such that M_1 and the message $M'_1 = (c_8^0, \ldots, c_{14}^0, c_{15}^0 + 16)$ lead to partially colliding hash values $H_1 = (\overline{c}\,_8^4, \ldots, \overline{c}\,_{15}^4)$ and $H'_1 = (\overline{c}\,_8^4, \ldots, \overline{c}\,_{15}^4 + 16)$.

2.2 Construction of a full collision using a partial collision

We now show how to find a 128-bit message $M_2 = (c_8^0, \ldots, c_{15}^0)$ such that the previously obtained hash values H_1 and H'_1 (denoted in this section by $(c_0^0, \ldots, c_7^0)$ and $(c'_1^0, \ldots, c'_6^0, c'_7^0) = (c_1^0, \ldots, c_6^0, c_7^0 + 16)$) respectively lead to the same hash value H_2 (when combined with M_2) : $g(H_1, M_2) = g(H'_1, M_2)$.

Our technique for finding M_2 is quite similar to the one used for finding M_1 and M'_1. Let us denote by c_j^i (resp c'_j^i) ($0 \le i \le 4$, $0 \le j \le 15$) the intermediate variables of the calculations of $g(H_1, M_2)$ (resp $g(H'_1, M_2)$).

We search M_2 values such that $c_6^2 = c_8^2 = c_6^4 = c_8^4 = 0$. The propositions 1 and 1' suggest that the probability that the 16-uples $(c_0^4, \ldots, c_{15}^4)$ and $(c'_0^4, \ldots, c'_{15}^4)$ differ only by their components c_7^4 and c'_7^4 which implies that the probability to have $g(H_1, M_2) = g(H'_1, M_2)$ is quite substantial, approximatively $1/8$.

There are two main steps for the search of M_2 :

<u>Step 1</u> : Selection of $c_8^0, c_{10}^0, c_{12}^0, c_{14}^0, c_9^0$.

An arbitrary (e.g random) value is taken for c_{14}^0. The values of $c_8^0, c_{10}^0, c_{12}^0$ are deduced from c_{14}^0 by solving the following linear system :

$$\begin{cases} c_{14}^1 = 0 & (3) \\ c_0^1 = -1 & (4) \\ c_8^1 = -2^8 & (5) \end{cases}$$

A preliminary calculation, where c_9^0, c_{11}^0 and e_{15}^0 are set to the temporary value 0 and c_{13}^0 is set to the temporary value 14, is made. The obtained value of c_6^2, denoted by δ, is kept.

<u>Proposition 5</u> : If $c_8^0, c_{10}^0, c_{12}^0, c_{14}^0$ are solutions of (3), (4), (5) and if in addition the values

$e_9^0 = p-\delta, c_{11}^0 = 0, c_{13}^0 = 14, e_{15}^0 = 0$ lead to intermediate values such that : c_1^2 mod 16 is not in

$\{9,11,13,15\}$; c_2^2 mod 16 is not in $\{9,11,13,15\}$; $c_3^2 \equiv 9$ mod 16; e_4^2 mod 16 is not in $\{9,11,13,15\}$;

c_5^2 mod 16 is in $\{0,6,14\}$, then if we fix the value $c_9^0 = p-\delta$, for any value of $c_{13}^0 \equiv 14$ and for any value

of e_{15}^0 such that $c_{15}^0 \equiv 0$ we have :

$$e_{14}^1 = 0 \; ; \qquad c_0^2 = 0 \; ; \qquad c_6^2 = 0 \; ; \qquad c_8^2 = 0 \; .$$

<u>Proof</u> : The proof of this proposition is easy. Finding the $c_8^0, c_{10}^0, c_{12}^0, c_{14}^0$ and c_9^0 values satisfying the conditions of the above proposition is quite easy, and requires the trial of a few hundreds c_{14}^0 values.

<u>Step 2</u> : Selection of $c_{11}^0, c_{13}^0, e_{15}^0$

The values of $c_8^0, c_{10}^0, c_{12}^0, e_{14}^0, c_9^0$ are taken from Step 1 ; these values are assumed to realize the conditions of the above proposition.

An arbitrary (e.g random) value is taken for c_{11}^0. A preliminary calculation is made, using the selected c_{11}^0 value and the temporary values $c_{13}^0 = 14; c_{15}^0 = 0$. The corresponding values of c_{12}^2 and c_8^3 are kept.

Based on these preliminary calculations, we "correct" the temporary value of c_{13}^0 by a quantity Δc_{13}^0 and

we also consider new values $c_{15}^0 + \Delta c_{15}^0$ for c_{15}^0. The variation Δc_{13}^0 is selected in such a way that for any

Δc_{15}^0 value satisfying $\Delta c_{15}^0 \equiv 0$, the new value $e_8^3 + \Delta e_8^3$ of e_8^3 is equal to -2^8 with a substantial probability.

Proposition 6 : If $e_{12}^2 \neq 0$ and $\dfrac{-2^8 - e_8^3}{2^{4.4.7} \, c_{12}^2} \equiv 0$ and e_j^2 mod 16 is not in $\{13,15\}$ for $1 \leq j \leq 11$ then for

any variation $\Delta c_{15}^0 \equiv 0$ on c_{15}^0 such that $c_{15}^2 + \Delta c_{15}^0 < p$ and $c_{15}^4 + \Delta c_{15}^0 < p$, the variation

$\Delta c_{13}^0 = \dfrac{-2^8 - e_8^3}{2^{4.4.7} \, c_{12}^2}$ on the c_{13}^0 value leads to the following new values :

$c_{14}^1 + \Delta c_{14}^1 = 0$; $c_0^2 + \Delta c_0^2 = 0$; $c_6^2 + \Delta c_6^2 = 0$; $c_8^2 + \Delta c_8^2 = 0$; $c_8^3 + \Delta c_8^3 = -2^8$.

We performed a number n_1 of trials of step 1. For each successful trial of step 1, we made a large

number n_2 of trials of c_{11}^0 values at step 2. For those c_{11}^0 values satisfying the conditions of the above

proposition, we made a large number n_3 of trials of new c_{15}^0 values such that $\Delta c_{15}^0 \equiv 0$. The probability

that the trial of a new Δc_{15}^0 value leads to intermediate variables satisfying the four equations $c_6^2 = 0$; $c_8^2 = 0$;

$c_6^4 = 0$; $e_8^4 = 0$ is basically the probability that randomly tried c_6^4 and c_5^4 values satisfy $c_6^4 = 0$ and $c_5^4 \equiv 6$; the

order of magnitude of this probability is therefore 2^{-20}.

Moreover, the probability that a message M_2 satisfying the four equations $c_6^2 = 0$; $c_8^2 = 0$; $c_6^4 = 0$; $c_8^4 = 0$ leads

to a full collision $g(H_1, M_2) = g(H_1, M_2)$ is basically the probability that none of the c_{i-3} mod 16 indices

occurring in the calculation of c_0^2 to e_{15}^2 and of c_0^4 to c_{15}^4 takes the value 15, which is close to 1/8. So in

summary approximatively 2^{23} partial computations of the g function are necessary to obtain a message M_2

giving a full collision.

2.3 Implementation details

The above attack method was implemented using a non-optimized Pascal program. The search for a collision took a few hours on a SUN3 workstation and less than an hour on a SPARC workstation. We provide in annex the detail of the intermediate calculations for two colliding messages $M_1 M_2$ and $M'_1 M_2$, of two 128-bit blocks each.

Note that for many other values M''_1 of the form $(e_0^0, \dots, e_{15}^0 + k.16)$ (k : an integer) of the first 128-bit block, the message $M''_1 M_2$ leads to the same hash value as $M_1 M_2$: the observed phenomenon is in fact a multiple collision.

3 Conclusions

The attack described in this paper takes advantage of the two following weaknesses of the FFT-Hashing algorithm :

- the influence of the term $c_{c_{i-3}}$ in the recurrence $c_i := c_i + c_{i-1} e_{i-2} + c_{c_{i-3}} + 2^i$ (mod p) on the security of the algorithm is rather negative (see for example the method to obtain $c_6^2 = 0$ (or $c_8^2 = 0$) at step 1 of Section 2.2).

- as mentioned in Section 1.3, the diffusion introduced by the four steps of the algorithm is quite limited. In particular, the FT_8 Fourier transform acts only on half of the intermediate values $(e_0, \dots, e_{15})$, namely the 8 values $e_0, e_2, \dots, e_{14}$.

This suggests that quite simple modifications might result in a substantial improvement of the security of the FFT-Hashing algorithm.

4 Acknowledgements

The autors are greateful to Jacques BURGER (SEPT PEM, 42 rue des Coutures, BP 6243, 14066 CAEN, France) for the Sparc implementation as well as useful discussions.

5 References

[1] : C.P. SCHNORR; FFT-Hashing : An Efficient Cryptographic Hash Function; July 15, 1991 (This paper was presented at the rump session of the CRYPTO'91 Conference, Santa Barbara, August, 11-15, 1991)

[2] : DAEMEN - BOSSELAERS - GOVAERTS - VANDEWALLE : Announcement made at the rump session of the ASIACRYPT '91 Conference, Fujiyoshida, Japan, November 11-14, 1991)

ANNEX

FIRST MESSAGE   M = M1 M2   with

```
M1 =  F95A  807A  26A   0    440   365E  0     0
      1537  5202  3284  358  5D1C  959E  6D6B  75E0
M2 =
```

calculation of H1 :

```
H0 =  123   4567  89AB  CDEF  FEDC  BA98  7654  3210
M1 =  F95A  807A  26A   0    440   365E  0     0

step 1:  10000  4567  4F72  CDEF  B84C  BA98  D98A  3210
         FB30   807A  F62E        3677  365E        0

step 2:  0      4569  4F76  1DD1  6CEA  F49C  1DB9  7D13
         ADDC   156   5AFE  CD52  A692  158A  4626  B808

step 1:  CFA9   4569  2466  1DD1  2F1A  F49C  F3D7  7D13
         B305   156   3057  CD52  5A7   158A  0     B80B

step 2:  0      456B  F1BC  91E1  64F8  F6D2  FB99  A787
         7DCA   CDE2  4508  3BE5  8F64  E23C  9BBA  5806

H1 =  7DCA  CDE2  4508  3BE5  8F64  E23C  98BA  5BE6
```

calculation of H2 :

```
H1 =  7DCA  CDE2  4508  3BE5  8F64  E23C  98BA  5BE6
M2 =  1537  5202  3284  358   5D1C  959E  6D6B  75E0

step 1:  10000  CDE2  4508  3BE5  3E13  E23C  418A  5BE6
         FF01   5202  9B04  358   EF0   959E  0     75E0

step 2:  0      CDE4  C5C2  17A9  6501  6370  0     2A49
         0      5402  F306  99A5  8BB5  9A6E  38EF  73A9

step 1:  E26B   CDE4  8B79  17A9  E6CC  6370  E7C2  2A49
         FF01   CD5   CD5   99A5  37CB  9A6E  7FF2  73A9

step 2:  5551   E84C  4E20  EA99  C82F  9BB6  0     9E72
         0      AB53  5EF5  27D8  9554  995   983F  89CF

H2 =  0  AB53  5EF5  27D8  9554  995  983F  89CF
```

HASHED MESSAGE : 0 AB53 5EF5 27D8 9554 995 983F 89CF

SECOND MESSAGE   M = M1 M2   with

```
M1 =  F95A  807A  26A   0    440   365E  0     10
      1537  5202  3284  358  5D1C  959E  6D6B  75E0
M2 =
```

calculation of H1 :

```
H0 =  123   4567  89AB  CDEF  BA90  7654  3210
M1 =  F95A  807A  26A   0    440   365E  0     10

step 1:  10000  4567  4F72  CDEF  B84C  BA98  D98A  3210
         FB30   807A  F62E        3677  365E        10

step 2:  0      4569  4F76  1DD1  6CEA  F49C  1DB9  7D13
         ADDC   156   5AFE  CD52  A692  158A  4626  B81B

step 1:  CFA9   4569  2466  1DD1  2F1A  F49C  F3D7  7D13
         B305   156   3057  CD52  5A7   158A  0     B81B

step 2:  0      456B  F1BC  91E1  64F8  F6D2  FB99  A787
         7DCA   CDE2  4508  3BE5  8F64  E23C  9BBA  5BF6

H1 =  7DCA  CDE2  4508  3BE5  8F64  E23C  98BA  5BF6
```

calculation of H2 :

```
H1 =  7DCA  CDE2  4508  3BE5  8F64  E23C  98BA  5BF6
M2 =  1537  5202  3284  358   5D1C  959E  6D6B  75E0

step 1:  10000  CDE2  C5BE  3BE5  3E13  E23C  418A  5BF6
         FF01   5202  9B04  358   EF0   959E  0     75E0

step 2:  0      CDE4  C5C2  17A9  6501  6370  0     2A59
         0      5402  F306  99A5  8BB5  9A6E  38EF  73A9

step 1:  E26B   CDE4  8B79  17A9  E6CC  6370  E7C2  2A59
         FF01   5402  99A5  99A5  37CB  9A6E  7FF2  73A9

step 2:  5551   E84C  4E20  EA99  C82F  9BB6  0     9E82
         0      AB53  5EF5  27D8  9554  995   983F  89CF

H2 =  0  AB53  5EF5  27D8  9554  995  983F  89CF
```

HASHED MESSAGE : 0 AB53 5EF5 27D8 9554 995 983F 89CF

FFT–Hash II, Efficient Cryptographic Hashing

C.P. Schnorr

Fachbereich Mathematik / Informatik, Universität Frankfurt

Abstract. We propose an efficient algorithm that hashes messages of arbitrary bit length into an 128 bit hash value. The algorithm is designed to make the production of a pair of colliding messages computationally infeasible. The algorithm performs a discrete Fourier transform and a polynomial recursion over a finite field. Each hash value in $\{0,1\}^{128}$ occurs with frequency at most 2^{-120}. This hash function is an improved variant of the algorithm FFT–hash I presented in the rump session of CRYPTO'91.

1 The hash algorithm

Overview. We present a novel design for a cryptographic hash algorithm. It may serve as an alternative to the MD4 / MD5 algorithm of Rivest. These novel hash algorithms are not based on any encryption scheme. We need to have more cryptographic hash algorithms since improved methods of cryptoanalysis may exhibit more weaknesses in the proposed hash functions. Our design goal is to make it impossible to produce a collision using less than 2^{64} operations. Our algorithm can easily be implemented in software using addition and multiplication either modulo 2^{16} or modulo 2^8. It uses the discrete Fourier transform in order to diffuse information in an ideal way. We also use a polynomial transformation of high degree over a finitefield. Such transformations generate local randomness, see Niederreiter, Schnorr (1992), this proceedings.

Padding the message. Let the message be given as a bit string $m_1 m_2 \ldots m_t$ of t bits. The message must be padded so that its length in bits becomes a multiple of 128. We recommend to append to the message a single "1" bit followed by a suitable number of "0" bits followed by the binary representation of t. Let the padded message $M_1 \| M_2 \ldots \| M_n$ consist of n blocks $M_1, \ldots, M_n$ that each is 128 bits long.

The initial value. H_0 is given in hexadecimal as

$$H_0 = 0123\ 4567\ 89ab\ cdef\ fedc\ ba98\ 7654\ 3210 \quad \in \{0,1\}^{128} \ .$$

R.A. Rueppel (Ed.): Advances in Cryptology - EUROCRYPT '92, LNCS 658, pp. 45-54, 1993.

Algorithm for the hash function h

INPUT $M = M_1 \| M_2 \cdots \| M_n \in \{0,1\}^{n \cdot 128}$ (a padded message)

     $H_i := g(H_{i-1}, M_i)$ for $i = 1, \ldots, n$

OUTPUT $h(M) := H_n$

Requirements for function $g : \{0,1\}^{256} \to \{0,1\}^{128}$.

We wish to make the function h *collision-free*. This means that it is infeasible to find distinct messages $M_1 \| M_2 \cdots \| M_n$, $M_1' \| M_2' \cdots \| M_{n'}'$ such that $H_n = H_{n'}'$. Specifically the construction of two colliding messages should require about 2^{64} steps which is the time bound for the birthday attack.

To achieve this goal the following problems must be infeasible to solve for given $H, H' \in \{0,1\}^{128}$.

Problem 1   Find message blocks $M, M' \in \{0,1\}^{128}$ such that $g(H, M) = g(H', M')$.

Problem 2   Find $M \in \{0,1\}^{128}$ such that $g(H', M) = H$.

Description of the function $g : \{0,1\}^{256} \longrightarrow \{0,1\}^{128}$

Let p be the prime $p = 65537 = 2^{16}+1$. We represent elements in $\mathbb{Z}_p = \mathbb{Z}/p\mathbb{Z}$ by the residues in the interval $[0, 2^{16}]$. We associate with a double byte $(x_1, \ldots, x_{16}) \in \{0,1\}^{16}$ the integer $\sum_{i=1}^{16} x_i 2^{i-1} \in \mathbb{Z}_p$. Identification of a bit string in $\{0,1\}^{16}$ with its corresponding integer yields natural inclusions $\{0,1\}^{16} \subset \mathbb{Z}_p$, $\{0,1\}^{16n} \subset \mathbb{Z}_p^n$ for all n. Conversely we associate with $a \in [0, 2^{16}]$ the integer a', $a' = a(\bmod\, 2^{16})$ in $[0, 2^{16} - 1]$, corresponding to a double byte in $\{0,1\}^{16}$. Here $a = 2^{16}$ and $a = 0$ both yield $a' = 0$.

Due to the inclusion $\{0,1\}^{256} \subset \mathbb{Z}_p^{16}$ it is sufficient to compute a function $g : \mathbb{Z}_p^{16} \longrightarrow \{0,1\}^{128}$. We will use the discrete Fourier transform $FT_8 : \mathbb{Z}_p^8 \longrightarrow \mathbb{Z}_p^8$ with the primitive root 2^4 of order 8. We have

$$FT_8(a_0, \ldots, a_7) = (b_0, \ldots, b_7) \qquad \text{with}$$

$$b_i = \sum_{j=0}^{7} 2^{4ij} a_j \;(\bmod\, p) \quad \text{for } i = 0, \ldots, 7 \,.$$

Algorithm for $g : \mathbb{Z}_p^{16} \longrightarrow \{0,1\}^{8 \cdot 16}$

INPUT $(e_0, \ldots, e_{15}) \in \{0,1\}^{16 \cdot 16} \subset \mathbb{Z}_p^{16}$

    1. FOR $i = 0, \ldots, 15$ DO $e_i := e_i + e_{i-1}^* e_{i-2}^* + e_{i-3} + 2^i \ (\mathrm{mod}\, p)$

    where $e^* = e$ if $e \neq 0$ and $e^* = 1$ if $e = 0$.

    (The indices $i, i-1, i-2, i-3$ are taken modulo 16)

    2. $(e_0, e_2, \ldots, e_{14}) := \mathrm{FT}_8(e_0, e_2, \ldots, e_{14})$

    $(e_1, e_3, \ldots, e_{15}) := \mathrm{FT}_8(e_1, e_3, \ldots, e_{15})$

    3. FOR $i = 0, \ldots, 15$ DO $e_i := e_i + e_{i-1}^* e_{i-2}^* + e_{i-3} + 2^i \ (\mathrm{mod}\, p)$

OUTPUT $e_i \,(\mathrm{mod}\, 2^{16})$ for $i = 8, \ldots, 15$

Comparison to the algorithm FFT-Hash I presented in the rump session of CRYPTO'91.
The previous version of the above algorithm g performed instead of Steps 1 – 3 the following steps:

REPEAT Steps 1,2 twice

    1. $(e_0, e_2, \ldots, e_{14}) := \mathrm{FT}_8(e_0, e_2, \ldots, e_{14})$
    2. FOR $i = 0, \ldots, 15$ DO $e_i := e_i + e_{i-1}e_{i-2} + e_{e_{i-3}} + 2^i(\mathrm{mod}\, p)$

END loop

This version has been broken by BOSSELAER and DAEMEN and independently by M. GIRAULT. A weakness of this algorithm is that the information contained in an input element e_j with $8 \leq j \leq 15$ must in some cases not be diffused to all final elements $e_0, \ldots, e_{15}$. E.g. if j is odd then e_j has no impact on the Fourier transform in Step 1. Moreover if $e_{j-1} = e_{j+1} = 0$ holds just before the execution of $e_j := e_j + e_{j-1}e_{j-2} + e_{e_{j-3}} + 2^j \,(\mathrm{mod}\, p)$ in Step 2 then the first round of Steps 1 and 2 will diffuse e_j only to those e_i for which $e_{i-3} = j$ holds during the execution of $e_i := e_i + e_{i-1}e_{i-2} + e_{e_{i-3}} + 2^i(\mathrm{mod}\, p)$. A critical point is that the transformation is almost linear in e_i if e_{i-1} and e_{i+1} are 0. The new function g does not suffer from this weakness.

Counter measures against the Bosselaer–Daemen/Girault attack
Each of the measures 1, 2 and 3, when taken separately, protects against the above attack, measure 4 slows the attack down.

1. Replace the term $e_{e_{i-3}}$ in FFT–Hash I by e_{i-3}

2. Replace the FT_8 on the elements e_i with even index i by a full Fourier transform FT_{16} on all 16 elements

3. Eliminate zero–multipliers in the polynomial recursion, e.g. replace the recursion in Step 2 of FFT–Hash I by

$$e_i := e_i + e_{i-1}^* e_{i-2}^* + e_{e_{i-3}} + 2^i \pmod{p} ,$$

where $e^* = e$ if $e \neq 0$ and $e^* = 1$ if $e = 0$.

4. Invert the order of Step 1 and Step 2 in FFT–Hash I.

In FFT–Hash II we have essentially combined all these four counter measures. Step 2 of this algorithm performs a full Fourier transform except for the stage that mixes even and odd elements. We think that this stage is not needed because this mixing is done in Steps 1 and 3.

The function g is easy to invert. This is because Steps 1 – 3 can easily be inverted. We can complement the output elements $e'_8, \ldots, e'_{15}$ by any elements $e_0, \ldots, e_7$. We obtain a corresponding input for g by inverting Steps 1 – 3 on $e_0, \ldots, e_7, e'_8, \ldots, e'_{15}$. If this inversion yields elements in the interval $[0, 2^{16}-1]$, we have found an inverse image of $(e'_8, \ldots, e'_{15})$ for g. Since g is easy to invert it is not collision–free. The fast inversion algorithm for function g does not necessarily constitute a weakness for the hash function h since it does not produce inverse images having the prefix $H_0 \in \{0, 1\}^{128}$.

Design principles for the function g. The polynomial recursion in Steps 1 and 3 is at both ends of the procedure. It forms a shell for the linear transformation in Step 2. Each of Steps 1 and 3 generates a polynomial transformation so that e_i^{new} is a polynomial of degree at least F_{i+3} in $e_0^{\text{old}}, \ldots, e_{15}^{\text{old}}$ where F_{i+3} is the $i+3$-th Fibbonacci number. Thus the degree of e_{15}^{new} is at least $F_8 = 2584$. We believe that the problems 1 and 2 are intractable in worst case for this polynomial transformation. That is it seems difficult to enforce a particular pattern for the second half of the output of this transformation by manipulating only the first half of the input. The function g is a polynomial transformation with the property that all output coordinates are polynomials of degree at least 229976 in the input coordinates.

We use the Fourier transform to mix in a perfect way the information contained in the numbers $e_0, \ldots, e_{15}$. It is of interest that the function g transforms message bits and hash bits in a similar way.

The cost for evaluating **g**.

Evaluating the Fourier transform FT_8 requires $3 \cdot 8$ additions and $3 \cdot 8$ shifts modulo p. The two evaluations of FT_8 cost 48 additions modulo p and 48 shifts modulo p. Steps 1 and 3 require 64 additions and 32 multiplications modulo p. We neglect the costs for adding the bit 2^i and for setting up the FOR–loop. We obtain the following total costs

$$112 \text{ additions } \bmod p$$

$$32 \text{ multiplications } \bmod p$$

$$\underline{48 \text{ shifts } \bmod p}$$

$$192 \text{ operations } \bmod p \, .$$

Variations of the function g *preserving the invertability*. The provable properties of Section 2 rely on the invertability of the Steps $1 - 3$ in the algorithm for
g. It is interesting to study a class of functions g with this invertability. We can
replace Steps 1 and 3 of the algorithm for g by the following step

$$e_i := e_i + f(e_0, \ldots, e_{i-1}, e_{i+1}, \ldots, e_{15}, i) \bmod p \text{ for } i = 0, \ldots, 15$$

where f is an arbitrary function. For any such function f the Steps 1,2,3 in
the algorithm for g remain invertible. It may be possible that there are better
choices for the function f enhancing security and efficiency than our choice
$f(e_0, \ldots, e_{i-1}, e_{i+1}, \ldots, e_{15}, i) = \bar{e}_{i-1}\bar{e}_{i-2} + e_{i-3} + 2^i$. Alternatively we can also
use the recursion

$$e_i := e_i \oplus f(e_0, \ldots, e_{i-1}, e_{i+1}, \ldots, e_{15}, i) \bmod p \text{ for } i = 0, \ldots, 15$$

where f is an arbitrary function and $\oplus : \mathbb{Z}_p^2 \to \mathbb{Z}_p$, $(x, y) \mapsto x \oplus y$ is the bitwise
addition which is defined as follows. Let $x, y \in \mathbb{Z}_p$ have binary representations
$(x_0, \ldots, x_{16})$ and $(y_0, \ldots, y_{16})$, i.e. $x = \sum_i 2^i x_i$, $y = \sum_i 2^i y_i$. Then we put
$x \oplus y = \sum_{i=0}^{16} 2^i (x_i + y_i \bmod 2) \bmod p$. It is of interest to note that the function
$\oplus$ is *associative*. The associativity of $\oplus$ implies that Steps 1 and 3 in the
algorithm for g remain invertible.

Transforming g *into a function that is most likely collision-free*. That there is a
simple heuristic that transforms g into a function $\bar{g}$ that is most likely collision–
free. We obtain $\bar{g}$ from g by inserting a Step 4 into the computation that adds
the input elements e_i^{inp} to e_i for $i = 8, \ldots, 15$, i.e.

Step 4. $e_i := e_i + e_i^{\text{inp}} (\bmod p)$ for $i = 8, \ldots, 15$.

2 Provable properties of the function g

We show in the next proposition that no information is lost during the computation of g . Moreover this computation cannot create any redundancies.

Proposition 1. *Steps 1 – 3 act as bijective transformations on the configurations* $(e_0, \ldots, e_{15}) \in \mathbb{Z}_p^{16}$.

Proof . The Fourier transform $FT_8 : \mathbb{Z}_p^8 \to \mathbb{Z}_p^8$ is bijective. Its inverse is given by

$$FT_8^{-1}(b_0, \ldots, b_7) = (a_0, \ldots, a_7)$$

$$a_i = 8^{-1} \sum_{j=0}^{7} 2^{-4ij} b_j \pmod{p} \quad \text{for } i = 0, \ldots, 7 .$$

The inverse is correct since 2^4 is a primitive root of order 8 in $\mathbb{Z}_p$ i.e. $2^{4 \cdot 8} = 1 \bmod p$, $2^{4 \cdot 4} = -1 \bmod p$.
Moreover a single step

$$e_i^{\text{new}} := e_i + e_{i-1}^* e_{i-2}^* + e_{i-3} + 2^i \pmod{p}$$

can be inverted as

$$e_i := e_i^{\text{new}} - e_{i-1}^* e_{i-2}^* - e_{i-3} - 2^i \pmod{p}$$

□

By Proposition 1 no collision of messages is possible within the evaluation of the function g, i.e. distinct messages yield distinct configurations in the same step. If the input $(e_0, \ldots, e_{15})$ for g is uniformly distributed over $\mathbb{Z}_p^{16}$ then the final configuration $(e_0, \ldots, e_{15})$ in the program for g is also uniformly distributed over $\mathbb{Z}_p^{16}$.

Next we study the probability distribution of $g(H, M)$ when (H, M) ranges uniformly over $\{0, 1\}^{256}$. We first prove an upper bound for the probability of any output value for g. Moreover we show that the distribution of $g(H, M)$ is close to the uniform distribution on $\{0, 1\}^{128}$. The following theorem is interesting since it excludes the construction of collisions for g corresponding to most likely outputs. Note that it is impossible to verify the statement of the theorem by empirical testing.

Theorem 2. *If the pair* (H, M) *is uniformly distributed over* $\{0, 1\}^{256}$ *then we have for all* $a \in \{0, 1\}^{128}$ *that* $\text{prob}\, [g(H, M) = a] \leq 2^{-120} e^{-2^{-13}} \approx 2^{-120}$.

Proof. Let out : $\mathbb{Z}_p^{16} \to \{0,1\}^{128}$ be the function that associates with a final configuration $(e_0, \ldots, e_{15}) \in \mathbb{Z}_p^{16}$ the output $(e'_8, \ldots, e'_{15}) \in \{0,1\}^{128}$ of g. For every $a = (e'_8, \ldots, e'_{15})$ we have

$$\# \text{out}^{-1}(a) \le (2^{16} + 1)^8 2^t$$

where $t = \#\{i | 8 \le i \le 15, e'_i = 0\}$. Since pairwise distinct inputs (H, M) for g are mapped into pairwise distinct final configurations we see that

$$\text{prob}\,[g(H, M) = a] \le (2^{16} + 1)^8 2^t 2^{-256}$$
$$= 2^{-128+t}(1 + 2^{-16})^8 \approx 2^{-128+t} e^{-2^{-13}} \approx 2^{-128+t} \,,$$

where the probability space is the set of all $(H, M) \in \{0, 1\}^{256}$. $\qquad\square$

Let X, Y be probability distributions over some finite set S. We measure the distance of X and Y by the 1-norm

$$\|X - Y\|_1 := \sum_{s \in S} \left| \text{prob}_x(s) - \text{prob}_y(s) \right| \,.$$

We have that $0 \le \|X - Y\|_1 \le 2$ and $\|X - Y\|_1 = 0$ if and only if $X = Y$. The norm satisfies the triangular inequality:

$$\|X - Y\|_1 \le \|X - Z\|_1 + \|Z - Y\|_1 \,. \tag{1}$$

The norm cannot increase by the application of a function f:

$$\|f(X) - f(Y)\|_1 \le \|X - Y\|_1 \,. \tag{2}$$

Here $f(X)$ is the probability distribution on Image (f) given as

$$\text{prob}_{f(X)}(t) = \sum_{f(s)=t} \text{prob}_X(s) \,.$$

Let U_S denote the uniform distribution on the set S. If $S' \subset S$ we have

$$\|U_{S'} - U_S\| = 2\left(1 - \frac{\#S'}{\#S}\right) \,. \tag{3}$$

Theorem 3. $\|g\,(U_{\{0,1\}^{256}}) - U_{\{0,1\}^{128}}\|_1 \le 10^{-3}$, *i.e. if* (H, M) *ranges uniformly over* $\{0, 1\}^{256}$ *the 1-norm distance of* $g(H, M)$ *from the uniform distribution on* $\{0, 1\}^{128}$ *is at most* 10^{-3}.

Proof. Let $\bar{g} : \mathbb{Z}_p^{16} \longrightarrow \mathbb{Z}_p^{16}$ be the bijective function that associates to an initial configuration the corresponding final configuration in the computation of g. We have that

$$\bar{g}\left(U_{\mathbb{Z}_p^{16}}\right) = U_{\mathbb{Z}_p^{16}} . \tag{4}$$

We see from (2) and (3) that

$$\left\| \bar{g}\left(U_{\mathbb{Z}_p^{16}}\right) - \bar{g}\left(U_{\{0,1\}^{256}}\right) \right\|_1 \leq \left\| U_{\mathbb{Z}_p^{16}} - U_{\{0,1\}^{256}} \right\|_1$$

$$\leq 2\left(1 - \left(\frac{2^{16}}{2^{16}+1}\right)^{16}\right) \approx 2\left(1 - \left(1 - \frac{1}{2^{16}+1}\right)^{16}\right) \tag{5}$$

$$\approx 2\left(1 - e^{-16/2^{16}}\right) \approx 4.88 \cdot 10^{-4} .$$

Let out $: \mathbb{Z}_p^{16} \longrightarrow \{0,1\}^{128}$ be the function that associates with a final configuration $(e_0, \ldots, e_{15}) \in \mathbb{Z}_p^{16}$ the output $(e_8', \ldots, e_{15}') \in \{0,1\}^{128}$. We have that out $(U_{\{0,1\}^{256}}) = U_{\{0,1\}^{128}}$. We finally get

$$\left\| g\left(U_{\{0,1\}^{256}}\right) - U_{\{0,1\}^{128}} \right\|_1$$

$$= \left\| \text{out } \bar{g}\left(U_{\{0,1\}^{256}}\right) - \text{out}\left(U_{\{0,1\}^{256}}\right) \right\|_1$$

$$\overset{(2)}{\leq} \left\| \bar{g}\left(U_{\{0,1\}^{256}}\right) - U_{\{0,1\}^{256}} \right\|_1$$

$$\overset{(1),(4)}{\leq} \left\| \bar{g}\left(U_{\{0,1\}^{256}}\right) - \bar{g}\left(U_{\mathbb{Z}_p^{16}}\right) \right\|_1 + \left\| U_{\mathbb{Z}_p^{16}} - U_{\{0,1\}^{256}} \right\|_1$$

$$\overset{(5)}{\leq} 2 \cdot 4.88 \cdot 10^{-4} < 10^{-3} .$$

$\square$

We next consider the discrete Fourier transform FT_8. If a_j ranges uniformly over $\mathbb{Z}_p$ and the a_k for $k \neq j$ are all fixed then each component b_i of $(b_0, \ldots, b_7) = FT_8(a_0, \ldots, a_7)$ ranges uniformly over $\mathbb{Z}_p$ for $i = 0, \ldots, 7$. This holds because no coefficient 2^{4ij} of FT_8 is zero in $\mathbb{Z}_p$. We next consider pairs of components.

Lemma 4. *Let $0 \leq i, \bar{\imath}, j, \bar{\jmath} \leq 7$ and $(j - \bar{\jmath})(i - \bar{\imath}) \neq 0 \pmod{8}$. If $(a_j, a_{\bar{\jmath}})$ ranges uniformly over $\mathbb{Z}_p^2$ and the a_k for $k \notin \{j, \bar{\jmath}\}$ are all fixed then the pair $(b_i, b_{\bar{\imath}})$ from $(b_0, \ldots, b_7) = FT_8(a_0, \ldots, a_7)$ ranges uniformly over $\mathbb{Z}_p^2$.*

Proof. Consider the matrix $[16^{ij}]_{0 \le i,j \le 15}$ representing the linear transformation FT_8. It is sufficient to prove that the 2×2 matrix

$$\begin{bmatrix} 16^{ij}, & 16^{i\bar{j}} \\ 16^{\bar{i}j}, & 16^{\bar{i}\bar{j}} \end{bmatrix}$$

corresponding to rows $i, \bar{i}$ and columns $j, \bar{j}$ of this matrix is regular. The determinant of this matrix is

$$16^{ij + \bar{i}\bar{j}} - 16^{i\bar{j} + \bar{i}j} = 16^{ij + \bar{i}\bar{j}} \left(1 - 16^{(\bar{i} - i)(j - \bar{j})} \right) .$$

This determinant is zero modulo p iff $(i - \bar{i})(j - \bar{j}) = 0 (\mathrm{mod}\, 8)$. $\square$

We see from Lemma 4 that independent pairs of input components $(a_j, a_{\bar{j}})$ for FT_8 yield independent pairs of output components $(b_i, b_{\bar{i}})$ for most $(i, \bar{i}, j, \bar{j})$, in particular for all $i, \bar{i}$ with $i - \bar{i}$ odd. The condition that $i - \bar{i}$ is odd seems to be sufficient for our purpose since any two output coordinates $e_k^{\mathrm{new}}, e_{\bar{k}}^{\mathrm{new}}$ with $k \ne \bar{k}$ of Step 2

$$e_k^{\mathrm{new}} := e_k + e_{k-1}^* e_{k-2}^* + e_{k-3} + 2^k \ (\mathrm{mod}\, p)$$

depend on two distinct output components $e_{2i}, e_{2\bar{i}}$ of

$$(e_0, e_2, \ldots, e_{14}) = FT_8(\bar{e}_0, \bar{e}_2, \ldots, \bar{e}_{14})$$

where $i - \bar{i}$ is odd. If e.g. k and $\bar{k}$ are even then $e_k^{\mathrm{new}}, e_{\bar{k}}^{\mathrm{new}}$ depend on $e_{2i}, e_{2\bar{i}}$ with $i = k/2$ and with $\bar{i} = \bar{k}/2$ and $\bar{i} = \bar{k}/2 + 1$.

Lemma 4 can be extended from pairs $(i, \bar{i})$, $(j, \bar{j})$ to triples, quadruples of components and so on. Since independence of components is unlikely to cancel out later on in the computation of g this shows that the function g has highly desirable statistical properties.

Appendix : A program for h

$M_1, M_2, \ldots, M_n \in \{0,1\}^{128}$ with $M_j = M_{j,0}\|M_{j,1}\cdots\|M_{j,7}$ and $M_{j,i} \in \{0,1\}^{16}$.

(Let $i = i_0 2^2 + i_1 2 + i_2 \in [0,7]$ with $i_j \in \{0,1\}$ and let $\mathrm{rev}(i) = i_2 2^2 + i_1 2 + i_0$ be the number i in reversed notation. Let $i(j,0)$ ($i(j,1)$, resp.) be obtained from i by setting $i_j := 0$ ($i_j := 1$, resp.). We abbreviate $e_i = e_i^{(0)}$ and $e_i' = e_i(\mathrm{mod}\,2^{16})$. Lower indices i of $e_i^{(j)}$ are taken modulo 16.)

INITIATION $(e_8, e_9, \ldots, e_{15}) := 0123456789abcdeffedcba9876543210 \in (\mathbb{Z}_p)^{18}$

(in hexadecimal notation)

FOR $k = 1, \ldots, n$ DO

FOR $i = 0, \ldots, 7$ DO $[e_i := e_{i+8}',\ e_{i+8} := M_{k,i}]$

FOR $i = 0, \ldots, 15$ DO

$$e_i := e_i + e_{i-1}^* e_{i-2}^* + e_{i-3} + 2^i \bmod (2^{16} + 1)$$

where $e^* = e$ if $e \neq 0$ and $e^* = 1$ if $e = 0$

FOR $j = 0, 1, 2$ FOR $i = 0, \ldots, 7$ DO

$$e_{2i}^{(j+1)} := e_{2\cdot i(j,0)}^{(j)} + 16^{\mathrm{rev}(i)2^{2-j}} e_{2\cdot i(j,1)}^{(j)} \bmod (2^{16} + 1)$$

$$e_{2i+1}^{(j+1)} := e_{2\cdot i(j,0)+1}^{(j)} + 16^{\mathrm{rev}(i)2^{2-j}} e_{2\cdot i(j,1)+1}^{(j)} \bmod (2^{16} + 1)$$

FOR $i = 0, \ldots, 15$ DO $[e_{2i} := e_{2\cdot\mathrm{rev}(i)}^{(3)},\ e_{2i+1} := e_{2\cdot\mathrm{rev}(i)+1}^{(3)}]$

FOR $i = 0, \ldots, 15$ DO

$$e_i := e_i + e_{i-1}^* e_{i-2}^* + e_{i-3} + 2^i \bmod (2^{16} + 1)$$

END for k

OUTPUT $e_8', \ldots, e_{15}'$.

Hash Functions Based on Block Ciphers

Xuejia Lai and James L. Massey

Signal and Information Processing Laboratory
Swiss Federal Institute of Technology
CH–8092 Zürich, Switzerland

Abstract. Iterated hash functions based on block ciphers are treated. Five attacks on an iterated hash function and on its round function are formulated. The wisdom of strengthening such hash functions by constraining the last block of the message to be hashed is stressed. Schemes for constructing m-bit and $2m$-bit hash round functions from m-bit block ciphers are studied. A principle is formalized for evaluating the strength of hash round functions, viz., that applying computationally simple (in both directions) invertible transformations to the input and output of a hash round function yields a new hash round function with the same security. By applying this principle, four attacks on three previously proposed $2m$-bit hash round functions are formulated. Finally, three new hash round functions based on an m-bit block cipher with a $2m$-bit key are proposed.

1 Introduction

This paper is intended to provide a rather rounded treatment of hash functions that are obtained by iterating a round function. Section 2 examines the possible attacks on such iterated hash functions, considers relations between the security of an iterated hash function and the security of its hash round function, and points out the wisdom of strengthening the hash function by constraining the last block of the message to be hashed.

In Section 3, we consider hash round functions constructed from secret-key block ciphers. In particular, we consider the problems of constructing m-bit hash round functions and $2m$-bit hash round functions from m-bit block ciphers. A principle is formalized for evaluating the strength of hash round functions, viz., that applying computationally simple (in both directions) invertible transformations to the input and output of a hash round function yields a new hash round function with the same security. To demonstrate this principle, we present four attacks on three previously proposed $2m$-bit hash round functions. Finally, three new hash round functions based on an m-bit block cipher with a $2m$-bit key are proposed.

R.A. Rueppel (Ed.): Advances in Cryptology - EUROCRYPT '92, LNCS 658, pp. 55-70, 1993.
© Springer-Verlag Berlin Heidelberg 1993

2 Iterated hash functions and attacks

A *hash function* is an easily implementable mapping from the set of all binary sequences of some specified minimum length or greater to the set of binary sequences of some fixed length. In cryptographic applications, hash functions are used within digital signature schemes and within schemes to provide data integrity (e.g., to detect modification of a message).

An *iterated hash function* is a hash function Hash($\cdot$) determined by an easily computable function $h(\cdot, \cdot)$ from two binary sequences of respective lengths m and l to a binary sequence of length m in the manner that the message $M = (M_1, M_2, ..., M_n)$, where M_i is of length l, is hashed to the *hash value* $H = H_n$ of length m by computing recursively

$$H_i = h(H_{i-1}, M_i) \qquad i = 1, 2, ..n, \tag{1}$$

where H_0 is a specified *initial value*. We will write $H = \text{Hash}(H_0, M)$ to show explicitly the dependence on H_0. The function h will be called the *hash round function*. Such a recursive construction of hash functions has been called the "meta-method" by Merkle [13], see also [4, 15]. For message data whose total length in bits is not a multiple of l, one can apply deterministic "padding" [7, 13] to the message to be hashed by (1) to increase the total length to a multiple of l.

For iterated hash functions, we distinguish the following five attacks:

1. **Target attack:** Given H_0 and M, find M' such that $M' \neq M$ but $\text{Hash}(H_0, M') = \text{Hash}(H_0, M)$.

2. **Free-start target attack:** Given H_0 and M, find H_0' and M' such that $(H_0', M') \neq (H_0, M)$ but $\text{Hash}(H_0', M') = \text{Hash}(H_0, M)$.

3. **Collision attack:** Given H_0, find M and M' such that $M' \neq M$ but $\text{Hash}(H_0, M') = \text{Hash}(H_0, M)$.

4. **Semi-free-start collision attack:** Find H_0, M and M' such that $M' \neq M$ but $\text{Hash}(H_0, M') = \text{Hash}(H_0, M)$.

5. **Free-start collision attack:** Find H_0, H_0', M and M' such that $(H_0', M') \neq (H_0, M)$ but $\text{Hash}(H_0', M') = \text{Hash}(H_0, M)$.

Remark. In applications where H_0 is specified and fixed, attacks 2, 4 and 5 are not "real attacks". This is because the initial value H_0 is then an integral part of the hash function so that a hash value computed from a different initial value will not be accepted. However, if the sender is free to choose and/or to change H_0, attacks 2, 4 and 5 can be real attacks, depending on the manner in which the hash function is used. Note that the free-start and semi-free-start attacks are never harder than the attacks where H_0 is specified in advance.

For an m-bit hash function, brute-force target attacks, in which one randomly chooses an M' until one hits the "target" $H = \text{Hash}(H_0, M)$, require about 2^m computations of hash values. It follows from the usual "birthday argument" that brute-force collision attacks require about $2^{m/2}$ computations of hash values. In particular, for hash round functions with $l \geq m$ so that all 2^m hash values can be reached with one-block messages, brute-force target attacks require about 2^m computations of the round function h while brute-force collision attacks require about $2^{m/2}$ computations of the round function h. We will say that the computational security of the hash function is *ideal* when there is no attack substantially better than brute force.

In the following discussion, we consider some relations between the security of an iterated hash function and the strength of its hash round function. By an *attack on the hash round function* we mean an attack in which all the involved messages contain only *one* block. For example, a target attack on the round function h reads: given H_0 and M_1, find M_1' such that $M_1' \neq M_1$ but $h(H_0, M_1') = h(H_0, M_1)$. Once a target attack on the round function yields M_1', then, by "attaching" the message blocks $M_2, ..., M_n$ of the given message to M_1', one obtains success in a target attack on the iterated hash function. Similar arguments hold also for other types of attacks.

Proposition 1. *For an iterated hash function, any attack on its round function implies an attack of the same type on the iterated hash function with the same computational complexity.*

It should be noted that the converse of the statement of Theorem 1 is not true in general. There may be attacks on the iterated hash function that are easier than attacks on the round function alone, as the following three examples show.

Example 1 (Long message attack.) *For an m-bit iterated hash function, given an n-block message $M = (M_1, M_2, ..., M_n)$, there is a target attack which takes about*

$$C = \begin{cases} \frac{2^m}{n} + n & \text{for } n \leq 2^{m/2} \\ 2 \times 2^{m/2} & \text{for } n > 2^{m/2} \end{cases}$$

computations of the round function. [Essentially the above result for $n \leq 2^{m/2}$ is due to Winternitz [23].]

Proof. First we consider the case $n < 2^{m/2}$. For the given M, we compute $H_i = h(H_{i-1}, M_i)$ for $i = 1, .., n$ and store these values. Then we compute $H^* = h(H_0, M_1^*)$ repetitively with randomly chosen M_1^*. After computing $\frac{2^m}{n}$ values for H^*, the probability that $H^* = H_i$ for some $i, 1 \leq i \leq n$, is

$$1 - [(1 - 2^{-m})^n]^{\frac{2^m}{n}} = 1 - (1 - 2^{-m})^{2^m} \approx 1 - e^{-1} \approx 0.63,$$

which shows that fewer than $\frac{2^m}{n}$ computations of round function will usually suffice. The message $M' = (M_1^*, M_{i+1}, ..., M_n)$ hashes to the same value H as the message M, and total number of computations of the round function is about $\frac{2^m}{n} + n$. The probability that $M' = M$ is negligible.

For $n > 2^{m/2}$, we compute and store only $H_1, H_2, ..., H_{2^{m/2}}$. Then $2^{m/2}$ random choices of M_1^* will yield a "match" of some H^* with some H_i, $1 \leq i \leq 2^{m/2}$, with probability about 0.63. $\qquad \square$

For an iterated hash function, one can always do the following "trivial" free-start attacks.

Example 2 (Trivial free-start attacks.) *Consider a message $M = (M_1, M_2)$ that hashes to H with initial value H_0. Then, for the initial value $H_1 = h(H_0, M_1)$, the "truncated" message $M' = M_2$ hashes also to the value $H = h(H_1, M_2)$. That is, a free-start target attack can always be done if the message contain more than one block. Similarly, one can do a trivial free-start collision attack.*

The following attack using a "fixed-point" of the hash round function was proposed in [16].

Example 3. (A trivial semi-free-start collision attack based on a 'fixed point'.) *If the hash round function h has a recognizable "fixed point", i.e., if one can somehow find (H, M) such that $H = h(H, M)$, then there is a trivial semi-free-start collision attack since, starting with the initial value $H_0 = H$, the "different" messages $M = M$ and $M' = (M, M)$ both hash to the same value H.*

Note that in the trivial free-start and semi-free-start attacks and in the "long-message" attack described in the above three examples, one breaks the iterated hash function without breaking its round function. Such attacks are based on the fact that, for an iterated hash function of the form (1), the attacker can take advantage of the fact that a falsified message can have a *length different* from that of the given genuine message. This problem can be overcome by the following strengthening of iterated hash functions, which was proposed independently by Merkle[13] and by Damgaard[4]:

Merkle-Damgaard Strengthening (MD-strengthening) *For the iterated hash function, specify that the last block M_n of the "message" $M = (M_1, M_2, ..., M_n)$ to be hashed must represent the length of the "true message" in bits, i.e., the length of the unpadded portion of the first $n - 1$ blocks.*

Using arguments similar to those in [4, 13, 17], one can show that:

Proposition 2. *Against a free-start (target or collision) attack, an iterated hash function with MD-strengthening, $Hash_{MD}$, has roughly the same computational security as its hash round function.*

In the previous discussions we have considered the security of an iterated hash function and the security of its round function against an attack of the same type. Now we consider how to relate "non-real" free-start target attacks to "real" target attacks. The following result shows that, for an iterated hash function, when a "random inverse" of the hash round function can be found with less than the ideal maximum of about 2^m computations, then there always exists a target attack on the hash function that is better than the brute-force target attack.

Proposition 3. (A meet-in-the-middle target attack by "working backwards".) *Let $Hash_{MD}$ be an m-bit iterated hash function with MD-strengthening and with round function h. If, for most H in the range of h, it takes about 2^s*

computations of h to find a new solution (H', M') of $H = h(H', M')$ for which H' appears to be essentially randomly chosen and if the unconstrained portion of messages contains at least two blocks, i.e., $n-1 \geq 2$, then there exists a target attack on Hash_{MD} that takes about $2 \times 2^{\frac{m+s}{2}}$ computations of h.

Proof. For given M and H_0, let the results of the first two iterations be

$$H_1 = h(H_0, M_1), \qquad H_2 = h(H_1, M_2).$$

We show how to find two message blocks (M_1', M_2') that hash to H_2 by a "meet-in-the-middle" attack. Then replacing the first two blocks (M_1, M_2) in the given message M by (M_1', M_2'), we obtain a message M' of the same length as, but different from, M that hashes to the same H.

First, we compute $G_1 = h(H_0, M_1')$ for $2^{\frac{m+s}{2}}$ randomly chosen M_1''s; then we find $2^{\frac{m-s}{2}}$ pairs (G_1', M_2') such that $H_2 = h(G_1', M_2')$ and G_1' appears essentially randomly chosen. The attack succeeds if some G_1 and some G_1' take on the same value. Thus, the attack succeeds with probability

$$1 - [(1 - 2^{-m})^{2^{\frac{m+s}{2}}}]^{2^{\frac{m-s}{2}}} = 1 - (1 - 2^{-m})^{2^m} \approx 1 - e^{-1} \approx 0.63,$$

as follows from the facts that the probability of choosing M_1' so that G_1 will not equal G_1' is $1 - 2^{-m}$, that there are $2^{\frac{m+s}{2}}$ independent chances to choose M_1' so that G_1 will "miss" a particular G_1', and there are $2^{\frac{m-s}{2}}$ independently chosen values of G_1' to miss. Both the "forwards" computation for computing values of G_1 and the "backwards" computation for computing values of G_1' take $2^{\frac{m+s}{2}}$ computations of the round function h. $\qquad\square$

The method used in the above proof of attacking an iterated hash function by "working backward" [1, 22] has been used to attack several proposed iterated hash functions [15, 22]. The above result shows that if the hash round function does not have ideal computational security against a free-start target attack, then the iterated hash function cannot achieve ideal computational security against a target attack. Proposition 2, together with the argument used to prove Proposition 3, implies:

Proposition 4. *Suppose that the unconstrained portion of messages must contain at least two blocks, i.e., $n - 1 \geq 2$. Then an iterated hash function with MD-strengthening, $\text{Hash}_{\text{MD}}(\cdot)$, has ideal computational security against a target attack if and only if its hash round function $h(\cdot, \cdot)$ has ideal computational security against a free-start target attack.*

Proof. Suppose the round function h has ideal computational security against a free-start target attack. Then Proposition 2 shows that $\text{Hash}_{\text{MD}}(\cdot)$ has the same ideal security against a free-start target attack. But a target attack without free start is no easier than a free-start target attack so that $\text{Hash}_{\text{MD}}(\cdot)$ also has ideal computational security against a target attack.

Conversely, if for an m-bit hash round function h, a free-start target attack takes less than 2^m computations, then Proposition 3 implies a target attack on Hash$_{MD}$ with less than 2^m computations. □

From the above two propositions, we see that MD-strengthening creates secure iterated hash functions from secure round functions. In particular, the trivial free-start and semi-free-start attacks and the long-message target attack in the above examples *cannot* be used to attack an iterated hash function with MD-strengthening. Such considerations suggest an obvious implementation principle for iterated hash functions, viz., that *iterated hash functions should be used only with MD-strengthening*. In the following discussion, whenever the security of an iterated hash function is considered, we always mean the security of the hash function with MD-strengthening.

Because of Proposition 4 and Proposition 2 and because one generally desires that the hash function be strong enough to provide protection against free-start attacks, the problem of constructing secure hash functions reduces to the problem of constructing hash round functions that are secure against free-start attacks, which will be considered in the next section.

3 Hash round functions based on block ciphers

In the following discussion, we consider schemes for constructing hash round functions from a block cipher. In what follows, we write $Y = E_Z(X)$, for an m-bit block cipher E with k-bit key, to mean that the m-bit ciphertext Y is computed from the m-bit plaintext X and k-bit key Z. Based on the discussion in the last section, we consider only attacks on the hash round function or equivalently, attacks on the iterated hash function with MD-strengthening.

3.1 Some m-bit hash round functions

Davies-Meyer (DM) scheme: The DM-scheme was proposed independently by Davies and by Meyer, cf. [5, 11, 22]. This scheme can be used with any block cipher. The message block M_i that is hashed in each step of this scheme has length l equal to the key length k of the block cipher, i.e., $l = k$. The hash round function is given by

$$h(H_{i-1}, M_i) = E_{M_i}(H_{i-1}) \oplus H_{i-1} \qquad (2)$$

and is illustrated in Fig.1 where here and hereafter $\oplus$ denotes bit-by-bit modulo-two addition.

The DM-scheme with MD-strengthening is generally considered to be secure in the sense that, if the block cipher has no known weakness, then no attack better than the brute-force attacks is known, i.e., the free-start target attack on h takes about 2^m computations and the free-start collision attack on h takes about $2^{m/2}$ computations. In particular, with MD-strengthening, none of the attacks mentioned in the three examples of the last section can be effectively used against an iterated hash function based on the DM-scheme. The DM-scheme is currently under consideration as an ISO standard [7].

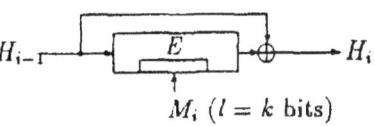

Figure 1: *The hash round function of the DM-scheme. The small box indicates the key input to the block cipher.*

A proposed m-bit hash round function using a block cipher with m-bit block and $2m$-bit key: This method is based on a block cipher with block-length m and key-length $k = 2m$. For example, one could use the block cipher PES [8] or its improved version IPES [9]. For such a cipher with $k = 2m$, we will write $Y = E_{Z_a, Z_b}(X)$ to mean that the m-bit ciphertext is computed from the m-bit plaintext X and two m-bit subkeys Z_a and Z_b. The proposed hash round function is given by

$$h(H_{i-1}, M_i) = E_{H_{i-1}, M_i}(H_{i-1})$$

and is illustrated in Fig. 2. We have been unable to find an attack on this hash

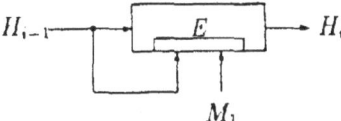

Figure 2: *A proposed m-bit hash function based on an m-bit block cipher with a $2m$-bit key.*

function better than the brute force attack when the underlying block cipher has no known weakness.

3.2 Construction of $2m$-bit hash round functions

When the block length m of a block cipher is 64 (which is the case for many practical block ciphers), one can obtain a 64-bit iterated hash function by using the DM-scheme. The "brute-force" collision attack on any 64-bit hash function has complexity about 2^{32}, which is certainly too small in many applications. Thus, several efforts [2, 13, 14, 18, 20] have been made to construct a $2m$-bit hash function based on an m-bit block cipher by modifying the (apparently secure) DM-scheme. This will be considered in the following sections.

3.3 A principle for evaluating hash round functions and four attacks on three $2m$-bit hash round functions

In this section, we point out an obvious (once the 5 attacks have been formulated) but useful principle for evaluating the security of a hash round function, viz. that *applying any simple (in both directions) invertible transformations to the input and to the output of the hash round function yields a new hash round function with the same security as the original one.* [A similar principle has been used by Meier and Staffelbach in [12] to classify nonlinearity criteria for cryptographic functions]. For example, for a block cipher with block length equal to key length, it follows from this principle that the hash round function (2) of the DM-scheme has the same security as the following hash round function proposed in [11]

$$h(H_{i-1}, M_i) = E_{H_{i-1}}(M_i) \oplus M_i,$$

since this hash round function differs from that in (2) only by a "swapping" of the input blocks H_{i-1} and M_i.

To demonstrate this principle, we present four "meet-in-middle" attacks on three $2m$-bit hash round functions based on an m-bit block cipher with an m-bit key. The basic purpose of these three schemes is to construct a $2m$-bit hash function based on an m-bit block cipher by modifying the (apparently secure) DM-scheme (2). We now show that these $2m$-bit hash round functions are in fact weaker than the m-bit hash round function of the DM-scheme. More precisely, for each scheme, we present a free-start target attack that takes only about $2^{m/2}$ (instead of the ideal maximum 2^{2m}) computations of the round function. [Recall that the free-start target attack on the m-bit hash round function in the DM-scheme has complexity 2^m.]

3.3.1 The Preneel-Bosselaers-Govaerts-Vandewalle (PBGV) scheme.

The PBGV scheme was proposed in [18]. In this scheme, which uses an m-bit block cipher with an m-bit key, a $2m$-bit hash value $H = (H_n, G_n)$ is computed from a $2mn$-bit message $(L_1, N_1, L_2, N_2, ..., L_n, N_n)$ and a $2m$-bit initial value (H_0, G_0). In each round, two new m-bit values H_i and G_i are computed from the two previous m-bit values H_{i-1} and G_{i-1} and from the two m-bit message blocks L_i and N_i as follows:

$$\begin{aligned} H_i &= E_{L_i \oplus N_i}(H_{i-1} \oplus G_{i-1}) \oplus L_i \oplus H_{i-1} \oplus G_{i-1} \\ G_i &= E_{L_i \oplus H_{i-1}}(N_i \oplus G_{i-1}) \oplus N_i \oplus H_{i-1} \oplus G_{i-1} \end{aligned} \tag{3}$$

for $i = 1, 2, \ldots, n$.

The round function for the PBGV-scheme produces the output pair (h, g) from the inputs (h_0, g_0, l, n) in the manner

$$\begin{aligned} h &= E_{l \oplus n}(h_0 \oplus g_0) \oplus l \oplus h_0 \oplus g_0 \\ g &= E_{l \oplus h_0}(n \oplus g_0) \oplus n \oplus h_0 \oplus g_0. \end{aligned} \tag{4}$$

By applying the simple and simply inverted transformations

$$(h, g) \longrightarrow (h, f) = (h, h \oplus g) \tag{5}$$

on the output and

$$(h_0, g_0, l, n) \longrightarrow (h_0', g_0', l', n') = (h_0 \oplus g_0, g_0 \oplus n, l \oplus n, n), \tag{6}$$

on the input, we obtain the round function illustrated in Fig.3 that computes (h, f) from the input (h_0', g_0', l', n') in the manner

$$\begin{aligned} h &= E_{l'}(h_0') \oplus l' \oplus n' \oplus h_0' \\ f &= E_{l' \oplus h_0' \oplus g_0'}(g_0') \oplus E_{l'}(h_0') \oplus l'. \end{aligned} \tag{7}$$

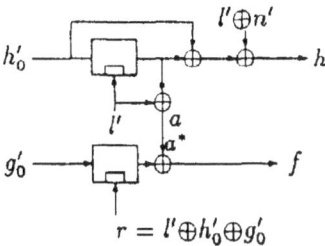

$$r = l' \oplus h'_0 \oplus g'_0$$

Figure 3: *The transformed function used to attack the PBGV round function.*

Because the transformations (5) and (6) are both easy to compute and easy to invert, it follows from our principle that an attack on the round function (7) has the same complexity as an attack on the round function (4).

A free-start target attack on the PBGV round function with complexity about $2^{m/2}$: In this attack, we show how to find a "random inverse" of (7), i.e., we show how, for given (h, f), to find (h'_0, g'_0, l', n') satisfying (4) for which (h'_0, g'_0) appears randomly chosen.

1. Choose an arbitrary constant c_0.

2. For the given h, compute $a = E_{l'}(h'_0) \oplus l'$ for $2^{m/2}$ randomly chosen values of h'_0 and corresponding l' such that $h'_0 \oplus l' = c_0$.

3. For the given f, compute $a^* = E_r(g'_0) \oplus f$ for $2^{m/2}$ randomly chosen values of g'_0 and corresponding r such that $g'_0 \oplus r = c_0$.

The probability that some a and some a^* take on the same value is about 0.63. For such $(g'_0, r, a = a^*, h'_0, l')$, we obtain a solution (h'_0, g'_0, l', n') for (7) by computing $n' = a \oplus l' \oplus h'_0 \oplus l' \oplus h$. ☐

[A recent result of Preneel [19] gives a free-start target attack on the PBGV round function that requires only the computation of one decryption with the block cipher.]

A target attack on the PBGV round function with complexity about 2^m: In this attack, we find, for the given (h_0, g_0) and (h, g), a message block (l, n) satisfying (4). We will use the notation of Fig.3.

From (5) and (6), we see that (h, f) and h'_0 are determined by the given (h_0, g_0) and (h, g). We randomly choose l', then compute

$$a = E_{l'}(h'_0) \oplus l',$$

$$n' = a \oplus h'_0 \oplus h,$$

$$r = l' \oplus h'_0 \oplus g'_0 = l' \oplus h'_0 \oplus g_0 \oplus n'$$

and

$$g'_0 = D_r(a \oplus f),$$

where $D_z(y)$ denotes the result of deciphering y with key z.

After 2^m such computations, $g'_0 \oplus n'$ will take on the given value g_0 with probability 0.63. Then using (5) and (6), we obtain a solution (l, n) for (4). ☐

3.3.2 The first Quisquater-Girault (QG-I) scheme.

The QG-I scheme was proposed in the Abstracts from Eurocrypt'89 [20]. It also appeared in a draft ISO standard [6], see also [15]. However, this scheme was dropped from the recent version of the draft ISO standard CD10118 [7]. [In unpublished work, Coppersmith pointed out to its inventors some weakness of this scheme [21]. In the subsequent Proceedings paper [21], a "weaker" round function was used, but with additional functional strengthening.] Similarly to the PBGV-scheme discussed above, the QG-I scheme is based on an m-bit block cipher with an m-bit key. A $2m$-bit hash value (H_n, G_n) is computed from a $2mn$-bit message $(L_1, N_1, L_2, N_2, ..., L_n, N_n)$ and a $2m$-bit initial value (H_0, G_0). In each round, two new m-bit values H_i and G_i are computed from the two previous m-bit values H_{i-1} and G_{i-1} and from the two m-bit message blocks L_i and N_i as follows:

$$
\begin{aligned}
W_i &= E_{L_i}(G_{i-1}\oplus N_i)\oplus N_i\oplus H_{i-1}\\
H_i &= W_i\oplus G_{i-1}\\
G_i &= E_{N_i}(W_i\oplus L_i)\oplus H_{i-1}\oplus G_{i-1}\oplus L_i
\end{aligned}
\tag{8}
$$

for $i = 1, 2, \ldots, n$.

The round function of the QG-I scheme produces the output pair (h, g) from the input (h_0, g_0, l, n) in the manner

$$
\begin{aligned}
h &= E_l(g_0\oplus n)\oplus n\oplus h_0\oplus g_0\\
g &= E_n\left(E_l(g_0\oplus n)\oplus n\oplus h_0\oplus l\right)\oplus h_0\oplus g_0\oplus l.
\end{aligned}
\tag{9}
$$

We will consider the pair $(h, f) = (h, h\oplus g)$ illustrated in Fig.4 and defined by

$$
\begin{aligned}
h &= E_l(g_0\oplus n)\oplus n\oplus h_0\oplus g_0\\
f = h\oplus g &= E_n\left(E_l(g_0\oplus n)\oplus n\oplus h_0\oplus l\right)\oplus E_l(g_0\oplus n)\oplus l\oplus n.
\end{aligned}
\tag{10}
$$

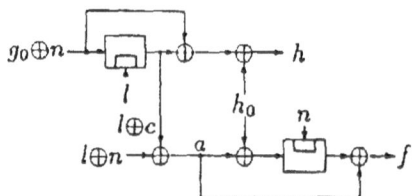

Figure 4: *The pair (h, f) used in the attack on the QG-I scheme.*

A free-start target attack on the QG-I scheme with complexity about $2^{m/2}$: In the following we show that, for any given (h, f), one can find, in about $2^{m/2}$ decrypting computations for the block cipher, a solution (h_0, g_0, l, n) satisfying (10) by a "meet-in-the-middle" attack.

We will use the notation shown in Fig.4. Let c be a fixed m-tuple.

1. Randomly choose values for a and choose n such that $a \oplus n = c$. Then, for the given value of f, compute $h'_0 = a \oplus D_n(a \oplus f)$. Repeat this process $2^{m/2}$ times to obtain $2^{m/2}$ values for (h'_0, n) with randomly chosen values for h'_0.

2. Randomly choose l and compute $h^*_0 = h \oplus (l \oplus c) \oplus D_l(l \oplus c)$. In $2^{m/2}$ computations, one obtains $2^{m/2}$ values for (h^*_0, l) with randomly chosen values for h^*_0.

Note that both h'_0 and h^*_0 are m-bit blocks so that some h'_0 and some h^*_0 obtained as above will take on the same value with probability about 0.63. Thus, we can find (h'_0, h^*_0, l, n) such that $h'_0 = h^*_0$. (Note that the constraint that $l \oplus c \oplus l \oplus n = a$ is automatically satisfied.) From the obtained (l, n), compute $g_0 = D_l(l \oplus c) \oplus n$. Then the resulting (h_0, g_0, l, n) is the desired solution. □

3.3.3   The LOKI Double Block Hash (DBH) function.

The block cipher LOKI, proposed in [2], is a DES-like 64-bit block cipher with a 64-bit key. In [2], a 128-bit iterated Double Block Hash (DBH) function based on the cipher LOKI was proposed, but this scheme can in fact be used for any m-bit block cipher with an m-bit key. In LOKI DBH, a $2m$-bit hash value (H_n, G_n) is computed from a $2mn$-bit message $(L_1, N_1, L_2, N_2, ..., L_n, N_n)$ and a $2m$-bit initial value (H_0, G_0). In each round, two new m-bit values H_i and G_i are computed from the two previous m-bit values H_{i-1} and G_{i-1} and from the two current m-bit message blocks L_i and N_i as follows:

$$
\begin{aligned}
W_i &= E_{L_i \oplus G_{i-1}}(G_{i-1} \oplus N_i) \oplus N_i \oplus H_{i-1} \\
H_i &= W_i \oplus G_{i-1} \\
G_i &= E_{N_i \oplus H_{i-1}}(W_i \oplus L_i) \oplus H_{i-1} \oplus G_{i-1} \oplus L_i
\end{aligned}
\tag{11}
$$

for $i = 1, 2, \ldots, n$.

The LOKI DBH round function was derived from the hash round function of the QG-I scheme (8) by the bitwise addition modulo 2 of the previous hash value blocks (H_{i-1} and G_{i-1}) to the current message blocks (L_i and N_i) to obtain the key inputs for the two LOKI encryptions. This was done to avoid some attacks derived from the 'weak key' of the underlying cipher. By applying our security evaluation principle, we obtain the following free-start target attack on the LOKI DBH round function that has complexity only about $2^{m/2}$.

The round function for the LOKI DBH produces the output pair (h, g) from the input (h_0, g_0, l, n) in the manner

$$
\begin{aligned}
h &= E_{l \oplus g_0}(g_0 \oplus n) \oplus n \oplus h_0 \oplus g_0 \\
g &= E_{n \oplus h_0}(E_{l \oplus g_0}(g_0 \oplus n) \oplus n \oplus h_0 \oplus l) \oplus h_0 \oplus g_0 \oplus l.
\end{aligned}
\tag{12}
$$

By applying the transformation

$$
(h, f) = (h, h \oplus g)
\tag{13}
$$

on the LOKI DBH output pair (h, g) and applying the transformation

$$
(h_0, g_0, l', n') = (h_0, g_0, l \oplus g_0, n \oplus g_0)
\tag{14}
$$

on the LOKI DBH inputs (h_0, g_0, l, n), we obtain the function illustrated in Fig.5 that computes (h, f) from the inputs (h_0, g_0, l', n') in the manner

$$
\begin{aligned}
h &= E_{l'}(n') \oplus n' \oplus h_0 \\
f &= E_{n' \oplus h_0 \oplus g_0}(h \oplus l') \oplus h \oplus l' \oplus h_0.
\end{aligned}
\tag{15}
$$

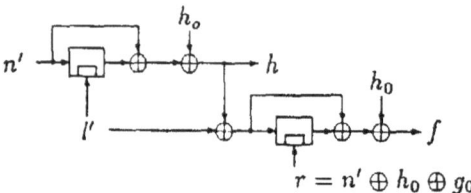

$$r = n' \oplus h_0 \oplus g_0$$

Figure 5: *The new function used to attack the LOKI DBH round function.*

A free-start target attack on the LOKI DBH with complexity about $2^{m/2}$: In the following, we show that, for any given (h, f), one can find, in about $2 \times 2^{m/2}$ encrypting computations for the block cipher, a solution for (h_0, g_0, l, n) satisfying (10) by a "meet-in-the-middle" attack.

Because the transformations (13) and (14) are both easy to compute and easy to invert, it follows from our principle that finding a solution (h_0, g_0, l, n) of (12) for a given (h, g) is computationally the same as finding a solution (h_0, g_0, l', n') of (15) for a given (h, f). This can be done in about $2 \times 2^{m/2}$ encryptions as we now show.

1. Choose an arbitrary value for l'.

2. For the given h and the chosen l', compute $h_0 = h \oplus n' \oplus E_{l'}(n')$ for $2^{m/2}$ randomly chosen values of n'.

3. For the given h, f and the chosen l', compute $h_0^* = E_r(h \oplus l') \oplus h \oplus l' \oplus f$ for $2^{m/2}$ randomly chosen values of r $(= n \oplus h_0 \oplus g_0)$.

The probability that some h_0 and some h_0^* take on the same value is about 0.63. For $h_0 = h_0^*$, by computing $g_0 = r \oplus n' \oplus h_0$, we obtain a solution (h_0, g_0, l', n') for (15). $\quad\square$

Remark. We have given three free-start target attacks on three hash round functions in this section. The "real" target attacks (with specified initial value) will usually be more difficult. For example, when m is 64 bits, a target attack on the 128-bit hash function LOKI DBH obtained by combining the above attack with the attack used in the proof of Theorem 3 will take about $2^{\frac{128-32}{2}} = 2^{80}$ computations. A similar conclusion holds also for the QG-I scheme hash function.

3.4 Complexity of known attacks on $2m$-bit hash functions

We consider here some known 128-bit iterated hash functions based on two uses of an $m = 64$-bit block cipher with key-length $k = 64$ or $k = 56$ in each round. All these schemes can be considered as slight modifications of the 64-bit DM-scheme hash round function. The complexities of known attacks on these hash functions are listed in Table 1. We assume that all the iterated hash functions are used with MD-strengthening and that the underlying block cipher has no known weakness (such as weak keys).

$h(\cdot,\cdot)$	PBGV	GQ-I	LOKI-DBH	Merkle [12]	M-S [13]	ideal
(m,k) [1]	(64,64)	(64,64)	(64,64)	(64,56)	(64,56)	(64,k)
target	2^{64} [2]	2^{80} [5]	2^{80} [9]	2^{112}	2^{81} [14]	2^{128}
f-s target	$o(1)$ [3]	2^{32} [6]	2^{32} [10]	2^{112}	2^{54} [15]	2^{128}
collision	2^{32} [3]	2^{64}	2^{64}	2^{56}	2^{54}	2^{64}
semi-f-s col.	2^{32} [3]	2^{32} [7]	2^{64}	2^{56}	2^{54}	2^{64}
f-s coll.	$o(1)$ [4]	$o(1)$ [8]	2^{32} [11]	2^{56}	2^{27} [16]	2^{64}
$\mathrm{leng}(M_i)$	128	128	128	7	64	l [17]

↝ 1: m: block-length, k: key-length of the underlying cipher;

↝ 2: see last section;

↝ 3: recent results of Preneel [19];

↝ 4: a free-start collision attack is no harder than a free-start target attack;

↝ 5: from the free-start target attack [6] and Proposition 3;

↝ 6: see last section;

↝ 7,8: see [16];

↝ 9,10: same as ↝ 5,6;

↝ 11: same as ↝ 4;

↝ 12: Merkle's scheme [13]: hash-code is of length 112 bits; this scheme appears to have ideal security; however, each round can 'digest' only 7 bits of message;

↝ 13: Meyer-Schilling's scheme [14]: 128-bit hash code, but round output has length 108 bits;

↝ 14,15: each round output (two blocks) has length 108 bits; a free-start target attack on one (54-bit) block takes about 2^{54} computations; then use Proposition 3; see also [14];

↝ 16: collision is achieved on one (54-bit) block.

↝ 17: see next section.

Table 1: *Complexity of known attacks on some hash round functions.*

3.5 Proposed schemes for block ciphers with $k = 2m$

The study of previously proposed hashing schemes (see Table 1) suggests that it is difficult, if not impossible, to build a $2m$-bit hash round function with ideal computational security that can "digest" in each round at least m bits of message by two uses of an m-bit block cipher with an m-bit key. However, if an m-bit block cipher with a $2m$-bit key is available, then there are more possibilities to construct a possibly secure $2m$-bit hash round function. In the following, we propose two $2m$-bit hash round functions that use an m-bit block cipher with a $2m$-bit key and that appear to be secure.

68

Tandem DM: We refer to our first proposed $2m$-bit hash function as the *Tandem DM* scheme because it is based on cascading two DM-schemes as in (2). The round function of the Tandem DM scheme is shown in Fig.6. In each iteration, two new

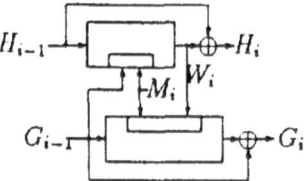

Figure 6: *The Tandem DM $2m$-bit hash round function based on an m-bit block cipher with a $2m$-bit key.*

m-bit values H_i and G_i are computed from the two previous m-bit values H_{i-1} and G_{i-1} and from an m-bit message block M_i as follows:

$$W_i = E_{G_{i-1},M_i}(H_{i-1})$$
$$H_i = W_i \oplus H_{i-1}$$
$$G_i = G_{i-1} \oplus E_{M_i,W_i}(G_{i-1}).$$

Abreast DM We next propose the *Abreast* DM scheme in which two DM-schemes are used side-by-side. The hash round function is illustrated in Fig.7. In each round, two new m-bit values (H_i, G_i) are computed from the two previous m-bit values (H_{i-1}, G_{i-1}) and from an m-bit message block M_i as follows:

$$H_i = H_{i-1} \oplus E_{G_{i-1},M_i}(H_{i-1})$$
$$G_i = G_{i-1} \oplus E_{M_i,H_{i-1}}(\overline{G}_{i-1})$$

where $\overline{G}$ denotes the bit-by-bit complement of G.

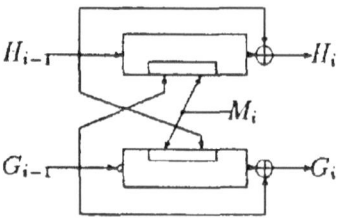

Figure 7: *The Abreast DM $2m$-bit hash round function based on an m-bit block cipher with a $2m$-bit key. The circle indicates that the input to the lower encrypter is bitwise complemented.*

Remarks: 1. The Tandem DM and the Abreast DM schemes were constructed on the following consideration. The round function h consists of two subfunctions h_1 and h_2:

$$(H_i, G_i) = h(H_{i-1}, G_{i-1}, M_i) = [h_1(H_{i-1}, G_{i-1}, M_i), h_2(H_{i-1}, G_{i-1}, M_i)],$$

both of which have the same inputs. Thus, to attack h (in a free-start target or free-start collision attack) implies that one must attack both h_1 and h_2 simultaneously. If the subfunctions h_1 and h_2 are so 'different' that an attack on one subfunction provides no help in attacking the other subfunction and if both h_1 and h_2 are equivalent (in the sense of security) to the apparently secure DM-scheme, then we can expect that an attack on h will have complexity equal to the product of the complexities of the attacks on h_1 and on h_2. In the proposed Tandem DM and Abreast DM schemes, the subfunctions h_1 and h_2 are chosen to be as "different" as possible.

2. The Abreast DM scheme gives a $2m$-bit hash function that is at least as strong as the m-bit DM-scheme. [This is true also for the Meyer-Schilling scheme [7, 14].]

3. Our investigations to this point have shown no weakness in either of these two new proposed $2m$-bit hash round functions, i.e., we have been unable to find any attacks better than brute-force attacks when the underlying cipher is assumed to have no weakness. We should point out, however, that our Tandem DM and Abreast DM schemes use two m-bit block encryptions for each block of m message bits in order to compute a final hash value of length $2m$ bits.

Acknowledgements

The authors are grateful to Prof. F. Piper, Dr. R. Rueppel and the anonymous referee for informing them about recent developments related to ISO standard CD10118. The authors would like in particular to thank B. Preneel for his many useful comments on this paper. This research was supported by the Swiss Commission for the Advancement of Scientific Research, Research Grant KWF 2146.1.

References

[1] S. G. Akl, "On the Security of Compressed Encodings", Advances in Cryptology-CRYPTO'83, Proceedings, pp. 209-230, Plenum Press, New York, 1984.

[2] L. Brown, J. Pieprzyk and J. Seberry, "LOKI – A Cryptographic Primitive for Authentication and Secrecy Applications", Advances in Cryptology – AUSCRYPT'90, Proceedings, LNCS 453, pp. 229-236, Springer-Verlag, 1990.

[3] Data Encryption Standard, FIPS PUB 46, National Tech. Info. Service, Springfield, VA, 1977.

[4] I. B. Damgaard, "A Design Principle for Hash Functions", Advances in Cryptology-CRYPTO'89, LNCS 435, pp. 416-427, Springer-Verlag, 1990.

[5] R. W. Davies and W. L. Price, "Digital Signature – an Update", Proc. International Conference on Computer Communications, Sydney, Oct 1984, Elsevier, North-Holland, pp. 843-847, 1985.

[6] I.S.O. DP 10118, Hash-functions for Digital Signatures, I.S.O., April 1989.

[7] ISO/IEC CD 10118, Information technology - Security techniques - Hash-functions, I.S.O., 1991.

[8] X. Lai and J. L. Massey, "A Proposal for a New Block Encryption Standard", Advances in Cryptology-EUROCRYPT'90, Proceedings, LNCS 473, pp. 389-404, Springer-Verlag, Berlin, 1991.

[9] X. Lai, J. L. Massey and S. Murphy, "Markov Ciphers and Differential Cryptanalysis", Advances in Cryptology-EUROCRYPT'91, Proceedings, LNCS 547, pp. 17-38, Springer-Verlag, Berlin, 1991.

[10] S. M. Matyas, "Key Processing with Control Vectors", Journal of Cryptology, Vol. 3, No. 2, pp. 113–136, 1991.

[11] S. M. Matyas, C. H. Meyer and J. Oseas, "Generating Strong One-way Functions with Cryptographic Algorithm", IBM Technical Disclosure Bulletin, Vol. 27, No. 10A, pp. 5658-5659, March 1985.

[12] W. Meier, O. Staffelbach, " Nonlinearity Criteria for Cryptographic Functions", Advances in Cryptology - EUROCRYPT'89, Proceedings, LNCS 434, pp. 549-562, Springer-Verlag, 1990.

[13] R. C. Merkle, "One Way Hash Functions and DES", Advances in Cryptology-CRYPTO'89, Proceedings, LNCS 435, pp. 428-446, Springer-Verlag, 1990.

[14] C. H. Meyer and M. Schilling, "Secure Program Code with Modification Detection Code", Proceedings of SECURICOM 88, pp. 111-130, SEDEP.8, Rue de la Michodies, 75002, Paris, France.

[15] C. J. Mitchell, F. Piper and P. Wild, "Digital Signatures", Contemporary Cryptology (Ed. G. Simmons), pp. 325-378, IEEE Press, 1991.

[16] S Miyaguchi, K. Ohta and M. Iwata, "Confirmation that Some Hash Functions Are Not Collision Free", Advances in Cryptology-EUROCRYPT'90, Proceedings, LNCS 473, pp. 326-343, Springer-Verlag, Berlin, 1991.

[17] M. Naor and M. Yung, "Universal One-way Hash Functions and Their Cryptographic Applications", Proc. 21 Annual ACM Symposium on Theory of Computing, Seattle, Washington, May 15-17, 1989, pp. 33-43.

[18] B. Preneel, A. Bosselaers, R. Govaerts and J. Vandewalle, "Collision-free Hashfunctions Based on Blockcipher Algorithms." Proceedings of 1989 International Carnahan Conference on Security Technology, pp. 203-210.

[19] Private communication, B. Preneel to X. Lai, June 1992.

[20] J. J. Quisquater and M. Girault, "2n-bit Hash Functions Using n-bit Symmetric Block Cipher Algorithms", Abstracts of EUROCRYPT'89.

[21] J. J. Quisquater and M. Girault, "2n-bit Hash Functions Using n-bit Symmetric Block Cipher Algorithms", Advances in Cryptology-EUROCRYPT'89, Proceedings, LNCS 434, pp. 102-109, Springer-Verlag, Berlin, 1990.

[22] R. S. Winternitz, "Producing One-Way Hash Function from DES", Advances in Cryptology-CRYPTO'83, Proceedings, pp. 203-207, Plenum Press, New York, 1984.

[23] R. S. Winternitz, "A Secure One-way Hash Function Built from DES", Proc. 1984 IEEE Symposium on Security and Privacy, Oakland, 1984, pp. 88-90.

Differential Cryptanalysis Mod 2^{32} with Applications to MD5

Thomas A. Berson

Anagram Laboratories

P.O. Box 791

Palo Alto, CA 94301, USA

Abstract

We introduce the idea of differential cryptanalysis mod 2^{32} and apply it to the MD5 message digest algorithm. We derive a theory for differential cryptanalysis of the circular shift function. We demonstrate a high-probability differentials which leave the message digest register unchanged for each of MD5's four rounds, and explain how more such differentials may be calculated.

1 Introduction

Differential cryptanalysis is a method which analyses the effect of particular differences in plaintext pairs on the differences of the resulting ciphertext pairs. Since differential cryptanalysis was first explained by Biham and Shamir [BS1] [BS2] it has been applied with success, sometimes with devastating success, to cryptosystems including DES [BS1], [BS2], Feal [BS3], N-Hash [BS3], PES [LMM], Snefru [BS4] Khafre [BS4], REDOC-II [BS4], LOKI [BS4] [Knud] [BKPS], and Lucifer [BS4]. A common element in these cryptosystems which makes them susceptible to differential cryptanalysis is their heavy use of the exclusive-or operation (denoted by $\oplus$ or XOR), which is equivalent to vector addition mod 2, to introduce confusion in combining partial results or in combining key with data. Differential cryptanalysis is able to reduce or negate the cryptographic effects of these XORs by considering the differences in ciphertexts which arise from operating the cryptosystem on pairs of plaintexts chosen so that they are at a fixed distance (mod 2) from one another. The appropriate fixed distance changes from cryptosystem to cryptosystem, and is derived by analysis of the structure of the cryptosystem.

Designers of cryptosystems published since the rise of differential cryptanalysis have sought to avoid its sting by reducing or avoiding the use of XOR in its traditional roles. The message digest algorithm MD5 [Riv] [RD] is an example of such a post-differential cryptosystem. MD5 is designed to be fast on 32-bit machines, and employs addition mod 2^{32} (denoted by +) to achieve confusion.

Unfortunately for designers, there is nothing which binds differential cryptanalysis to any particular algebraic group. In this paper we apply differential cryptanalysis mod 2^{32} to MD5.

R.A. Rueppel (Ed.): Advances in Cryptology - EUROCRYPT '92, LNCS 658, pp. 71-80, 1993.

So far as possible, we will follow the notation of Biham and Shamir [BS2] and Rivest and Dusse [RD]. In particular:

$Y*$, Y': At any point during the operation of the MD5 algorithm on pairs of messages, Y and $Y*$ are the values of the two executions of the algorithm, and Y' is defined to be $Y' = Y - Y*$ mod 2^{32}, $0 \leq Y' \leq 2^{32} - 1$.

$\bar{M}$: the input message, b bits in length. b is any non-negative integer.

n_x: A hexidecimal number is denoted by the subscript x.

2   The MD5 Message Digest Algorithm

The MD5 message digest algorithm takes as input a message of arbitrary length and produces as output a 128-bit message digest. MD5 is intended by its authors for use in digital signature applications. These applications require that it be computationally infeasible to produce two messages having the same digest, for it is the digest which is signed, not the original message.

2.1   Overall Structure

The computation of the message digest of $\bar{M}$ by MD5 is carried out in five stages.

Stage 1. The message $\bar{M}$ is padded (extended) so that its length in bits is congruent to 448 modulo 512. The padding consists of a single "1" bit followed by as many "0" bits as necessary to reach the desired length.

Stage 2. A 64-bit representation of b, the length of $\bar{M}$ before padding, is appended to the results of Stage 1. The resulting message M has a length which is an exact multiple of 512 bits. Equivalently, this message has a length which is an exact multiple N of 16 32-bit words.

Stage 3. A four-word register MD is used to compute the message digest. This is initialized to a constant.

Stage 4. The message M is processed in consecutive *blocks* of 16 words, M_1, M_2, ... , M_{N-1}, M_N. The processing of each block consists of four *rounds*, each of which consists of 16 *steps*. We will have much more to say about rounds and steps below.

Stage 5. Register MD now contains the calculated message digest. This is output.

In seeking two equivalent messages we will work within a single 16-word block. We would like the freedom to alter every word of that block independently of any other. The structure of M_N, and possibly of M_{N-1}, is constrained by Stage 1 and Stage 2 processing, these are therefore not easy blocks to attack. We can focus our attack on any other block or blocks. For purposes of this paper we will attack a single block $M_{a,a\neq N,N-1}$ and will hold all other blocks in M_1, ... , M_{N-2} identical in M and $M*$.

2.2 Block Processing

The message digest register, MD, begins in a specified constant state MD_0 and is updated during the processing of each block. Its final state MD_N is the value assigned to MD5(m).

The processing of the jth block involves four *round functions*, FF, GG, HH, and II, as follows:

$$MD_j = MD_{j-1} + II\Big(M_j, HH\big(M_j, GG\big(M_j, FF\big(M_j, MD_{j-1}\big)\big)\big)\Big)$$

The round functions are similar to one another in structure. The message digest register is treated as a four-element shift register, with each element being one word wide. The elements are referred to as A, B, C, and D. Each round consists of 16 steps of this register.

At each step, $A = B + ((A + f(B,C,D) + x[s] + t) <<< k)$, where f is an *auxiliary function* which varies from round-to-round; $x[s]$ is a word chosen from M_j; s, t, and k are parameters of the step; and $<<< k$ signifies a k-bit left circular shift of a word. Note that each step involves four + operations, one $<<<$ operation, and one auxiliary function.

The auxiliary functions each take three 32-bit words as input and produce one 32-bit word as output. They are bit-wise parallel, which is to say that each bit of the output word depends only on the corresponding bits of the input words. The auxiliary functions f are defined in Table 1, where $\bar{v}$ denotes the bit-wise complement of v.

Table 1. Auxiliary functions.

Round	f	$f(X,Y,Z)$
FF	F	$XY \vee \bar{X}Z$
GG	G	$XZ \vee Y\bar{Z}$
HH	H	$X \oplus Y \oplus Z$
II	I	$Y \oplus (X \vee \bar{Z})$

3 The Cryptanalytic Problem

Message digest algorithms present two related cryptanalytic challenges. The simpler of these is to find two messages with the same digest. Rivest and Dusse conjecture that the difficulty of doing this for MD5 is on the order of 2^{64} operations. The other challenge is to find any message with a given digest. Rivest and Dusse conjecture that the difficulty of this feat for MD5 is on the order of 2^{128} operations.

We will attack the simpler problem, that of finding two messages $m \neq m^*$ such that MD5(m) = MD5(m^*). Note that under the simplistic assumption that the cryptanalytic difficulty of MD5 is uniformly distributed across its sixty-four steps, we will need to succeed at each step with probability > 0.5 in order to do better than Rivest's conjecture. This is a daunting prospect. Throughput is another measure of MD5's difficulty under differential cryptanalysis. MD5 produces only 2 output bits per step. DES produces 4.

The overwhelming number of + operations per step provides the motivation to attempt differential cryptanalysis mod 2^{32}. Analysis of the + operation is hard working mod 2; but it is easy working mod 2^{32}, and leads to output differences with probability = 1. On the other hand, analysis of the f functions is very hard mod 2^{32}. There is no theory to help, and a complete simulation is beyond the size of available computer memory. We must content ourselves with approximations. Table 2 summarizes the trade-offs facing the cryptanalyst.

Table 2. Cryptanalytic trade-offs within an MD5 step.

op	#	DCA mod 2	DCA mod 2^{32}
+	4	Hard analysis.	Easy analysis. See §3.1. Output differences with probability = 1.
<<<	1	Easy analysis. Output differences with probability = 1.	Non-trivial analysis. See §3.2. For any input difference only four output differences are possible, at least one has probability greater than or equal to 0.25.
f	1	Easy analysis. Low weight input differences lead to high probability output differences.	Hard analysis. See §3.3. Requires approximations. High probability output differences.

3.1 Differential Cryptanalysis of the Add Function

Let $x+c=z$, and $x^*+c=z^*$, then $z'=z-z^*=x-x^*=x'$, with probability = 1. Similarly, where $x+y=z$, and $x^*+y^*=z^*$, then $z'=z-z^*=x-x^*+y-y^*=x'+y'$, with probability = 1.

3.2 Differential Cryptanalysis of the Circular Shift

Theorem 1. If $i \in Z$ and $x \in R$, then $\lfloor i+x \rfloor = i + \lfloor x \rfloor$.

Theorem 2. If $a, b, m \in Z$, then

$$\left\lfloor \frac{a}{m} \right\rfloor - \left\lfloor \frac{a-b}{m} \right\rfloor = \begin{cases} \left\lfloor \dfrac{b}{m} \right\rfloor + 1, & a \bmod m < b \bmod m; \\ \left\lfloor \dfrac{b}{m} \right\rfloor, & a \bmod m \geq b \bmod m. \end{cases}$$

Let x be a 32-bit word and $CLS_k(x)$ denote the circular left shift of x by k places. Let $n=2^{32}$ and $m=2^{32-k}$. We can then write $CLS_k(x)$ as the sum of two quantities

$$CLS_k(x) = 2^k x + \left\lfloor \frac{x}{m} \right\rfloor \bmod n.$$

The quantity $2^k x$ is simply a left shift of x by k places, with zeroes filled at the right.

The quantity $\left\lfloor \frac{x}{m} \right\rfloor$ is just right shift of x by 32-k places, with zeroes filled at the left.

Recall that $x' = x - x * \bmod n$, $0 \le x' \le n-1$. What does the mod n difference, $z' = CLS_k(x) - CLS_k(x*) \bmod n$, look like? Assume $0 \le x, x' < n$. Then $z' = CLS_k(x) - CLS_k(x - x' \bmod n) \bmod n =$

$$2^k x + \left\lfloor \frac{x}{m} \right\rfloor - 2^k(x - x' \bmod n) - \left\lfloor \frac{x - x' \bmod n}{m} \right\rfloor \bmod n =$$

$$2^k x' + \left\lfloor \frac{x}{m} \right\rfloor - \left\lfloor \frac{x - x' \bmod n}{m} \right\rfloor \bmod n.$$

This can be divided, for further analysis, into two cases:

1. $x \ge x' \Rightarrow x - x' \ge 0 \Rightarrow x - x' \bmod n = x - x'$.
2. $x < x' \Rightarrow x - x' < 0 \Rightarrow x - x' \bmod n = x - x' + n$.

How many times does each case occur? If x' is held constant while x is varied over all possible n values, Case 1 occurs when x takes the $n - x'$ values $x = x', x'+1, \ldots, n-1$. Case 2 occurs when x takes the x' values $x = 0, 1, \ldots, x' - 1$. We will consider the two cases separately.

Case 1. $x \ge x'$

Let $x = q_1 m + r_1$, $x' = q_2 m + r_2$, $0 \le r_1, r_2 < m$. Then $\left\lfloor \frac{x}{m} \right\rfloor = q_1$, $\left\lfloor \frac{x'}{m} \right\rfloor = q_2$ and

$$\left\lfloor \frac{x - x'}{m} \right\rfloor = q_1 - q_2 + \left\lfloor \frac{r_1 - r_2}{m} \right\rfloor. \text{ By Theorem 2,}$$

$$\left\lfloor \frac{x}{m} \right\rfloor - \left\lfloor \frac{x-x'}{m} \right\rfloor = \begin{cases} q_2, & \text{if } \left\lfloor \dfrac{r_1-r_2}{m} \right\rfloor = 0; \\[3mm] q_2+1, & \text{if } \left\lfloor \dfrac{r_1-r_2}{m} \right\rfloor = -1. \end{cases}$$

The quantity $\left\lfloor \dfrac{r_1-r_2}{m} \right\rfloor = -1$ when $r_1 < r_2$. How often does this occur? There are $2^k - q_2 - 1$ m-long sub-intervals in $[x', n-1]$. In each of these, $r_1 < r_2$ for the first r_2 values of x. So the event in question occurs $r_2(2^k - q_2 - 1)$ times.

Case 2. $x < x'$

Let x and x' be defined as in Case 1. We are now examining $\left\lfloor \dfrac{x}{m} \right\rfloor - \left\lfloor \dfrac{x-x'+n}{m} \right\rfloor \bmod n$ for $0 \le x < x < x' < n$. Note that $n/m = 2^k$, so

$$\left\lfloor \frac{x}{m} \right\rfloor - \left\lfloor \frac{x-x'+n}{m} \right\rfloor = \left\lfloor \frac{x}{m} \right\rfloor - \left\lfloor \frac{x-x'}{m} \right\rfloor - 2^k \bmod n =$$

$$\begin{cases} q_2 - 2^k, & \text{if } \left\lfloor \dfrac{r_1-r_2}{m} \right\rfloor = 0; \\[3mm] q_2 - 2^k + 1, & \text{if } \left\lfloor \dfrac{r_1-r_2}{m} \right\rfloor = -1. \end{cases}$$

We are again interested in how often $r_1 < r_2$ occurs. There are q_2 m-long sub-intervals in $[0, x'-1]$. In each of these, $r_1 < r_2$ for the first r_2 values of x. The final sub-interval is only r_2 long, and $r_1 < r_2$ throughout this final sub-interval. So the event in question occurs $r_2(q_2+1)$ times.

We have proven the following

Theorem 3. Select x and x^* at random from the integers in $[0, n-1]$ so that $x' = x - x^* \bmod n$. Let $z' = CLS_k(x) - CLS_k(x^*) \bmod n$, and $x' = qm + r, 0 \le r < m$. Then

$$Pr(z') = \frac{1}{n} \begin{cases} n - x' - r(2^k - 1 - q), & \text{if } z' = q + x'2^k \bmod n; \\ r(2^k - 1 - q), & \text{if } z' = q + 1 + x'2^k \bmod n; \\ r(q+1), & \text{if } z' = q - 2^k + 1 + x'2^k \bmod n; \\ x' - r(q+1), & \text{if } z' = q - 2^k + x'2^k \bmod n; \\ 0, & \text{otherwise.} \end{cases}$$

Note that this generalizes to other word sizes.

3.3 Differential Cryptanalysis of the Step Feedback Functions

The step feedback functions f present a difficult problem. In general, where $w = f(X,Y,Z)$ and $w^* = f(X^*,Y^*,Z^*)$, we are interested in $w' = w - w^*$ mod $2^{32} = f(X,Y,Z) - f(X^*,Y^*,Z^*)$ mod 2^{32}.

A question which arises is, given X', Y', and Z', what can be said about w'? At present, nothing much is known either about the values of w' or their associated probabilities. We do not know how to do a mod 2^{32} theoretical analysis of these functions, and the memory required for a complete simulation is not available.

One hunch is that certain values of X', Y', and Z', for example those with low weight, may be likely to lead to values of w' with high probability. Another hunch is that it may be worthwhile to allow only one of X', Y', and Z' to take on a non-zero value. Monte Carlo evaluation within such restricted spaces begins to be feasible, and may lead to useful results.

In the meanwhile, we observe that the differential cryptanalyses mod 2^{32} and mod 2 are identical when we restrict X', Y', and Z' to differ in only the high-order bit. Analysis of the f s mod 2 is straightforward.

4 Cryptanalysis of the Rounds

We can now apply the analytic tools developed in §3 to the rounds of MD5. Recall that each round consists of 16 steps. Fig. 1 is a schematic representation of the calculations which make up a step. The constant additions are omitted as they make no difference from a differential point of view. The inputs have been rotated one element to the right for clarity. Table 3 contains data from an example analysis of the FF round. The notation used in Fig.1 and Table 3 is:

A, B, C, D: elements of the message digest shift register MD.

w: the output of the step auxiliary function f.

x: a word chosen from the current message block M_j. Specified in [RD].

z: the output of the circular left shift function CLS_k. [RD] specifies the values of k.

p, q: intermediate values included to clarify the illustration.

i: a step number. V_i indicates the value of variable v during step i. In the case of A_i, B_i, C_i, and D_i this is to be interpreted as the value of these message register elements at the conclusion of step i.

Our objective is to find two message blocks M_a and M_a^* such that $FF(M_a) = FF(M_a^*)$. We chose, arbitrarily, to work toward the end of the round. The situation at the end of step 11 is $A'_{11}, B'_{11}, C'_{11}, D'_{11} = 0$. Our objective will be met if we can introduce some difference, and then remove it, so that $A'_{16}, B'_{16}, C'_{16}, D'_{16} = 0$.

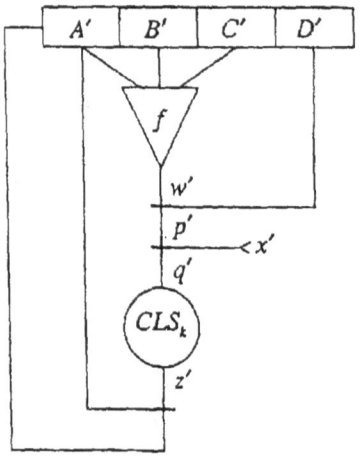

Figure 1. Schematic of an MD5 step, showing only those operations which impact the differential analysis.

Step 12. We can analyze the FF round auxiliary function F only in the case that A', B', and C' are restricted to the high order bit. So our immediate goal is to choose x'_{12} which leads to $A'_{12} = 2^{31}$. Examination of Fig. 1 and Table 3 shows that this reduces to choosing an x' such that $2^{31} = CLS_{22}(x')$. An important insight is that $z = CLS_k(x)$ is a permutation, whose inverse is $x = CLS_{n-k}(z)$. So it is possible to use the theory we developed in §3.2 to calculate the x' we are looking for. There are two values with high probability, we choose 2^9, whose probability is 1/2.

Step 13. We would now like to know the possible values and corresponding probabilities of $w' = F(X,Y,Z) - F(X*,Y,Z)$. Carries have no effect when input differences are restricted to the high-order bit. The step auxiliary functions can then be evaluated as Boolean functions on a single bit. This straightforward, either symbolically or by enumeration. For the function at hand, the h.o. bit of $w' = 0$ or 1, each with probability of 1/2. We choose the 0. Now our goal is to choose an x'_{13} such that $A'_{13} = 0 \Rightarrow z'_{13} = 2^{31}$. Working with the inverse of CLS_7 we choose $x' = 2^{24} \Rightarrow z'_{13} = 2^{31}$ with probability 1/2.

Table 3. Step-by-step analysis of round FF.

i	k_i	w'_i	p'_i	x'_i	q'_i	z'_i	prob	A'_i	B'_i	C'_i	D'_i
11							1	0	0	0	0
12	22	0	0	2^9	2^9	2^{31}	$1 \cdot \frac{1}{2}$	2^{31}	0	0	0
13	7	0	0	2^{24}	2^{24}	2^{31}	$\frac{1}{2} \cdot \frac{1}{2}$	0	2^{31}	0	0
14	12	0	0	0	0	0	$\frac{1}{2} \cdot 1$	0	0	2^{31}	0
15	17	0	0	0	0	0	$\frac{1}{2} \cdot 1$	0	0	0	2^{31}
16	22	0	2^{31}	2^{31}	0	0	$1 \cdot 1$	0	0	0	0

$$\Pi = 2^{-5}$$

Step 14. Now we need to know the values and probabilities of $w' = F(X,Y,Z) - F(X,Y*,Z)$. Working as in Step 12, we calculate the h.o. bit of $w' = 0$ or 1, each with probability of 1/2. Again we choose the 0. Our goal is now to choose an x'_{14} such that $A'_{14} = 0 \Rightarrow z'_{14} = 0$. Thus x'_{14} is trivially 0, and leads to $z'_{14} = 0$ with probability 1.

Step 15. Now we need to know the values and probabilities of $w' = F(X,Y,Z) - F(X,Y,Z*)$. Working as in Step 12, we calculate the h.o. bit of $w' = 0$ or 1, each with probability of 1/2. Again we choose the 0. Our goal is now to choose an x'_{15} such that $A'_{15} = 0 \Rightarrow z'_{15} = 0$. As in the previous step, x'_{15} is trivially 0, which leads to $z'_{15} = 0$ with a probability of 1.

Step 16. By inspection, $w' = 0$ with probability 1. Our goal is to choose an x'_{16} such that $A'_{16} = 0$. Notice that $p'_{16} = 2^{31}$. If we select $x'_{16} = 2^{31}$ then $q'_{16} = 0$, since all additions are mod 2^{32}. Since $q'_{16} = 0$ then $A'_{16} = 0$ with probability 1. We have reached our goal: $A'_{16}, B'_{16}, C'_{16}, D'_{16} = 0$. This completes the analysis of round FF.

The procedure just described has also been applied to rounds GG, HH, and II. The results for all four rounds are summarized in Table 4.

It should be clear that each time one of the equiprobable w's was chosen, we could have as easily chosen the other, which would have lead to a different differential. What may be less clear is that the second most probable z' often occurs with probability $1/2 - \varepsilon$, and also leads to useful differentials. A last comment is that the differentials exhibited here can be slid as far forward in a round as the round's first step.

Table 4. Example message block differentials M'_a for which *round function*$(M'_a, MD) = MD$.

word	FF	GG	HH	II
0	0	0	0	0
1	0	0	0	0
2	0	0	2^{31}	0
3	0	0	0	0
4	0	0	0	2^{25}
5	0	0	0	0
6	0	0	$2^{32} - 2^8$	0
7	0	0	0	0
8	0	2^{11}	0	0
9	0	0	78 00 00 00$_x$	2^{31}
10	0	0	0	0
11	2^9	0	0	2^{31}
12	2^{24}	2^{31}	2^{31}	0
13	0	2^{26}	0	2^{10}
14	0	0	0	0
15	2^{31}	0	2^{31}	0
probability	2^{-5}	2^{-2}	2^{-2}	2^{-4}

The challenge remains to find a single M'_a which makes all four rounds of MD5 ineffective simultaneously.

References

[BKPS] Lawrence Brown, Matthew Kwan, Josef Pieprzyk and Jennifer Seberry, "Improving Resistance to Differential Cryptanalysis and the Redesign of LOKI," in *Asiacrypt '91 Abstracts*, pp. 25-30.

[BS1] Eli Biham and Adi Shamir, "Differential Analysis of DES-like Cryptosystems," in *Advances in Cryptology -- Crypto '90*, pp. 2-21.

[BS2] Eli Biham and Adi Shamir, "Differential Analysis of DES-like Cryptosystems," *Journal of Cryptology* (1991) 4:1, pp. 3-72.

[BS3] Eli Biham and Adi Shamir, "Differential Analysis of FEAL and N-Hash," in *Advances in Cryptology -- Eurocrypt '91*, pp. 1-16.

[BS4] Eli Biham and Adi Shamir, "Differential Analysis of Snefru, Khafre, REDOC-II, LOKI and Lucifer," in *Advances in Cryptology -- Crypto '91*.

[Knud] Lars Ramkilde Knudsen, "Cryptanalysis of LOKI," in *Asiacrypt '91 Abstracts*, pp. 19-24.

[LMM] Xeujia Lai, James L. Massey and Sean Murphey, "Markov Ciphers and Differential Cryptanalysis," in *Advances in Cryptology -- Eurocrypt '91*, pp. 17-38.

[RD] R. Rivest and S. Dusse, "The MD5 Message-Digest Algorithm," Network Working Group Internet Draft, RSA Data Security Inc., 10 July 1991.

[Riv] Ronald Rivest, "MD5", presentation at Crypto '91 rump session.

A New Method for Known Plaintext Attack of FEAL Cipher

Mitsuru Matsui   Atsuhiro Yamagishi

Computer & Information Systems Laboratory

Mitsubishi Electric Corporation

5-1-1, Ofuna, Kamakura, Kanagawa 247, Japan

E-mail : matsui@mmt.isl.melco.co.jp

Abstract

We propose a new known plaintext attack of FEAL cipher. Our method differs from previous statistical ones in point of deriving the extended key in definite way. As a result, it is possible to break FEAL-4 with 5 known plaintexts and FEAL-6 with 100 known plaintexts respectively. Moreover, we show a method to break FEAL-8 with 2^{15} known plaintexts faster than an exhaustive search.

1 Introduction

FEAL cipher [SM, MKOM] is a block cipher algorithm which is designed for software implementations on 8 or 16 bit microprocessors. The most recent version of FEAL cipher is announced as FEAL-NX, where N is the number of rounds and X denotes an optional 128 bit key input.

As for known plaintext attacks of FEAL cipher, Biham and Shamir have shown FEAL-8 is breakable with 2^{38} known plaintexts using differential cryptanalysis [BS], and Tardy-Corfdir and Gilbert have presented a statistical method to break FEAL-4 with 1000 known plaintexts and FEAL-6 with 20000 known plaintexts respectively [TG].

In this paper, we propose a new technique of a known plaintext attack of FEAL cipher. Our method is a kind of meet-in-the-middle attack with a partial exhaustive search; hence we can derive the extended key directly and definitely. We have made

R.A. Rueppel (Ed.): Advances in Cryptology - EUROCRYPT '92, LNCS 658, pp. 81-91, 1993.

computer experiments to attack FEAL cipher with up to seven rounds. As for FEAL-8, we have estimated the computational complexity to derive the key.

The main results in this paper are as follows. The experiments were implemented with C and assembly language programs on HP9425 workstation computer (68040/25MHz).

- FEAL-4 is breakable with 5 known plaintexts in 6 minutes.

- FEAL-6 is breakable with 100 known plaintexts in 40 minutes.

- FEAL-7 is breakable with 2^{14} known plaintexts in 170 hours.

- FEAL-8 is breakable with 2^{15} known plaintexts faster than an exhaustive search for 64-bit keys and with 2^{28} known plaintexts as fast as an exhaustive search for 50-bit keys.

2 Preliminaries

We use the following notations throughout this paper.

P : The plain text.
C : The corresponding cipher text.
P_H, C_H : The left 32 bit data of P and C respectively.
P_L, C_L : The right 32 bit data of P and C respectively.
$A[i]$: The i-th bit of A.
$A[i, j, ..., k]$: The XORed value of the i-th, j-th,.., and k-th bits of A.
$A[i \sim j]$: The $j - i + 1$ bit data consisting of the i-th, $i+1$-th,.., and j-th bits of A.

For convenience of the following chapters, we define modified F-function $O_R = mF_R(I_R, K_R)$ as figure 1, where R denotes the round.
Then one has easily

$$I[0] = O[2, 8] \oplus K[0], \tag{1}$$
$$I[8] = O[2, 8, 10, 16] \oplus K[0, 8], \tag{2}$$
$$I[16] = O[10, 18, 26] \oplus K[16, 24], \tag{3}$$
$$I[24] = O[16, 26] \oplus K[24]. \tag{4}$$

Consequently, for three round algorithm with modified F-function as figure 2, we obtain the following useful relations which hold for any plaintext P and the corresponding ciphertext C by tracing the bold line in figure 2:

$$P_H[2,8] \oplus P_L[0] \oplus C_H[2,8] \oplus C_L[0] = K_1[0] \oplus K_3[0] \oplus K_4[2,8], \tag{5}$$

$$P_H[2,8,10,16] \oplus P_L[8] \oplus C_H[2,8,10,16] \oplus C_L[8] = \tag{6}$$
$$K_1[0,8] \oplus K_3[0,8] \oplus K_4[2,8,10,16],$$

$$P_H[10,18,26] \oplus P_L[16] \oplus C_H[10,18,26] \oplus C_L[16] = \tag{7}$$
$$K_1[16,24] \oplus K_3[16,24] \oplus K_4[10,18,26],$$

$$P_H[16,26] \oplus P_L[24] \oplus C_H[16,26] \oplus C_L[24] = \tag{8}$$
$$K_1[24] \oplus K_3[24] \oplus K_4[16,26].$$

We note that the left side of each equation is a constant value for an attacker. In the following chapters, we will construct similar constant functions for each cipher so that the number of related key bits is as small as possible.

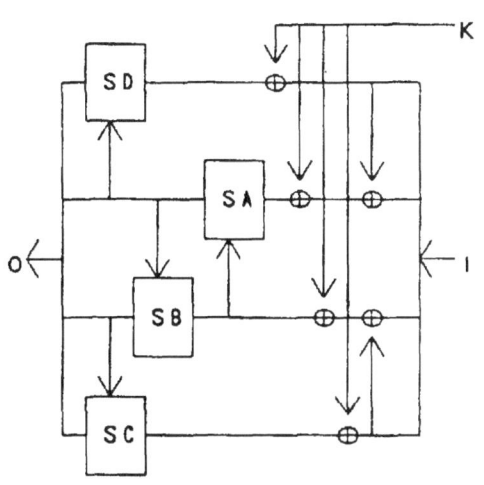

$$S A(x, y) = S C(x, y) = ROL2 (x+y+1 (mod256))$$
$$S B(x, y) = S D(x, y) = ROL2 (x+y (mod256))$$

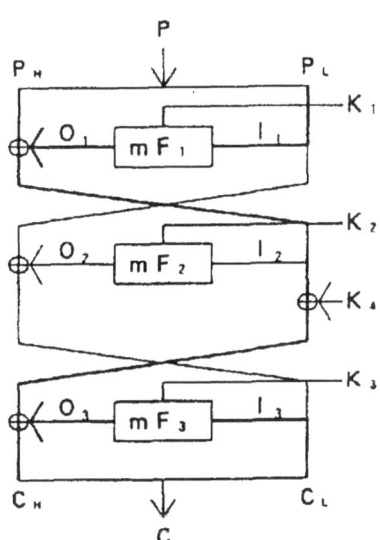

Figure 1 :   Modified F function   Figure 2 :   Three round algorithm

3 Principle of the attack

First, we construct explicitly a function $g(x, y, z)$ which satisfies the following conditions:

- The size of z is sufficiently small.

- There exists $\tilde{K}$ which depends only on the extended key, then $g(P, C, \tilde{K})$ is a constant value for any plaintext P and the corresponding ciphertext C.

- For any fixed $K \neq \tilde{K}$, there exist plaintexts P, P' and the corresponding ciphertexts C, C' such that $g(P, C, K) \neq g(P', C', K)$.

Once we succeed in obtaining g, then we are able to make an exhaustive search for $\tilde{K}$; namely, for all possible K we verify that g always outputs a constant value for given known plaintexts and the corresponding ciphertexts. Since this verification is expected to fail in almost cases except the correct $\tilde{K}$, we will obtain only several candidates for $\tilde{K}$ if we have sufficiently many known plaintexts. By repeating this method using various functions, we can reach whole key bits finally.

Although this attack is effective when the number of rounds is small, it is generally difficult to find a function g so that the size of z is sufficiently small. Then next, we try to reduce the substantial size of z in g by selecting convenient plaintexts to attack. This is realized by controlling the spread of carry bits of the addition in S-boxes.

Our method determines the extended key directly and definitely, moreover the computational complexity generally decreases as the number of known plaintexts increases. In subsequent chapters we will describe the results of our computer experiments about the number of known plaintexts and the breaking time for each cipher with the independent extended key bits.

4 Attack of FEAL-4

We start to rewrite FEAL-4 as figure 3. Then the original extended key is corresponding to $K_1, K_2, .., K_6$ in figure 3 one-to-one and linearly. Hence in this chapter we describe the method to derive these keys from given known plaintexts.

First, applying the equation (7) from the second round to the fourth round in figure 3, we obtain

$$P_H[10, 16, 18, 26] \oplus P_L[10, 18, 26] \oplus C_H[10, 16, 18, 26] \oplus C_L[16] \oplus$$
$$m F_1(P_H \oplus P_L, K_1)[16] = T, \tag{9}$$

where T is independent of P and C.

Now we can easily see that the key bits which influence the left side of the equation (9) are $K_1[8 \sim 14]$ and $K_1[16 \sim 22]$, in which $K_1[14]$ and $K_1[22]$ are only XORed in the left side. Hence by transposing these two bits to the right side, we may suppose the essential key bits in the left side are $K_1[8 \sim 13]$ and $K_1[16 \sim 21]$. We call these bits the effective key bits in the equation (9).

Therefore, we obtain candidates for $K_1[8 \sim 13]$ and $K_1[16 \sim 21]$ through an exhaustive search for 12-bit keys by checking that the left side of the equation (9) gives a constant value for every known plaintext and the corresponding ciphertext.

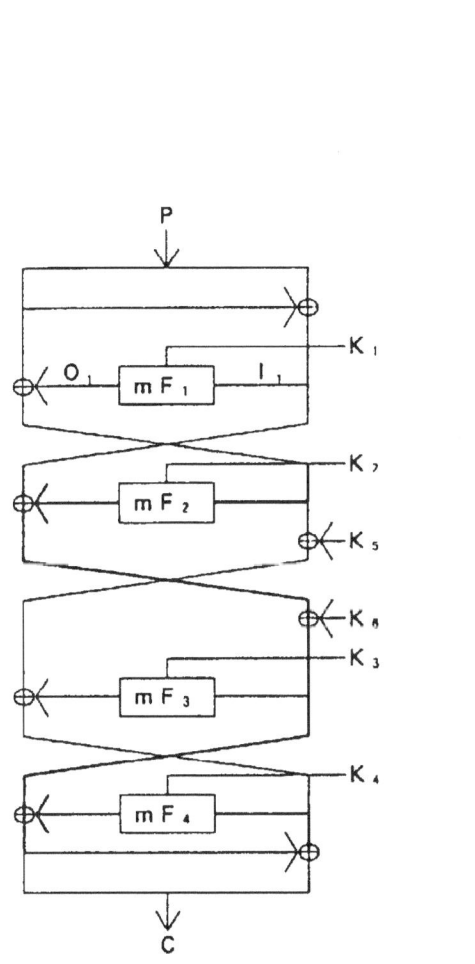

Figure 3: FEAL-4 cipher

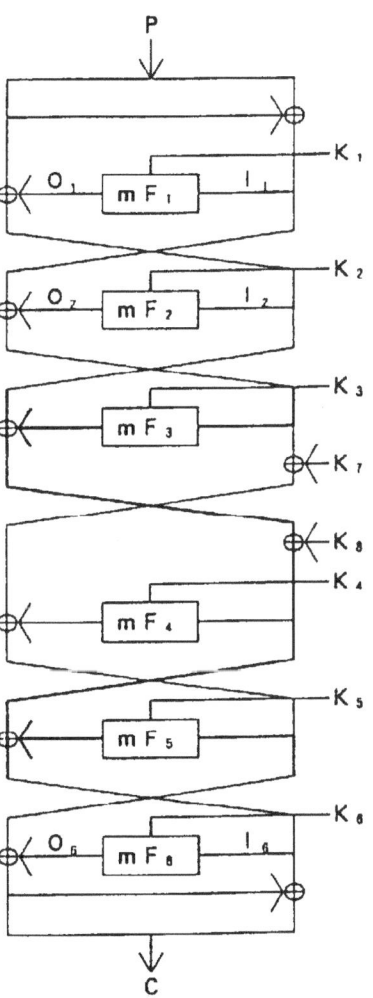

Figure 4: FEAL-6 cipher

Similarly applying the equations (6), (8) and (9) in figure 3, one has candidates for $K_1[0 \sim 5]$ and $K_1[8 \sim 29]$ finally. It is easy to derive remaining bits, so we omit the detail.

In our computer experiments to derive $K_1, K_2, .., K_6$ completely, it takes 2 seconds with 10 known plaintexts and 350 seconds with 5 known plaintexts respectively. Our program uses 30KB memory in running.

5   Attack of FEAL-6

We also start to rewrite FEAL-6 as figure 4. In this case $K_1, K_2, .., K_8$ are not independent; for example we may suppose

$$K_1[j] = K_3[j],$$
$$K_4[j] = K_6[j] \quad (0 \leq j \leq 7 , \ 24 \leq j \leq 31), \tag{10}$$

though we do not need these relations afterward.

Now applying the equation (7) from the third round to the fifth round in figure 4, we obtain

$$P_H[10, 18, 26] \oplus C_H[10, 16, 18, 26] \oplus C_L[10, 18, 26] \oplus$$
$$mF_2(P_H \oplus mF_1(P_H \oplus P_L, K_1), K_2)[16] \oplus$$
$$mF_6(C_H \oplus C_L, K_6)[16] = T. \tag{11}$$

Then we easily see that the effective key bits in the equation (11) are the following 48 bits:

$$K_1[0 \sim 3], K_1[8 \sim 27], K_i[8 \sim 13], K_i[16 \sim 21] \quad (i = 2, 6). \tag{12}$$

However, it is computationally heavy to solve this equation using the same search method as FEAL-4. Hence we try to reduce the number of the effective key bits. First assume

$$I_6[5, 13, 21, 29] \oplus K_6[13, 21] = 0, \tag{13}$$

then the carry bit from the 5-th to the 6-th bit in the addition of S-box S_A in mF_6 can be denoted as

$$I_6[5, 13] \oplus K_6[13]. \tag{14}$$

This shows we may ignore the influence of $K_6[8 \sim 12]$ and $K_6[16 \sim 20]$ in the left side of the equation (11) under the assumption (13). Similarly assuming

$$I_2[5, 13, 21, 29] \oplus K_2[13, 21] = 0, \tag{15}$$

we can eliminate the influence of $K_2[8 \sim 12]$ and $K_2[16 \sim 20]$, and hence the effective key bits in the equation (11) are reduced to the following 26 bits:

$$K_1[0 \sim 3], K_1[8 \sim 27], K_2[13, 21], K_6[13, 21]. \tag{16}$$

In fact the following search algorithm leads us to obtain candidates for these bits.

Step 1 Select 26-bit data $K_1[0 \sim 3], K_1[8 \sim 27], K_2[13, 21], K_6[13, 21]$.

Step 2 Calculate I_2 and I_6 from given plaintexts and the corresponding ciphertexts.

Step 3 Check that the left side of the equation (11) gives a constant value for every plaintext which satisfies the equations (13) and (15).

Step 4 If the check is correct, then let the 26-bit data be a candidate. Otherwise, the selection in step 1 is wrong.

Subsequently, we can reach other key bits using a similar method to breaking FEAL-4, however we omit the detail. In our computer experiments using randomly generated 100 known plaintexts, it takes 37 minutes to derive $K_1, K_2, .., K_8$ completely. Our program uses 100KB memory in running.

Moreover, we can solve the key faster by adding more relations and known plaintexts. For example, assume

$$I_1[5, 13, 21, 29] \oplus K_1[13, 21] = 0, \tag{17}$$

$$I_1[0, 8, 16, 24, 26] \oplus K_1[8, 16, 26] = 1, \tag{18}$$

$$I_1[6, 14, 22, 30] \oplus K_1[14, 22] = 0, \tag{19}$$

$$I_1[0, 2, 5, 6, 8, 13, 14, 22, 30] \oplus K_1[2, 8, 13, 14, 22] = 0, \tag{20}$$

which are introduced in order to cut off the spread of each carry bit at the 5-th bit of S_A, the 2-nd bit of S_C, the 6-th bit of S_A and the 2-nd bit of S_D in mF_1 respectively. Then the effective key bits in the equation (11) are reduced to the following 18 bits:

$$K_1[2 \sim 3], K_1[8 \sim 11], K_1[13], K_1[14, 22], K_1[15, 22, 23],$$
$$K_1[16 \sim 19], K_1[21], K_1[26 \sim 27], K_2[13, 21], K_6[13, 21]. \tag{21}$$

After the calculation of these 18 bits, we can reach whole 26 bits by repeating the previous method without equations (17) $\sim$ (20). In this case, the derivation of $K_1, K_2, .., K_8$ takes 5 minutes using 500 known plaintexts and 32 minutes using 200 known plaintexts respectively.

6 Attack of FEAL-7

The principle of breaking FEAL-7 is the same as FEAL-6. Now our fundamental equation is

$$mF_2(P_H \oplus mF_1(P_H \oplus P_L, K_1), K_2)[16] \oplus$$
$$mF_6(C_H \oplus mF_7(C_H \oplus C_L, K_7), K_6)[16] \oplus$$
$$P_H[10, 18, 26] \oplus C_H[10, 18, 26] = T. \tag{22}$$

In this case we use known plaintexts which satisfy (13),(15),(17) $\sim$ (20) and subsequent four relations simultaneously:

$$I_7[5, 13, 21, 29] \oplus K_7[13, 21] = 0, \tag{23}$$
$$I_7[0, 8, 16, 24, 26] \oplus K_7[8, 16, 26] = 1, \tag{24}$$
$$I_7[6, 14, 22, 30] \oplus K_7[14, 22] = 0, \tag{25}$$
$$I_7[0, 2, 5, 6, 8, 13, 14, 22, 30] \oplus K_7[2, 8, 13, 14, 22] = 0. \tag{26}$$

Then the effective key bits in the equation (22) are the following 34 bits:

$$K_i[2 \sim 3], K_i[8 \sim 11], K_i[13], K_i[14, 22], K_i[15, 22, 23],$$
$$K_i[16 \sim 19], K_i[21], K_i[26 \sim 27], K_2[13, 21], K_6[13, 21] \quad (i = 1, 7). \tag{27}$$

We have described a FEAL-7 breaking program to derive the extended key with 2^{14} randomly generated known plaintexts. This program uses 700KB memory in executing. It takes 170 hours to complete our attack.

7 Attack of FEAL-8

Our algorithm to attack FEAL-8 with independent keys $K_1, K_2, .., K_{10}$ (figure 5) is to decipher by one round using K_8, and then to apply the attack of FEAL-7. Since this computation is beyond our computer's power, in this chapter we try to estimate the computational complexity of our attack.

First, we note the choices of K_8 are 2^{30} since $K_8[7]$ and $K_8[31]$ are only XORed in mF_8. Hence these 2 bits may be included in other key bits. For example, we may modify the keys as follows:

$$K_i[1] = K_i[1] \oplus K_8[7] \quad (i = 5, 7, 10),$$
$$K_i[9] = K_i[9] \oplus K_8[7] \quad (i = 5, 7),$$
$$K_i[17] = K_i[17] \oplus K_8[31] \quad (i = 5, 7),$$
$$K_i[25] = K_i[25] \oplus K_8[31] \quad (i = 5, 7, 10),$$
$$K_8[7] = K_8[31] = 0. \tag{28}$$

Consequently, the total time required to break FEAL-8 is estimated to be 2^{30} times as much time as FEAL-7.

Next, we have constructed a simple key search program of FEAL-8 using assembly language in order to compare our method with exhaustive key search. As a result, the breaking time of FEAL-7 using our method is almost same as exhaustive search for 34-bit keys of FEAL-8. This shows the computational time to attack FEAL-8 with 2^{14} known plaintexts is estimated to be the same as exhaustive search for 64-bit keys.

For the rest of this chapter, we try to reduce the computational complexity. Now assume

$$O_8[16] = 0. \tag{29}$$

Then we note $K_8[24]$ can be also included in other key bits. Namely, we may modify the key as follows:

$$
\begin{aligned}
K_i[18] &= K_i[18] \oplus K_8[24] \quad (i = 5, 7), \\
K_i[26] &= K_i[26] \oplus K_8[24] \quad (i = 5, 7, 10), \\
K_8[24] &= 0.
\end{aligned}
\tag{30}
$$

This reduces the choices for K_8 to 2^{29}. Nevertheless, since the ciphertexts which satisfy the equation (29) are half of whole ones, we need twice as many known plaintexts to attack FEAL-7 with the same efficiency.

This fact also holds on $O_8[17], O_8[18], .., O_8[22]$ and $O_8[8], O_8[9], .., O_8[14]$ in this order. Namely, by assuming the equations

$$
\begin{aligned}
O_8[i] &= 0 \quad (17 \leq i \leq 22), \\
O_8[i] &= 1 \quad (8 \leq i \leq 14),
\end{aligned}
\tag{31}
$$

we can reduce 13 more bits for key search, though we need 2^{13} times as many plaintexts. Then similar key modifications are possible; for example, as for $O_8[8]$ we may suppose

$$
\begin{aligned}
K_i[2] &= K_i[2] \oplus K_8[0] \quad (i = 5, 7, 10), \\
K_i[10] &= K_i[10] \oplus K_8[0] \quad (i = 5, 7), \\
K_8[0] &= 0.
\end{aligned}
\tag{32}
$$

This shows the breaking time of FEAL-8 with 2^{28} known plaintexts is the same as exhaustive search for 50-bit keys. In fact, we can use the following algorithm to carry out the attack of FEAL-8:

Step 1 Select a candidate of 16-bit data $K_8[8 \sim 23]$, and then let $K_8[j] = 0$
$(0 \leq j \leq 7,\ 24 \leq j \leq 31)$.

Step 2 Decipher by one round using the key K_8 and given ciphertexts.

Step 3 Attack the FEAL-7 using known plaintexts and the corresponding ciphertexts which satisfy (29) and (31).

Step 4 If $K_8[8 \sim 23]$ is the correct value, the attack of step 3 succeeds. Otherwise it fails.

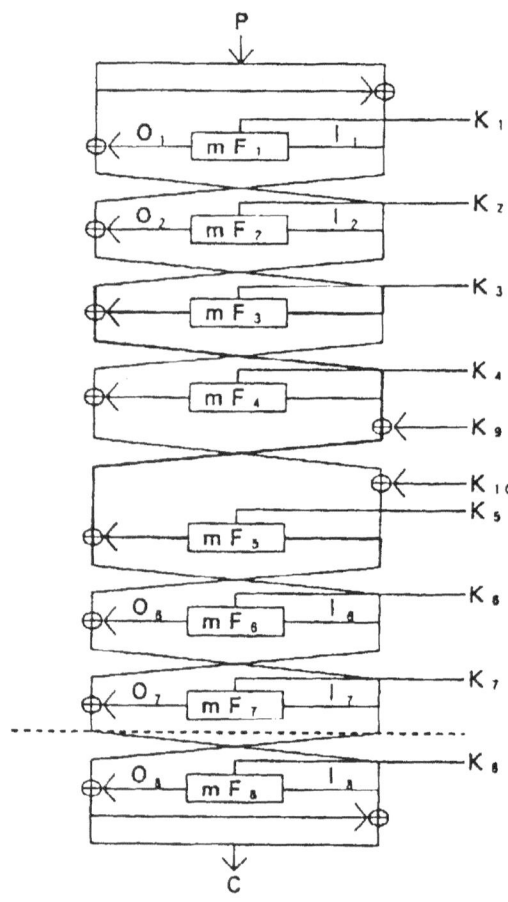

Figure 5 : FEAL-8 cipher

8 Conclusion

We have introduced a new method to attack FEAL cipher with up to eight rounds. Our attack derives the extended key directly and definitely without any assumption of the key schedule algorithm. Moreover we have proposed a method to reduce the computational time at the cost of the number of known plaintexts. As for FEAL-8, it is still possible to reduce the breaking time with more known plaintexts. We will discuss it in the full paper.

Acknowledgement

The authors would like to thank Kouichi Sakurai and Tohru Inoue very much for their many helpful comments and many hours of discussion about this work.

References

[SM]   A. Shimizu and S. Miyaguchi, "Fast Data Encipherment Algorithm FEAL," *Advances in Cryptology, Proceedings of EUROCRYPT '87*, pp.267, 1987.

[MKOM] S. Miyaguchi, S. Kurihara, K. Ohta and H. Morita, "Expansion of FEAL cipher," *NTT Review Vol.2 No.6* , pp.116-127, 1990.

[BS]   E. Biham and A. Shamir, "Differential Cryptanalysis of FEAL and N-Hash," *Extended Abstract, Proceedings of EUROCRYPT '91*, 1991.

[TG]   A. Tardy-Corfdir and H. Gilbert, "A known plaintext attack of FEAL-4 and FEAL-6," *Proceedings of Crypto '91*, 1991.

On the construction of highly nonlinear permutations

Kaisa Nyberg

Finnish Defence Forces
University of Helsinki
(on leave)

1. Introduction

Highly nonlinear permutations play an important role in the design of cryptographic transformations such as block ciphers, hash functions and stream ciphers. The substitution boxes of DES are relatively small in dimension and they can be generated by testing randomly chosen functions for required design criteria. Security may be increased by the use of substitution transformations of higher dimensions. But when the dimensions grow larger, analytic construction methods become necessary.

In this paper a general methodology is developed to construct permutations of a vector space over a finite field such that the nonlinearity of both the permutation itself and its inverse can be kept in control. The nonlinearity measure used is based on the Hamming distance from the set of affine functions. For quadratic functions there is a close relationship with this nonlinearity measure and the number of the so called linear structures of the function. This approach leads to a necessary and sufficient condition under which a transformation of F_q^n (n odd, $q = 2^d$, d odd), with quadratic coordinate functions, is a highly nonlinear permutation with equally highly nonlinear inverse.

Finally, we shall apply our general methodology to give a general construction of which the cubing permutation is a special case.

It was observed by Pieprzyk [6] that the coordinate functions of the cubing permutation in $GF(2^n)$, n odd, are of high nonlinearity, when considered with respect to a self-dual normal basis in $GF(2^n)$ over $GF(2)$. His measure of nonlinearity is weaker than the one given in the present work, since it only takes into account the coordinate functions of the permutation. Our nonlinearity measure involves all nontrivial linear combinations of the coordinate functions of the permutation and allows a rigorous proof of the fact that the inverse permutation is of the same nonlinearity.

The permutations of $GF(2^n)$ constructed in §4 have the property that their coordinate functions as well as the coordinate functions of their inverses are all of the same large distance from the set of affine functions independently of the choices for the bases in the input and and output spaces. This degree of nonlinearity only depends on over which subfield $GF(2^n)$ is considered as a linear space.

Current address: Prinz Eugenstraße 18/6, A-1040 Wien, Austria

R.A. Rueppel (Ed.): Advances in Cryptology - EUROCRYPT '92, LNCS 658, pp. 92-98, 1993.
© Springer-Verlag Berlin Heidelberg 1993

2. The nonlinearity measure

Let $F = F_q$ be a finite field with q elements and consider a function $f : F^n \to F$.

DEFINITION 1. *The nonnegative integer*

$$\mathcal{N}(f) = \min_{u \in F^n, v \in F} \#\{x \in F^n \mid f(x) \neq u^t x + v\}$$

is the Hamming distance of f from affine functions.

It is easily seen that $\mathcal{N}(f)$ is independent of the choice of the basis in the linear space F^n over F.

LEMMA 1. *For all $u \in F^n$, $u \neq 0$*

$$\mathcal{N}(f) \leq (q - 1)q^{n-1} = \#\{x \in F^n \mid u^t x \neq 0\}.$$

By the help of this lemma the third equality in the following definition can be established.

DEFINITION 2. *The nonlinearity of a vector function $f : F^n \to F^m$ is*

$$\begin{aligned}
\mathcal{N}(f) &= \min_{w \in F^m, w \neq 0} \mathcal{N}(w^t f) \\
&= \min_{u \in F^n, w \in F^m, v \in F, w \neq 0} \#\{x \in F^n \mid w^t f(x) \neq u^t x + v\} \\
&= \min_{u \in F^n, w \in F^m, v \in F, u \neq 0 \text{ or } w \neq 0} \#\{x \in F^n \mid w^t f(x) \neq u^t x + v\}
\end{aligned}$$

PROPOSITION 1. *The nonlinearity $\mathcal{N}(f)$ of $f : F^n \to F^m$ is invariant under linear permutations of the input space F^n and also under linear permutations of the output space F^m.*

This measure of nonlinearity has the following property of symmetry.

THEOREM 1. *Let $f : F^n \to F^n$ be a permutation. Then $\mathcal{N}(f^{-1}) = \mathcal{N}(f)$.*

PROOF:

$$\begin{aligned}
\mathcal{N}(f^{-1}) &= \min_{u, w \in F^n, v \in F, u \neq 0 \text{ or } w \neq 0} \#\{y \in F^n \mid w^t f^{-1}(y) \neq u^t y + v\} \\
&= \min_{u, w \in F^n, v \in F, u \neq 0 \text{ or } w \neq 0} \#\{x \in F^n \mid w^t x \neq u^t f(x) + v\} \\
&= \mathcal{N}(f).
\end{aligned}$$

The following result will be used later.

PROPOSITION 2. *Let* $f : F^n \to F$ *have nonlinearity* $\mathcal{N}(f)$. *Then the function* $g : F^{n+1} \to F$

$$(x_1, x_2, \ldots, x_n, x_{n+1}) \mapsto f(x_1, x_2, \ldots, x_n) + x_{n+1}$$

has nonlinearity $q\mathcal{N}(f)$.

PROOF:

$$\mathcal{N}(g) = \min_{u \in F^{n+1}, v \in F} \#\{x \in F^{n+1} \mid g(x) \neq u^t x + v\}$$

$$= \min_{u \in F^n, u_{n+1} \in F, v \in F} \sum_{i=0}^{q-1} \#\{x \in F^n, x_{n+1} = i \mid f(x) + i \neq u^t x + i u_{n+1} + v\}$$

$$\geq \sum_{i=0}^{q-1} \min_{u \in F^n, v_i \in F} \#\{x \in F^n \mid f(x) \neq u^t x + v_i\}$$

$$= \sum_{i=0}^{q-1} \mathcal{N}(f) = q\mathcal{N}(f),$$

and this lower bound is obtained by the choice $u_{n+1} = 1$.

The linear behaviour of a function can also be measured by the number of its linear structures.

DEFINITION 3. *A vector* $w \in F^n$ *is called a* linear structure *of a function* $f : F^n \to F$ *if* $f(x + w) - f(x)$ *is constant* $(= f(w) - f(0))$ *as* $x \in F^n$ *varies.*

It was shown in [2] that if F is a prime field, then the linear structures form a linear subspace on which the restriction of the function is linear. This does not hold in general for arbitrary finite fields. In the next section it shown however that the linear structures of a quadratic function of finitely many variables over any field form a linear space whose dimension determines the Hamming distance from linear functions given in Definition 1.

3. Quadratic functions

Let

$$f(x_1, x_2, \ldots, x_n) = \sum_{i,j} a_{ij} x_i x_j$$

be a quadratic form of n indeterminates over a finite field F with q elements. Then after fixing a basis in F^n we can consider f as a function, a quadratic polynomial, from F^n to F of the form

$$f(x) = f(x_1, x_2, \ldots, x_n) = x^t A x,$$

where $A = (a_{ij})$. Two quadratic forms $f(x) = x^t A x$ and $g(x) = x^t B x$ are called equivalent if they represent the same quadratic form, i.e., there is a linear permutation (a change of basis) C, such that $A = C^t B C$, or what is the same, $g(Cx) = f(x)$.

PROPOSITION 3. *Let $f(x) = x^t Ax$ be a quadratic form of n indeterminates over F. Then the linear structures of f form a linear subspace of dimension $\mathcal{X}(f) = n - rank(A + A^t)$.*

PROOF: We have

$$f(x + w) - f(x) = (x + w)^t A(x + w - x^t Ax$$
$$= x^t Aw + w^t Ax + w^t Aw$$
$$= x^t (A + A^t)w + f(w).$$

Hence w is a linear structure of f if and only if $(A + A^t)w = 0$.

The following result is a consequence of Theorem 6.30 in [4] and the preceeding proposition.

PROPOSITION 4. *Let n be odd and $q = 2^d$ and $f(x) = x^t Ax$ be a quadratic form in F^n with $rank(A + A^t) = r$. Then r is even and f is equivalent to*

$$x_1 x_2 + x_3 x_4 + \cdots + x_{r-1} x_r + L(x_1, x_2, ..., x_n)$$

where L is a linear form of n indeterminates.

The quadratic form $f(x_1, x_2, \ldots, x_n) = x_1 x_2 + x_3 x_4 + \cdots + x_{n-1} x_n$ (for an even n) is a perfect nonlinear function from F^n to F, that is, for every fixed $w \in F^n$ the difference $f(x + w) - f(x)$ obtains each value in F equally many times. Hence it is also a bent function, if F is a prime field with q elements, and the distance of f to the set of affine functions is the maximum

(1) $$\mathcal{N}(f) = (q - 1)(q^{n-1} - q^{\frac{n}{2}-1})$$

(see [5], Theorem 3.3). It is straightforward to check that if f is considered over $F = GF(2^d)$ then the formula (1) also holds with $q = 2^d$.

Let us remark that the quadratic functions of n variables over $GF(2)$ belong to the class of partially bent functions ([1], [7]). By definition due to C. Carlet a Boolean function is partially bent if the product of the numbers of the nonzeros of the autocorrelation function and the nonzeros of the Walsh transform obtain the absolute lower bound 2^n. So partially bent functions are optimal in this sense. But since linear functions are contained in the class of partially bent functions, also high linearity has to be required of functions to be used in cryptography. For a quadratic function f this means that $\mathcal{X}(f)$ should be as small as possible. For n odd the minimum of $\mathcal{X}(f)$ is 1.

Summarizing the results of Propositions 2, 3 and 4 we obtain the following

THEOREM 2. *Let* $F = GF(2^d)$ *and* $q = 2^d$. *Then every quadratic form in* F^n,

$$f(x) = x^t A x$$

with $\operatorname{rank}(A + A^t) = r$ *has the distance*

$$\mathcal{N}(f) = q^{n-r}(q-1)(q^{r-1} - q^{\frac{r}{2}-1})$$

from the set of affine functions.

Observe that for n odd a nondegenerate quadratic form over $GF(2^d)$ is balanced (obtains each value in F equally many times). Conversely, if $x^t A x$ is balanced and $\operatorname{rank}(A + A^t) = n - 1$, then it is nondegenerate.

The special quadratic form that we shall make use of in our construction is

$$x_1 x_2 + x_2 x_3 + x_3 x_4 + \cdots + x_{n-1} x_n + x_n x_1 = x^t R x,$$

where $R : F^n \to F^n$ is the linear permutation

$$R : (x_1, x_2, \ldots, x_n) \mapsto (x_2, x_3, \ldots, x_n, x_1),$$

i.e., the cyclic shift of the coordinates. By using the general substitution algorithm of Lemma 6.29 of [4] it is easy to verify that $x^t R x$ in an odd number n of indeterminates over $Gf(2^d)$ is equivalent to $x_1 x_2 + x_3 x_4 + \cdots + x_{n-2} x_{n-1} + x_n$.

The main result of [6], which has had a strong impact on the present work, is the observation that

$$Tr(x^3) = x^t R x, \quad x \in GF(2^n),$$

with respect to a self-dual normal basis in $GF(2^n)$, for n odd. Indeed, our construction contains the cubing permutation as a special case. By replacing R by R^i, $i = 1, 2, \ldots, n-1$, in the construction in §4, we obtain classes of equally highly nonlinear permutations where the permutations $x \to x^{2^i+1}$, $i = 1, 2, \ldots, n-1$, are as special cases. Let us recall that these are exactly the permutations on which the public key cryptosystem C^* proposed in [4] is based.

4. The construction

We combine Theorems 1 and 2 to obtain the following method for constructing permutations with desired distance from linear functions.

THEOREM 3. *Let* $F = GF(2^d)$ *and* $q = 2^d$. *Then the function* $f = (f_1, f_2, \ldots, f_n) : F^n \to F^n$ *with quadratic coordinate functions* f_k, $k = 1, \ldots, n$, *is a permutation of* F^n *with*

$$\mathcal{N}(f) = \mathcal{N}(f^{-1}) \geq q^{n-r}(q-1)(q^{r-1} - q^{\frac{r}{2}-1})$$

if and only if every nontrivial linear combination of the coordinate functions $f_1, f_2, \ldots, f_n$ *is a balanced quadratic form* $\mathbf{x}^t \mathbf{C} \mathbf{x}$ *with* $\operatorname{rank}(\mathbf{C}^t + \mathbf{C}) \geq r$.

The condition of the theorem on the coordinate functions can be tested for in low dimensions. In what follows we shall give an analytic construction, which is feasible also in large dimensions.

Let n and d be odd positive integers and $n \geq 3$. Then $GF(2^{nd})$ is an n-dimensional linear space over $\mathbf{F} = GF(2^d)$. Let $e_1, e_2, \ldots, e_n$ be a basis in $\mathbf{F}^n = GF(2^{nd})$ over $\mathbf{F}$. Then the matrix

$$
\mathbf{E}(e_1, e_2, \ldots, e_n) = \begin{pmatrix}
e_1 & e_2 & \cdots & e_n \\
e_1^2 & e_2^2 & \cdots & e_n^2 \\
e_1^4 & e_2^4 & \cdots & e_n^4 \\
\vdots & \vdots & \ddots & \vdots \\
e_1^{2^{n-1}} & e_2^{2^{n-1}} & \cdots & e_n^{2^{n-1}}
\end{pmatrix}
$$

is a nonsingular matrix over $\mathbf{F}$ (see Corollary 2.38 [4]). Choose $\alpha_1, \alpha_2, \ldots, \alpha_n \in GF(2^{nd})$ such that their cubes $\alpha_1^3, \alpha_2^3, \ldots, \alpha_n^3$ are linearly independent over $GF(2^d)$. This is possible since cubing is a permutation in $GF(2^{nd})$ if nd is odd.

Set

$$
\mathbf{E}_k = \mathbf{E}(\alpha_k e_1, \alpha_k e_2, \ldots, \alpha_k e_n) \text{ and } \mathbf{B}_k = \mathbf{E}_k^t \mathbf{R} \mathbf{E}_k,
$$

$k = 1, 2, \ldots, n$. Then the ij^{th} entry of $\mathbf{B}_k$ equals

$$
Tr_{\mathbf{F}}(\alpha_k^3 e_i e_j^2) \in GF(2^d).
$$

Let $c_k \in GF(2^d)$, $k = 1, 2, \ldots, n$ not all equal to 0. Then the ij^{th} entry of $\sum_k c_k \mathbf{B}_k$ is equal to

$$
Tr_{\mathbf{F}}\left(\sum_k c_k \alpha_k^3 e_i e_j^2\right) = Tr_{\mathbf{F}}(\gamma^3 e_i e_j^2)
$$

for some $\gamma \in GF(2^{nd})$, $\gamma \neq 0$. Hence

$$
(2) \qquad \sum_k c_k \mathbf{B}_k = \mathbf{C}^t \mathbf{R} \mathbf{C},
$$

where $\mathbf{C} = \mathbf{E}(\gamma e_1, \gamma e_2, \ldots, \gamma e_n)$.

Now $\operatorname{rank}(\mathbf{R} + \mathbf{R}^t)$ is equal to the odd number of the indeterminates minus 1 over any field over which the nondegenerate quadratic form $\mathbf{x}^t \mathbf{R} \mathbf{x}$ is considered. Since $\mathbf{B}_k = \mathbf{E}_k^t \mathbf{R} \mathbf{E}_k$, where $\mathbf{E}_k$ is a nonsingular matrix, it then follows that $\operatorname{rank}(\mathbf{B}_k^t + \mathbf{B}) = n - 1$ and the quadratic form $f_k(\mathbf{x}) = \mathbf{x}^t \mathbf{B}_k \mathbf{x}$ is nondegenerate and hence balanced, $k = 1, 2, \ldots, n$. Due to the identity (2) the same holds for every linear combination (over $GF(2^d)$) of $f_1, f_2, \ldots, f_n$. Hence it follows from Theorem 3 that the function $\mathbf{f} = (f_1, f_2, \ldots, f_n)$ is a permutation in $GF(2^{nd}) = GF(2^d)^n$ with nonlinearity

$$
\mathcal{N}(\mathbf{f}) = \mathcal{N}(\mathbf{f}^{-1}) = \mathcal{N}(f_k) = q(q-1)(q^{n-2} - q^{\frac{n-1}{2}-1}),
$$

where $q = 2^d$.

Acknowledgements. The most part of this paper was written while the author was visiting Prof. J. L. Massey at ETH Zürich. His kind hospitality and interest to my work is gratefully acknowledged. I would also like to thank Prof. Z. Wan and Dr. S. Hellberg for pointing out errors in an earlier version of the paper.

REFERENCES

1. C. Carlet, *Partially-bent functions*, Codes, Designs and Cryptography (to appear).

2. X. Lai, *Linear structures of functions over prime fields*, Unpublished preprint (1990).

3. R. Lidl and H. Niederreiter, "Finite fields.," Encyclopedia of Mathematics and its applications, Vol. 20. Addison-Wesley, Reading, Massachusetts, 1983.

4. T. Matsumoto and H. Imai, *Public quadratic polynomial-tuples for efficient signature-verification and message-encryption*, Advances in Cryptology Eurocrypt '88. Lecture Notes in Computer Science, Springer-Verlag, 1989.

5. K. Nyberg, *Constructions of bent functions and difference sets*, Advances in Cryptology – Proceedings of Eurocrypt '90. Lecture Notes in Computer Science **473**, Springer-Verlag, 1991.

6. J. Pieprzyk, *On bent permutations*, Technical Report CS91/11; The University of New South Wales, Department of Computer Science. Presented at the International Conference on Finite Fields, Coding Theory and Advances in Communications and Computing, Las Vegas, 1991.

7. B. Preneel et al., *Cryptographic properties of quadratic Boolean functions*, International Conference on Finite Fields, Coding Theory and advances in Communications and Computing, Las Vegas, 1991; *Boolean functions satisfying higher order propagation criteria*, Advances in Cryptology - Eurocrypt '91. Lecture Notes in Computer Science **547**, Springer-Verlag, 1991.

The One-Round Functions of the DES Generate the Alternating Group

Ralph Wernsdorf

SIT Gesellschaft für Systeme der Informationstechnik mbH

O-1252 Grünheide (Mark), Germany

Charlottenstraße 7

Abstract: In each of the 16 DES rounds we have a permutation of 64-bit-blocks. According to the corresponding key-block there are 2^{48} possible permutations per round. In this paper we will prove that these permutations generate the alternating group. The main parts of the paper are the proof that the generated group is 3-transitive, and the application of a result from P. J. Cameron based on the classification of finite simple groups. A corollary concerning n-round functions generalizes the result.

1 Introduction

In each of the 16 DES-rounds a permutation of 64-bit-blocks is carried out /NBS 77/. According to the corresponding key-block there are 2^{48} possible permutations per round. In the following we will answer the question which group they generate.

A question like this is important from the cryptographic point of view. If the generated group is "too small" in a certain sense, then the algorithm might be vulnerable to cryptanalytic attacks (see for example /KRS 88/ and /RM 85/).

Several publications are concerned with group theoretic properties of the DES or DES-like ciphers:

- Coppersmith/Grossman in /CG 75/ and Even/Goldreich in /EG 83/ derived general results on "DES-like functions". The one-round permutations of DES form a subset of those functions. Further research in the direction of "DES-like permutations" is done by Pieprzyk/Zhang in /PZ 90/. The permutations defined there also generate the alternating group.

R.A. Rueppel (Ed.): Advances in Cryptology - EUROCRYPT '92, LNCS 658, pp. 99-112, 1993.
© Springer-Verlag Berlin Heidelberg 1993

- Reeds/Manferdelli in /RM 85/ and Chaum/Evertse in /CE 86/ exclude the existence of several classes of nontrivial linear factors.

- Group theoretic properties of the 16-round DES-cipher are subject of papers written by Kaliski/Rivest/Sherman /KRS 88/ and Simmons/Moore /SM 87/. The results support the hypothesis that the corresponding group is not "small".

The main result obtained in this paper is stated in Theorem 1. It will be proved that the 2^{48} one-round permutations generate the alternating group, i.e. that the generated group is "large". An essential part of the proof is done in Section 3. There the 3-transitivity of the group is derived from some computational results concerning properties of the S-boxes. In Section 4 we complete the proof of Theorem 1. For this purpose we apply some propositions of P. J. Cameron /Cam 81/ based on the classification of finite simple groups. Corollary 4 shows that the result also holds for n-round functions with independent subkeys.

2 Notations

$\forall (m, n) \in N^2$: $\overline{m, n} := \{m, m+1, ..., n\}$ for $m \leq n$.

$\forall m \in N$: $V_m := \{0, 1\}^m$ (m-dimensional vector space over $\{0, 1\}$)

$\forall a \in V_{2m}$: $a_L := (a_1, a_2, ..., a_m)$

$\forall a \in V_{2m}$: $a_R := (a_{m+1}, a_{m+2}, ..., a_{2m})$ $\quad (a = (a_L, a_R))$.

$\langle \Pi \rangle$:= the permutation group generated by the set Π of permutations.

$A_{2^{64}}$:= the alternating group on V_{64}.

$S_{2^{64}}$:= the symmetric group on V_{64}.

We consider the set of functions F_k: $V_{32} \times V_{32} \rightarrow V_{32} \times V_{32}$:

$\forall k \in V_{48} \, \forall a \in V_{32} \, \forall b \in V_{32}$: $F_k(a, b) := (b, a \oplus S(k \oplus EPb))$,

where $E: V_{32} \rightarrow V_{48}$, $P: V_{32} \rightarrow V_{32}$ and $S: V_{48} \rightarrow V_{32}$ are defined according to /NBS 77/.

The functions F_k represent permutations on V_{64} and describe one round of the DES-algorithm, if we follow an equivalent description of the DES algorithm given in /DDF 84/ on page 183. (This modification does not influence the group theoretical properties considered here.)

The main object of our interest will be the group G:

$$G := \langle \{F_k \in S_{2^{64}} \mid k \in V_{48}\} \rangle.$$

Further we will use the following notations:

$d \quad := (0, 0, ..., 0, 1, 0, ..., 0) \in V_{64}$
$\qquad\quad \underset{1\ \ 2\ \ ...\ \ 30\ \ 31\ \ 32\ \ ...\ \ 64}{}$

$d' \quad := (0, 0, ..., 0, 1, 0, ..., 0) \in V_{48}$
$\qquad\quad \underset{1\ \ 2\ \ ...\ \ 21\ \ 22\ \ 23\ \ ...\ \ 48}{}$

$G_0 \quad := \{g \in G \mid g(0) = 0\}$ - stabilizer of zero

$G_{0,d} \quad := \{g \in G \mid g(0) = 0 \wedge g(d) = d\}$ - stabilizer of zero and d

$M \quad := \{(k, k') \in V_{48}^2 \mid k \neq k' \wedge S(k) = S(k')\}$

$M_{d'} \quad := \{(k, k') \in V_{48}^2 \mid k \neq k' \wedge S(k) = S(k') \wedge S(k \oplus d') = S(k' \oplus d')\}$

$\forall\ (k, k') \in M\ \ \forall\ (a, b) \in V_{32}^2:$

$F_{k,k'}^L (a, b) \quad := F_{k'}^{-1}(F_k(a, b)) = (a \oplus S(k \oplus EPb) \oplus S(k' \oplus EPb), b);$

$F_{k,k'}^R (a, b) \quad := F_{k'}(F_k^{-1}(a, b)) = (a, b \oplus S(k \oplus EPa) \oplus S(k' \oplus EPa)).$

(Obviously we have: $\forall\ (k, k') \in M:$ $(F_{k,k'}^L \in G_{0,d} \wedge F_{k,k'}^R \in G_0)$ and

                    $\forall\ (k, k') \in M_{d'}:$ $(F_{k,k'}^L \in G_{0,d} \wedge F_{k,k'}^R \in G_{0,d}).)$

3 The 3-Transitivity of the Group G

Two elementary properties of G are stated in Lemma 1:

Lemma 1:

a) $G \subseteq A_{2^{64}}$ /EG 83/.

b) G is transitive on V_{64}.

Lemma 2 shows the way we will pursue to prove the 3-transitivity of G. Its proof can be derived from theorem 9.1. in the book of H. Wielandt /Wie 64/.

Lemma 2:

If $G_{o,d}$ is transitive on $V_{64} \setminus \{0, d\}$ and G_o is transitive on $V_{64} \setminus \{0\}$, then G is 3-transitive on V_{64}.

To prepare the next steps we consider the following linear subspaces:

$\forall j \in \overline{1,8} \ \forall y \in V_6 \setminus \{0\}$:

$$U^j (y) := L\{S_j ((k_i)^{6j}_{i=6j-5} \oplus y) \oplus S_j ((k_i')^{6j}_{i=6j-5} \oplus y) \,|\, (k, k') \in M\}$$

$$U^j_{d'} (y) := L\{S_j ((k_i)^{6j}_{i=6j-5} \oplus y) \oplus S_j ((k_i')^{6j}_{i=6j-5} \oplus y) \,|\, (k, k') \in M_{d'}\}.$$

($L :=$ linear subspace of V_4 spanned by the given subset of V_4; $S_j: V_6 \to V_4$ denotes the j-th S - box; $(k_i)^{6j}_{i=6j-5}$ denotes the vector $(k_{6j-5}, k_{6j-4}, ..., k_{6j})$.)

The propositions (a) - (d) of Lemma 3 were obtained by a computer program:

Lemma 3:

(a) $\forall j \in \overline{1,8} \setminus \{4\} \ \forall y \in V_6 \setminus \{0\}: U^j (y) = U^j_{d'} (y) = V_4$.

(b) $\forall y \in V_6 \setminus \{(0, 0, 0, 0, 0, 0), (1, 0, 1, 1, 1, 1)\}: U^4 (y) \neq \{(0, 0, 0, 0)\}$.

(c) $U^4 ((1, 1, 1, 1, 0, 1)) = U^4_{d'} ((1, 1, 1, 1, 0, 1)) = V_4$.

(d) $\forall y \in V_6 \setminus \{(0, 0, 0, 0, 0, 0), (0, 0, 0, 1, 0, 0), (1, 0, 1, 0, 1, 1), (1, 0, 1, 1, 1, 1)\}$:

$$U^4_{d'} (y) \neq \{(0, 0, 0, 0)\}.$$

Notations:

- "$\sim$" denotes the following equivalence:

$$\forall (x, y) \in V^2_{64}: (x \sim y \longleftrightarrow \exists g \in G_{o,d}: g(x) = y).$$

- Further we denote

$$e := (\underset{1}{1}, \underset{2}{1}, ..., \underset{9}{1}, \underset{10}{0}, \underset{11}{1}, ..., \underset{32}{1}) \in V_{32}.$$

Lemma 4:

$$\forall\, (z, z') \in V_{32}^2 \colon (e, z) - (e, z').$$

Proof:

The proof uses propositions (a) and (c) of Lemma 3.

Let $(z, z') \in V_{32}^2$ be arbitrarily fixed.

Property (a) of Lemma 3 includes the equality $U_{d'}^1\,((1, 1, 1, 1, 1, 1)) = V_4$; hence the vector $(z_1 \oplus z_1', z_2 \oplus z_2', z_3 \oplus z_3', z_4 \oplus z_4')$ is a linear combination of vectors

$$S_1((k_i^1)_{i=1}^6 \oplus (1, 1, 1, 1, 1, 1)) \oplus S_1((k_i^2)_{i=1}^6 \oplus (1, 1, 1, 1, 1, 1)),$$

where $(k^1, k^2) \in M_{d'}, (k_i^1)_{i=1}^6 \ne (k_i^2)_{i=1}^6$.

For the corresponding (k^1, k^2) we set:

$$\forall\, i \in \overline{7,48} \colon k_i^1 = k_i^2 = 0.$$

By carrying out the corresponding permutations F_{k^1,k^2}^R we finally obtain:

$$(e, z) \sim (e, (z_1', z_2', z_3', z_4', z_5, z_6, ..., z_{32})).$$

Analogous, from $U_{d'}^2\,((1, 1, 1, 1, 1, 1)) = V_4$ (see (a)) we obtain:

$$(e, (z_1', z_2', z_3', z_4', z_5, z_6, ..., z_{32})) \sim (e, (z_1', z_2', z_3', z_4', z_5', z_6', z_7', z_8', z_9, z_{10}, ..., z_{32})).$$

The continuation of these considerations (where the equality $U_{d'}^4\,((1, 1, 1, 1, 0, 1))$ $= V_4$ follows from (c) and not from (a)) finally yields the statement of Lemma 4.

∎

Lemma 5:

$$\forall\, a \in V_{64} \setminus \{0, d\}\, \exists\, z \in V_{32} \colon a - (e, z).$$

Proof:

Let $a \in V_{64} \setminus \{0, d\}$ be arbitrarily fixed.

At first we show:

$$\exists\, a' \in V_{64} \setminus \{0, d\}: (a' \sim a \wedge \exists\, i \in \overline{1,32} \setminus \{2, 5, 10, 18, 26, 31\}: a'_i = 1). \qquad (1)$$

If $\exists\, i \in \overline{1,32} \setminus \{2, 5, 10, 18, 26, 31\}: a_i = 1$, then we immediately obtain (1) $(a' := a)$.

If not:

- If we have $\exists\, i \in \overline{33,64}: a_i = 1$,

   then because of (a), (b), the properties $\forall\, (k^1, k^2) \in M: F^{L}_{k^1,k^2} \in G_{o,d}$
   and

   $$\forall\, b \in V_{32}: ([EPb]_{17} = [EPb]_{19} \wedge [EPb]_{24} = [EPb]_{26}) \qquad (2)$$

   there exists a pair $(k^1, k^2) \in M$ and

   an index $j \in \overline{1,32} \setminus \{2, 5, 10, 18, 26, 31\}$ such that $[F^{L}_{k^1,k^2}(a)]_j = 1$.

   We fix $a' := F^{L}_{k^1,k^2}(a)$.

- If we have $\forall\, i \in \overline{33,64}: a_i = 0$,

   then because of (2), (d) and $a \notin \{0, d\}$ we obtain:

   $$\exists\, j \in \overline{33,64}\ \exists\, (k^1, k^2) \in M_{d'}: [F^{R}_{k^1,k^2}(a)]_j = 1.$$

Hence, this case is traced back to the case: $\exists\, i \in \overline{33,64}: a_i = 1$.

Therefore the proof of (1) is complete.

We fix a vector $a' \in V_{64} \setminus \{0, d\}$ according to (1).

As a next step we prove:

$$\exists\, a'' \in V_{64} \setminus \{0, d\}: (a'' \sim a' \wedge \forall\, i \in \overline{1,32} \setminus \overline{13,16}: a''_i = e_i). \qquad (3)$$

If $\forall\, i \in \overline{1,32} \setminus \overline{13,16}: a'_i = e_i$, then we immediately obtain (3).

If not, then we choose an index $j \in \overline{1,32} \setminus \{2, 5, 10, 18, 26, 31\}$ with $a'_j = 1$ according to (1).

Property (a) implies:

$$\exists\, a^0 \in V_{64} \setminus \{0, d\}: (a^0 - a' \wedge [a^0]_L = [a']_L \wedge \forall\, i \in I(j): [a^0]_{32+i} = 1), \quad (4)$$

where the sets $I(j)$ are defined in table 1.

We fix $a^0 \in V_{64} \setminus \{0, d\}$ according to (4).

Because of (a) we obtain:

$$\exists\, a^1 \in V_{64} \setminus \{0, d\}: (a^1 \sim a^0 \wedge [a^1]_R = [a^0]_R \wedge \forall\, i \in J(j): a_i^1 = e_i),$$

where the sets $J(j)$ are defined according to table 2.

In the case of $j \in \{1, 6, 9, 14, 16, 17, 19, 21, 22, 25, 29, 32\}$ property (3) holds with $a'' := a^1$. In the other case, (3) follows by carrying out the same

procedure for a^1. At this we can take an index $j \in \overline{1,32} \setminus \{2, 5, 10, 18, 26, 31\}$ with $a_j^1 = 1$ and the property $j \in \{1, 6, 9, 14, 16, 17, 19, 21, 22, 25, 29, 32\}$.

That is possible because of:

$$\forall\, j \in \overline{1,8}: \overline{4j\text{-}3, 4j} \cap \{1, 6, 9, 14, 16, 17, 19, 21, 22, 25, 29, 32\} \neq \varnothing.$$

This completes the proof of (3).

We fix $a'' \in V_{64} \setminus \{0, d\}$ according to (3). Then we can prove:

$$\exists\, z \in V_{32}: a'' \sim (e, z). \tag{5}$$

We have: $\forall\, i \in \overline{1,32} \setminus \overline{13,16}: a_i'' = e_i$.

From this besides (a) and $\overline{13,16} \cap \{2, 5, 10, 18, 26, 31\} = \varnothing$ we get:
$\exists\, z \in V_{32}: (a'' \sim ([a'']_L, z) \wedge z_{10} = 0 \wedge z_2 = z_5 = z_{18} = z_{26} = z_{31} = 1)$.
We fix such a vector $z \in V_{32}$. Because of (c) the equivalence $([a'']_L, z) \sim (e, z)$ holds.
Therefore the proof of (5) is complete.

Considering the chain $a \sim a' \sim a'' \sim (e, z)$ we obtain the proposition of Lemma 5 from (1), (3) and (5).

j	$I(j)$
16 or 25	$\overline{1,4} \cup \overline{29,32}$
7 or 20	$\overline{1,4}$
21 or 29	$\overline{1,8}$
12 or 28	$\overline{5,8}$
1 or 17	$\overline{5,12}$
15 or 23	$\overline{9,12}$
8 or 24	$\overline{17,20}$
14 or 32	$\overline{17,24}$
3 or 27	$\overline{21,24}$
9 or 19	$\overline{21,28}$
13 or 30	$\overline{25,28}$
6 or 22	$\overline{25,32}$
4 or 11	$\overline{29,32}$

Table 1. Definition of $I(j)$

j	$\overline{1,32} \setminus (\overline{13,16} \cup J(j))$
1,6,9,14,16,17,19,21,22,25,29,32	$\varnothing$
7 or 20	$\overline{1,4} \cup \overline{25,28}$
12 or 28	$\overline{5,8} \cup \overline{21,24}$
15 or 23	$\overline{1,4} \cup \overline{9,12}$
8 or 24	$\overline{17,20} \cup \overline{29,32}$
3 or 27	$\overline{21,24}$
13 or 30	$\overline{17,20} \cup \overline{25,28}$
4 or 11	$\overline{9,12} \cup \overline{29,32}$

Table 2. Definition of $J(j)$

Corollary 1:

$$G_{o,d} \text{ is transitive on } V_{64} \setminus \{0, d\}.$$

Proof:

Let $(a, a') \in (V_{64} \setminus \{0, d\})^2$. By Lemma 4 and Lemma 5 we get:

$$\exists (z, z') \in V_{32}^2 : (a \sim (e, z) \sim (e, z') \sim a').$$

Corollary 2:

$$G_o \text{ is transitive on } V_{64} \setminus \{0\}.$$

Proof:

Because of Corollary 1 it suffices to show that

$$\exists g \in G_o : g(d) \neq d \quad (g(d) \neq 0 \text{ because of } g \in G_o).$$

Let $(k, k') \in M \setminus M_{d'}$. $(M \setminus M_{d'} \neq \emptyset$; for example for $k = 0$ and

$$k' := (\underset{1}{0}, \underset{2}{0}, ..., \underset{19}{0}, \underset{20}{1}, \underset{21}{0}, \underset{22}{0}, \underset{23}{1}, \underset{24}{1}, \underset{25}{0}, ..., \underset{48}{0})$$

we have $S(k) = S(k')$ and $S(k \oplus d') \neq S(k' \oplus d')$.)

$$\Rightarrow F_{k,k'}^R (d) = (d_L, d_R \oplus S(k \oplus d') \oplus S(k' \oplus d')) \neq d.$$

Since we know $F_{k,k'}^R \in G_0$, the proof is complete.

∎

From Lemma 2, Corollary 1, and Corollary 2 we immediately obtain:

Corollary 3:

$$G \text{ is 3-transitive on } V_{64}.$$

4 Proof of the main Theorem

The proposition of Lemma 6 will help to complete the proof of the main theorem.

Lemma 6:

$$\exists (k, k') \in M \colon |\{x \in V_{64} \mid F_{k,k'}^R (x) = x \}| = 5 \cdot 2^{59}.$$

Proof:

We fix the following pair $(k, k') \in M$:

$$k \quad := (\underset{1}{0}, \underset{2}{0}, ..., \underset{12}{0}, \underset{13}{1}, \underset{14}{0}, ..., \underset{48}{0}) \in V_{48} \text{ and}$$

$$k' \quad := (\underset{1}{0}, \underset{2}{0}, ..., \underset{17}{0}, \underset{18}{1}, \underset{19}{0}, ..., \underset{48}{0}) \in V_{48}.$$

Then the permutation $F_{k,k'}^R$ exactly has the following set of fixed points:

$$\{(x_L, x_R) \in V_{64} \mid S(k \oplus EPx_L) \oplus S(k' \oplus EPx_L) = 0\} \; =$$

$$= \{(x_L, x_R) \in V_{64} \mid S_3(((k \oplus EPx_L)_j)_{j=13}^{18}) = S_3(((k' \oplus EPx_L)_j)_{j=13}^{18})\} =$$

$$= \{(x_L, x_R) \in V_{64} \mid (((EPx_L)_j)_{j=13}^{18}) \in \{(0,0,0,0,0,0), (1,0,0,0,0,1), (0,0,0,1,1,0),$$
$$(1,0,0,1,1,1), \quad (0,0,1,0,0,1), \quad (1,0,1,0,0,0), \quad (0,1,1,0,0,1),$$
$$(1,1,1,0,0,0), (0,1,1,1,0,1), (1,1,1,1,0,0)\}\}.$$

Thus the permutation $F_{k,k'}^R$ exactly has $10 \cdot 2^{26} \cdot 2^{32} = 5 \cdot 2^{59}$ fixed points.

∎

Theorem 1:

The following equality holds: $G = A_{2^{64}}$.

Proof:

Suppose $G \neq A_{2^{64}}$.

G is 3-transitive. Therefore we can apply proposition 5.2. of /Cam 81/ implying that G has a unique minimal normal subgroup which is Abelian or simple.

- Suppose that this normal subgroup is simple. Following the table on page 8 of /Cam 81/, which is based on the classification of finite simple groups (see also /CC 91/, p. 462), we obtain that 2^{64} has the form $\frac{q^2 - 1}{q - 1} = q + 1$, where q is a prime or a power of a prime. But because $2^{64} - 1$ is neither a prime nor a power of a prime, we get a contradiction.

- If this normal subgroup is Abelian, then G is "similar" to a subgroup of the affine group $Aff(V_{64})$ (see /Rob 82/, pp. 192-193).

  Two permutation groups H and H' on X and X', respectively, are called similar, if there exist an isomorphism $\alpha: H \to H'$ and a bijection $\beta: X \to X'$ with the property:

  $$\forall h \in H \; \forall x \in X: \; \beta(h(x)) = (\alpha(h))(\beta(x)) \qquad \text{(see /Rob 82/, p. 32)}.$$

That means that each element of G has the same cycle representation as a certain element of $Aff(V_{64})$. Particularly the number of fixed points of each element of G must be an element of the set $\{0, 2^0, 2^1, ..., 2^{64}\}$, because the fixed points of an affine mapping (if there are any) form an affine subspace of V_{64}.

From Lemma 6 we know that there exists an element of G with exactly $5 \cdot 2^{59}$ fixed points. Thus, also in the Abelian case we get a contradiction.

Hence, the supposition $G \neq A_{2^{64}}$ is wrong and the theorem is proved.

■

Corollary 4:

The groups $G_{(n)}$ generated by the n - round functions of the DES (with independent subkeys) are equal to $A_{2^{64}}$ ($n = 2, 3, ...$).

Proof (sketch):

Obviously $G_{(n)}$ is a transitive subgroup of G.

Besides this, all permutations $F_{k,k'}^R$ and $F_{k,k'}^L$ also belong to $G_{(n)}$

(consider $(F_k \circ F_k \circ ... \circ F_k) \circ (F_{k'} \circ F_k \circ F_k \circ ... \circ F_k)^{-1}$ $\qquad = F_k \circ F_{k'}^{-1}$

and $\quad (F_k \circ F_k \circ ... \circ F_k)^{-1} \circ (F_k \circ F_k \circ ... \circ F_k \circ F_{k'})$ $\qquad = F_k^{-1} \circ F_{k'}$).

Because the given proofs of Lemma 4, Lemma 5, Corollary 2 and Lemma 6 make use of no other group elements than $F_{k,k'}^R$ and $F_{k,k'}^L$, we can prove $G_{(n)} = A_{2^{64}}$ in the same way as we proved $G = A_{2^{64}}$.

■

5 Conclusions

Theorem 1 gives an answer to an open question formulated for example by Pieprzyk/Zhang in /PZ 90/ ("Are the DES generators complete?"). The result shows that the structure of the one-round DES-permutations and the current S-boxes do not restrict the number of possible permutations attainable by composition. Since the generated alternating group $A_{2^{64}}$ is a large simple group and primitive on V_{64} we can exclude several imaginable cryptanalytic "shortcuts" of the DES-algorithm.

Though the proofs in this paper are based on various special properties of the permutations it is possible to find certain other S-boxes such that after replacement we obtain the same final result as in Theorem 1.

Finally, with regard to Corollary 4 one may expect that the set of 16-round DES-cipher-permutations generates the alternating group, too. This, however, remains an open question, because the influence of the DES key-scheduling must be taken into account.

Acknowledgement

The author would like to thank the members of the SIT GmbH for their help.

References

/Cam 81/ Cameron, P. J.: "Finite Permutation Groups and Finite Simple Groups"
Bull. London Math. Soc., 13, 1981, 1-22

/CC 91/ Cameron, P. J.; Cannon, J.: "Fast Recognition of Doubly Transitive Groups"
Journal of Symbolic Computation, 12, Nr. 4&5, 1991, 459-474

/CE 86/ Chaum, D.; Evertse, J. H.: "Cryptanalysis of DES with a reduced number of rounds; Sequences of linear factors in blockciphers"
Proc. CRYPTO '85, Lect. Notes Comp. Sci., 218, 1986, 192-211

/CG 75/ Coppersmith, D.; Grossman, E.: "Generators for certain alternating groups with applications to cryptography"
Journal of Applied Mathematics, 29, Nr. 4, 1975, 624-627

/DDF 84/ Davio, M.; Desmedt, Y.; Fosseprez, M. et al.: "Analytical Characteristics of the DES"
Proc. CRYPTO '83, Plenum Press, New York and London, 1984, 171-202

/EG 83/ Even, S.; Goldreich, O.: "DES-like functions can generate the alternating group"
IEEE Transactions on Information Theory, IT-29, Nr. 6, 1983, 863-865

/KRS 88/ Kaliski, B. S.; Rivest, R. L.; Sherman, A. T.: "Is the Data Encryption Standard a Group? (Results of Cycling Experiments on DES)"
Journal of Cryptology, 1, Nr. 1, 1988, 3-36

/NBS 77/   National Bureau of Standards: "Data Encryption Standard"
            FIPS PUB 46, Washington, 1977

/PZ 90/   Pieprzyk, J.; Zhang, X. M.: "Permutation Generators of Alternating
            Groups"
            Proc. AUSCRYPT '90, Lect. Notes Comp. Sci., 453, 1990, 237-244

/RM 85/   Reeds, J. A.; Manferdelli, J. L.: "DES has no per round linear factors"
            Proc. CRYPTO '84, Lect. Notes Comp. Sci., 196, 1985, 377-389

/Rob 82/   Robinson, D. J. S.: "A Course in the Theory of Groups"
            Springer, New York, Heidelberg, Berlin, 1982

/SM 87/   Simmons, G. J.; Moore, J. H.: "Cycle structure of the DES for keys
            having palindromic (or antipalindromic) sequences of round keys"
            IEEE Transactions on Software Engineering, SE-13, Nr. 2, 1987,
            262-273

/Wie 64/   Wielandt, H.: "Finite Permutation Groups"
            Academic Press, New York, London, 1964

CORRELATION VIA LINEAR SEQUENTIAL CIRCUIT APPROXIMATION
OF COMBINERS WITH MEMORY

Jovan Dj. Golić

Institute of Applied Mathematics and Electronics, Belgrade
School of Electrical Engineering, University of Belgrade
Bulevar Revolucije 73, 11001 Beograd, Yugoslavia

Abstract: Correlation properties of a general binary combiner with an arbitrary number of memory bits are analyzed. It is shown that there exists a pair of certain linear functions of the output and input, respectively, that produce correlated binary sequences. An efficient procedure, based on a linear sequential circuit approximation, is developed for finding such pairs of linear functions. The result may be a basis for a divide and conquer correlation attack on a stream cipher generator consisting of several linear feedback shift registers combined by a combiner with memory.

I. INTRODUCTION

A common way to combine several linear feedback shift registers (LFSRs) in a pseudorandom sequence generator for cryptographic or spread-spectrum applications is by a memoryless function. Siegenthaler [7] has shown that such structures are not resistant against divide and conquer correlation attacks and has introduced the corresponding concept of correlation immunity of Boolean functions [8]. This concept has been further developed in [9] using the Walsh transform of Boolean functions. According to the Xiao-Massey lemma [9] it follows that the output of a Boolean function is correlated to at least one linear function of its inputs. Given such a linear function, fast cryptanalytic algorithms for the LFSRs initial states reconstruction might also be feasible

This research was supported by the Science Fund of Serbia, grant #0403, through Institute of Mathematics, Serbian Academy of Arts and Sciences.

R.A. Rueppel (Ed.): Advances in Cryptology - EUROCRYPT '92, LNCS 658, pp. 113-123, 1993.
© Springer-Verlag Berlin Heidelberg 1993

(for basic principles, see [2]). Their efficiency is a monotonic function of the corresponding correlation coefficient. Moreover, it has been shown in [3] that the sum of the squares of the correlation coefficients to all the linear functions of the inputs is equal to one.

In order to increase the resistance against correlation attacks one can use Boolean functions with memory, see [5], [6]. Correlation properties of combiners with one bit memory have been investigated in [4]. The sum of the squares of the correlation coefficients to all the linear functions of input sequences is derived in [4]. It follows that Boolean functions with one bit memory exist whose output is correlated to none of the linear functions of input sequences. However, in this case the sum of two successive outputs is shown to be correlated to at least one linear function of the input sequences.

In this paper, we study the correlation properties a general Boolean function with an arbitrary number, M, of bits of memory. We show that there exists a linear function of at most M+1 successive outputs that is correlated to a linear function of at most M+1 successive inputs. In addition, we propose an efficient procedure for finding such pairs of linear functions, given an arbitrary combiner with memory. The procedure is based on a linear sequential circuit approximation of a nonlinear combiner with memory.

II. GENERAL BINARY COMBINER WITH MEMORY

A general binary combiner with an arbitrary number M of memory bits is a sequential circuit defined by

$$\underset{\sim}{s}_t = G(\underset{\sim}{x}_{t-1}, \underset{\sim}{s}_{t-1}), \quad t \geq 1 \tag{1}$$

$$y_t = f(\underset{\sim}{x}_t, \underset{\sim}{s}_t), \quad t \geq 0 \tag{2}$$

where $G: GF(2)^{N+M} \longrightarrow GF(2)^M$ is a next-state vector Boolean function, $f: GF(2)^{N+M} \longrightarrow GF(2)$ is an output Boolean function, $\underset{\sim}{s}_t = (s_{1t}, \ldots, s_{Mt})$

is a state vector at time t. $\underset{\sim}{s}_0$ is an initial state, $\{\underset{\sim}{x}_t\} = \{(x_{1t}, \ldots, x_{Nt})\}$ is an N-dimensional vector input sequence consisting of N binary sequences $\{x_{it}\}$, $1 \leq i \leq N$, and $\{y_t\}$ is the output binary sequence.

In order to study the correlation properties of this combiner we assume that the input sequences are mutually independent, balanced (uniformly distributed), and independent random binary sequences. A necessary condition to be satisfied for cryptographic applications is that the output sequence is also balanced and independent. It is not difficult to see that the output sequence is balanced and independent iff the output function $\underset{\sim}{f}(\underset{\sim}{x}, \underset{\sim}{s})$ is balanced for each $\underset{\sim}{s}$.

III. CORRELATION PROPERTY OF A GENERAL COMBINER WITH MEMORY

Consider an arbitrary vector Boolean function $\underset{\sim}{f}: GF(2)^n \longrightarrow GF(2)^m$. The vector function $\underset{\sim}{f}$ consists of m component Boolean functions, that is, $\underset{\sim}{f} = (f_1, \ldots, f_m)$. Let $\underset{\sim}{z} = \underset{\sim}{f}(\underset{\sim}{x}, \underset{\sim}{y})$, $\underset{\sim}{x} \in GF(2)^{n_1}$, $\underset{\sim}{y} \in GF(2)^{n_2}$, $n = n_1 + n_2$. Assume that $\underset{\sim}{x}$ and $\underset{\sim}{y}$ are statistically independent balanced random variables. Throughout this paper, we for simplicity assume the same notation for a random variable and its values. We study the statistical dependence between $\underset{\sim}{z}$ and $\underset{\sim}{x}$. In accordance with the Xiao-Massey lemma [9], one can establish the following two properties.

Property 1: A vector Boolean function $\underset{\sim}{f}(\underset{\sim}{x})$ is balanced iff all the nonzero linear combinations of its component functions are balanced.

Property 2: A vector Boolean function $\underset{\sim}{f}(\underset{\sim}{x}, \underset{\sim}{y})$ is statistically independent of $\underset{\sim}{x}$ (respectively, is balanced for each $\underset{\sim}{x}$) iff each nonzero linear combination of the component functions of $\underset{\sim}{f}$ is statistically independent of each nonzero (is balanced for each) linear function of $\underset{\sim}{x}$.

Property 2 can be further developed by using the following simple property.

Property 3: A Boolean function $f(\underset{\sim}{x})$ and a balanced Boolean function (for example, a nonzero linear function) $g(\underset{\sim}{x})$ are statistically independent iff their sum is balanced.

Thus we obtain the following.

Property 4: A vector Boolean function $\underset{\sim}{f}(\underset{\sim}{x}, \underset{\sim}{y})$ is statistically independent of $\underset{\sim}{x}$ (respectively, balanced for each $\underset{\sim}{x}$) iff none of the nonzero linear combinations of the component functions of $\underset{\sim}{f}$ can be expressed as the sum of a nonzero linear function (linear function) of $\underset{\sim}{x}$ and a nonbalanced Boolean function of $(\underset{\sim}{x}, \underset{\sim}{y})$.

Now, consider a general binary combiner with M bits of memory described by (1) and (2). From (1) and (2) we have

$$(y_t, y_{t-1}, \ldots, y_{t-M}) = F(\underset{\sim}{x}_t, \underset{\sim}{x}_{t-1}, \ldots, \underset{\sim}{x}_{t-M}, \underset{\sim}{s}_{t-M}), \quad t \geq M \quad (3)$$

where F is the corresponding vector Boolean function $GF(2)^{N(M+1)+M} \longrightarrow GF(2)^{M+1}$, which can be expressed as a composition of f and G. Input sequence $\{\underset{\sim}{x}_t\}_{t=0}^{\infty}$ is assumed to be balanced and independent random sequence, whereas $\underset{\sim}{s}_0$ is a given initial state. It is easy to see that the output sequence $\{\underset{\sim}{y}_t\}_{t=0}^{\infty}$ is also balanced and independent iff F is balanced for each $\underset{\sim}{s}_{t-M}$, that is, if $f(\underset{\sim}{x}, \underset{\sim}{s})$ is balanced for each $\underset{\sim}{s}$.

Using properties 3 and 4 one can prove the following result.

Theorem: Let the output function $f(\underset{\sim}{x}, \underset{\sim}{s})$ of a general binary combiner with M bits of memory be balanced for each $\underset{\sim}{s}$. Then, there exists a linear function $L_{\underset{\sim}{\omega}}(y_t, y_{t-1}, \ldots, y_{t-M})$ effectively depending on y_t that can be expressed as the sum of a linear function $L_{\underset{\sim}{w}}(\underset{\sim}{x}_t, \underset{\sim}{x}_{t-1}, \ldots, \underset{\sim}{x}_{t-M})$ effectively depending on $\underset{\sim}{x}_t$ and a nonbalanced Boolean function $\epsilon(\underset{\sim}{x}_t, \underset{\sim}{x}_{t-1}, \ldots, \underset{\sim}{x}_{t-M}, \underset{\sim}{s}_{t-M})$ statistically independent of $\underset{\sim}{s}_{t-M}$, for all $t \geq M$.

The main point in the proof is that the function F as a balanced (M+1)-dimensional vector Boolean function can not be balanced for each $(\underset{\sim}{x}_t, \underset{\sim}{x}_{t-1}, \ldots, \underset{\sim}{x}_{t-M})$, since $\underset{\sim}{s}_{t-M}$ has dimension only M. The theorem essentially states that for a general binary combiner with M memory bits there exists a nonzero linear function of at most M+1 successive outputs that is correlated to a nonzero linear function of at most M+1 successive inputs. Given these two linear functions, it is possible to apply the standard cryptanalytic methods [7], [2], for example, developed for memoryless combiners, in order to reconstruct the initial states of all those LFSRs involved by the linear function of the inputs. Note that for M=1, the theorem states that either the output or the sum of two successive outputs is correlated to a nonzero linear function of at most two successive inputs. This is in accordance with the results obtained in [4].

IV. LINEAR APPROXIMATION OF A GENERAL BINARY COMBINER WITH MEMORY

The correlation theorem for a general binary combiner with memory, given in the previous section, asserts the existence of a pair of certain linear functions of the output and the input that are statistically dependent, which is a basis for divide and conquer correlation attacks. However, the problem remains how to find such a pair. Exhaustive search would require balance checkings of $2^{M(N+1)}(2^N-1)$ Boolean functions of $N(M+1)$ variables, which is intractable for large M or N, even if we use the Walsh transform which requires $O((NM+N)2^{NM+M+N})$ time complexity.

We propose an efficient procedure having $O((M+1)(M+N)2^{M+N})$ time complexity at worst, which very likely leads to the desired solution. The procedure is based on the linear approximation of the output and the next-state functions of a binary combiner with memory, see Fig.1.

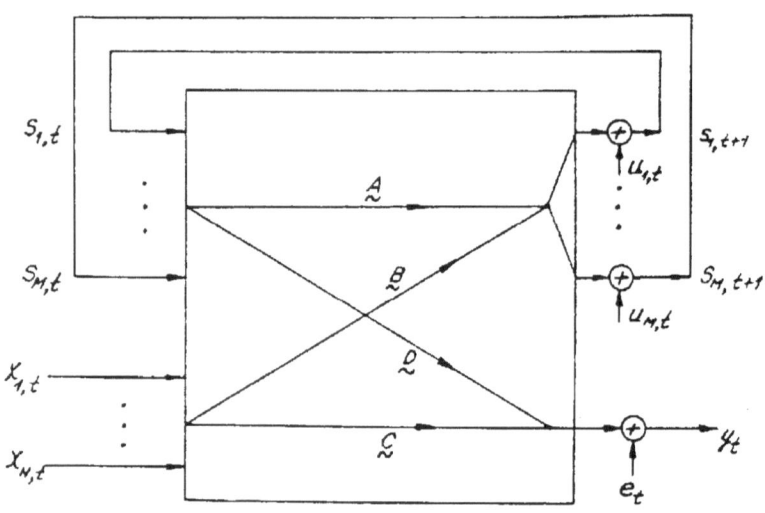

Fig.1. Linear sequential circuit approximation.

First, decompose the output function f and each of the component functions of the next-state function $G = (g_1, \ldots, g_M)$ into the sum of a linear function and a nonbalanced Boolean function. This decomposition is always possible due to the correlation properties of a memoryless combiner. If the function being decomposed is balanced, then the linear function is nonzero and, according to property 3, statistically independent of the nonbalanced function. If the function being decomposed is nonbalanced, then one can choose the linear function to be zero.

After the decomposition, the basic equations (1) and (2) using the matrix notation become

$$\underset{\sim}{s}_t = \underset{\sim}{A} \ \underset{\sim}{s}_{t-1} + \underset{\sim}{B} \ \underset{\sim}{x}_{t-1} + \underset{\sim}{u}(\underset{\sim}{x}_{t-1} \cdot \underset{\sim}{s}_{t-1}), \qquad t \geq 1 \tag{4}$$

$$y_t = \underset{\sim}{C} \ \underset{\sim}{x}_t + \underset{\sim}{D} \ \underset{\sim}{s}_t + e(\underset{\sim}{x}_t \cdot \underset{\sim}{s}_t), \qquad t \geq 0 \tag{5}$$

where the vectors are regarded as one-column matrices, $\underset{\sim}{A} = \left[a_{ij}\right]_{MxM}$, $\underset{\sim}{B} = \left[b_{ij}\right]_{MxN}$, $\underset{\sim}{C} = \left[c_{ij}\right]_{1xN}$, and $\underset{\sim}{D} = \left[d_{ij}\right]_{1xM}$ are binary matrices, and e and each of the component functions of $\underset{\sim}{u} = (u_1, \ldots, u_M)$ are nonbalanced Boolean functions.

Second, consider (4) and (5) as the basic equations of a general binary linear combiner with memory, that is, of a binary linear sequential circuit, LSC, (see [1], for example), formally assuming that $\{\underset{\sim}{u}_t\}_{t=0}^{\infty} = \{\underset{\sim}{u}(\underset{\sim}{x}_t \cdot \underset{\sim}{s}_t)\}_{t=0}^{\infty}$ and $\{e_t\}_{t=0}^{\infty} = \{e(\underset{\sim}{x}_t \cdot \underset{\sim}{s}_t)\}_{t=0}^{\infty}$ are input sequences to this LSC. Let us call this binary LSC a linear approximation of a general binary combiner with memory. Then find a solution to (4) and (5) using the generating function representation (often called the formal power series or the D-transform) of binary sequences [1]. Namely, let $\underset{\sim}{S} = \Sigma_{t=0}^{\infty} \underset{\sim}{s}_t z^t$, $\underset{\sim}{X} = \Sigma_{t=0}^{\infty} \underset{\sim}{x}_t z^t$, $\underset{\sim}{U} = \Sigma_{t=0}^{\infty} \underset{\sim}{u}_t z^t$, $E = \Sigma_{t=0}^{\infty} e_t z^t$, and $Y = \Sigma_{t=0}^{\infty} y_t z^t$ denote the corresponding generating functions. Then (4) and (5) result in

$$\underset{\sim}{S} = z \ \underset{\sim}{A} \ \underset{\sim}{S} + z \ \underset{\sim}{B} \ \underset{\sim}{X} + z \ \underset{\sim}{U} + \underset{\sim}{s}_0 \tag{6}$$

$$Y = \underset{\sim}{C} \ \underset{\sim}{X} + \underset{\sim}{D} \ \underset{\sim}{S} + E. \tag{7}$$

The solution to (6) and (7) is

$$Y = (\underset{\sim}{C} - z \ \frac{\underset{\sim}{D} \ (z\underset{\sim}{A} - \underset{\sim}{I})^{adj} \ \underset{\sim}{B}}{\det \ (z\underset{\sim}{A} - \underset{\sim}{I})}) \ \underset{\sim}{X} - \frac{\underset{\sim}{D} \ (z\underset{\sim}{A} - \underset{\sim}{I})^{adj}}{\det \ (z\underset{\sim}{A} - \underset{\sim}{I})} \ (z\underset{\sim}{U} + \underset{\sim}{s}_0) + E \tag{8}$$

where det $(z\underset{\sim}{A}-\underset{\sim}{I}) = \varphi(z)$ is a nonzero polynomial in z of degree at most rang $\underset{\sim}{A} \leq M$, which is the reciprocal of the characteristic polynomial of $\underset{\sim}{A}$, and the elements of the matrix $(z\underset{\sim}{A}-\underset{\sim}{I})^{adj}$ are polynomials in z of degree at most M-1. Accordingly, (8) can be put in the form

$$Y = \frac{1}{\varphi(z)} \sum_{i=1}^{N} h_i(z)X_i + \frac{1}{\varphi(z)} \sum_{j=1}^{M} p_j(z)(zU_j + s_{j0}) + E, \qquad (9)$$

with the polynomials in z satisfying deg $\varphi(z) \leq$ rang $\underset{\sim}{A} \leq M$, $\varphi(0)=1$, deg $h_i(z) \leq M$, $1 \leq i \leq N$, and deg $p_j(z) \leq M-1$, $1 \leq j \leq M$, which is equivalent to

$$\varphi(z)Y = \sum_{i=1}^{N} h_i(z)X_i + \sum_{j=1}^{M} p_j(z)(zU_j + s_{j0}) + \varphi(z)E. \qquad (10)$$

Letting $\varphi(z) = \sum_{k=0}^{M} \varphi_k z^k$, $h_i(z) = \sum_{k=0}^{M} h_{ik} z^k$, $1 \leq i \leq N$, $p_j(z) = \sum_{k=0}^{M-1} p_{jk} z^k$, $1 \leq j \leq M$, (10) in time domain reduces to

$$\sum_{k=0}^{M} \varphi_k y_{t-k} = \sum_{i=1}^{N} \sum_{k=0}^{M} h_{ik} x_{i,t-k} + \sum_{j=1}^{M} \sum_{k=0}^{M-1} p_{jk} u_j(\underset{\sim}{x}_{t-1-k}, \underset{\sim}{s}_{t-1-k}) +$$
$$+ \sum_{k=0}^{M} \varphi_k e(\underset{\sim}{x}_{t-k}, \underset{\sim}{s}_{t-k}), \quad t \geq M. \quad (11)$$

Now, if by using the next-state function, $\underset{\sim}{s}_{t-k}$ is expressed in terms of $(\underset{\sim}{x}_{t-k-1}, \ldots, \underset{\sim}{x}_{t-M}, \underset{\sim}{s}_{t-M})$, $0 \leq k \leq M$, (11) reduces to

$$\sum_{k=0}^{M} \varphi_k y_{t-k} = \sum_{i=1}^{N} \sum_{k=0}^{M} h_{ik} x_{i,t-k} + \epsilon(\underset{\sim}{x}_t, \ldots, \underset{\sim}{x}_{t-M}, \underset{\sim}{s}_{t-M}), \quad t \geq M, \quad (12)$$

which is the desired composition. However, it seems impossible to prove in general that the noise function ϵ is a nonbalanced function. Namely, $u_j(\underset{\sim}{x}_{t-1-k}, \underset{\sim}{s}_{t-1-k})$, $1 \leq j \leq M$, $0 \leq k \leq M-1$, and

$e(x_{t-k}, s_{t-k})$, $0 \leq k \leq M$, are nonbalanced provided that s_{t-k} is balanced, $0 \leq k \leq M$. If s_{t-k} is expressed in terms of $(x_{t-k-1}, \ldots, x_{t-M}, s_{t-M})$, $0 \leq k \leq M$, it might not be balanced any more, since we then assume that s_{t-M} is balanced. Therefore, it is not impossible that u_j as a function of $(x_{t-1-k}, \ldots, x_{t-M}, s_{t-M})$, $1 \leq j \leq M$, $0 \leq k \leq M-1$, or e as a function of $(x_{t-k}, \ldots, x_{t-M}, s_{t-M})$, $0 \leq k \leq M$, become balanced. On the other hand, it is also not impossible that $\epsilon(x_t, \ldots, x_{t-M}, s_{t-M})$ as the sum of nonbalanced, not necessarily statistically independent, functions becomes balanced. Nevertheless, it appears highly unlikely that $\epsilon(x_t, \ldots, x_{t-M}, s_{t-M})$ is balanced.

Note that (9) is a basic equation which can be further modified. For example, one can remove from $\varphi(z)$ all the factors that are common to all $p_j(z)$, $1 \leq j \leq M$, and $h_i(z)$, $1 \leq i \leq N$, provided they exist, and thus obtain and use $\varphi'(z)$.

V. CORRELATION ATTACK

We now analyze the discrimination potential of (11), that is, (12) regarding the statistical reconstruction of the input sequences, provided that the noise function ϵ is nonbalanced. Suppose that each of the input sequences $\{x_{it}\}$ is generated by a LFSR with feedback polynomial $F_i(z)$, $1 \leq i \leq N$. Also suppose that $F_i(z)$, $1 \leq i \leq N$, are pairwise coprime. Then, using (12) one can statistically reconstruct all the input sequences $\{x_{it}\}$, $1 \leq i \leq N$, such that $h_{ik} \neq 0$ for at least one k, $0 \leq k \leq M$, and that $F_i(z)$ does not divide $h_i(z)$. It is important to note that for the statistical procedure one need not know how much ϵ is nonbalanced, although the knowledge of the overall correlation coefficient, that is, the correlation coefficient between ϵ and the zero function would enable the calculation of the necessary length of the observed output segment. However, one may also take a modified approach that leads to the reconstruction of individual input sequences. Namely, if we multiply both sides of (10) by the product $F_i'(z)$, of degree L_i', of $F_j(z)$ for all $j \neq i$ that effectively appear in (12), then (10) reduces to

$$F_i'(z)\varphi(z)Y = F_i'(z)h_1(z)X_1 + F_i'(z) \sum_{j=1}^{M} p_j(z)(zU_j + s_{j0}) + F_i'(z)\varphi(z)E +$$
$$+ \Delta_i(z). \qquad (13)$$

with $\Delta_i(z)$ being a polynomial of degree at most $M + L_i' - 1$, which results in a decomposition equation analogous to (12), with $M_i' = M + L_i') M$ instead of M. If the resulting noise function ϵ' is nonbalanced and $F_i(z)$ does not divide $h_i(z)$, the reconstruction of $\{x_{it}\}$ is possible in principle.

Similarly, one can conjecture that the linear function of the output sequence in the generating function domain defined by $\varphi(z) \prod_{i=1}^{N} F_i(z) Y$ is with high probability nonbalanced.

Note that the described decomposition procedure may lead to efficient reconstruction only if the absolute value of the correlation coefficient between the resulting noise function and the zero function is sufficiently large. Since the noise function is expressed as the sum of individual noise functions, the absolute value of the correlation coefficient is in general much smaller than for memoryless combiners. It appears reasonable that in order to maximize the absolute value of the overall correlation coefficient one should approximate the output function and the components of the next-state function by the linear functions with maximum absolute values of correlation coefficients. If we use the Walsh transform, determination of such functions requires $O((M+1)(M+N)2^{M+N})$ time complexity, assuming that most of the functions effectively depend on all $M+N$ input and state variables. If many of them depend on just a subset of the input and state variables, the time complexity can be considerably smaller.

VI. CONCLUSION

In this paper, we study the correlation properties of a general binary combiner with an arbitrary number, M, of memory bits. We show that there exists a pair of correlated linear functions of at most $M+1$ successive output and input bits, respectively, which is a generalization of a result from [4] regarding the binary combiners with one bit of memory. We also develop and analyze an efficient procedure for finding such pairs of linear functions. The procedure is based on a linear sequential circuit approximation of a nonlinear combiner with memory. It is still an open question whether there exist other such procedures. The result may be a basis for a divide and conquer correlation attack on a stream cipher generator consisting of several linear feedback shift registers combined by a combiner with memory. In general, it also applies to an arbitrary synchronous finite-state machine as well.

REFERENCES

[1]   A. Gill, *Linear Sequential Circuits*. McGraw-Hill, 1966.

[2]   W. Meier, O. Staffelbach, "Fast correlation attacks on certain stream ciphers", *Journal of Cryptology*, Vol. 1 (3), pp. 159-176, 1989.

[3]   W. Meier, O. Staffelbach, "Nonlinearity criteria for cryptographic functions", *Advances in Cryptology - - EUROCRYPT '89, Proceedings*, LNCS, Vol. 434, pp. 549-562, Springer-Verlag, 1990.

[4]   W. Meier, O. Staffelbach, "Correlation properties of combiners with memory in stream ciphers", *Advances in Cryptology - EUROCRYPT '90, Proceedings*, LNCS, Vol. 473, pp. 204-213, Springer-Verlag, 1991.

[5]   R.A. Rueppel, *Analysis and Design of Stream Ciphers*. Springer-Verlag, 1986.

[6]   R.A. Rueppel, "Correlation immunity and the summation generator", *Advances in Cryptology - CRYPTO '85, Proceedings*, LNCS, pp. 260-272, Springer-Verlag, 1986.

[7]   T. Siegenthaler, "Decrypting a class of stream ciphers using ciphertext only", *IEEE Trans. Comput.*, Vol. C-34, pp. 81-85, Jan. 1985.

[8]   T. Siegenthaler, "Correlation-immunity of nonlinear combining functions for cryptographic applications", *IEEE Trans. Inform. Theory*, Vol. IT-30, pp. 776-780, Sept. 1984.

[9]   G.Z. Xiao, J.L. Massey, "A spectral characterization of correlation-immune combining functions", *IEEE Trans. Inform. Theory*, Vol. IT-34, pp. 569-571, May 1988.

CONVERGENCE OF A BAYESIAN ITERATIVE ERROR-CORRECTION PROCEDURE ON A NOISY SHIFT REGISTER SEQUENCE

Miodrag J. Mihaljević and Jovan Dj. Golić

Institute of Applied Mathematics and Electronics, Belgrade
School of Electrical Engineering, University of Belgrade
Bulevar Revolucije 73, 11001 Beograd, Yugoslavia

ABSTRACT: Convergence of an algorithm for a linear feedback shift register initial state reconstruction using the noisy output sequence, based on a bitwise Bayesian iterative error-correction procedure and different weight parity-checks, is analyzed. It is proved that the self-composition of the Bayes error probability converges to zero if and only if the noise probability is less than a critical value expressed in terms of the numbers of parity-checks. An alternative approach to the critical noise estimation based on the residual error-rate after each iterative revision is also discussed.

Key words: Cryptanalysis, Decoding, Shift registers, Fast correlation attack, Algorithms, Convergence.

I. INTRODUCTION

Many of the published keystream generators are based on binary linear feedback shift registers (LFSRs) combined by a memoryless function. Such a generator is called a combination generator. A weakness of a combination generator for stream ciphers is demonstrated in [1]. In [2]-[8], various algorithms for the efficient realization of the attack are proposed and analyzed. The main underlying ideas for these algorithms are based on the iterative error-correction [9]. The algorithms are iterative procedures with two main phases in each iteration. In the first phase, a criterion

This research was supported by Science Fund of Serbia, grant #0403, through Institute of Mathematics, Serbian Academy of Arts and Sciences.

for the second phase is bit-by-bit calculated, using the parity-checks corresponding to the considered bit, and in the second phase a bit-by-bit decision (error-correction) is made. Experimental convergence analysis of these algorithms is given in [2], [4], and [8], whereas a probabilistic approach through the convergence of the error-rate self-composition, having origins in [9], is developed in [3], [5] and [7] for the majority and threshold decision rules, respectively, in the error-correction phase. An equivalent convergence representation for the Bayesian decision rule in error-correction is derived in [6].

In this paper, we consider an iterative algorithm employing the parity-checks of different weights and the Bayesian decision rule in error-correction for each bit, assuming that the error-rate from the previous iteration is used as the noise probability in the current one. We analyze the convergence of the Bayes error probability self-composition, which, as in [3], [5], and [9], may be regarded as an indicator of the iterative error-correction convergence. An alternative approach to the convergence consideration based on the residual error-rate after each iterative revision is also suggested.

II. PROBLEM STATEMENT

Denote by $\{x_n\}_{n=1}^{N}$ an output segment of a LFSR of length L. In a statistical model, a binary noise sequence $\{e_n\}_{n=1}^{N}$ is assumed to be a realization of a sequence of i.i.d. binary variables $\{E_n\}_{n=1}^{N}$ such that $Pr(E_n=1) = p_n$, $n=1,2,\ldots,N$. Let $\{z_n\}_{n=1}^{N}$ be a noisy version of $\{x_n\}_{n=1}^{N}$ defined by

$z_n = x_n \oplus e_n$, $n=1,2,\ldots,N$.

where $\oplus$ denotes the modulo 2 addition, in GF(2).

We consider a reconstruction of the LFSR initial state given the segment $\{z_n\}_{n=1}^{N}$, provided that the feedback polynomial and p ($p_n=p$, $n=1,2,\ldots,N$) are known, using an algorithm based on iterative error-correction.

Suppose that a set of orthogonal parity-checks related to the n-th bit is generated in an appropriate way (see [2]-[4] and [7]) n=1,2,...,N.

Let $No_n(w)$ denote the number of parity-checks of weight w for the n-th bit, which involve exactly $w+1$ bits, and let $s_n(w)$ be the number of satisfied parity-checks among them, n=1,2,...,N. Let Ω denote a set of possible weights for each bit. In the statistical model, for every n=1,2,...,N, $s_n(w)$ is a realization of the integer stochastic variable $S_n(w)$, $w=1,2,...,L$. $Pr(E_n,\{S_n(w)\}_{w=1}^L)$ is the joint probability of the variables E_n and $S_n(w)$, $w=1,2,...,L$, and $Pr(E_n|\{S_n(w)\}_{w=1}^L)$ is the corresponding posterior probability, n=1,2,...,N. In the statistical model, for every n=1,2,...,N, $\theta_n = [s_n(w)]_{w\in\Omega}$ is a realization of the stochastic multi-dimensional vector variable $\tilde{\theta}_n = [S_n(w)]_{w\in\Omega}$. Then, it can be shown that the characteristic ratio of posterior probabilities

$$q_n(\theta_n) = Pr(E_n=1|\tilde{\theta}_n=\theta_n) \; / \; [1 - Pr(E_n=1|\tilde{\theta}_n=\theta_n)] \tag{1}$$

is given by the following lemma.

Lemma 1: For given p, $[No_n(w)]_{w\in\Omega}$, and an observed $\theta_n = [s_n(w)]_{w\in\Omega}$, the posterior probability quotient is given by

$$q_n(\theta_n) = \frac{p}{1-p} \prod_{w\in\Omega} \left[\frac{1 + (1-2p)^w}{1 - (1-2p)^w}\right]^{No_n(w)-2s_n(w)} , \quad n=1,2,...,N . \tag{2}$$

Note that the product (2) effectively contains only those terms for which $No_n(w) \geq 1$.

The main purpose of this paper is the theoretic convergence analysis of the following algorithm [8], which can be viewed as a modification/simplification of the algorithms [2]-[4].

A L G O R I T H M :

Input : The noisy sequence $\{z_n\}_{n=1}^{N}$.

Initialization: $i=0$, $p^{(1)}=p$, $I=const \rangle 1$, $c=const \geq 1$.

Step 1: Set $i \rightarrow i+1$. If $i \rangle I$ go to Step 6.

Step 2: Calculate $\Theta_n = [s_n(w)]_{w \in \Omega}$. $n=1,2,\ldots,N$.

Step 3: For $n=1,2,\ldots,N$, calculate $q_n(\Theta_n,p^{(i)})$ and
$P_n^{(i)} = q_n(\Theta_n,p^{(i)}) / [1 + q_n(\Theta_n,p^{(i)})]$ where $q_n(\Theta_n,p^{(i)})$
stands for $q_n(\Theta_n)$ corresponding to the noise probability
$p^{(i)}$.

Step 4: If $q_n(\Theta_n,p^{(i)}) \rangle c$, set $z_n \rightarrow z_n \oplus 1$, $P_n^{(i)} \rightarrow 1-P_n^{(i)}$,
$n=1,2,\ldots,N$.

Step 5: Calculate $p^{(i+1)} = (1/N) \sum_{n=1}^{N} P_n^{(i)}$. If $p^{(i+1)} \langle p^{(i)}$ go to

Step 1.

Step 6: Set $\hat{x}_n \rightarrow z_n$, $n=1,2,\ldots,N$, and stop the procedure.

Output: The reconstructed sequence $\{\hat{x}_n\}_{n=1}^{N}$.

Besides the direct analysis of the algorithm convergence to the original sequence (meaning that $\hat{x}_n = x_n$, $n=1,2,\ldots,N$), its convergence could be considered through the convergence of the sequence $\{p^{(i)}\}_{i=1}^{\infty}$ (in Step 5 $p^{(i)}$ is the expected relative number of errors in $\{z_n\}_{n=1}^{N}$ after the $(i-1)$-th iteration step, given $\{\Theta_n\}_{n=1}^{N}$), or through the convergence of the expected value sequence $\{P_B^{(i)}\}_{i=1}^{\infty}$, where $P_B^{(i)}$ denotes the Bayes probability of error after the i-th iteration (average value of $p^{(i)}$ over $\{\Theta_n\}_{n=1}^{N}$, when $c=1$). Since the direct analysis of the algorithm convergence seems to be intractable, in this paper we analyze the convergence of $\{P_B^{(i)}\}_{i=1}^{\infty}$ and $\{p^{(i)}\}_{i=1}^{\infty}$.

III. SELF-COMPOSITION OF THE BAYES ERROR PROBABILITY

When $c = 1$ in the ALGORITHM Step 4, we have the Bayesian approach to the error-correction. When the Bayesian decision rule is employed, for a given n-th bit of noise, both the conditional and average probabilities of decision error are minimized, and are given by

$$P_B(n,\theta_n) = \min\{Pr(E_n=0|\theta_n), Pr(E_n=1|\theta_n)\} \quad . \tag{3}$$

$$P_B(n) = \sum_{\theta_n} P_B(n,\theta_n) \, Pr(\tilde{\theta}_n=\theta_n) \quad . \tag{4}$$

respectively, with

$$Pr(\tilde{\theta}=\theta) = p \prod_{w\in\Omega} \binom{No(w)}{s(w)} p_w^{s(w)} (1-p_w)^{No(w)-s(w)} +$$

$$(1-p) \prod_{w\in\Omega} \binom{No(w)}{s(w)} (1-p_w)^{s(w)} p_w^{No(w)-s(w)} \quad , \tag{5}$$

where $p_w = [1-(1-2p)^w]/2$, $w \in \Omega$, and subscript n in (5) is omitted for simplicity. After certain algebraic manipulations, one can obtain the following form for the Bayes probability of error:

$$P_B(n) = p - \sum_{\theta_n:\, q_n(\theta_n)\geq 1} q_n(\theta_n) \, Pr(\tilde{\theta}_n=\theta_n) \, \frac{q_n(\theta_n) - 1}{q_n(\theta_n) + 1} \quad , \tag{6}$$

where $Pr(\tilde{\theta}_n=\theta_n)$ is given by (5) and $q_n(\theta_n)$ by Lemma 1.

Without loss of generality, suppose that $No_n(w) = No(w)$, $n=1,2,\ldots,N$, $w \in \Omega$. This can be obtained by clipping of the original sequence $\{No_n(w)\}_{n=1}^{N}$ to the minimum value. Under this assumption, Lemma 1 and (6) yield that the Bayes error probability $P_B(n)$ is the same for all n , that is, $P_B(n) = P_B$, $n=1,2,\ldots,N$.

Consequently, the self-composition of the Bayes error probability is defined as the recursion

$$p^{(i)} = p^{(i-1)} - f(p^{(i-1)}) \quad , \qquad i=1,2,\ldots \quad . \tag{7}$$

$$f(P) = \sum_{\theta:\, q(\theta,P)\rangle 1} Pr(\tilde{\theta}=\theta) \, \frac{q(\theta,P) - 1}{q(\theta,P) + 1} \quad , \qquad P \in [0, 0.5] \quad . \tag{8}$$

where $p^{(0)} = p \leq 0.5$, p being the initial noise probability, and $q(\theta,P)$ stands for $q_n(\theta_n)$ corresponding to the noise probability P.

IV. CONVERGENCE ANALYSIS

The convergence of the self-composition of the probability of error for the majority decision error-correction rule was analyzed in [3], [5]. A condition for the threshold error-correction is given in [7], although the proof is not quite precise. In this section, we consider the convergence of the Bayes error probability self-composition, which, as in [3], [5], may be regarded as an indicator of the iterative error-correction convergence.

First note that $f(P)$ is a continuous nonnegative function on the segment $[0,0.5]$ such that $f(P=0) = f(P=0.5) = 0$. We now prove three lemmas and a theorem giving the necessary and sufficient conditions for (7) to converge to zero. When $No(1) = 1$ and $No(w) = 0$, $w=2,3,\ldots,L$, it follows that $P^{(i)} = p$, $i=1,2,\ldots$. In all what follows this case is simply called the degenerate one. Note that lemma 4 essentially contains the result that is missing in [7].

Lemma 2: The recursion (7) converges to 0 if and only if

$$f(P) > 0 \quad , \quad P \in (0, p] \quad . \tag{9}$$

Proof: Since $f(P)$ is a nonnegative function not greater than P, the sequence $\{P^{(i)}\}_{i=1}^{\infty}$ is nonnegative and nonicreasing and, hence, it converges to a limit $P^* \in [0, p]$ such that $f(P^*) = 0$. Consequently, it is straightforward to show that $P^* = 0$ if and only if (9) is true. Q.E.D.

Lemma 3: For each $P \in (0, 0.5)$, we have that $f(P) > 0$ if and only if $q(\theta, P) > 1$ for $\theta = \underline{0} = [0,0,\ldots,0]$, that is, when all the parity-checks are unsatisfied. Otherwise, $f(P) = 0$.

Proof: First, note that, except in the degenerate case, for all $\theta \neq \underline{0} = [0\ 0\ \ldots\ 0]$ we have

$$q(\underline{0}, P) > q(\theta, P) \quad , \qquad P \in (0,0.5) \quad .$$

because, according to (2), the inequality

$$[\frac{1 + (1-2P)^w}{1 - (1-2P)^w}]^{No(w)} > [\frac{1 + (1-2P)^w}{1 - (1-2P)^w}]^{No(w)-2s(w)} \quad .$$

holds if $s(w) > 0$, for any $w=1,2,\ldots,L$.

On the other hand, according to (5) $Pr(\tilde{\Theta}=\theta) > 0$ for all θ . Therefore, in view of (8), it follows that a necessary and sufficient condition for $f(P) > 0$ is that a θ exists such that $q(\theta, P) > 1$. However, the first formula in the proof implies that this is equivalent to $q(\underline{0}, P) > 1$.   Q.E.D.

Lemma 4: Let $Q(P)$ be the function defined by

$$Q(P) = \frac{P}{1 - P} \prod_{w\in\Omega} [\frac{1 + (1-2P)^w}{1 - (1-2P)^w}]^{No(w)} \quad , \quad P \in (0, 0.5] . \quad (10)$$

For $\Omega = \{1\}$ and $No(1) = 1$. $Q(P) = 1$, $P \in (0, 0.5]$. For $\Omega = \{1\}$ and $No(1) > 1$, $Q(P) > 1$, $P \in (0, 0.5)$, and $Q(0.5) = 1$. Finally, for $\Omega \neq \{1\}$ a critical value $P_0 \in (0, 0.5)$ exists such that $Q(P) > 1$ for $0 < P < P_0$, $Q(P_0) = 1$. $Q(P) < 1$ for $P_0 < P < 0.5$, and $Q(0.5) = 1$.

Proof: First note that $Q(P)$ is a positive and continuous function such that $Q(0.5)=1$. It can be shown that the first derivative of $Q(P)$ is

$$Q'(P) = Q(P) [\frac{1}{P (1-P)} - \sum_{w\in\Omega} \frac{4 No(w) w (1-2P)^{w-1}}{1 - (1-2P)^{2w}}] , \quad P\in(0,0.5). \quad (11)$$

which, using a substitution $P = (1-x)/2$, becomes

$$Q'((1-x)/2) = \frac{4 Q((1-x)/2)}{|\Omega| (1-x^2)} F(x) . \quad (12)$$

where

$$F(x) = \sum_{w\in\Omega} [1 + \frac{|\Omega|No(w)w (x^{w+1} - x^{w-1})}{1 - x^{2w}}] , \quad x\in(0,1) . \quad (13)$$

and $|\Omega|$ is the cardinality of Ω . The zeroes of $Q'(x)$ on $(0,1)$ are thus determined by $F(x)$. So, we proceed by analyzing $F(x)$.

When $\Omega = \{1\}$ and $No(1) = 1$. $F(x) = 0$ for all $x \in (0,1)$. meaning that $Q(P) = 1$. $P \in (0, 0.5]$. When $\Omega = \{1\}$ and $No(1) > 1$, $F(x) < 0$ for all $x \in (0,1)$. Bearing in mind that $Q(P) > 0$, $P \in (0, 0.5]$, it then follows that when $\Omega = \{1\}$ and $No(1) > 1$, $Q(P)$ is a decreasing function on $(0, 0.5]$ such that $Q(0.5) = 1$, which implies the lemma statement.

Assume now that $\Omega \neq \{1\}$. The first derivative of $F(x)$ on $(0,1)$ is

$$F'(x) = \sum_{w \in \Omega} - \frac{|\Omega| No(w) w \, x^{w-2}}{(1 - x^{2w})^2} [(w-1) - (w+1)x^2 + (w+1)x^{2w} - (w-1)x^{2w+2}].$$

(14)

According to (14), we now analyze the following functions on $(0,1)$:

$$\phi_w(x) = (w-1) - (w+1)x^2 + (w+1)x^{2w} - (w-1)x^{2w+2} \quad , \quad w \in \Omega \setminus \{1\} \quad . \quad (15)$$

The first derivative of $\phi_w(x)$ is

$$\phi_w'(x) = 2(w+1) \, x \, [-1 + w \, x^{2w-2} - (w-1) \, x^{2w}] \quad . \tag{16}$$

Proceeding in the same manner, we finally consider the following functions on $(0,1)$:

$$\varphi_w(x) = -1 + w \, x^{2w-2} - (w-1) \, x^{2w} \quad , \qquad w \in \Omega \setminus \{1\} \quad . \tag{17}$$

The first derivative of $\varphi_w(x)$ is

$$\varphi_w'(x) = 2 \, w \, (w-1) \, x^{2w-3} \, [1 - x^2] \quad . \tag{18}$$

According to (18), $\varphi_w'(x) \cdot x \in (0,1)$, is a positive function, for each $w \in \Omega \setminus \{1\}$. Consequently, using (12)-(18) and the fact that $Q(P) > 0$, $P \in (0, 0.5)$, it is not difficult to obtain that $Q'(P)$ has exactly one zero, P^*, on $(0,0.5)$, and that $Q'(P)$ is a negative function for $0 < P < P^*$, and a positive one for $P^* < P < 0.5$. Finally, from $Q(0+) = \infty$, $Q(0.5) = 1$, and the established characteristics of $Q'(P)$ it follows that for $\Omega \neq \{1\}$, a $P_0 \in (0, 0.5)$ exists such that $Q(P) > 1$ for $0 < P < P_0$, $Q(P_0) = 1$, $Q(P) < 1$ for $P_0 < P < 0.5$, and $Q(0.5) = 1$. Q.E.D.

Theorem 1: The self-composition of the Bayes error probability converges to 0 for $0 < p < P_0$ and is in any iteration step equal to p for $P_0 \leq p \leq 0.5$. The critical value P_0 is equal to the unique value of $P \in (0, 0.5)$ such that

$$\frac{P}{1 - P} \prod_{w \in \Omega} \left[\frac{1 + (1-2P)^w}{1 - (1-2P)^w}\right]^{No(w)} = 1 \quad . \tag{19}$$

if $\Omega \neq \{1\}$. For $\Omega = \{1\}$ and $No(1) > 1$, $P_0 = 0.5$, and for $\Omega = \{1\}$ and $No(1) = 1$, $P_0 = 0$.

Proof: For $\Omega = \{1\}$ and $No(1) = 1$ the proof is trivial. For

$\Omega \neq \{1\}$ and $\Omega = \{1\}$. $No(1) > 1$ the existence of P_0 is established by Lemma 4. Then, according to Lemma 4, we have that $Q(P) > 1$ for any $0 < P < P_0$. By Lemma 3, $Q(P) > 1$ implies that $f(P) > 0$, $0 < P < P_0$. Finally, by Lemma 2, the self-composition converges to 0 if $0 \leq p < P_0$.

On the other hand, in view of Lemma 3, we have that $f(P) = 0$ for $P_0 \leq P \leq 0.5$, so that (7) yields that the self-composition equals p in any iteration step. Q.E.D.

Note that by virtue of Lemma 4, Theorem 1 essentially states that for $p > 0$ the self-composition of the Bayes error probability converges to zero if and only if

$$\frac{p}{1-p} \prod_{w\in\Omega} \left[\frac{1+(1-2p)^w}{1-(1-2p)^w}\right]^{No(w)} > 1 . \tag{20}$$

which is the desired convergence condition.

Finally, some examples related to Lemma 4 and Theorem 1 are presented. Denote by W the number of feedback tapes on a given LFSR. The function $Q(P)-1$, when $N=10^5$, 10^6 , $\Omega \subseteq \{1,2,\ldots,L\}$ and $\Omega \subseteq \{1,2,\ldots,W\}$ such that $No(w) \geq 1$, $w \in \Omega$, and $\Omega = \{W\}$, assuming that $\{No(w)\}_{w=2}^{L}$ stands for the average values determined by the approach presented in [4], is for $L=40$, $W=14$ displayed in Table I.

According to Theorem 1, the nonacceptable noise P_0 is the value of $P \in (0, 0.5)$ such that $Q(P)-1 = 0$. So defined nonacceptable noise is the minimum noise-rate above which the algorithm is bound to fail. Some values of P_0 are given in Table II.

The presented numerical examples are self-explanatory and provide the quantitative illustrations of the analytical results.

Table I: The function $Q(P)-1$, for $L=40$, $W=14$, when $N=10^5$, 10^6,
$\Omega \subseteq \{1,2,\ldots,L\}$, $\Omega \subseteq \{1,2,\ldots,W\}$, and $\Omega = \{W\}$.

P	$\Omega \subseteq \{1,2,\ldots,L\}$		$\Omega \subseteq \{1,2,\ldots,W\}$		$\Omega = \{W\}$	
	$N=10^5$	$N=10^6$	$N=10^5$	$N=10^6$	$N=10^5$	$N=10^6$
0.01	)1000	)1000	)1000	)1000	)1000	)1000
0.02	)1000	)1000	)1000	)1000	)1000	)1000
0.03	)1000	)1000	)1000	)1000	)1000	)1000
0.04	)1000	)1000	)1000	)1000	)1000	)1000
0.05	)1000	)1000	)1000	)1000	)1000	)1000
0.06	)1000	)1000	)1000	)1000	)1000	)1000
0.07	)1000	)1000	)1000	)1000	)1000	)1000
0.08	)1000	)1000	)1000	)1000	)1000	)1000
0.09	)1000	)1000	)1000	)1000	)1000	)1000
0.10	)1000	)1000	)1000	)1000	)1000	)1000
0.11	)1000	)1000	)1000	)1000	)1000	)1000
0.12	)1000	)1000	)1000	)1000	99.9	861.4
0.13	)1000	)1000	)1000	)1000	13.1	60.8
0.14	)1000	)1000	)1000	)1000	2.6	8.9
0.15	)1000	)1000	)1000	)1000	0.425	1.8
0.16	)1000	)1000	)1000	)1000	-0.234	0.204
0.17	)1000	)1000	)1000	)1000	-0.488	-0.310
0.18	)1000	)1000	)1000	)1000	-0.602	-0.517
0.19	)1000	)1000	)1000	)1000	-0.656	-0.612
0.20	)1000	)1000	)1000	)1000	-0.682	-0.656
0.21	)1000	)1000	713.1	)1000	-0.691	-0.676
0.22	)1000	)1000	48.1	)1000	-0.691	-0.681
0.23	99.5	)1000	7.4	)1000	-0.684	-0.679
0.24	8.9	)1000	1.6	)1000	-0.674	-0.670
0.25	1.5	)1000	0.275	)1000	-0.660	-0.658
0.26	0.151	)1000	-0.190	)1000	-0.645	-0.644
0.27	-0.264	341.3	-0.382	58.9	-0.628	-0.627
0.28	-0.422	18.9	-0.469	7.5	-0.610	-0.609
0.29	-0.488	2.8	-0.508	1.6	-0.591	-0.591
0.30	-0.513	0.532	-0.521	0.280	-0.571	-0.571
0.31	-0.517	-0.080	-0.521	-0.150	-0.550	-0.550
0.32	-0.510	-0.298	-0.512	-0.321	-0.529	-0.529
0.33	-0.496	-0.385	-0.497	-0.394	-0.507	-0.507
0.34	-0.479	-0.419	-0.479	-0.422	-0.485	-0.485
0.35	-0.458	-0.425	-0.458	-0.426	-0.461	-0.461
0.36	-0.435	-0.417	-0.435	-0.418	-0.437	-0.437
0.37	-0.412	-0.401	-0.412	-0.402	-0.413	-0.413
0.38	-0.386	-0.381	-0.386	-0.381	-0.381	-0.387
0.39	-0.360	-0.357	-0.360	-0.357	-0.361	-0.361
0.40	-0.333	-0.331	-0.333	-0.332	-0.333	-0.333
0.41	-0.305	-0.304	-0.305	-0.304	-0.305	-0.305
0.42	-0.276	-0.275	-0.276	-0.275	-0.276	-0.276
0.43	-0.246	-0.245	-0.246	-0.245	-0.246	-0.246
0.44	-0.214	-0.214	-0.214	-0.214	-0.214	-0.214
0.45	-0.182	-0.182	-0.182	-0.182	-0.182	-0.182
0.46	-0.148	-0.148	-0.148	-0.148	-0.148	-0.148
0.47	-0.113	-0.113	-0.113	-0.113	-0.113	-0.113
0.48	-0.077	-0.077	-0.077	-0.077	-0.077	-0.077
0.49	-0.039	-0.039	-0.039	-0.039	-0.039	-0.039
0.50	0.000	0.000	0.000	0.000	0.000	0.000

Table II: The values of nonacceptable noise P_0 when $N=10^5$, 10^6 .
$\Omega \subseteq \{1,2,\ldots,L\}$, $\Omega \subseteq \{1,2,\ldots,W\}$, $\Omega = \{W\}$ for $L=40$,
$W=14$ and $L=60$, $W=18$.

	P_0					
	$\Omega\subseteq\{1,2,\ldots,L\}$		$\Omega\subseteq\{1,2,\ldots,W\}$		$\Omega=\{W\}$	
	$N=10^5$	$N=10^6$	$N=10^5$	$N=10^6$	$N=10^5$	$N=10^6$
$L=40$, $W=14$	0.25	0.30	0.24	0.29	0.15	0.16
$L=60$, $W=18$	0.17	0.20	0.13	0.17	0.11	0.12

V. ALTERNATIVE APPROACH TO THE CONVERGENCE ANALYSIS

The critical noise established in Theorem 1 is the noise above
which the algorithm for iterative error-correction is bound to fail,
because there is no complementation in Step 4. However, the
convergence to zero of the Bayes error probability below the critical
noise essentially relies on the fact that with nonzero probability
all the parity-checks related to a bit could be unsatisfied. But in
most cases this probability is extremely small, which makes this
critical noise overly optimistic, when regarded as the noise below
which the algorithm is successful. Accordingly, we now take a more
realistic approach dealing with the convergence of the sequence of
residual error-rates $\{p^{(i)}\}_{i=1}^{\infty}$ (see Step 5).

Starting from the ALGORITHM Step 5, it can be shown that the
residual error-rate can be put in the following form.

Lemma 5:

$$p^{(i)} = p^{(i-1)} - \sum_{\theta:\ q(\theta,p^{(i-1)})>1} \frac{m^{(i)}(\theta)}{N} \frac{q(\theta,p^{(i-1)}) - 1}{q(\theta,p^{(i-1)}) + 1} \quad . \quad (21)$$

where q is defined by Lemma 1, and $m^{(i)}(\theta)$ is the number of
indices n such that $\theta_n = \theta$ in the i-th iteration step.

Note that the expected value of $m^{(i)}(\theta) / N$ over $\tilde{\theta}$, is $Pr(\tilde{\theta}=\theta)$, which yields the Bayes error probability. Since $m^{(i)}(\theta)$ depends on i the recursion (21) is not a self-composition. Therefore, we can not take the same approach as in section IV. It is easy to see that $\{p^{(i)}\}_{i=1}^{\infty}$ is a positive nonincreasing sequence which remains constant for $i > j$ if for some j $p^{(j)} = p^{(j-1)}$. A desirable convergence property is that $p^{(j)} = p^{(j-1)}$ can occur only if $\theta = [No(w)]_{w \in \Omega}$, that is, when all the parity-checks are satisfied. It appears very difficult to derive exact necessary and sufficient conditions for this to happen. Instead, we take a heuristic approach based on the following three lemmas.

Lemma 6: $p^{(i)} < p^{(i-1)}$ if and only if $m^{(i)}(\theta) > 1$ and $q(\theta, p^{(i-1)}) > 1$ for at least one θ.

Lemma 7: For each θ such that $s(w) < No(w)/2$, $w \in \Omega$, $q(\theta, p) > 1$ implies $q(\theta, P) > 1$ for all $0 \leq P \leq p$.

Lemma 8: For each $p \in (0, 0.5)$, $q(\theta, p) > 1$ implies $q(\theta', p) > 1$ for all $\theta' \leq \theta$, where the inequality is defined componentwise.

Note that Lemma 6 corresponds to Lemma 3, whereas Lemma 7 can be proved in a similar way as Lemma 4. Lemma 8 is a simple consequence of Lemma 1.

In view of Lemmas 5 - 8, we come to the following more realistic estimations of the critical noise, P_0^* and P_0^{**}, below which the algorithm is very likely capable of correcting all the errors in the noised sequence. P_0^* and P_0^{**} are the solutions, in $(0, 0.5)$, to the equation

$$\frac{P}{1 - P} \prod_{w \in \Omega} \left[\frac{1 + (1-2P)^w}{1 - (1-2P)^w} \right]^{No(w)-2s(w)} = 1 . \tag{22}$$

assuming that

− $s(w) = s^*(w)$, $w \in \Omega$, are the elements of an arbitrary

θ - vector such that $[1 - \mathrm{Pr}(\tilde{\theta} \le \theta^*)]^N < \epsilon$ where $\epsilon \cong 0$, and that so-defined P_0^* is close to maximum given ϵ.

and

- $\quad s(w) = s^{**}(w)$, $w \in \Omega$, where $\theta^{**} = [s^{**}(w)]_{w \in \Omega} = \bar{\theta} = \Sigma \, \theta \, \mathrm{Pr}(\tilde{\theta}=\theta)$.
  $\quad \theta$

for P_0^* and P_0^{**} . respectively. It follows that for reasonably small ϵ

$$P_0 > P_0^* > P_0^{**} . \tag{23}$$

where P_0 is given by Theorem 1. In connection with P_0^* , one can also define the critical noise probability $\bar{P}_0^*$ as the solution, in $(0, 0.5)$, to (22) with $\theta = [s(w)]_{w \in \Omega}$ being an arbitrary vector such that $\mathrm{Pr}(\tilde{\theta} = \theta) > 1/N$, and that $\bar{P}_0^*$ is close to maximum.

VI. CONCLUSION

A cryptanalytic problem of a LFSR initial state reconstruction using a noisy output sequence is considered. The paper is dedicated to the convergence analysis of the self-composition of the Bayes probability of error, which is an indicator of the iterative error-correction procedure convergence relevant for the cryptanalysis. It is proved that a critical value of the noise probability exists below which the error self-composition converges to zero, and above which it remains equal to the initial noise probability. This critical value, expressed in terms of the numbers of parity-checks, is the noise-rate above which the iterative error-correction fails. An alternative, more realistic, estimation of the critical noise below which the iterative error-correction procedure is successful, based on the convergence of the residual error-rate sequence, is also given.

REFERENCES

[1] T.Siegenthaler, "Decrypting a Class of Stream Ciphers Using Ciphertext Only", *IEEE Trans. Comput.*, vol. C-34, pp.81-85, Jan. 1985.

[2] W.Meier, O.Staffelbach, "Fast Correlation Attacks on Certain Stream Ciphers", *Journal of Cryptology*, vol.1, pp.159-176, 1989.

[3] K.Zeng, M.Huang, "On the Linear Syndrome Method in Cryptanalysis", *Advances in Cryptology - CRYPTO '88, Lecture Notes in Computer Science*, vol.405, pp.469-478, Springer-Verlag, 1990.

[4] M.Mihaljević, J.Golić, "A Fast Iterative Algorthm for a Shift Register Initial State Reconstruction Given the Noisy Output Sequence", *Advances in Cryptology - AUSCRYPT '90, Lecture Notes in Computer Science*, vol.453, pp.165-175, Springer-Verlag, 1990.

[5] K.Zeng, C.H.Yang, T.R.N.Rao, "An Improved Linear Syndrome Algorithm in Cryptanalysis with Applications", *Proc. CRYPTO '90.*

[6] M.Živković, "An Analysis of Linear Recurrent Sequences over the Field GF(2)", Ph.D. thesis, Belgrade University, 1990.

[7] V.Chepyzhov, B.Smeets, "On a Fast Correlation Attack on Stream Ciphers", *Advances in Cryptology - EUROCRYPT '91, Lecture Notes in Computer Science*, vol.547, pp.176-185, Springer-Verlag, 1991.

[8] M.Mihaljević, J.Golić, "A Comparison of Cryptanalytic Principles Based on Iterative Error-Correction", *Advances in Cryptology - EUROCRYPT '91, Lecture Notes in Computer Science*, vol.547, pp.527-531, Springer-Verlag, 1991.

[9] R.G.Gallager, "Low-Density Parity-Check Codes", *IRE Trans. Inform. Theory*, vol. IT-8, pp.21-28, Jan. 1962.

Suffix trees and string complexity

Luke O'Connor   Tim Snider

Department of Computer Science
University of Waterloo, Ontario, Canada, N2L 3G1
email: ljpoconn@watmath.uwaterloo.ca

Abstract

Let $s = (s_1, s_2, \ldots, s_n)$ be a sequence of characters where $s_i \in Z_p$ for $1 \leq i \leq n$. One measure of the complexity of the sequence s is the length of the shortest feedback shift register that will generate s, which is known as the maximum order complexity of s [17, 18]. We provide a proof that the expected length of the shortest feedback register to generate a sequence of length n is less than $2 \log_p n + o(1)$, and also give several other statistics of interest for distinguishing random strings. The proof is based on relating the maximum order complexity to a data structure known as a suffix tree.

1 Introduction

A common form of stream cipher are the so-called running key ciphers [4, 9] which are deterministic approximations to the one time pad. A running key cipher generates an ultimately periodic sequence $s = (s_1, s_2, \ldots, s_n)$, $s_i \in Z_p$, $1 \leq i \leq n$, for a given seed or key K. Encryption is performed as with the one time pad, using s as the key stream, but perfect security is no longer guaranteed. Considerable effort has been devoted to developing algorithms for generating sequences s that are pseudorandom [11, 13, 19, 28]. The purpose of such work is to define sequences that are efficiently generated and satisfy one, or possibly several, measures of randomness for finite strings. Let Ω_p^n be the set of all sequences s of length n where $s_i \in Z_p$ for $1 \leq i \leq n$. A *statistic* is a function $\alpha : \Omega_p^n \to \mathbb{R}$ which measures some property of a sequence, such as the length of the longest gap for binary sequences, or the distribution of the binary derivative [14]. If the

R.A. Rueppel (Ed.): Advances in Cryptology - EUROCRYPT '92, LNCS 658, pp. 138-152, 1993.
© Springer-Verlag Berlin Heidelberg 1993

distribution of α can be computed, in particular its expectation $\mathbf{E}(\alpha)$, α may be used to distinguish between random and nonrandom sequences by discarding those sequences s for which $\alpha(s)$ deviates significantly from the mean. If there are several statistics $\alpha_1, \alpha_2, \ldots, \alpha_j$ available for which the expectations are known, the more likely we are to detect nonrandom sequences. A collection of statistical tests for randomness is given in Knuth [21].

A notion attributed to Kolmogorov [22] characterizes the randomness of a sequence s as the encoded length of the smallest Turing machine program to produce s. Unfortunately, the Kolmogorov complexity of a sequence is not computable in general [23, §2.5], and consequently, the model of computation must be simplified in order to obtain a computable complexity measure. *Finite state machines* [4, 11] (FSM) are a class of automatons that consist of a finite set of states $Q = \{q_1, q_2, \ldots, q_m\}$, and a transition function $\delta : Q \rightarrow Q$. There is also an output function $\Delta : Q \rightarrow A$ which outputs a character from the alphabet A on each transition. The function δ is the 'program' associated with a FSM, and in this case, when executed will cause an infinite sequence of characters to be printed. If the state sequence of a FSM M after t transitions is $q_1', q_2', \ldots, q_t'$, then the output of M will be $\Delta(q_1'), \Delta(q_2'), \ldots, \Delta(q_t')$. We will informally say that the size of an FSM M, or the length of its description, is defined as the size of the information required to identify state q_i, plus the space required to store δ, denoted as $|\delta|$.

A FSM is a special instance of a deterministic finite automaton (DFA) [15] where δ only depends on the current state, rather than also depending on a current input symbol. It is clear that DFAs with the ability to write output symbols, known as Moore machines, can mimic any FSM. Alternately, a FSM is a Moore machine which prints the same output string for every input string w.

Feedback shift registers (FSR) are a special class of FSMs which have much practical import as they can be directly implemented in hardware, and are fundamental to the design of most digital circuitry. A FSR consists of m stages, or memory cells, $x_1, x_2, \ldots, x_m$ and a feedback polynomial $f(X) \in Z_2[x_1, x_2, \ldots, x_m]$. A state transition in a FSR corresponds to a shift of the register contents ($x_i = x_{i-1}$, $2 \leq i \leq m$), and the assignment $x_1 = f(x_1, x_2, \ldots, x_m)$. The size or description of each state in an FSR is then m, the size of the shift register, and $|f(X)|$ is the cost of storing $f(X)$: it follows that the size of the machine is $m + |f(X)|$. With respect to FSRs, the complexity of a sequence s is given as the smallest FSR that generates s.

As with the class of Turing machines, we are now left with the problem of actually determining m and $f(X)$ for a given sequence s, or the shortest program which describes the sequence. A further attraction of FSRs is that when $f(X)$ is restricted to be a linear polynomial (degree at most 1), the celebrated Massey-Berlekamp algorithm [24] can be used to determine m and $f(X)$ in time which is a polynomial function of the sequence length. Such a machine is known as a linear feedback shift register (LFSR). The number of stages required to generate a given sequence $s = (s_1, s_2, \ldots, s_n)$ with a linear feedback polynomial is known as the *linear complexity* or *linear span* of a sequence, and will be denoted as $L(s)$. For example, if $s = 010101110010$ then $L(s) = 5$, and the corresponding feedback polynomial is $f(X) = x_3 + x_4$. The following theorem is due to Rueppel [28].

Theorem 1.1 Let $s = (s_1, s_2, \ldots, s_n)$ where $s_i \in Z_2$ for $1 \le i \le n$. Then assuming the uniform distribution on Ω_2^n, the expected linear span $E(L(s))$ of a binary sequence is

$$E(L(s)) = \frac{n}{2} + \frac{4 + [n \text{ is odd}]}{18} + O(1). \tag{1}$$

$\square$

Our original notion was to find the smallest program to generate a given sequence, with respect to some machine class. The actual program size of a LFRS is $m + |f(X)|$, but the linear span is typically measured as simply m, since the number of terms in $f(X)$ is bounded by m. The size of a LFSR is then at most $2m + 1$, where we have included a constant term in $f(X)$.

A natural extension of these ideas is to permit the feedback function $f(X)$ of the FSR to be a polynomial of arbitrary degree. We reiterate that the linear case is attractive as the linear span can be directly computed, but there has been little work in computing spans of higher order with the exception of Chan and Games [7]. Let $F_k(s)$ be the span of the sequence $s = s_1, s_2, \ldots, s_n$ where $f(X)$ is a polynomial of degree at most k, $0 \le k \le n$. For a fixed sequence length n it follows that

$$F_n(s) \le F_{n-1}(s) \le \ldots \le F_1(s) \le F_0(s) \tag{2}$$

where $F_1(s) = L(s)$ is the linear span of s. The inequalities in eq. (2) state that the size of the shift register may decrease as we allow the degree of $f(X)$ to increase. We may illustrate this possible reduction in memory by considering de Bruijn sequences [8]. Chan and Games [7] have studied the quadratic span of these sequences, and their results show a large difference between the linear and quadratic spans. For example, the de Bruijn sequences of length $63 = 2^6 - 1$ have an expected linear span of at least 62, while the expected quadratic span is at most 12.

There will always exist a pair (n_0, n_1) where $1 \le n_0, n_1 \le n$ such that $F_i(s) = n_0$ for $n_1 \le i \le n$, which states that at some point the size of the shift register cannot be reduced any further by permitting higher order feedback polynomials. Let $M^*(\mathbf{s})$ be the *span* of s, defined as the length of the shortest FSR generates s. The span gives a lower bound on the amount of information that must be stored in the states of a FSR to generate a given sequence. We note that for a de Bruijn sequence of length $2^n - 1$ the span is exactly n. Chan and Games [7] have commented that in general, determining the span of an arbitrary sequence s appears to be difficult given the nonlinearities involved. Surprisingly, there are several efficient nonalgebraic algorithms for determining the span of a sequence s. Later in this paper we will prove the following theorem.

Theorem 1.2 Let $\mathbf{s} = (s_1, s_2, \ldots, s_n)$ where $s_i \in Z_p$ for $1 \leq i \leq n$. Then assuming the uniform distribution on Ω_p^n, the expected span $E(M^*\mathbf{s})$ of s is then

$$E(M^*(\mathbf{s})) = 2\log_p n + o(1).$$

$\square$

Similar results to this theorem have been proven by Arratia, Gordon and Waterman [3], Apostolico and Szpankowski [2], Jansen [17] and Maurer [25]. In this paper we will prove Theorem 1.2 by showing that $M^*(\mathbf{s})$ is equivalent to the height of a data structure known as a suffix tree [26]. From this equivalence it will also follow that $M^*(\mathbf{s})$ can be computed in $O(n)$ time for sequences of length n. Also from this characterization there are several other statistics of interest that can be computed which can be used to distinguish random from nonrandom sequences.

The results of theorems 1.1 and 1.2 indicate that the length of the shift register to generate a sequence is influenced dramatically by the degree of the feedback polynomial: the expected linear span of binary sequence of length n is roughly $\frac{n}{2}$, while the expected span for the same sequence is approximately $2\log n$. For example, if s is a binary sequence of length 10^6, then the expected linear span is approximately $500,000$ while the expected span is less than 50. The striking difference between the linear span and the span of a sequence can be accounted for by considering the space required to store the feedback polynomial $f(X)$. As noted for the linear case, the size of the feedback polynomial is bounded by the linear span itself. But for the case where $f(X)$ is unrestricted, the size of $f(X)$ may be an exponential function of the span. Thus the dramatic saving in memory cells for the shift register with arbitrary feedback is explained by encoding more information about the sequence into the feedback polynomial.

The paper is organized as follows. In §2.1 we review previous work on modeling sequence complexity with FSRs using nonlinear feedback. In §2.2 we introduce the suffix tree, give its relation to maximum order complexity, and present some asymptotic properties of suffix trees. In §3 we examine the expected size of a feedback polynomial, both theoretically and experimentally, and prove that the expected size of a feedback polynomial is exponential in the length of the FSR.

2 Computing the span of a sequence

2.1 Previous work

Chan and Games [7] have presented an algorithm for computing the quadratic span of a sequence. The algorithm determines the coefficients of the feedback polynomial by repeatedly solving systems of linear equations. A feedback polynomial $f(X)$ over the m indeterminates $x_1, x_2, \ldots, x_m$ is quadratic if it can be written in the form

$$f(X) = \sum_{i=1}^{m} \sum_{j=1}^{m} a_{i,j} x_i x_j \tag{3}$$

where $a_{i,j} \in Z_2$ for $1 \leq i, j \leq m$, $i \neq j$. For example, if $s = s_1, s_2, \ldots, s_8$ is a sequence of length 8, then the coefficients of the feedback polynomial for a shift register of length 3 are given as the solution to

$$
\begin{pmatrix}
s_1 & s_2 & s_1 s_2 & s_3 & s_1 s_3 & s_2 s_3 \\
s_2 & s_3 & s_2 s_3 & s_4 & s_2 s_4 & s_3 s_4 \\
s_3 & s_4 & s_3 s_4 & s_5 & s_3 s_5 & s_4 s_5 \\
s_4 & s_5 & s_4 s_5 & s_6 & s_4 s_6 & s_5 s_6 \\
s_5 & s_6 & s_5 s_6 & s_7 & s_5 s_7 & s_6 s_7
\end{pmatrix}
\begin{pmatrix}
a_1 \\
a_2 \\
a_{1,2} \\
a_3 \\
a_{1,3} \\
a_{2,3}
\end{pmatrix}
=
\begin{pmatrix}
s_4 \\
s_5 \\
s_6 \\
s_7 \\
s_8
\end{pmatrix}. \tag{4}
$$

If there is no solution to the linear system in eq. (4), then the quadratic span of s necessarily exceeds 3. The Chan and Games algorithm, as well as the Massey-Berlekamp algorithm, is an *on-line* algorithm, which means that the algorithm processes the sequence one character at a time (left to right) and computes the span for the portion of the sequence that has been read. Let the span of the first k characters of s be n_k, such that $f_k(X)$ is the current feedback polynomial. If $s_{k+1} \neq f(s_{k-n_k+1}, \ldots, s_k)$, then a *discrepancy* is said to occur. When a discrepancy occurs the length of the shift register may have to be incremented. For the linear span, a change in the current span only occurs if $n_k < \frac{k}{2}$, and if this is the case, then the new linear span is given as $(k+1) - n_k$ [28]. Part of the efficiency of the Massey-Berlekamp algorithm is the knowledge of the correct increment to be made when a discrepancy occurs. For the quadratic case, the appropriate increment to be made when a discrepancy occurs is not known, and must be found by searching an interval of possible increments. The Chan and Games algorithm is very general, and can in fact be used to compute $F_k(s)$ for $0 \leq k \leq n$. Perhaps for this reason, Games and Chan commented that determining the span of a sequence is a difficult problem, as their algorithm becomes less efficient for higher degree feedback polynomials.

A result similar to Theorem 1.2 was also presented in the thesis of Jansen [17]. Jansen has shown that $M^*(s)$ can be characterized as a property of the Directed Acyclic Word Graph (DAWG) [5] for a sequence s. The DAWG for a sequence s is a finite automaton that recognizes all substrings or subwords of s, or accepts the language $L_s = \{u \mid vuz = s\}$. For this reason, the DAWG is also called the subword automaton. It can be shown that if the longest path in the DAWG for s from the start vertex to a vertex of outdegree at least 2 is k, then $M^*(s) = k + 1$ [17]. While we may determine the span of s from the DAWG for s, there is no obvious way to determine any analytical information concerning the span from considering the DAWG since enumerating automatons is difficult in general. Jansen proved several of his results using statistical and combinatorial arguments which did not refer to the DAWG (see §3.4 of the thesis). Some properties of the DAWG are presented in [6, 12].

2.2   Pattern-matching algorithms

Pattern-matching algorithms are concerned with methods for finding and/or retrieving a pattern or substring w from a given piece of text Y [1]. If there are to be repeated searches on the text Y, then Y may be preprocessed and stored in a particular data structure to make searches of Y more efficient. Aho et al. [1] present a data structure which is useful to solve the following three pattern-matching problems: (a) given text Y and pattern w, find all occurrences of w in Y ; (b) given text Y, determine the longest repeated substring of Y; (c) given two texts Y_1 and Y_2, determine the longest string that is a substring of both Y_1 and Y_2. For this paper, it will be the solution of problem (b) that is of interest.

The data structure they presented is called a *position tree*, or later, a *suffix tree* [26]. Let $w = w_1, w_2, \ldots, w_n, \$$ be a string of length $n + 1$ such that $\$$ is a unique character that only appears in position $n + 1$. Then every position in w is uniquely identified by at least one substring of w, namely $w_i, w_{i+1}, \ldots, w_n, \$$, which are known as the *suffixes* of w. Let $v(i)$ be the shortest substring of w that uniquely identifies position i of w. Consider inserting $v(1), v(2), \ldots, v(n)$ into a tree $T(w)$, which is called the *suffix tree* for

w. The tree $T(\mathbf{w})$ will have n leaves, corresponding to the first n positions of **w**, and the edges of the tree are labeled so that the path from the root to the leaf representing position i is $v(i)$ for $1 \leq i \leq n$. The height of a suffix tree $h(T(\mathbf{w}))$ is the length of the longest path from the root to some leaf in the tree.

Example 2.1 The suffix tree for s = 010101110010$ is given in Figure 1. □

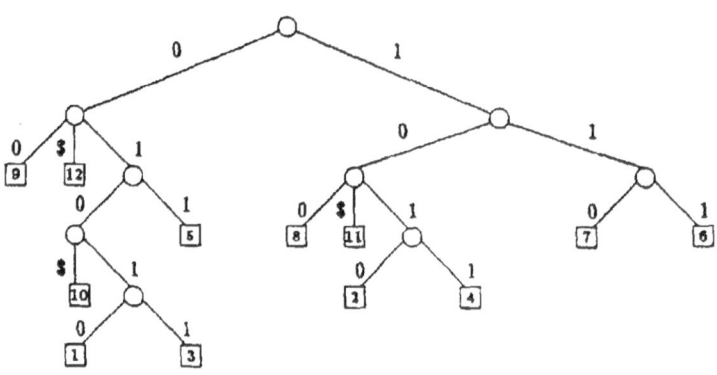

Figure 1: Suffix tree for 010101110010$

The span of a sequence s is greater than k when there exist two substrings $s^1 = s_i, s_{i+1}, \ldots, s_{i+k-1}$ and $s^2 = s_j, s_{j+1}, \ldots, s_{j+k-1}$ such that $s^1_{i+h} = s^2_{j+h}$, $0 \leq h < k$, and $s_{i+k} \neq s_{j+k}$. We will say that the substring $s' = s_i, s_{i+1}, \ldots, s_{i+k-1}$ occurs in s with *two different successor characters*. Thus no function $f(x_1, x_2, \ldots, x_k)$ of k variables can generate the sequence s since

$$f(s_i, s_{i+1}, \ldots, s_{i+k-1}) = s_{i+k} \neq s_{j+k} = f(s_j, s_{j+1}, \ldots, s_{j+k-1}). \tag{5}$$

For the string **w**, let LRS(**w**) be the longest repeated substring of **w**, and let |LRS(**w**)| be the length of the LRS of **w**. Then observe that $|\text{LRS}(\mathbf{w})| \leq h(T(\mathbf{w})) - 1$. From Figure 1, for s = 010101110010$, LRS(s) = 0101 and |LRS(s)| = 4. In this case $|\text{LRS}(\mathbf{w})| = h(T(\mathbf{w})) - 1$ which is not true in general. In fact, the difference between the span of a finite sequence and the length of its longest common substring can differ dramatically. Consider the $(n+1)$-bit string

$$s = 1\underbrace{0, 0, \ldots, 0}_{n-1}\$ \tag{6}$$

for which $|LRS(s)| = n - 2$, while the span $M(s) = 1$ (where we assume that the shift register does not have to generate \$ as it is only appended to terminate the suffixes). The span of a sequence is then the longest string s' that occurs in s with two different successor characters β_1, β_2 such that $\beta_1, \beta_2 \notin \{\$\}$. This relation is formally proved next.

Theorem 2.1 The span of s is equal to the longest path from the root of $T(s)$ to a leaf u, where the parent of u has at least two children whose edges are not labeled by \$.

Proof. Let the length of the path from the root to u be k, such that the path represents the substring $s_i, s_{i+1}, \ldots, s_{i+k-1}$ from s. Since the parent of u has two children whose edges are not labeled by the \$ character, then the substring $s_i, s_{i+1}, \ldots, s_{i+k-1}$ occurs twice in s with different successor characters, both distinct from \$. Further, for any $j, 1 \le j \le k - 1$, there is a substring of length j, namely $s_{i+j}, s_{i+j+1}, \ldots, s_{i+k-1}$, with different successor characters, distinct from \$. Thus no FSR of length shorter than k can generate s, and it follows that $M(s) \ge k$.

It remains to prove that $M(s) \le k$. Let u' be any leaf in $T(s)$ of maximum depth $k' > k$. Since $u' \ne u$, then the parent of u' must be an internal node of degree 2, with one child whose edge is labeled as \$. W.l.o.g, let u' be the child that has the edge labeled \$. If u' occurs in position i of s then $v(i) = s_i, s_{i+1}, \ldots, s_n, \$$. Thus if the character \$ is deleted from s, then the string $v'(i) = s_i, s_{i+1}, \ldots, s_n$ occurs at least twice in s, and either has a unique successor character, or no successor character. It follows that s can be generated with a FSR of length $k' - 1$.

By repeating the above argument for all leaves at depths $k' - 1, k' - 2, \ldots, k + 1$, it follows that s can be generated by a FSR of length less than $(k + 1) - 1 = k$. Thus $M^*(s) = k$ which completes the proof. □

In general we may assume that the height of the suffix tree is an upper bound on the maximum order complexity. The experiments of Jansen [17] suggest that this bound is tight in the expected case. The possible discrepancy between $|LRS(s)|$ and $M^*(s)$, as shown in eq. (6) and formalized in Theorem 2.1, is due to the fact that s is finite. A string s is said to be a semi-infinite string, or a *sistring* [12], if $s = s_1, s_2, s_3, \ldots$ and thus extends infinitely to the right. The ith suffix $v(i)$ of s is defined as the substring $v(i) = s_i, s_{i+1}, s_{i+2}, \ldots$, for $i \ge 1$, and a suffix tree $T(s^n)$ can be built from the first $n \ge 1$ suffixes of s. The analysis of suffix trees by Apostolico and Szpankowski [2] is done using sistrings but is directly adapted to the case of finite strings.

Recall that for a sequence s, the suffix tree $T(s)$ for s is a tree containing strings $v(1), v(2), \ldots, v(n)$, where $v(i)$ is the smallest string that identifies position i in s. The strings $v(1), v(2), \ldots, v(n)$ are not *independent* as they may overlap. The (search) tree that results from inserting n independent strings $v'(1), v'(2), \ldots, v'(n)$ is a digital search tree called a *trie*. The properties of digital search trees have been extensively studied [10, 20, 27, 30], and in particular, the expected height of a random trie built with n strings drawn from an alphabet with p symbols is $2\log_p n + o(\log n)$. The properties of $T(s^a)$ are similar (asymptotically equivalent) to the properties of a suffix tree built using a string of length n [16].

The structure of a suffix tree depends upon the overlaps that exist between the given suffixes of a string. For suffixes $v(i)$ and $v(j)$, $1 \le i \ne j \le n$, let V_{ij} be defined as the length of the longest prefix that is common to both $v(i)$ and $v(j)$. Then we may define the following statistics:

$$H_n = \max_{1 \le i,j \le n} V_{ij} \tag{7}$$

$$h_n = \min_{1 \le i \le n} \max_{1 \le j \le n} V_{ij} \tag{8}$$

$$D_n = \sum_{i=1}^{n} \frac{\max_{1 \le j \le n} V_{ij}}{n}. \tag{9}$$

Then H_n is called the nth-height, which corresponds to the height of a suffix tree built on n suffixes; h_n is called the nth-shallowness, and D_n is the average depth of the suffix tree. We note that each of these quantities can be computed directly as the suffix tree for a sequence can be constructed in $O(n)$ time using linear $O(n)$ space [26]. Further assume that the characters of s are drawn randomly and independently from Z_p, such that the jth character of s is p_i with probability q_i for $0 \le i \le p - 1$. If $q_i = p^{-1}$ then the strings occur with uniform probability, which is known as the Bernoulli model [2].

Recall that $f(n) \sim g(n)$ if and only if $f(n) = g(n) + o(1)$.

Theorem 2.2 (Apostolico and Szpankowski [2]) Assuming the Bernoulli model, for large n, $E(H_n) \sim 2\log_p n$, $E(h_n) \sim \log_p n$, $E(D_n) \sim \log_p n$. ◻

For binary sequences Jansen [17] has computed the span of all sequences up to length 22, and the empirical average was close to $2\log_2 n$, which agrees with $E(H_n)$. Other estimates of $E(H_n)$ have also been obtained, for example [3], but the derivation from the suffix tree seems the most robust. In fact, Apostolico and Szpankowski [2] are still able to compute the expected values of H_n, h_n and D_n when the Bernoulli model is not assumed. For example, when the q_i are not equal, expected value of H_n becomes

$$E(H_n) = \frac{2}{\ln q_{max}^{-1}} \ln n + c \tag{10}$$

where ln is the natural logarithm, q_{max} the maximum of the q_i, and c a constant.

3 Properties of feedback polynomials

In this section we will restrict ourselves to considering binary sequences, because such sequences are of the most practical importance. Let the feedback polynomial of a FSR be $f(X)$. It then follows that $f(X) \in Z_2[x_1, x_2, \ldots, x_m]$, which is the set of all polynomial expressions involving the indeterminates $x_1, x_2, \ldots, x_m$. The polynomials of $Z_2[x_1, x_2, \ldots, x_m]$ correspond to boolean functions $f : \{0, 1\}^m \to \{0, 1\}$. Conversely, any m-bit boolean function f can be represented as a polynomial $Q_f(X) \in Z_2[x_1, x_2, \ldots, x_m]$, which is known as the Ring Sum Expansion (RSE) [29] or Algebraic Normal Form (ANF) [28] of f. Let $Z_2^A[x_1, x_2, \ldots, x_m]$ be the set of 2^{2^m} ANF polynomials, which are exactly those polynomials that remain in $Z_2[x_1, x_2, \ldots, x_m]$ when the elements of the ring are are simplified according to $x_i = x_i^2$, $1 \leq i \leq m$.

We are interested in two quantities associated with feedback polynomials: the size, or number of terms, and the degree. For $f(X) \in Z_2^A[x_1, x_2, \ldots, x_m]$, let $T(f(X))$ be the number of terms in $f(X)$, and let $\deg(f(X))$ be the degree of $f(X)$. Then the space required to store $f(X)$ is bounded by $n \cdot T(f(X))$.

Theorem 3.1 Assuming the uniform distribution on $Z_2^A[x_1, x_2, \ldots, x_m]$

$$E(\deg(f, m)) = m - \frac{1}{2} + O\left(\frac{1}{2^m}\right)$$

Proof. Let $B(m, k) = \sum_{0 \leq j \leq k} \binom{m}{k}$ be the sum of the first $k + 1$ binomial coefficients, $0 \leq k \leq m$. Then

$$
\begin{aligned}
E(\deg(f, m)) &= \sum_{0 \leq i \leq m} i \cdot \Pr(Q_f(X) \text{ has degree } i) \\
&= \sum_{0 \leq i \leq m} i \cdot \frac{2^{B(m, i-1)} \cdot (2^{\binom{m}{i}} - 1)}{2^{2^m}} \\
&= 2^{-2^m} \cdot \left[\sum_{1 \leq i \leq m} i \cdot 2^{B(m, i)} - i \cdot 2^{B(m, i-1)} \right] \\
&= 2^{-2^m} \cdot \left[m \cdot 2^{2^m} - 2^{2^m - 1} - \sum_{0 \leq i \leq m-2} 2^{B(m, i)} \right] \\
&= m - \frac{1}{2} + O\left(\frac{1}{2^m}\right).
\end{aligned}
$$

$\square$

Thus we expect all taps of the FSR to be involved in the feedback polynomial. Further it follows from the binomial theorem that $E(T(f(X))) = 2^{m-1}$. Thus a random polynomial in $Z_2^A[x_1, x_2, \ldots, x_m]$ has a number of terms exponential in m, and a degree approximately m. Again consider the example of computing the linear span and span of of sequence of length $n = 10^6$. The total storage for the LFSR is approximately 750,000 bits, assuming half the coefficients in $f(X)$ are nonzero. Then from Theorem 1.2, the smallest shift register will have approximately $2 \log n$ stages, and from Theorem 3.1, assuming that the feedback polynomial is drawn at random from $Z_2^A[x_1, x_2, \ldots, x_{2\log n}]$, $f(X)$ will have $2^{2\log n - 1} > 2^{39}$ terms, which is far larger than the length of the sequence!

We note that the feedback polynomial to generate a given sequence is not unique in general. Let the span of a sequence s be m, and let $f(X)$ be a feedback polynomial. There are 2^m m-bit strings, and let $k, 1 \leq k \leq 2^m$, of these strings occur as substrings in s. It then follows that there are in fact $2^{2^m - k}$ polynomials $g(X)$ that agree with $f(X)$ on the k input strings found in s. We may still speak of *the* feedback polynomial $f(X)$ associated with every sequence s by defining $f(X)$ to be the polynomial which generates s, and for which $f(X) + 1$ has the smallest number of roots.

3.1 Experimental results

For this experiment we generated random bit sequences using the C library function *srandom*. The span for each sequence was calculated by comparing suffixes and a feedback function generated for the sequence. The feedback function was simplified to give a count of the number of terms using MAPLE™. This was applied to 100,000 32-bit sequences. The mean span was 8.65 with a standard deviation of 1.59. The following table shows the number of terms in the feedback function for different length spans.

4 Conclusion

We have shown that the suffix tree provides an alternate characterization of the maximum order complexity of a sequence. With this characterization it is then possible to accurately determine distribution of the maximum order complexity of a sequence, and also several other statistics. Also we have given some evidence to indicate that the linear complexity of a sequence is expected to be less than the maximum order complexity when both the length of the shift register and the size of the feedback polynomial are taken into account. One open problem is then to determine that order k of a feedback polynomial for which $F_k(s) + f(X)$ is minimized for sequences s of length n.

Span	# of cases	Mean # of terms	Std. Dev.
5	11	15.82	3.24
6	2871	30.30	4.84
7	20917	53.86	11.80
8	29822	92.47	11.80
9	22228	154.25	66.08
10	12485	251.68	140.59
11	6328	415.13	292.70
12	2887	666.99	599.33
13	1339	1078.58	1103.69
14	638	1713.08	2199.19
15	256	2679.77	4304.51
16	109	4335.17	9380.74
17	60	14028.02	32298.27
18	30	24725.47	64266.20
19	11	10025.73	14349.16
20	5	12698.00	11095.74
21	0	0.00	0.00
22	2	560.00	464.00
23	1	874.00	0.00

Table 1: Number of terms for different spans

ACKNOWLEDGEMENTS

We would like to thank Alfredo Viola, Richard Games, and the Eurocrypt referees for their comments on earlier versions of the paper.

References

[1] A. V. Aho, J. E. Hopcroft, and J. D. Ullman. *The Design and Analysis of Computer Algorithms.* Addison–Wesley Publishing Company, 1974.

[2] A. Apostolico and W. Szpankowski. Self-alignments in words and their applications. Technical Report CDS-TR-732, Purdue University, 1987.

[3] R. Arratia, L. Gordon, and M. Waterman. An extreme value theory for sequence matching. *The Annals of Statistics*, 14(3):971–993, 1986.

[4] H. Beker and F. Piper. *Cipher Systems.* Wiley, 1982.

[5] A. Blumer, J. Blumer, A. Ehrenfeucht, D. Haussler, and R McConnell. Linear size finite automata for the set of all subwords of a word: outline of results. *Bulletin of the European Association of Theoretical Computer Science*, 21:12–20, 1983.

[6] A. Blumer, E. Ehrenfeucht, and D. Haussler. Average sizes of suffix trees and DAWGs. *Discrete Applied Mathematics*, 24:37–45, 1989.

[7] A. Chan and R. Games. On the quadratic spans of periodic sequences. *IEEE Transactions on Information Theory*, IT-36(4):822–829, 1990.

[8] N. G. de Bruijn. A combinatorial problem. *Nederl. Akad. Wetensch. Proc*, 49:754–758, 1946.

[9] D. E. R. Denning. *Cryptography and Data Security.* Addison–Wesley Publishing Company, 1982.

[10] L. Devroye. A probabilistic analysis of the height of tries and of the complexity if triesort. *Acta Informatica*, 21:229–232, 1984.

[11] S. Golumb. *Shift Register Sequences.* Aegean Park Press, 1982.

[12] G. H. Gonnet and R. Baeza-Yates. *Handbook of Algorithms and Data Structures.* Addison-Wesley, Second Edition, 1991.

[13] E. J. Growth. Generation of binary sequences with controllable complexity. *IEEE Transactions on Information Theory*, 17(3):288–296, 1971.

[14] H. Gustafson, E. Dawson, and W. Caellie. Comparison of block ciphers. *Advances in Cryptology, AUSTCRYPT 90, Lecture Notes in Computer Science, vol. 453, J. Seberry and J. Piepryzk eds., Springer-Verlag*, pages 208–220, 1990.

[15] J. Hopcroft and J. Ullman. *An introduction to automata, languages and computation*. Addison–Wesley Publishing Company, 1979.

[16] P. Jacquet and W. Szpankowski. Autocorrelation on words and its applications: analysis of suffix trees by string-ruler approach. preprint, 1990.

[17] C. J. A. Jansen. *Investigations on Nonlinear Streamcipher systems: Construction and Evaluation methods*. PhD thesis, Philips, USFA BV, 1989.

[18] C. J. A. Jansen and D. Boekee. The shortest feedbck shift register that can generate a sequence. *Advances in Cryptology, CRYPTO 89, Lecture Notes in Computer Science, vol. 218, G. Brassard ed., Springer-Verlag*, pages 90–99, 1990.

[19] E. L. Key. An analysis of the structure and complexity of nonlinear binary sequence generators. *IEEE Transactions on Information Theory*, 22(6):732–736, 1976.

[20] D. E. Knuth. *The Art of Computer Programming : Volume 3, Sorting and Searching*. Addsion Wesley, 1973.

[21] D. E. Knuth. *The Art of Computer Programming : Volume 2, Seminumerical Algorithms*. Addsion Wesley, 1981.

[22] A. N. Kolmogorov. Three approaches to the quantitative definition of definition. *Problems in Information Transmission*, 1(1):1–7, 1965.

[23] M. Li and P. M. B. Vitanyi. Two decades of applied Kolmogorov complexity. Technical Report CS-R8813, Centre for Mathematics and Computer Science, April, 1988.

[24] J. L. Massey. Shift-register synthesis and BCH decoding. *IEEE Transactions on Information Theory*, 15:122–127, 1969.

[25] U. M Maurer. Asymptotically tight bounds on the number of cycles in generalized de Bruijn-Good graphs. to appear in Discrete Applied Mathematics.

[26] E. M. McCreight. A space-economical suffix tree construction algorithm. *Journal of the ACM*, 23(2):262–272, 1976.

[27] M. Regnier. On the average height of trees in digital search and dynamic hashing. *Information Processing Letters*, 13(2):64–66, 1981.

[28] R. A. Rueppel. *Design and Analysis of Stream Ciphers*. Springer–Verlag, 1986.

[29] J. Savage. *The Complexity of Computing*. John Wiley, 1976.

[30] W. Szpankowski. On the analysis of the average height of a digital search tree: another approach. Technical Report CSD-TR-646, Purdue University, 1986.

Attacks on Protocols for Server-Aided RSA Computation

Birgit Pfitzmann

Institut für Informatik
Universität Hildesheim
Marienburger Platz 22
W-3200 Hildesheim, FRG
pfitzb@informatik.uni-hildesheim.de

Michael Waidner

Institut für Rechnerentwurf
und Fehlertoleranz
Universität Karlsruhe, Postfach 6980
W-7500 Karlsruhe 1, FRG
waidner@ira.uka.de

Abstract. On Crypto '88, Matsumoto, Kato, and Imai presented protocols to speed up secret computations with insecure auxiliary devices. The two most important protocols enable a smart card to compute the secret RSA operation faster with the help of a server that is not necessarily trusted by the card holder.

It was stated that if RSA is secure, the protocols could only be broken by exhaustive search in certain spaces. Our main attacks show that much smaller search spaces suffice. These attacks are passive and therefore undetectable.

It was already known that one of the protocols is vulnerable to active attacks. We show that this holds for the other protocol, too. More importantly, we show that our attack may still work if the smart card checks the correctness of the result; this was previously believed to be an easy measure excluding all active attacks.

Finally, we discuss attacks on related protocols.

1   Introduction

1.1   The Model

Smart cards are often considered as appropriate for carrying out secret cryptographic computations for individual owners. ISO standards for smart cards, however, emphasize flexibility in the physical sense more than flexibility regarding computations. Only rather small chips can therefore be used. Hence, the computing abilities are limited. In particular, at least if the smart card is equipped with a general-purpose CPU, the speed does not suffice for asymmetric algorithms, such as signing a message with RSA [RSA_78]. (Special-purpose designs exist nowadays [QuWB_91, WaQu_91]. Nevertheless, the protocols considered in the following are still to be used in practical systems [KaSh_90].)

In most applications, the smart card communicates directly with a device with much larger computing abilities, such as a point-of-sale terminal. Such a device will be called a *server* in the following. The basic idea of [MaKI_90] (and previous Japanese publications by the same authors) was to use the computing power of the server to help the smart card. This is complicated by the fact that the owner of the smart card need not trust the server.

The question whether an untrusted server can help a less powerful device with a secret computation can be seen as a general theoretical problem, too. A similar problem has been considered in [Feig_86, AbFK_89]. There, the server has unrestricted computational power and needs this power even in the correct protocol. In contrast, here the server is restricted to

R.A. Rueppel (Ed.): Advances in Cryptology - EUROCRYPT '92, LNCS 658, pp. 153-162, 1993.

polynomial-time computations both in the correct protocol and in the attacks. (Otherwise, it could break RSA anyway.)

1.2 Overview over Protocols and Attacks

The main existing server-aided protocols enable a smart card to compute RSA signatures faster with the help of the server. They all share the same basic structure. In Section 2, we describe the first two protocols, RSA-S1 and RSA-S2 from [MaKI_90], and sketch the remaining ones [MaKI_90, QuSo_91, LaYH_91]. The performance of some protocols has been further considered in [KaSh_90].

In [MaKI_90], active attacks are not considered, and it is claimed that if RSA is secure, the best possible passive attacks are brute force search in certain search spaces. We will show that this is not correct, and that much smaller search spaces are sufficient. These attacks can be countered by increasing the system parameters. However, one must carefully consider at what point one loses the advantage over direct computation.

The attacks also work for all protocol variants except that from [QuSo_91], which is provably secure against passive attacks if RSA is secure. However, its drawback is large communication overhead, which makes it impractical for smart cards with a standard interface.

It had already been noticed that the scheme RSA-S2 is vulnerable to active attacks, see a remark in [QuSo_91] and a complete description in [ShKa_90].[1] In both cases, it is proposed that the smart card should check the result, i.e., the RSA signature, before it outputs it to the server. This is possible if the public RSA exponent is small. It is claimed in [ShKa_90] that this countermeasure excludes all possible active attacks. However, we establish a new active attack that requires stronger countermeasures. Additionally, the new attack also works for RSA-S1 and the protocol from [QuSo_91].

We also show that another protocol from [MaKI_90], used to solve modular equations, is not secure. Finally, we make some remarks about another protocol in [QuSo_91].

Note that server-aided protocols for testing RSA signatures with small public exponent exist, too [QuSo_91, Bos_92].

1.3 Other Security Considerations

Note that in the example of ISO standard smart cards, the basic assumptions of the protocols considered here are a little inconsistent: Such smart cards have neither a keyboard nor a display. Consequently, the owner enters secret data, such as a PIN, via the server, which is considered as insecure, and the owner cannot check that the correct message is signed. Thus, no real security can be achieved in this scenario. (And if one deviates from the ISO standards by adding keyboards and displays, one can also use devices with more computational power, e.g., [PrCh_89, BaEi_90].)

---

[1] Two further references containing active attacks and countermeasures, which we unfortunately cannot read, are: Tsutomu Matsumoto, Hideki Imai: How to Ask and Verify Oracles for Speeding Up Secret Computations, Part 1 and 2 (in Japanese); IEICE Technical Reports (Institute of Electronics, Information and Communication Engineers) 89/45 (1989) 21-28, ISEC89-4, and 89/145 (1989) 13-20, IT89-24.

Remark: It has been argued that one might build in a display without deviating from the standard otherwise, and solve the PIN problem with measures such as in [Malm_91]. Although this approach is interesting (in particular against the threat that someone is watched while typing a PIN), its effectiveness against repeated use of a smart card with the same server, with parameter sizes acceptable for human users, remains to be shown.

2 The RSA Protocols

The smart card wants to compute $y = x^d \bmod n$, where n is the product of two large primes p and q and d is a secret exponent. Both basic schemes, RSA-S1 and RSA-S2, have two parameters, M and L.

2.1 RSA-S1

As a precomputation for all following signatures, the smart card (or a trusted larger device) breaks down the secret exponent d: It randomly generates an integer vector $D = (d_1,..., d_M)$ and a binary vector $F = (f_1,..., f_M)$ with

$$d = f_1 d_1 + ... + f_M d_M \bmod \lambda(n) \tag{1}$$

and $Weight(F) \leq L$. Here, $Weight$ denotes the Hamming weight. The computation of the signature on an actual message x proceeds as follows:

1. The smart card sends n, D, and x to the server.
2. For $i := 1,..., M$, the server computes

$$z_i := x^{d_i} \bmod n$$

and sends $Z := (z_1,..., z_M)$ back to the smart card.
3. The smart cards obtains y by multiplying the z_i's for which $f_i = 1$.

Thus, the smart card only needs $L - 1$ multiplications, and the communication is $2(M + 1)$ numbers.

2.2 RSA-S2

RSA-S2 is to improve on RSA-S1 by use of the Chinese remainder theorem, similar to [QuCo_82]. As a precomputation, the smart card breaks down d as

$$d = f_1 d_1 + ... + f_M d_M \bmod (p-1)$$
and
$$d = g_1 d_1 + ... + g_M d_M \bmod (q-1),$$

where $D = (d_1,..., d_M)$ is an integer vector again, and $F = (f_1,..., f_M)$ and $G = (g_1,..., g_M)$ are binary vectors with $Weight(F) + Weight(G) \leq L$.

Steps 1 and 2 are just like in RSA-S1.

In Step 3, the smart card obtains $y_p := y \bmod p$ as the product of the z_i's for which $f_i = 1$, and $y_q := y \bmod q$ as the product of the z_i's for which $g_i = 1$. Finally, it applies the Chinese remainder theorem.

2.3 Other Variants

First, one can also use the Chinese remainder theorem to speed up RSA-S1: The computation is exactly like that in RSA-S2 when F and G are equal. However, the security considerations are different for the two versions. Anyway, since the difference to RSA-S1 is only in the local computations of the smart card, the scheme is just as secure as RSA-S1.

So-called non-binary variants of RSA-S1 and RSA-S2 are obtained if the coefficients f_i and g_i may have other values than just 0 and 1 [MaKI_90, KaSh_90]. Of course, the values must still be very small integers so that the smart card needs just a few multiplications.

The remaining protocols vary the choice of the set of exponents, D:

In [LaYH_91], D is chosen so that the server can compute all the powers $x^{d_i} \bmod n$ more easily with one addition chain. (This is the second proposal in that paper; the first one only changes the local computation of the server.)

Last but not least, the only variant which makes a real security difference is that from [QuSo_91]: There, one and the same fixed set D is used for all smart cards. This scheme is obviously secure against passive attacks, i.e., attacks where the server always sends back correct data: The only information that the smart card gives the server is n, x, and the signature, just as if it computed the signature alone. Unfortunately, a set D which allows every possible secret exponent d to be expressed as in Formula (1) with a vector F of small weight is larger than special sets for special exponents. This increases the communication overhead. In the example in [QuSo_91], $|D| = 832$ for 512-bit exponents. With a standard ISO interface, i.e., 9600 bit/s, the communication would take more than 40 seconds. Hence, in spite of its security advantage, this variant cannot be used in several applications, and it still makes sense to consider the other variants further.

3   Passive Attacks

By "passive attacks" we mean attacks where the server never deviates from its protocol, i.e., it sends back the correct powers $x^{d_i} \bmod n$. Hence, no measures to prevent passive attacks are possible.

3.1   Passive Attack on RSA-S1

It is stated in [MaKI_90] that RSA-S1 could only be broken by searching the true value d via the exhaustion of

$$\sum_{i=1}^{L} \binom{M}{i} \tag{2}$$

possibilities. (That is, all the possible vectors F of weight $\leq L$.)

The following attack shows that the search space can be considerably smaller:

(1) For a message x where the signature $y = x^d$ is known, one first computes all the values $z_i = x^{d_i}$.

(2) Next, one computes all the products

$$y_F := \prod_{i=1}^{M} z_i^{f_i} \bmod n$$

for vectors F of only half the weight, i.e., with $Weight(F) \leq \lceil L/2 \rceil$.

(3) One also computes the values

$$y^*_F := y \, y_F^{-1} \bmod n$$

for the same vectors F.

(4) The values y_F are sorted, and the values y^*_F compared to them. Whenever a match is found, i.e., $y_{F_1} = y^*_{F_2}$, and F_1 and F_2 are disjoint (i.e., they do not have a 1 at the same position), the vector $F_1 + F_2$ is a candidate for the true secret vector F.

(5) If F is not uniquely determined by this procedure, one can test the remaining candidates by use of a second message x'.

The reason why this works is that the true vector F can be represented as the sum $F_1 + F_2$ of two disjoint vectors of weight $\leq \lceil L/2 \rceil$. The equation for the signature is

$$y = \prod_{i=1}^{M} z_i{}^{f_i} = \prod_{i=1}^{M} z_i{}^{f_{1,i} + f_{2,i}} = y_{F_1} y_{F_2} \bmod n.$$

Thus
$$y_{F_1} = y\, y_{F_2}{}^{-1} = y^*_{F_2} \bmod n.$$

Complexity. The number of values y_F is

$$N := \sum_{i=0}^{\lceil L/2 \rceil} \binom{M}{i},$$

and they can be computed with little more than one multiplication on average (if one stores intermediate results). The same holds for the values y^*_F, if one starts by computing the values $z_i{}^{-1}$. Sorting and searching take about $N \log(N)$ operations.

All this is considerably less than the number in Formula (2) (a bit larger than the square root).

3.2 Passive Attack on RSA-S2

It is said in [MaKI_90] that RSA-S2 could only be broken by searching the true value d via the exhaustion of

$$\sum_{j=1}^{L} \sum_{i=0}^{j} \binom{M}{i}\binom{M}{j-i} > \binom{M}{L/2}^2 \tag{3}$$

possibilities. (That is, all combinations of vectors F and G of total weight $\leq L$.) How two vectors are actually checked for correctness is described in [KaSh_90].

Again, we can reduce the search space considerably: One can search for one of the vectors F or G individually. Clearly, one of them must be of weight $\leq L/2$.

(1) For a message x where the signature $y = x^d$ is known, one first computes all the values $z_i = x^{d_i}$.

(2) Next, one computes all the products

$$y_F := \prod_{i=1}^{M} z_i{}^{f_i} \bmod n$$

for vectors F with $Weight(F) \leq L/2$.

(3) One also computes the values $v_F := \gcd(y_F - y, n)$. If one of them is neither 1 nor n, one has factored the modulus.

(4) Otherwise, at least one value v_F (and usually not many) will be n, and either the true F or the true G must be among the corresponding vectors. Normally, one will find $F = G$ in this case.

The reason why this works is that for the true F, $y_F = y_p = y \bmod p$, and for the true G, $y_G = y_q = y \bmod q$. Thus, not all the values v_F can be 1. Furthermore, except when $F = G$, y_F will usually not be congruent $y \bmod q$, too, and vice versa. In these cases, one can factor the modulus according to Step (3). Otherwise, one has usually found both F and G. In the remaining cases, a very small search space will remain, and anyway, it is enough for an attack to succeed in most cases.

Complexity. The number of values y_F is

$$N := \sum_{i=0}^{L/2} \binom{M}{i},$$

and, as above, they can be computed with a bit more than one multiplication on average. Since the greatest common divisor takes more time, one should initially compute it for the product of several differences $(y_F - y)$.

Again, the complexity is not much more than the square root of that described in Formula (3).

3.3   Passive Attacks on Other Variants

The attacks described easily generalize to the non-binary variants of RSA-S1 and RSA-S2. Furthermore, the protocol from [LaYH_91] is just a special case of the original protocols as far as security is concerned. Hence the attacks work for all known variants except the provably secure one from [QuSo_91].

4   Active Attacks

In active attacks, the server deviates from its protocol by sending back a wrong vector $Z' = (z'_1, \ldots, z'_M)$, instead of the powers $z_i = x^{d_i} \bmod n$. On the one hand, this makes active attacks more powerful, and they usually result in a total break of the system in a few steps. On the other hand, active attacks can often be detected or even prevented, in contrast to passive ones. Hence, the most important thing to know about active attacks is not how they work, but whether one exists and which countermeasures are effective.

4.1   Description of Attacks

Attacks on RSA-S1. As mentioned, we describe the first active attack on RSA-S1. The basic idea is to use Jacobi symbols: If the server sends back any vector $Z' = (z'_1, \ldots, z'_M)$, then the smart card outputs

$$y' := \prod_{i=1}^{M} z'_i{}^{f_i} \bmod n.$$

The server can compute Jacobi symbols modulo n. Let $(-1)^c$ be that of y', and let I be the subset of indices i where the Jacobi symbols of z'_i are -1. Since Jacobi symbols are multiplicative, the server knows

$$(-1)^c = \left(\frac{y'}{n}\right) = \prod_{i=1}^{M} \left(\frac{z'_i}{n}\right)^{f_i} = \prod_{i \in I} (-1)^{f_i}.$$

This yields

$$c = \sum_{i \in I} f_i \bmod 2.$$

Thus from each vector Z', the server obtains one linear equation about F in GF(2). As long as the server sends back correct data, this does not matter much, since it always obtains the same equation. However, in an active attack, it can choose just one z'_i with Jacobi symbol -1, and the rest with Jacobi symbol $+1$. Then it obtains the value of f_i directly. Thus, after M rounds of this attack, the server knows F and therefore d.

If L is much smaller than M, the server needs less than M rounds if it chooses the vectors Z so that the resulting linear equations form the parity check matrix of a code correcting up to L errors.

If the server is not likely to meet the same smart card often enough, it can also use the information obtained from some rounds of the active attack to speed up the passive attack.

A far more elegant attack has been found recently [Ande_92]: The server sends back a selection of small primes (or a blinded version thereof), factors the product that the smart card outputs (which is $< n$ if M is not too large) by trial division, and thus finds out F. However (see Section 4.2), this elegant attack is easily prevented by countermeasures that were previously proposed for other protocol variants, and which are needed against our attack, too, whereas our attack needs additional countermeasures. Hence, in practice, the additional elegance makes no difference, and the less elegant attack is even more dangerous.

The Attack on RSA-S2. The attack on RSA-S2 from [ShKa_90] uses the following fact: If the server changes the sign of just one of the values z_i, and if $f_i \neq g_i$, then the resulting value y' and the true signature y are significantly different square roots of the same value. The server does not know y. However, since the public exponent e is odd, the same holds for the values y'^e and $y^e = M$. In this case, the server can therefore factor the modulus by computing $\gcd(y'^e - y^e, n)$.

4.2 Discussion of Countermeasures

Why No Active Attack Can Be Ignored. The attitude towards active attacks in [MaKI_90] was to assume that the server would refrain from them. However, such an assumption about an untrusted server is unjustified unless such an attack would at least entail a severe risk. One risk might be that the owner of the smart card notices that the smart card outputs a wrong signature. However, if the owner obtains the signature at all, it obtains it through the server, and all the attacks described above have a variant where the server can output the correct signature.

With the one-round attacks [ShKa_90, Ande_92], this is clear since the server obtains the secret key at once. With the Jacobi symbol attack, the server chooses a with Jacobi symbol -1 modulo n. It first computes the correct vector Z and its Jacobi symbols. Where it wants the Jacobi symbol changed, it uses

$$z'_i := a \, z_i, \text{ otherwise } z'_i := z_i.$$

Thus the smart card's output y' is the product of the real signature y and a value a^x, where $x \leq L$ and $x \leq$ the number of factors a used. The server can find out the correct signature by searching among the few numbers $y' \, a^{-x}$. If the Jacobi symbol of -1 modulo n is -1, one would use $a = -1$. Then the signature y is $\pm y'$.

Hence each of the attacks implies that countermeasures in the smart card itself are needed.

How Much Does Signature Checking Help? The easiest countermeasure is that the smart card tests the resulting signature and only outputs it if it is correct [QuSo_91]. This restricts the protocol to RSA with small public exponents. It also means additional computation; however, this may be small compared to the L multiplications needed anyway.

In fact, this measure effectively excludes all one-round attacks, such as [ShKa_90, Ande_92].

It was even said in [ShKa_90] that it excludes all conceivable active attacks. However, multi-round attacks are still possible.

For instance, if the server chooses just one z'_i wrong each time, like in the Jacobi symbol attack, it can infer f_i from whether it receives a signature or not. By choosing all the values z'_i as $\pm z_i$, the server can obtain any linear equation about F again. Hence also the variant with error-correcting codes still works.

With RSA-S2, choosing one z'_i wrong reveals $(f_i \vee g_i)$, and changing more than one sign reveals different information.

Other Countermeasures. The only general countermeasure that would certainly exclude all active attacks would be to check the values sent by the server, instead of the result. However, this looks infeasible.

The next-best general solution (from a theoretical point of view) is that the smart card stops once and for all if it detects an attack. However, this may have practical disadvantages. In particular, since the smart card does not have a display, the client will only notice this fact during the next transaction with an honest server.

Instead, one could let the smart card continue and try to issue a warning through the next server it communicates with. However, this seems quite dangerous since someone might do all their shopping in one supermarket for quite a while.

In the special protocols considered here, an easier measure might be to change the vectors D and F with each signature (in addition to checking the result). In this case, the smart card must be able to generate random numbers quickly, and the procedure of breaking down d must definitely be implemented on the smart card itself. Then just one bit of information can be obtained about each F; this corresponds to a passive attack with slightly smaller parameters L and M. However, it is not clear if having several vectors D allows new passive attacks. Anyway, this measure seems impossible with the protocol from [QuSo_91].

5   Attacks on Related Protocols

There are two more types of server-aided secret computations in [MaKI_90].

The first type is matrix multiplication, where the matrices are to be kept secret. In the protocol, the server receives versions of the matrices where the rows and columns are permuted. It has been noticed in [MaKI_90] itself that quite a lot of information, such as the determinants, is not hidden by these operations. Thus this protocol does not provide secrecy in the cryptographic sense.

The second type is solving modular equations. The smart card has secret integers $a_0, \ldots, a_{m-1}$ and an integer k. (k was declared secret, too, but that must have been a slip of the pen.) It wants to know a solution x to the equation

$$a_0 + a_1 x + \ldots + a_{m-1} x^{m-1} + x^m = 0 \bmod k.$$

For this, it chooses a random number r and computes $b_{m-i} := a_{m-i} r^i$ for $i := 1, \ldots, m$, and sends k and the tuple $B = (b_0, b_1, \ldots, b_{m-1})$ to the server. The server computes a solution y to the equation with the coefficients B. The smart card can compute x from y as $x = y \, r^{-1} \bmod k$. This protocol is said to reveal nothing about the secret to the server. It does, however. For example, if $\{x_1, \ldots, x_u\}$ is the complete set of zeroes of the original polynomial, the server can compute $\{r \, x_1, \ldots, r \, x_u\}$ and therefore all the quotients $x_i \, x_j^{-1}$. As more easily computable functions of the original secret coefficients, the server can compute $a_{m-1}{}^i \, a_{m-i}{}^{-1} = b_{m-1}{}^i \, b_{m-i}{}^{-1}$.

In [QuSo_91], a special protocol for deciphering RSA-encrypted messages is also contained. First, this protocol assumes that the modulus n is secret from the server, which seems a rather strange assumption to make with a public-key cryptosystem. Secondly, if the server receives the decrypted messages (which is not as natural as that the server obtains the signatures, though), this scheme is vulnerable to active attacks, too: The smart card has chosen two additional primes r_1, r_2 and computed $n_1 := p r_1$ and $n_2 := q r_2$. It has also blinded the secret exponent as $\sigma_1 := d_1 + \rho_1(p-1)$ and $\sigma_2 := d_2 + \rho_2(q-1)$, where d_1 and d_2 must be the reductions of d modulo $p-1$ and $q-1$, resp., and $\rho_1 \leq p-2$ and $\rho_2 \leq q-2$ are random numbers. Now, together with a ciphertext C, it sends n_1, n_2, σ_1, and σ_2 to the server. The server should answer with $M_1 := C^{\sigma_1} \bmod n_1$ and $M_2 := C^{\sigma_2} \bmod n_2$. The smart card computes the message M by applying the Chinese remainder theorem to $M_1 \bmod p$ and $M_2 \bmod q$. If the server gives back the same value M_1 in two different protocol executions and receives the two results M and M', then $M - M'$ is a multiple of p. Thus, with high probability, $\gcd(n_1, M - M') = p$.

6 Conclusion

We have described several attacks on server-aided computation protocols, in particular, protocols for the computation of RSA signatures. Several of these attacks were previously declared impossible. None of the attacks on the signature protocols is disastrous, i.e., they can all be rendered ineffective by increasing the parameters or by performing additional tests. However, all these countermeasures cost time and may therefore annihilate the advantages of the server-aided approach.

Furthermore, we are by no means sure that our passive attacks are already optimal. In particular, one could try to exploit the obvious connection to modular knapsacks. (In contrast, better active attacks are of no practical importance, since the countermeasures needed so far exclude active attacks generally.)

Thus, like with all unproven cryptographic schemes, one should let server-aided computation undergo a lengthy evaluation phase. In this special case, the result is likely to be that by the time that the schemes are sufficiently well evaluated, smart cards that can compute RSA on their own are available for everyone. Other applications for these or similar protocols are not inconceivable, but one trades computation for communication.

Acknowledgement

We would like to thank Prof. Tsutomu Matsumoto for helping us to more literature about this subject, and Ross Anderson for an interesting discussion.

References

AbFK_89   Martin Abadi, Joan Feigenbaum, Joe Kilian: On Hiding Information from an Oracle; Journal of Computer and System Sciences 39/1 (1989) 21-50.

Ande_92   Ross Anderson: Personal communication, 26.5.1992; to be submitted to Electronics Letters.

BaEi_90   Paul Barrett, Raymund Eisele: The smart diskette – A universal user token and personal crypto-engine; Crypto '89, LNCS 435, Springer-Verlag, Heidelberg 1990, 74-79.

Bos_92   Jurjen Bos: Practical Privacy; Proefschrift, Technische Universiteit Eindhoven 1992.

Feig_86   Joan Feigenbaum: Encrypting Problem Instances, Or ..., Can You Take Advantage of Someone Without Having to Trust Him?; Crypto '85, LNCS 218, Springer-Verlag, Berlin 1986, 477-488.

KaSh_90   Shin-ichi Kawamura, Atsushi Shimbo: Performance Analysis of Server-Aided Secret Computation Protocols for the RSA Cryptosystem; The Transactions of The Institute of Electronics, Information and Communication Engineers IEICE, E73/7 (1990) 1073-1080.

LaYH_91   Chi-Sung Laih, Sung-Ming Yen, Lein Harn: Two Efficient Server-Aided Secret Computation Protocols Based on the Addition Sequence; Asiacrypt '91 – Abstracts, 270-274.

MaIm_91   Tsutomu Matsumoto, Hideki Imai: Human Identification Through Insecure Channel; Eurocrypt '91, LNCS 547, Springer-Verlag, Berlin 1991, 409-421.

MaKI_90   Tsutomu Matsumoto, Koki Kato, Hideki Imai: Speeding up Secret Computations with Insecure Auxiliary Devices; Crypto '88, LNCS 403, Springer-Verlag, Berlin 1990, 497-506.

PrCh_89   Wyn L. Price, Bernard Chorley: The Intelligent Token or 'Super-Smart' Card; SMART CARD 2000 (1987), North-Holland, Amsterdam 1989, 133-138.

QuCo_82   Jean-Jaques Quisquater, C. Couvreur: Fast Decipherment Algorithm for RSA Public-Key Cryptosystem; Electronics Letters 18/21 (1982) 905-907.

QuSo_91   Jean-Jaques Quisquater, Marijke De Soete: Speeding up Smart Card RSA Computation with Insecure Coprocessors; Proceedings Smart Cards 2000 (1989), North-Holland, Amsterdam 1991, 191-197.

QuWB_91   Jean-Jaques Quisquater, Dominique de Waleffe, Jean-Pierre Bournas: Corsair: A chip card with fast RSA capability; Proceedings Smart Cards 2000 (1989), North-Holland, Amsterdam 1991, 199-206.

RSA_78   Ronald L. Rivest, Adi Shamir, Leonard Adleman: A Method for Obtaining Digital Signatures and Public-Key Cryptosystems; Communications of the ACM 21/2 (1978) 120-126, reprinted: 26/1 (1983) 96-99.

ShKa_90   Atsushi Shimbo, Shin-ichi Kawamura: Factorisation attack on certain server-aided computation protocols for the RSA secret transformation; Electronics Letters 26/17 (1990) 1387-1388.

WaQu_91   Dominique de Waleffe, Jean-Jaques Quisquater: CORSAIR: A Smart Card for Public Key Cryptosystems; Crypto '90, LNCS 537, Springer-Verlag, Berlin 1991, 502-513.

Public-Key Cryptosystems with Very Small Key Lengths

Greg Harper, Alfred Menezes and Scott Vanstone

Dept. of Combinatorics and Optimization, University of Waterloo
Waterloo, Ontario, N2L 3G1, Canada

Abstract. In some applications of public-key cryptography it is desirable, and perhaps even necessary, that the key size be as small as possible. Moreover, the cryptosystem just needs to be secure enough so that breaking it is not cost-effective. The purpose of this paper is to investigate the security and practicality of elliptic curve cryptosystems with small key sizes of about 100 bits.

1 Introduction

It is sometimes convenient for the users of a public-key cryptosystem that the key sizes be as small as possible. For the applications we have in mind it is the public part of the key which should be relatively small. The size of the public key is typically difficult to control. For example, in the RSA system [23] the public key consists of the integers e and n, where n is the modulus. Although e can be chosen to be small there is not the same flexibility with the choice of n (n should be at least 512 bits in length). For the Diffie Hellman [7] and ElGamal schemes [8] based on discrete exponentiation in a finite field, the private key k, an integer, can be restricted but the public key α^k, α a generator of the field, is the size of the field (which should certainly be at least 2^{500}). For the Chor-Rivest knapsack [6], the public key is $(a_0, a_1, \ldots, a_{n-1})$, where $0 \leq a_i \leq p^m - 2$ (one choice suggested in [6] is $p = 197$, $m = 24$). The next two paragraphs describe some situations which come to mind where a small public key size might be desirable.

Consider the scenario where we have a small network where we would like to have secure e-mail exchange, or where we would like to have secure transmission of messages by fax. Rather than exchange public keys using certificates, key exchange is done verbally with authentication provided by voice recognition. If a symbol set consists of 32 alphanumeric characters represented by all 5-bit vectors then an n-bit key can be exchanged by representing it as an $\lceil n/5 \rceil$-symbol alphanumeric string. For n about 100 such a string is less than twice the length of most current international telephone numbers. Strings of this length would also be convenient for business cards, letterheads etc.

Consider also the following scenario: A software company places various programs on one distribution medium, however the purchaser can only access those programs he has paid for (the distribution medium could contain special purpose hardware that is tamper-proof for this purpose). If the user later wishes

R.A. Rueppel (Ed.): Advances in Cryptology - EUROCRYPT '92, LNCS 658, pp. 163-173, 1993.
© Springer-Verlag Berlin Heidelberg 1993

to purchase some of the other programs, he phones the company and places his request. The company in turn replies with the appropriate access information, which is digitally signed. The signature is verified by the user's terminal, and access is granted.

Most of the known public-key cryptosystems are totally insecure if the key size is restricted to about 100 bits. For example, since factoring 100-bit integers can be readily done on a microcomputer, the RSA system is insecure for keys of that size. The same holds true for systems whose security is based on the intractability of the discrete logarithm problem in a finite field, such as the ElGamal cryptosystem; recently La Macchia and Odlyzko [14] computed logarithms in the field F_p where p is a 192-bit prime, while Gordon and McCurley [9] were able to compute logarithms in $F_{2^{401}}$.

A good candidate that remains is the elliptic curve cryptosystem, which was first proposed in 1985 by N. Koblitz [11] and V. Miller [18]. The security of these systems is based on the difficulty of the logarithm problem in the elliptic curve group. If the curve is carefully chosen, then the only known attacks on the problem are the so-called square root attacks whose running times are proportional to the square root of the largest prime dividing the order of the group. The most efficient way known to implement this algorithm is the Pollard ρ-method [22] which requires very little storage.

The paper is organized as follows. We begin with a review of elliptic curve cryptosystems in Section 2. We outline Pollard's method in Section 3, and present some results of our experiments. In Section 4 we compare two ways of performing finite field arithmetic in software. Finally, in Section 5 we present the results of our software implementation of the elliptic curve cryptosystems. All implementations were done in the C-language on a SUN SPARCstation-2.

2 Elliptic Curve Cryptosystems

We will be concerned here with elliptic curves over fields of characteristic 2. Some recent work has been done on the implementation of cryptosystems based on these curves [12, 13, 16]. For an elementary introduction to elliptic curves consult [10].

We let $q = 2^n$. Let E be a non-supersingular elliptic curve defined over the field F_q. There are precisely $2(q-1)$ such curves, whose defining equations have the form

$$E : y^2 + xy = x^3 + ax^2 + b, \tag{1}$$

where $b \in F_q^*$, and $a \in \{0, \gamma\}$, $\gamma \in F_q$ being an element of trace 1. The set of solutions (x, y) in $F_q \times F_q$ to the equation (1), together with the special point at infinity, denoted $\mathcal{O}$, form an (additively written) abelian group denoted by $E(F_q)$. By Hasse's Theorem, the order of this group is $\#E(F_q) = q + 1 - t$, where $|t| \leq 2\sqrt{q}$, and hence $\#E(F_q) \approx q$. It is this group that is used instead of the multiplicative group of a finite field to implement the Diffie-Hellman key passing, and the ElGamal message passing and signature schemes as described in [11, 16].

The security of the elliptic curve schemes is based on the presumed difficulty of computing logarithms in E; namely the problem of finding the integer k given the points $P \in E(F_q)$ and $Q = kP$. We assume that the order of P (and thus also $\#E(F_q)$) is divisible by a large prime p. To avoid the recent attack of [15], the curve must also satisfy the condition that p does not divide $q^l - 1$ for all small l (i.e. for all $l \leq s$, where the discrete logarithm problem in F_{2^s} is considered intractable). If these criteria are met, then the best known attack is the combination of the Pollard ρ and Pohlig-Hellman methods.

To select an appropriate curve one may simply pick random curves over F_{2^n} and compute $\#E(F_{2^n})$ by Schoof's algorithm [24], until the conditions of the previous paragraph are met. Schoof's algorithm appears practical for $n \leq 155$ as demonstrated by the computations in [17]. An alternate method is to pick a curve defined over a small subfield K of F_{2^n}, count the number of points over K directly, and then lift the result to F_{2^n} by using the Weil Conjecture (see [10]).

3   Pollard ρ-Method

We outline the method for computing logarithms in an elliptic curve group which is a combination of the methods of Pohlig-Hellman [21] and Pollard [22].

Let $P \in E(F_q)$ and $Q = kP$. We assume that the order m of P is known as is its prime factorization $m = p_1^{e_1} p_2^{e_2} \cdots p_t^{e_t}$. We first show how to determine k modulo p^e, where p is one of $p_1, p_2, \ldots, p_t$. One may similarly compute k modulo $p_i^{e_i}$, $1 \leq i \leq t$, and then use the Chinese remainder theorem to recover k.

Let $z \equiv k \pmod{p^e}$, and write $z = \sum_{i=0}^{e-1} z_i p^i$, where $0 \leq z_i < p$. We have

$$\frac{m}{p} Q = \frac{km}{p} P = z\left(\frac{m}{p}P\right) = z_0\left(\frac{m}{p}P\right);$$

the latter two equations hold since $(m/p)P$ has order p. Using the method to be described below, we determine z_0, the logarithm of $(m/p)Q$ to the base $(m/p)P$, which has order p. To find z_1, we observe that

$$\frac{m}{p^2}(Q - z_0 P) = \frac{m}{p^2}Q - z_0\left(\frac{m}{p^2}P\right) = k\left(\frac{m}{p^2}P\right) - z_0\left(\frac{m}{p^2}P\right) = z_1\left(\frac{m}{p}P\right).$$

Again, by the method described below, we may then compute z_1, and continue.

We can now assume that the order of P is a prime p. If $R = (x, y)$, let $wt(R)$ denote the sum modulo 3 of the Hamming weights of the binary representations of x and y (with $wt(\mathcal{O}) = 0$). Define a sequence of points $\{R_i\}$ by $R_0 = \mathcal{O}$, and

$$R_{i+1} = Q + R_i, \ 2R_i, \ \text{or} \ P + R_i, \quad \text{if } wt(R_i) = 0, 1, \text{ or } 2 \text{ respectively.} \quad (2)$$

If the sequence $\{R_i\}$ behaves like a random sequence, then the least index j with $R_j = R_{2j}$ has expected value close to $1.0308\sqrt{p}$ [22]. We may find this j by Floyd's cycle-finding algorithm as follows. Define integer sequences $\{a_i\}, \{b_i\},$

by $a_0 = 0$, $b_0 = 0$, and $a_{i+1} = a_i + 1, 2a_i$ or $a_i \pmod{p}$, $b_{i+1} = b_i, 2b_i$ or $b_i + 1$ $\pmod{p}$, according to the three cases in (2). Observe that

$$R_i = a_i Q + b_i P \quad \text{for } i \geq 0.$$

We now compute the sequence $(R_i, a_i, b_i, R_{2i}, a_{2i}, b_{2i})$ for $i = 1, 2, 3, \ldots$ until we have $R_i = R_{2i}$. We then have $(a_i - a_{2i})Q = (b_{2i} - b_i)P$, which yields the congruence

$$(a_i - a_{2i})k \equiv (b_{2i} - b_i) \pmod{p}.$$

This congruence can then be solved for k.

A faster way to find indices i, j, $i \neq j$, with $R_i = R_j$ is to use Brent's cycle-finding algorithm [3], which is never slower and is on average 36% faster than Floyd's algorithm. Let the sequences $\{R_i\}$, $\{a_i\}$ and $\{b_i\}$ be defined as above. Brent's algorithm is the following.

> Let $Y := R_0$; $r := 1$; $i := 0$; *done* := **false**;
> **repeat** $X := Y$; $j := i$; $r := 2r$;
>   **repeat** $i := i + 1$; $Y := R_i$; *done* := $(X = Y)$
>   **until** *done* **or** $(i \geq r)$
> **until** *done*

When the algorithm terminates, $X = R_j$, $Y = R_i$, and $R_i = R_j$. We then have $(a_i - a_j)Q = (b_j - b_i)P$, which yields the congruence $(a_i - a_j)k \equiv (b_j - b_i) \pmod{p}$. This congruence is then solved for k. The expected value of i is $1.9828\sqrt{p}$, assuming that $\{R_i\}$ behaves like a random sequence [3].

For cryptographic applications, we only consider curves whose order is divisible by a large prime p. The running time of the algorithm is then dominated by the time to compute logarithms modulo this p. If we count elliptic curve additions as a basic step, then the expected running time is $1.9828\sqrt{p}$ curve additions with Brent's algorithm and $3.0924\sqrt{p}$ curve additions with Floyd's algorithm (since it takes 3 curve additions to compute the pair of points (R_{i+1}, R_{2i+2}) from (R_i, R_{2i})). Note that the amount of storage required by the algorithm is negligible.

We have implemented the Pohlig-Hellman and Pollard ρ-methods for computing logarithms in elliptic curves over F_{2^n}. Field elements are represented with respect to an optimal normal basis over F_2 (see Section 4). Tables 1, 2 and 3 give the average running times for computing logarithms in various curves over the fields $F_{2^{65}}$, $F_{2^{89}}$ and $F_{2^{105}}$.

4 Field Arithmetic

There are various methods for performing arithmetic in a finite field. The most popular seem to be through polynomial and normal basis representations. In this section we have attempted to compare software implementations for specific fields. It is extremely difficult to state with confidence which method is better

Order of curve	Number of digits in largest prime factor	Observed time (in minutes)
$2^3 \cdot 3 \cdot 79 \cdot 21755761 \cdot 894410947$	9	5
$2^2 \cdot 3^2 \cdot 3845557 \cdot 266494324513$	12	53
$2^2 \cdot 11 \cdot 2003 \cdot 418616258967271$	15	2625
$2^2 \cdot 7 \cdot 313 \cdot 4209663184776557$	16	5538

Table 1. Average times to compute logarithms in curves over $F_{2^{63}}$

Order of curve	Number of digits in largest prime factor	Observed time (in minutes)
$1360932^3 \cdot 7^2 \cdot 109 \cdot 419 \cdot 5441 \cdot 30779339$	8	1
$2^2 \cdot 3 \cdot 2309 \cdot 902977 \cdot 53353603 \cdot 463686011$	9	2
$2^3 \cdot 5 \cdot 23 \cdot 61 \cdot 1009 \cdot 2017 \cdot 135559 \cdot 39978503273$	11	11
$2^3 \cdot 3^2 \cdot 9049 \cdot 510843049 \cdot 1859726335343$	13	149
$2^2 \cdot 37 \cdot 467 \cdot 547 \cdot 942593 \cdot 17369186627491$	14	423
$2^3 \cdot 3 \cdot 71 \cdot 821 \cdot 646039 \cdot 684854210761471$	15	2268

Table 2. Average times to compute logarithms in curves over $F_{2^{89}}$

since any implementation is machine and code dependent. With this caveat we proceed.

We compare the speeds in software of the basic operations of multiplication, squaring, and inversion in the fields $F_{2^{105}}$ and $F_{2^{104}}$, the former represented with respect to an optimal normal basis over F_2, and the latter represented with respect to a polynomial basis over F_{2^8}.

Order of curve	Number of digits in largest prime factor	Observed time (in minutes)
$2^2 \cdot 11 \cdot 29 \cdot 43 \cdot 211 \cdot 421 \cdot 751 \cdot 1051 \cdot 15541 \cdot 25621 \cdot 26481841$	8	6
$2^3 \cdot 3^2 \cdot 457 \cdot 212913163 \cdot 1431658159 \cdot 4044445651$	10	17
$2^2 \cdot 7^4 \cdot 13^2 \cdot 1009 \cdot 131251 \cdot 173741 \cdot 1086211970677$	13	172
$2^3 \cdot 31 \cdot 71^2 \cdot 131 \cdot 211 \cdot 281 \cdot 3697 \cdot 1129982311077901$	16	3324

Table 3. Average times to compute logarithms in curves over $F_{2^{105}}$

4.1 Normal Basis Representation

A normal basis for $F_{2^{105}}$ over F_2 is a basis of the form

$$N = \{\beta, \beta^2, \beta^{2^2}, \ldots, \beta^{2^{104}}\}.$$

The basis is optimal if its multiplication table is as "simple" as possible; see [19] for more details and for an easy way to construct such a basis. Given any $\alpha \in F_{2^{105}}$, we can then express $\alpha = \sum_{i=0}^{104} c_i \beta^{2^i}$, where $c_i \in F_2$, and we write $\alpha = (c_0, c_1, c_2, \ldots, c_{104})$. In software, α is represented by a bit vector of length 105, i.e. on a 32-bit machine, α is stored in an array of unsigned integers of length 4, the last 23 bits of which are unused.

Addition of elements is achieved by simply XOR-ing the vector representations. Since

$$\alpha^2 = \sum_{i=0}^{104} c_i \beta^{2^{i+1}} = \sum_{i=0}^{104} c_{i-1} \beta^{2^i}$$

(with indices reduced modulo 105), squaring α is accomplished by a cyclic shift of its vector representation.

The most efficient way we know to compute the inverse of α is to first convert to a polynomial basis representation of $F_{2^{105}}$ using a precomputed change of basis matrix, compute the inverse using an efficient implementation of the extended Euclidean algorithm as described in [2, p.41], and then transform the result back to the normal basis representation.

It has been our experience that a software implementation of an optimal normal basis is more efficient that implementing an arbitrary normal basis. The field $F_{2^{105}}$ was chosen because it is about 100 bits and an optimal normal basis exists for it.

4.2 Polynomial Basis Representation

We choose to represent $F_{2^{104}}$ as a vector space over F_{2^8}, instead of over F_2 as is usually done. The polynomial

$$g(z) = 1 + z^2 + z^3 + z^7 + z^8$$

is a primitive polynomial over F_2, and so the elements of F_{2^8} are the set of polynomials $F_2[z]$ modulo $g(z)$. For efficiency, we store two tables "log" and "antilog" which are defined as follows:

$$\log[a] = i, \quad \text{where } z^i = \alpha \text{ and } 0 \leq i \leq 254$$

and

$$\text{antilog}[i] = a, \quad \text{where } z^i = \alpha \text{ and } 0 \leq i \leq 254.$$

(a is the binary vector representation of $\alpha \in F_{2^8}^*$.) Multiplication in F_{2^8} is then simply accomplished by table-lookup. For, if $a, b \in F_{2^8}^*$, then

$$a \cdot b = \text{antilog}[(\log[a] + \log[b]) \bmod 255].$$

We also store a table of inverses of elements in $F_{2^8}^*$.

The polynomial

$$f(x) = 1 + x + x^6 + x^7 + x^{13}$$

is irreducible over F_2, and thus also over F_{2^8} since $\gcd(8, 13) = 1$. Consequently, we may represent the elements of $F_{2^{104}}$ as polynomials in $F_{2^8}[x]$ of degree at most 12 modulo $f(x)$. Each coefficient of the polynomial is stored in a single computer word.

If $a = \sum_{i=0}^{12} a_i x^i$, $b = \sum_{i=0}^{12} b_i x^i \in F_{2^{104}}$, then we compute the product $c = a \cdot b$ by initializing c to 0, and then computing $a, ax, \ldots, ax^{12}$; after each stage, we add $b_i(ax^i)$ to c. Squaring is faster than a multiplication since $a^2 = \sum_{i=0}^{12} a_i^2 x^{2i}$. Finally, the inverse of a is computed by applying the extended Euclidean algorithm in $F_{2^8}[x]$ to find $s(x)$ such that $a(x)s(x) \equiv 1 \pmod{f(x)}$.

4.3 Timings

Arithmetic in the fields $F_{2^{105}}$ and $F_{2^{104}}$ was implemented. In Table 4 we give the running times for the operations of squaring, multiplication and inversion. The arithmetic in $F_{2^{104}}$ was observed to be faster than in $F_{2^{105}}$ in our implementations, and this is primarily due to our choice of representing $F_{2^{104}}$ as a vector space over F_{2^8}, the arithmetic in F_{2^8} being essentially for free.

	$F_{2^{105}}$	$F_{2^{104}}$
10,000 squarings	0.02	0.15
10,000 multiplications	7.6	1.95
10,000 inversions	15.3	9.46

Table 4. Times for field operations (in seconds)

5 Implementation of an Elliptic Curve Cryptosystem

For a curve with equation (1), the rules for adding points are the following. The point $\mathcal{O}$ serves as the identity element. Let $P = (x_1, y_1)$ and $Q = (x_2, y_2)$ be points on the curve. Then $-P = (x_1, x_1 + y_1)$. If $Q \neq -P$, then $P + Q = (x_3, y_3)$, where

$$x_3 = \begin{cases} \left(\dfrac{y_1 + y_2}{x_1 + x_2}\right)^2 + \dfrac{y_1 + y_2}{x_1 + x_2} + x_1 + x_2 + a_2 & P \neq Q \\[4mm] \dfrac{a_6}{x_1^2} + x_1^2 & P = Q \end{cases}$$

and

$$y_3 = \begin{cases} \left(\dfrac{y_1 + y_2}{x_1 + x_2} \right)(x_1 + x_3) + x_3 + y_1 & P \neq Q \\[2ex] x_1^2 + \left(x_1 + \dfrac{y_1}{x_1} \right) x_3 + x_3 & P = Q. \end{cases}$$

Note that two multiplications and an inversion are needed to add two distinct points, while a point can be doubled in 3 multiplications and an inversion (the other operations are relatively inexpensive). We should also note that inversions can, in general, be avoided by changing to projective coordinates at the expensive of doing more multiplications. Hardware implementations (see for example [1]) can take advantage of this whereas, in software, the affine representation appears to be superior.

Table 5 lists the running times of the elliptic curve operations for random curves over $F_{2^{105}}$ and $F_{2^{104}}$ (which are represented as described in Section 4).

	$F_{2^{105}}$	$F_{2^{104}}$
10,000 curve additions	33.9	14.4
10,000 curve doublings	38.6	16.2

Table 5. Times for elliptic curve operations (in seconds)

For each of the Diffie-Hellman key passing and ElGamal message passing and signature schemes, one has to compute kP for a random integer k. This is accomplished by the repeated doubling and adding algorithm. When $P \in E(F_{2^{104}})$, k will have 52 bits on average that are 1 in its binary representation. Hence computing kP requires 103 curve doublings and 51 curve additions on average, for an expected time of .24 seconds when working with $E(F_{2^{104}})$. This time is certainly tolerable for many applications, including the ones described in the Introduction.

If the cryptosystems are implemented in software, then there is additional storage available. In this case, the following method of [5] is more efficient for computing kP. We first precompute the points $P_i = 8^i P$ for $0 \leq i \leq 34$. Let $1 \leq k \leq \#E(F_{2^{104}})$. We write k in its base 8 representation as

$$k = k_0 + k_1 8 + k_2 8^2 + \cdots + k_{34} 8^{34}, \quad 0 \leq k_i \leq 7.$$

(This is an easy task given the binary representation of k.) The algorithm to compute kP is the following:

(i) Compute $B := \sum_{k_i = 7} P_i$, and set $A := B$.
(ii) For d from 6 to 1 do
$$B := B + \sum_{k_i = d} P_i$$
$$A := A + B$$

(iii) Output A

The method requires the storage of 35 points and needs 36 curve additions on average for random k. This reduces our time estimate for computing kP from .24 seconds to .052 seconds.

In the elliptic curve analogue of the ElGamal cryptosystem, a curve $E(F_q)$ and point $P \in E$ are public knowledge. Each user A has a public key aP, where integer a is the corresponding secret key. To send messages $m_1, m_2 \in F_q$ to A, user B selects a random integer k, and computes the points kP and $kaP = (\bar{x}, \bar{y})$. Finally, B sends $(kP, m_1\bar{x}, m_2\bar{y})$ to A. With our software implementation, the encryption rate is about 2 Kbits/sec.

The public key is an elliptic curve point which is $2n$ bits long (where $q = 2^n$). The key length can be shortened to $n + 1$ bits as follows. Observe first that the change of variables $(x, y) \longrightarrow (x, xz)$ transforms equation (1) to

$$z^2 + z = x + a + bx^{-2}. \tag{3}$$

Given the x-coordinate of a point $P = (\bar{x}, \bar{y})$, we can compute the right hand side of (3). Then (3) has precisely 2 solutions, namely z' and $z' + 1$, and these solutions can be easily found. We can then select the correct solution $\bar{z}$ (and hence reconstruct $\bar{y}$ as $\bar{y} = \bar{x}\bar{z}$) if we know the least significant bit of $\bar{z}$. Thus to transmit P it is sufficient to transmit $\bar{x}$ and the least significant bit of $\bar{y}/\bar{x}$.

If E is a curve defined over F_{2^8} with equation (1), then $\#E(F_{2^8})$ is even, and $226 \leq \#E(F_{2^8}) \leq 228$. If $\#E(F_{2^8}) = 242$, 250 or 254, then $\#E(F_{2^{104}})$ is divisible by a 29 digit prime. For example, the following curve

$$E \; : \; y^2 + xy = x^3 + zx^2 + (z^3 + z^4 + z^6 + z^7) \tag{4}$$

(with z defined as in Section 4.2), has $\#E(F_{2^8}) = 250$, and

$$\#E(F_{2^{104}}) = 2 \cdot 11^2 \cdot 8381160993244491690540 4513441,$$

where the last integer is a 29-digit prime.

Based on our experimental results of Section 3, computing even a single logarithm in $E(F_{2^{104}})$ using Pollard's ρ-method is computationally infeasible. For a very rough analysis of the security, observe that the Pollard ρ-method requires about 10^{15} elliptic curve operations to compute a single logarithm in the curve (4). A rough count shows that an elliptic curve operation takes at least 1000 word operations (i.e. operations such as exclusive-or's on 32-bit words). Hence the expected number of word operations to compute a logarithm is at least 10^{18}. This is roughly equal to the computing resources required to factor a 512-bit integer by the multipolynomial quadratic sieve, and for computing logarithms in the field $F_{2^{700}}$ by the index-calculus method (see [25]).

As noted by Brent [4], it does not seem possible to use parallelism to speed up the Pollard ρ-method. The Lambda method [22] is a method for computing logarithms given that the logarithm lies in some interval $[A, B]$. The running time of the method is $O(\sqrt{w})$, where $w = B - A$. If t processors are available, then to compute $\log_P Q$ where the order of P is a prime p, the interval $[0, p-1]$

is divided into t intervals of equal length, and the intervals are searched in parallel by the t processors. The expected number of elliptic operations for each processor is $\sqrt{p/t}$. For example, if 10,000 processors are available then for the curve (4) the expected number of elliptic curve operations is about 10^{13}. If this does not provide adequate security for the application at hand, then of course the underlying field can be replaced by a larger field $F_{2^{8r}}$, $r \geq 14$ (to use the polynomial basis representation of Section 4.2 we need to work in a field F_{2^n} where n is a multiple of 8).

6 Conclusions

We have demonstrated that implementing elliptic curve cryptosystems in F_{2^n} for $n \approx 100$ can be done efficiently in software on a workstation. Curves over $F_{2^{104}}$ can be selected so that by current best methods computing logarithms requires about 10^{15} elliptic curve operations (if additional security is required then, of course, n can be increased). It should be pointed out that each logarithm in such a system requires this amount of work. This is unlike the index-calculus method [20] where there are 2 phases. Once phase one is completed all other logarithms are relatively easy to find. Computing kP, where $P \in E(F_{2^{104}})$ and k is a random 104-bit integer, takes about .052 seconds. We can thus implement the ElGamal cryptosystem and achieve an encryption rate of 2 Kbits/sec in software. Public keys are only 105 bits in length. The code for this scheme is fairly compact and occupies about 40 Kbytes.

References

1. G. Agnew, R. Mullin and S. Vanstone, "An implementation of elliptic curve cryptosystems over $F_{2^{155}}$", preprint, 1992.
2. E. Berlekamp, *Algebraic Coding Theory*, McGraw-Hill, New York, 1968.
3. R. Brent, "An improved Monte Carlo factoring algorithm", *BIT*, **20** (1980), 176-184.
4. R. Brent, "Parallel algorithms for integer factorisation", in *Number Theory and Cryptography*, Cambridge University Press, 1990, 26-37.
5. E. Brickell, D. Gordon, K. McCurley and D. Wilson, "Fast exponentiation with precomputation", preprint, 1992.
6. B. Chor and R. Rivest, "A knapsack-type public key cryptosystem based on arithmetic in finite fields", *IEEE Transactions on Information Theory*, **34** (1988), 901-909.
7. W. Diffie and M. Hellman, "New directions in cryptography", *IEEE Transactions on Information Theory*, **22** (1976), 644-654.
8. T. ElGamal, "A public key cryptosystem and a signature scheme based on discrete logarithms", *IEEE Transactions on Information Theory*, **31** (1985), 469-472.
9. D. Gordon and K. McCurley, "Computation of discrete logarithms in $GF(2^n)$", presentation at Crypto '91, Santa Barbara, 1991.
10. N. Koblitz, *A Course in Number Theory and Cryptography*, Springer-Verlag, New York, 1987.

11. N. Koblitz, "Elliptic curve cryptosystems", *Mathematics of Computation*, **48** (1987), 203-209.

12. N. Koblitz, "Constructing elliptic curve cryptosystems in characteristic 2", *Advances in Cryptology: Proceedings of Crypto '90*, Lecture Notes in Computer Science, **537** (1991), Springer-Verlag, 156-167.

13. N. Koblitz, "CM-Curves with good cryptographic properties", *Advances in Cryptology: Proceedings of Crypto '91*, Lecture Notes in Computer Science, **576** (1992), Springer-Verlag, 279-287.

14. B. La Macchia and A. Odlyzko, "Computation of discrete logarithms in prime fields", *Designs, Codes and Cryptography*, 1 (1991), 47-62.

15. A. Menezes, T. Okamoto and S. Vanstone, "Reducing elliptic curve logarithms to logarithms in a finite field", *Proceedings of the 22nd Annual ACM Symposium on the Theory of Computing*, 80-89, 1991.

16. A. Menezes and S. Vanstone, "Elliptic curve cryptosystems and their implementation", submitted to *Journal of Cryptology*, 1991.

17. A. Menezes, S. Vanstone and R. Zuccherato, "Counting points on elliptic curves over F_{2^m}", to appear in *Mathematics of Computation*, 1992.

18. V. Miller, "Uses of elliptic curves in cryptography", *Advances in Cryptology: Proceedings of Crypto '85*, Lecture Notes in Computer Science, **218** (1986), Springer-Verlag, 417-426.

19. R. Mullin, I. Onyszchuk, S. Vanstone and R. Wilson, "Optimal normal bases in $GF(p^n)$", *Discrete Applied Mathematics*, **22** (1988/89), 149-161.

20. A. Odlyzko, "Discrete logarithms in finite fields and their cryptographic significance", *Advances in Cryptology – Proceedings of Eurocrypt '84*, Lecture Notes in Computer Science, **209** (1985), Springer-Verlag, 224-314.

21. S. Pohlig and M. Hellman, "An improved algorithm for computing logarithms over $GF(p)$ and its cryptographic significance", *IEEE Transactions on Information Theory*, **24** (1978), 106-110.

22. J. Pollard, "Monte Carlo methods for index computation (mod p)", *Mathematics of Computation*, **32** (1978), 918-924.

23. R. Rivest, A. Shamir and L. Adleman, "A method for obtaining digital signatures and public-key cryptosystems", *Communications of the ACM*, **21** (1978), 120-126.

24. R. Schoof, "Elliptic curves over finite fields and the computation of square roots mod p", *Mathematics of Computation*, **44** (1985), 483-494.

25. P. van Oorschot, "A comparison of practical public key cryptosystems", in *Contemporary Cryptology*, IEEE Press, 1992, 289-322.

Resource Requirements for the Application of Addition Chains
in Modulo Exponentiation

Jörg Sauerbrey

Andreas Dietel

Lehrstuhl für Datenverarbeitung
Technische Universität München
P. O. Box 20 24 20
W-8000 München 2
Germany

sy@ldv.e-technik.tu-muenchen.de

Abstract

Addition chains or sequences can be used to reduce the amount of multiplications to accomplish an exponentiation at the cost of more memory required. We examine known methods of exponentiations based on addition sequences and derive the parameters determining operation count and number of required registers for storing intermediate results. As a result an improved method is proposed to choose window distributions as a basis for using known addition sequence heuristics.

1. Introduction

A lot of cryptographic methods and protocols rely on the fast evaluation of powers modulo a large number n. One of the famous members of this class of methods is RSA. Exponentiation is usually based on modulo multiplications which can be broken down into additions. A lot of research work has been done to implement fast modulo multiplication [AliMar91, CuBoKa91, Even90, LipPos90, LuHaLH88, Montgo85, Moraga89, Morita90] and to speed up addition [KocHun90, Hwang79]. This paper deals with reducing the number of modulo multiplications for one modulo exponentiation. The aim is to obtain a computation rule for a specific exponent which leads to less multiplications than the usual methods. The effort for deriving this computation rule pays off, if one has to compute a lot of exponentiations with the same exponent. This holds for example for RSA, where the exponent is part of the key and subsequent encryptions use the same key. The computation rule could then be stored together with the key.

In this paper different computation rules are compared with regard to the number of multiplications and the amount of memory required to perform the exponentiations. We focus our investigation mainly on the amount of memory, because memory is a scarce resource for a VLSI implementation. We propose an improved method for finding computation rules, where the memory requirements are economic.

2. Exponentiation and Addition Chains

According to the rule $b^x b^y = b^{x+y}$ the computation of a power b^{x+y} corresponds to the problem of finding a sequence of increasing integers approaching the exponent. The sequence has to begin with 1 and every integer is the sum of two preceeding integers in the sequence. Such sequences are called addition chains. In general, an addition chain for n is a sequence of integers $(1=a_0, a_1, ... , a_r=n)$ with the property that $a_i = a_j + a_k$ for some k and $j, k \leq j < i$, for all $i=1, 2, ..., r$ ([Knuth69], p. 402).

R.A. Rueppel (Ed.): Advances in Cryptology - EUROCRYPT '92, LNCS 658, pp. 174-182, 1993.
© Springer-Verlag Berlin Heidelberg 1993

For example, the exponentiation b^x accomplished by repeated multiplications by b results in the addition chain $(1, 2, 3,..., x)$. The well known binary algorithm (repeated squaring), which is based upon the binary representation of x, defines an addition chain $(1, 2, ... , a_i, ... , x)$, where

$$a_{i-1} = \begin{cases} a_i - 1 & a_i \text{ odd} \\ a_i/2 & a_i \text{ even} \end{cases} \qquad (1)$$

We will show later that it is desirable to compute several predefined powers within the application of a single sequence of exponentiations. According to this requirement we define an addition sequence for $n_0,...,n_k$ as an addition chain, which at least contains the elements $n_0,...,n_k$. A star chain is an addition chain or sequence $(a_0,..,a_r)$, where each term a_i is the sum of a_{i-1} and a previous a_k.

We define the following: $l(n_0,n_1,...)$ is the length of the addition chain or sequence containing $n_0,n_1,...$ The length of the addition chain or sequence with the star property (star chain) is defined by $l^*(n_0,n_1,...)$. The Hamming weight $v(n)$ denotes the number of 1's in the binary representation of n. For $l(n)$ of the shortest addition chain it is known that $\log_2 n + \log_2 v(n) - 2.13 \le l(n)$. For addition sequences it is [Yao76]:

$$l(n_0, n_1, ..., n_k) \le \log n_k + c \sum_{i=0}^{k} \frac{\log n_i}{\log\log (n_i + 2)} \quad \text{where } c \text{ is a constant.} \qquad (2)$$

The star property may be suitable to reduce the amount of memory required to accomplish the exponentiation, but it does not guarantee that we get the shortest chain as $l(n) \le l^*(n)$ (Theorem of W. Hansen in [Knuth69], p. 413). For example the addition chain defined by equation (1) has the property, that only the value of the preceding step of the chain (a_{i-1}) and the value of b have to be stored for computing b^x. So using the binary algorithm, only two n bit registers are necessary to compute $b^x \bmod n$.

There are plenty of theoretical results and asymptotic bounds concerning different kinds of addition chains (see references in [BosCos89]), but much less practical hints for building and using addition chains. The main problem is, that computing the shortest addition chain is an NP-complete problem [DoLeSe81].

[McCart86] discusses the interaction between the efficiency of the basic multiplication algorithm and the addition chain used to compute b^x. If the cost of a multiplication is bound to the length of the operands, the multiplicative cost of evaluating b^x is minimized by using repeated squaring. But for modulo multiplications the cost of a multiplication is nearly constant. In this case it is clear, that the shortest chain for x will yield to the cheapest evaluation of b^x, in terms of number of multiplications. The binary algorithm does not define the shortest chain, as it can be shown for $x=15$. Here the shortest chain is $(1, 2, 3, 6, 12, 15)$, whereas the binary algorithm yields to $(1, 2, 3, 6, 7, 14, 15)$. However in most cases, saving of multiplications results in the necessity of storing more intermediate results.

[BosCos89] have shown some heuristics to compute addition sequences, which are mostly better than those derived from the standard binary method for exponentiation. With these heuristics it is possible to produce addition chains, which are on the average 21% shorter than those of the binary algorithms (using 512 bit exponents).

3. Known methods for the application of addition chains

Since even with these heuristics it is not feasible to compute an addition chain for an exponent x with length of 512 or 1024 bits, the goal of determining the computation rule is achieved in two steps. The first step is to reduce the computation of an addition chain for a large x to the computation of an addition sequence by choosing an appropriate set of numbers (window values) which are much smaller than x. The last step is to compute a 'good' addition sequence for these numbers.

There are different methods to accomplish the first step. They are all based on the m-ary method described in [Knuth69, p. 404]. If an exponentiation with the m-ary method has to be computed, x is rewritten as $x = d_0 m^t + d_1 m^{t-1} + \ldots + d_t$. This means that with $m = 2^k$ the binary representation of the exponent is divided in t windows with the width of k bit each and window values $d_i \in [0, m-1]$. The corresponding addition chain is as follows:

$$1,2,3,\ldots,(m-2),(m-1),$$

$$2d_0,4d_0,\ldots,md_0,(md_0+d_1),\qquad\qquad (3)$$

$$2(md_0+d_1),4(md_0+d_1),\ldots,m(md_0+d_1),(m^2d_0+md_1+d_2),$$

$$\ldots,(m^t d_0+m^{t-1}d_1+\ldots+d_t)$$

The first row serves to compute all possible window values (d_i), whereas each following row 'shifts' a new window value to the next window position in the binary representation of the exponent. Note that the width of the windows not necessarily have to be fixed.

There are several ways to decrease the length of the chain in order to reduce the number of multiplications for the exponentiation.

1. The last operation of a line of equation (3) correspondent to a particular window can be omitted if the window value is zero. If we take each string of one or more zeros of the binary representation of the exponent as a window, the LSBs of the remaining window values cannot be zero. Now the values of the first line of equation (3) decrease to the odd values between $1, \ldots, m-1$.

2. Compute only those values in the first line, which are used as d_i's in the following lines. This can be accomplished with an addition sequence algorithm. The maximum size of the windows can now be much larger, since we don't have to store all (odd) values of $[1, m-1]$ in a table.

3. The window distribution should be optimized. For example: the window values should be chosen, such that a short addition sequence can be constructed, with respect to the preceding point; there should be as few windows as possible; the same window values should appear more than once.

4. Choose the windows from right to the left such that many windows contain patterns of bits from windows standing to the right of them. This method is analog to the compression algorithm of Ziv and Lempel [ZivLem78]. If the values are stored in a tree, the derivation of new values is done easily using the values already computed. This method has been published in [Yacobi90] and is called compression method.

Of course, these reductions can be combined. The methods for defining windows have a great effect on the operation count, but their effect on the number of intermediate results to be stored (memory demand) is not obvious.

4. Effects on operation count and memory demand

The effect on the number of operations is rather clear. In detail there are four factors:

1. The length of the exponent x determining the number of squarings.

2. The length of the leftmost window (MSB-window), which reduces the number of squarings.

3. The number of non-zero windows minus one, which determines the number of multiplications. (Note: windows with the value of zero don't need a multiplication.)

4. The length of the addition sequence minus one, which defines the number of operations needed to compute the different window values. (Note: This is not the case for the method of [Yacobi90])

Figure 1 gives an example:

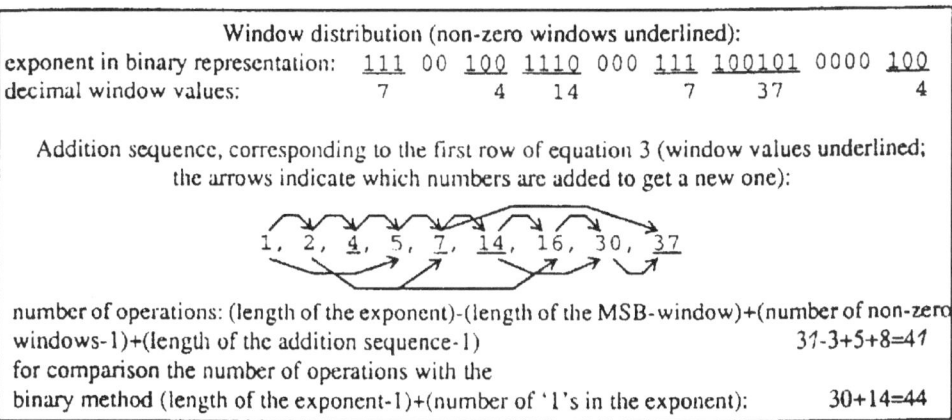

Figure 1: Window distribution and number of operations

The importance of examining the memory demand for computing b^x with the methods described in section 3 becomes obvious, if we look at the correspondent hardware realization.

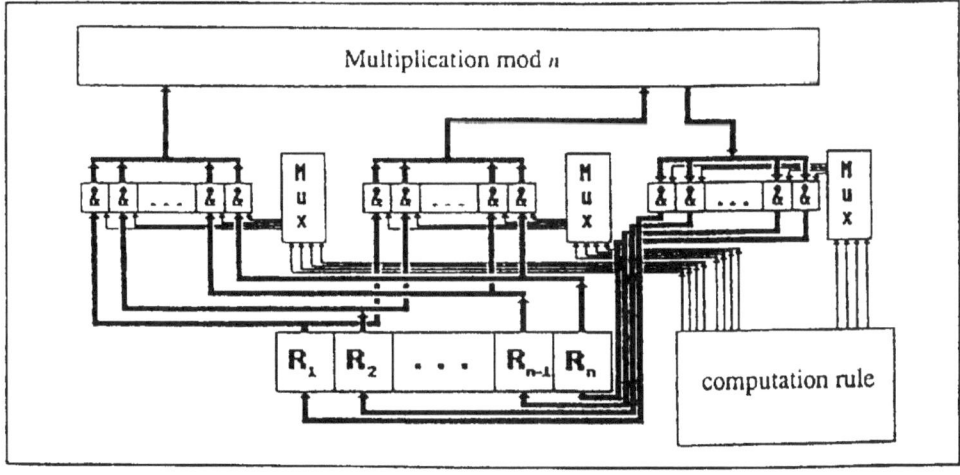

Figure 2: Hardware implementation of an exponentiation unit using addition chain methods

Figure 2 shows an example for the implementation of a hardware unit to compute b^x. There is a fixed number of registers (R_i) available for storing intermediate results. Initially R_1 is loaded with b. The box 'computation rule' is a controller selecting the correct source and destination registers R_i for each computation step. The computation rule for the actual x to accomplish the exponentiation is loaded into this block in advance. It implicitly contains the window distribution and the addition sequence, found by one of the algorithms above.

The effects of the computation rule on the required number of registers, cannot be summarized in a simple formula whereas the effects on the number of operations can be. The effect on the register count is investigated in the following points:

1. Does the window distribution contain non-adjacent windows sharing a common value?
   Once this value is computed, it has to to be stored for later use. The window value 4 in the example shown in figure 1 explains this point.

2. Is a window value necessary for the computation of the following window values?
   The value 14 in the addition sequence is an example for this point, because it is needed to create the value of 30.

3. Are the results, arising from the computation of the present window value necessary for the computation of the following window values?
   The value of 2 shows this aspect, because 2 is necessary to build 4, but has to be stored longer to create 16.

4. If traversed from left to right, how well are the windows sorted for increasing values?
   This point is important because we can start to compute the corresponding row of equation 3 once we have computed the corresponding window value within the addition sequence. If this value is not required any more we can drop it.

In the example of figure 1 at least 5 registers are needed to compute b^x. The content of the registers during the computation is as follows (figure 3):

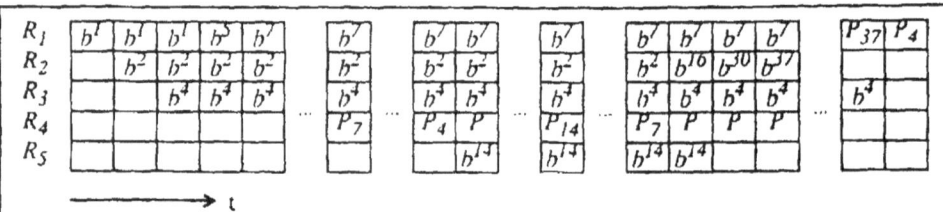

R_i: row shows occupation of register i over time
b^i: computed power of b^i using the addition sequence
...: build the new partial result P_i by squaring (see the corresponding row of equation 3)
P_i: new partial result built by squaring the old partial result and multiplication of b^i
P: unchanged partial result

Figure 3: Content of the registers during the computation

5. Operation count and memory requirements of different exponentiation methods

In this section we investigate three different exponentiation methods, which are combinations of the reduction methods discussed in section 3:

1. A combination of the standard m-ary method, reduction 2, and the addition sequence heuristics of [BosCos89], called 'modified m-ary method'.

2. A combination of the reductions 1, 2 and 3, suggested by [BosCos89]. We will refer to this method as 'optimized m-ary method'.

3. A new combination of reductions 1, 2 and 3, with a special emphasis on the reduction of the number of intermediate results to be stored, which we will call 'size oriented window distribution'.

The following figures show the average number of the required operations and registers for 100 exponentiations with randomly selected 512 bit exponents ('1' and '0' are distributed evenly). The effect of different window sizes on the number of operations and registers is presented.

Figure 4 shows the simulation results for the 'modified m-ary method'. On the horizontal axis the window size is given in bit.

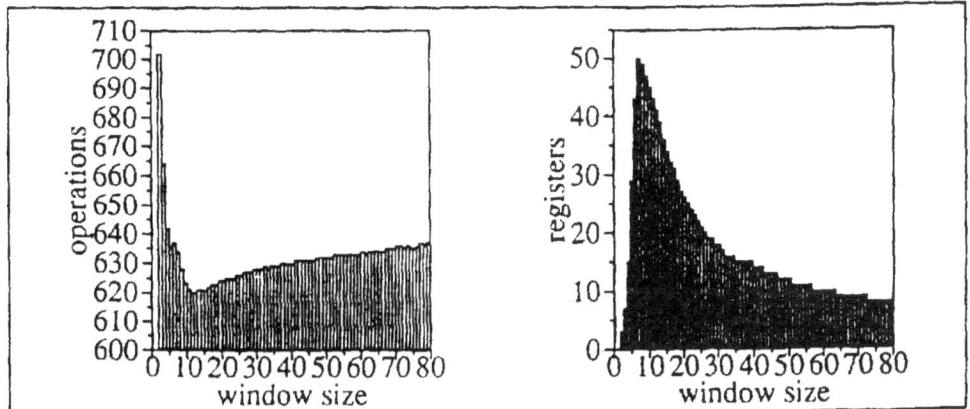

Figure 4: Number of operations and registers when applying the 'modified m-ary method'

Figure 5 shows simulation results for the 'optimized m-ary method'. The horizontal axis shows the maximal permitted window size in bits. In this algorithm the MSB-window has always the maximum size, thus reducing the number of squaring operations, due to the effect on the operation count as stated in section 4.

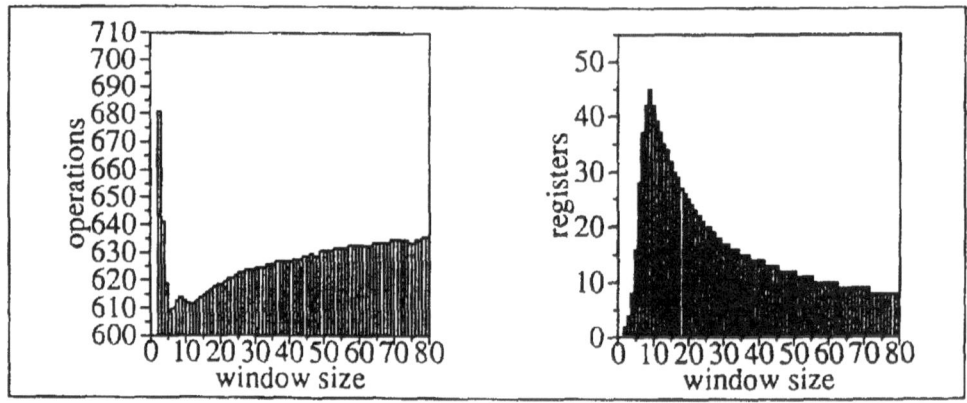

Figure 5: Number of operations and registers when applying the 'optimized *m*-ary method'

For comparison it should be recalled, that for 512 bit exponents the standard binary algorithm leads on the average to 767 operations and 2 required registers.

In both figures the number of operations has a distinct minimum. This is due to the fact that with increasing window sizes the number of windows decreases while the size of the MSB-window increases. However, when increasing the window sizes further, the addition chains for creating the window values become longer, because the employed heuristics are inefficient in identifying short sequences out of large numbers.

For both methods there is a rapid increase of the number of required registers with increasing window sizes, because of the exponential increase of the number of possible window values. This trend stops at some maximum value and the development reverses with further increasing window size. The reason for this effect is, that the bigger the window size the smaller the probability of the existence of all possible window values in the set of window values. For example, if the window size is 5 there are 16 possible window values (optimized *m*-ary method). The probability is very high that these window values exist more then once in the set of required window values of a 512 bit exponent. Thus a lot of different window values have to be stored for a long time, because they are needed later. With further increase of the window size, this probability decreases rapidly. Now only some of the existing window values have to be stored for later use.

The results of figures 4 and 5 illustrate the following problem: window sizes leading to good results for the number of operations require a lot of registers, and vice versa.

To reduce the memory requirements while keeping the number of operations low, we developed a new algorithm for selecting windows. The idea is to generate a window distribution with constantly increasing window values from left to right. This overcomes some reasons of storing values, because now window values appeare once and are sorted. However, this does not entirely eliminate the need for storing values (see section 4). The algorithm starts with a given size of the MSB-window and chooses new window values such that these values are greater than those of the windows to the left. Figure 6 shows an example of a window distribution generated with the 'size oriented window distribution'.

Example of a 'size oriented window distribution' (windows underlined; initial window size: 3):						
exponent in binary representation:	<u>111</u>	<u>001001</u>	<u>1100</u>	<u>001111</u>	<u>0010100</u>	0010...
decimal window values:	7	9	12	15	20	

Figure 6: Window distribution with the 'size oriented window distribution'

Figure 7 shows simulation results for the 'size oriented window distribution'. The horizontal axis shows the initial window size in bit.

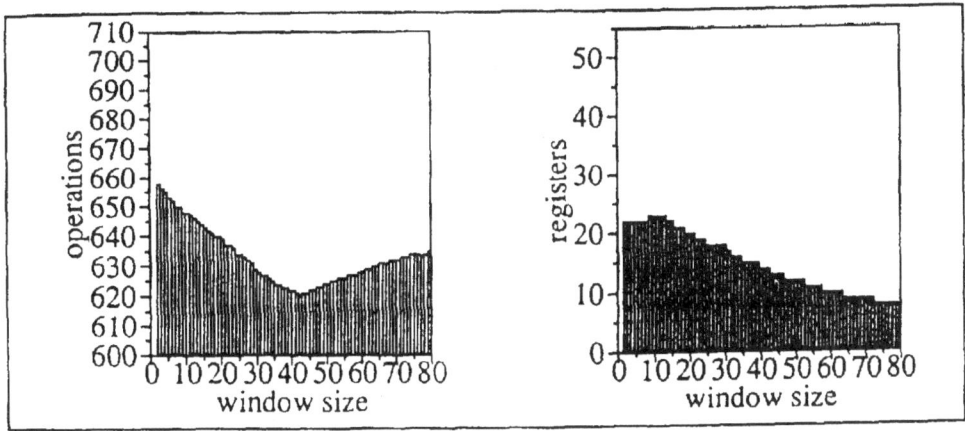

Figure 7: Number of operations and registers needed applying the
'size oriented window distribution'

The new algorithm actually provides a good compromise balancing the number of operations and the registers needed to accomplish the exponentiation. It is conceivable that an even better ratio can be achieved by a further improvement of the 'size oriented window distribution' algorithms and of the addition chain heuristics according to the effects explained in section 4.

6. Conclusion

We have examined different methods of exponentiation using addition chains. These methods have been compared with respect to the average number of multiplications and registers needed to accomplish exponentiations with randomly chosen exponents. The main factors influencing the operation count and memory requirements have been stated. A new method to choose window distributions (size oriented window distribution) has been proposed. It shows a better compromise between operation count and memory demand than previously known techniques.

7. References

[AliMar91]   Alia, Giuseppe; Martinelli, Enrico: "A VLSI Modulo m Multiplier", IEEE
                Transactions on Computers, Vol. 40, No. 7, p. 873-878, July 1991

[BosCos89]   Bos, Jurjen; Coster, Matthijs: "Addition Chain Heuristics", in Brassard, G. (Ed.):
                "Advances in Cryptology - Crypto '89", Proceedings (Lecture Notes in Computer
                Science 435), p. 400-407, Springer, 1989

[CuBoKa91]   Curiger, A.V.; Bonnenberg, H.; Kaeslin, H.: "Regular VLSI Architectures for
                Multiplication Modulo (2 exp n + 1)", IEEE Journal on Solid State Circuits, Vol. 26,
                No. 7, p. 990-994, July 1991

[DoLeSe81]   Downey, P.; Leong, B.; Sethi, R.: "Computing Sequences with Addition Chains",
                SIAM Journal on Computing, Vol. 3, No. 3, p. 638-646, August 1981

[Even90]   Even, Shimon: "Systolic Modular Multiplication", in Menezes, A.J.; Vanstone,
                S.A.(Eds.): "Advances in Cryptology - Crypto'90 Proceedings (Lecture Notes in
                Computer Science 537), p. 619-624, Springer, 1990

[Hwang79]   Hwang, Kai: "Computer Arithmetic: Principles, Architecture, and Design", John
                Wiley & Sons, New York, 1979

[Knuth69]   Knuth, Donald E.: "The Art of Computer Programming, Vol. 2: Seminumerical
                Algorithms", Addison-Wesley, Reading, Massachusetts, 1969

[KocHun90]   Koc, C. K.; Hung, C. Y.: "Multi-Operand Modulo Addition Using Carry Save
                Adders", Electronics Letters, Vol. 26, No. 6, p. 361-363, IEE, March 1990

[LipPos90]   Lippitsch, P.; Posch, K.C.; Posch, R.: "Multiplication As Parallel As Possible",
                Institute for Information Processing Graz, Report 290, October 1990

[LuHaLH88]   Lu, E.H.; Harn, L.;Lee, J.Y.; Hwang, W.Y.: "A Programmable VLSI Architecture for
                Computing Multiplication and Polynomial Evaluation Modulo a Positive Integer",
                IEEE Journal on Solid State Circuits, Vol. 23, No. 1, p. 204-207, February 1988

[McCart86]   McCarthy, D. P.: "Effect of Improved Multiplications Efficiency on Exponentiation
                Algorithms Derived from Addition Chains", Mathematics of Computations, Vol. 46,
                No. 174, p. 603/608, American Mathematical Society, April 1987

[Montgo85]   Montgomery, P. L.: "Modular Multiplication without Trial Division", Mathematics
                of Computation, Vol. 44, No. 170, p. 519-521, April 1985

[Moraga89]   Moraga, Claudio: "Design of a Modulo p Multiplier", International Journal on
                Electronics, Vol. 67, No. 5, p. 819-827, Taylor & Francis, 1989

[Morita90]   Morita, Hikaru: "A Fast Modular-multiplication Module for Smart Cards",
                Proceedings of AUSCRYPT '90 (Lecture Notes in Computer Science 453), p. 406-
                409, Springer, January 1990

[Yacobi90]   Yacobi, Y.: "Exponentiation Faster with Addition Chains", in Damgard, I.B. (Ed.):
                "Advances in Cryptology - EUROCRYPT '90", Proceedings (Lecture Notes in
                Computer Science 473), p. 222-229, Springer, 1990

[Yao76]   Yao, Andrew: "On the Evaluation of Powers", SIAM Journal on Computing, Vol. 5,
                No. 1, pp. 100-103, March 1976

[ZivLem78]   Ziv, Jacob; Lempel, Abraham: "Compression of Individuel Sequences via Variable-
                Rate Coding", IEEE Transactions on Information Theory, Vol. IT-24, No. 5, pp. 530-
                536, September 1978

Massively parallel elliptic curve factoring

B. Dixon[*], A. K. Lenstra[**]

Abstract. We describe our massively parallel implementations of the elliptic curve factoring method. One of our implementations is based on a new systolic version of Montgomery multiplication.

1. Introduction

The study of theoretical and practical aspects of factoring algorithms is of continuing interest for the analysis of various well known public-key cryptosystems. In this paper we describe two massively parallel implementations of the elliptic curve factoring method [5].

Usually one distinguishes two types of factoring algorithms, the *general purpose algorithms* whose expected run time depends solely on the size of the number n being factored, and the *special purpose algorithms* whose expected run time also depends on properties of the (unknown) factors of n. Examples of special purpose algorithms are *trial division, Pollard's rho method, Pollard's $p - 1$ method*, and the *elliptic curve method*. To find a factor p, trial division needs time approximately linear in p, Pollard's rho method is approximately linear in $\sqrt{p}$, and Pollard's $p - 1$ method is approximately linear in the largest prime factor of $p - 1$. Thus, in the worst case all these methods take exponential time. The elliptic curve method takes expected time $O((\log n)^2 L_p[\sqrt{2}])$ to find p, where

$$L_x[a] = \exp((a + o(1))\sqrt{\log x \log \log x}),$$

for $x \to \infty$. It follows that in the worst case $p \approx \sqrt{n}$ the method takes expected time $L_n[1]$, which is subexponential in n. Because their run time depends so strongly on the size of smallest factor of n, and only polynomially on the size of n itself, the performance of special purpose methods is usually measured by the size of the prime discovered.

From a security point of view, the above exponential time methods are not something to worry about, except that one sometimes takes the precaution to construct n such that Pollard's $p - 1$ method will not be unexpectedly lucky. Such precautions cannot be taken against the elliptic curve method. Whereas in Pollard's $p - 1$ method p will be discovered if $p - 1$ has only small factors, the elliptic curve method will find p if the method is so lucky

[*] Department of Computer Science, Princeton University, Princeton, NJ 08544, U.S.A,
email: bdd@princeton.edu.

[**] Bellcore, rm 2Q334, 445 South Street, Morristown, NJ 07960, U.S.A,
email: lenstra@bellcore.com.

R.A. Rueppel (Ed.): Advances in Cryptology - EUROCRYPT '92, LNCS 658, pp. 183-193, 1993.
© Springer-Verlag Berlin Heidelberg 1993

to hit upon a number *close to p* that has only small factors. Because the method consists of many independent trials, there is always the possibility that *p* will be discovered.

In this paper we consider two implementations of the elliptic curve method on a particular type of massively parallel computer, so-called *single instruction, multiple data* (SIMD) machines. As far as we know SIMD elliptic curve implementations have not been considered before, although massive parallelism is not new to this area. In [4], for instance, a distributed network implementation of the elliptic curve method is described, which may be viewed as a large scale *multiple instruction, multiple data* (MIMD) approach. We will see that SIMD machines, even though they are relatively cheap compared to MIMD machines, nevertheless can achieve an impressive elliptic curve performance*.

Using one of our implementations we have been able, for the first time for the elliptic curve method, to find a 40 digit factor**. The previous elliptic curve record was 38 digits, which occurred three times as far as we know. On the negative side, the elliptic curve method has also missed many smaller factors, in the 35 digit range, even after serious efforts. Using the run time estimates given above, one finds that finding a 60 digit factor can be expected to be more than 3000 times more difficult than finding a 40 digit factor. Given how much computing time has been invested to this date in the elliptic curve method, it seems safe to say that it is unlikely that with present day technology we will ever be able to discover factors of 60 or more digits using the elliptic curve method. Estimates of this type are, for instance, useful for the design of cryptosystems based on discrete logarithms modulo a composite modulus, like [8]. Finding 50 digit factors with the elliptic curve method is approximately 70 times more difficult than finding 40 digit factors; it is therefore not inconceivable that such factors might be found using more powerful machines.

The remainder of this paper is organized as follows. In Section 2 a short description of the SIMD machine that we use for our implementations is given. A very superficial

---

* It is the policy of Bellcore to avoid any statements of comparative analysis or evaluation of products or vendors. Any mention of products or vendors in this presentation or accompanying printed materials is done where necessary for the sake of scientific accuracy and precision, or to provide an example of a technology for illustrative purposes, and should not be construed as either a positive or negative commentary on that product or vendor. Neither the inclusion of a product or a vendor in this presentation or accompanying printed materials, nor the omission of a product or a vendor, should be interpreted as indicating a position or opinion of that product or vendor on the part of the presenter or Bellcore.

** In the mean time Dave Rusin from the University of Northern Illinois has found a 42 digit factor, using Peter Montgomery's elliptic curve program.

description of the elliptic curve method, and our first SIMD implementation are presented in Section 3. Section 4 describes our block-wise SIMD multi-precision modular integer arithmetic. The second elliptic curve implementation, which is based on this block-wise arithmetic, is presented in Section 5, along with some results concerning random generators with very long period, that were constructed using this second elliptic curve implementation.

2. The hardware

This section contains a short overview of the 16K MasPar, the massively parallel computer that we have used for the implementations to be described in this paper. Our description is *very* incomplete, and only covers those aspects of the machine that are referred to in the following sections. For a complete description of the MasPar we refer to the manuals, like [7].

The 16K MasPar is a SIMD machine, consisting of, roughly, a *front end*, an *array control unit* (ACU), and a 128×128 array of *processing elements* (PE array). Masks, or conditional statements, can be used to select and change a subset of active processors in the PE array, the so-called *active set*. The fact that it is a SIMD machine means that instructions are carried out sequentially, and that instructions involving parallel data are executed simultaneously by all processors in the active set, while the other processors in the PE array are idle. The instructions involving singular (i.e., non-parallel) data are executed either on the front end or on the ACU; for the purposes of our descriptions the front end and the ACU play the same role.

According to our rough measurements, each PE can carry out approximately $2 \cdot 10^5$ additions on 32 bit integers per second, and can be regarded as a 0.2 MIPS processor. Furthermore, each PE has 64KBytes of memory, which implies that the entire PE array has 1GByte of memory. Each processor can communicate with its north, northeast, east, southeast, south, southwest, west, and northwest neighbor, with toroidal wraparound. Actually, a processor can send data to a processor at any distance in one of these eight directions, with the possibility that all processors that lie in between also get a copy of the transmitted data. There is also a global router that allows any processor to communicate with any other processor, but we never needed it.

Each job has a size between 8K and 64K, reflecting the amount of PE-memory it uses. Only those jobs which together occupy at most 64K are scheduled in a round robin fashion, giving each job 10 seconds before it is preempted, while the others must wait. This means that jobs are never swapped out of PE-memory.

For our implementations we used the MasPar Parallel Application Language MPL, which is from our perspective, a simple extension of C.

3. Elliptic curve method

In this section we briefly describe our first SIMD implementation of the elliptic curve method. For a detailed description of the elliptic curve method, hints for its implementation and parameter choices, we refer to [3; 5; 10; 11]. For our purposes it suffices to know that the elliptic curve method consists of a number of independent trials. For each trial an elliptic curve E modulo n and a point x in a group G related to E are randomly selected. The group operation in G, which we will write multiplicatively, consists of several additions, subtractions, multiplications, and inversions of integers modulo n, and can be carried out in time $O((\log n)^2)$ per operation. Clearly, the group operation breaks down if some integer y for which $\gcd(n, y) \neq 1$ has to be inverted modulo n in the course of a group operation; in this case a factor of n has been found.

This is exploited as follows. Using the group operation, the point x is raised to a huge power k consisting of the product of all prime powers below a certain bound B_1. The trial is lucky if this computation cannot be completed because the group operation breaks down, since in this case a factor of n has been found. The trial fails if $x^k \in G$ has been computed successfully. If p is n's smallest prime divisor and $B_1 = L_p[\sqrt{1/2}]$, then each trial has probability $L_p[-\sqrt{1/2}]$ to factor n, as explained in [3; 5]. This implies that the number of independent trials needed to factor n can be expected to be $L_p[\sqrt{1/2}]$. One trial takes time $O((\log n)^2 B_1)$, from which the total expected run time $O((\log n)^2 L_p[\sqrt{2}])$ follows.

The computation of $x^k \in G$ is usually referred to as the first phase of the algorithm. In the second phase $x^{kq} \in G$ is computed for all primes q in $[B_1, B_2]$ for some bound B_2. This requires approximately an additional $\pi(B_2) - \pi(B_1)$ group operations, where $\pi(b)$ denotes the number of primes $\leq b$. It appears to be close to optimal to select B_1 and B_2 in such a way that the two phases take approximately the same amount of time, which makes B_2 an order of magnitude larger than B_1.

The optimal parameter choice for the elliptic curve method depends on the unknown factor p of n, and therefore cannot be known beforehand. A common approach is to do a few trials with fairly low B_1 (and corresponding B_2), upon failure a few trials with slightly larger B_1, and so on, until either the number is factored or the factoring attempt is aborted. The expected amount of work to find a certain factor usually varies only slightly: the success probability for t trials with bound B_1 is not dramatically different from the success probability for $c \cdot t$ trials with bound B_1/c, but only for c in a limited range, say $1/2 < c < 2$. The optimal parameter choices depend on the implementation as well. For instance, for the implementation described in [4], the optimal choices are $t = 300$ and $B_1 = 65000$ to find 25 digit factors with a 60% success probability, $t = 950$, $B_1 = 275000$ for 30 digits, and $t = 2300$, $B_1 = 1100000$ for 35 digits.

The trials of the elliptic curve method are independent, but any number of them can be carried out simultaneously. Since the sequence of operations involved in the computation of $x^k \in G$ depends only on the value of k, and *not* on the actual data to which the operations are applied, t elliptic curve trials can be carried out in parallel on a t processor SIMD machine. An exception occurs if one of the trials factors n, but in that case the process can be terminated. Using this elementary approach we ran 16K trials in parallel on the 16K MasPar, with each of the 16K PE's working on its own elliptic curve, uniquely generated using a single random seed. For this purpose the multi precision integer arithmetic used in [4] was adapted to the MasPar, in such a way that each PE operates independently on its own extended precision integers, simultaneous with the other PE's (in the active set). Thus we can carry out 16K elementary operations $(+, -, *, \text{quotient and remainder})$ on extended precision integers in parallel, without interprocessor communication.

Because inversions modulo n are slow, particularly so on a SIMD machine where the cost is determined by the processor that needs the most iterations, we kept track of the numerators and denominators (modulo n) of the group elements, without performing the inversions. But since these inversions are supposed to lead to the factorization of n, they cannot be avoided entirely. At regular intervals we therefore computed the product modulo n of all denominators (using 14 (i.e., $\log_2(16K)$) multiplications modulo n on the PE array), and computed the greatest common divisor g of the resulting product and n on the (much faster) front end. If g turns out to be > 1, the PE's can be inspected to see how many of them found a factor, or to refine the factorization if g is not prime. The latter usually happened only in the presence of several very small factors (up to 10 digits); larger factors are usually found on only one PE.

Addition and subtraction modulo n are easily made efficient in SIMD mode, but multiplication modulo n is more problematic. This is caused by the remainder computation modulo n, where the instruction stream is more dependent on the values involved. We used the so-called *Montgomery representation* throughout our program, because it allows a modular multiplication that is oblivious of the data involved, without affecting the addition or the subtraction; see Section 4 or [9].

Although these and various other improvements considerably enhanced the performance of our first implementation, there was not much we could do about the fact that a 0.2 MIPS PE is a fairly slow processor: it took about a day to complete 16K curves with $B_1 = 50000$ and $B_2 = 10^6$ for a 100 digit n. With these parameters one can almost guarantee that all factors up to 28 digits will be found, but this is not an optimal parameter choice. To find 28 digit factors far fewer curves with much larger B_1 would be better, whereas the optimal B_1 that corresponds to 16K trials is more than 10^7. The latter would be a good choice if one wants to look for 45 digit factors, but it would require an inordinate amount of time on this machine.

Consequently, this parallel elliptic curve program is not ideally suited for the present MasPar. For future generations of SIMD machines, however, our program might turn out to be useful: if future PE's run at speeds comparable to that of current workstations, then 16K parallel trials with matching bounds could be processed in at most a few days.

Given the current PE speed, the only way to get a better parallel elliptic curve program on the 16K MasPar seems to be to divide the work per curve over r PE's, for some r. This would decrease the number of trials by r and hopefully increase the speed per curve by the same factor, thus allowing parameter choices which are closer to optimal for factors in the 30 digit range. We achieved this by designing a multi precision integer arithmetic that is entirely different from the one mentioned above. This arithmetic will be described in the next section.

4. Block-wise modular arithmetic

In this section we describe an alternative integer arithmetic for SIMD machines, that essentially reconfigures the machine into a machine with fewer but faster processors, at least for arithmetic modulo a fixed integer n.

Let b be some small integer such that arithmetic operations on b bit integers can be carried out efficiently on a PE. Because multiplication of 32 bit integers with a 64 bit result is the most efficient multiplication that a PE can perform, we used $b = 30$, so that we could also add without overflow problems. Let r be the smallest integer with $2^{b \cdot r} > n$, where n is the odd number being factored. Each of the 128 rows of 128 PE's is divided into $u = \lfloor 128/(r+1) \rfloor$ disjoint *blocks* of $r + 1$ consecutive PE's, and $128 - (r+1) \cdot u$ idle

PE's. Thus, there are $128 * u$ blocks. The active set consists of the PE's that are contained in a block.

Suppose that the consecutive PE's in a block are numbered from 0 to r. If the ith PE contains a b bit integer $v_i \geq 0$, then $v_0, v_1, \ldots, v_r$ together represent the number $v = \sum_{i=0}^{r} v_i 2^{b \cdot i}$. Since we use the integers in $\{0, 1, \ldots, n-1\}$ to represent residue classes modulo n, we usually have that $v < n$ and therefore $v_r = 0$. This extra PE is used in the multiplication modulo n. Let $v = \sum_{i=0}^{r} v_i 2^{b \cdot i}$ and $w = \sum_{i=0}^{r} w_i 2^{b \cdot i}$ be two integers modulo n. To compute the sum $s = v + w$ the $r + 1$ PE's compute $\bar{s}_i = v_i + w_i$ and $c_i = \bar{s}_i / 2^b$, next c_i is sent by the ith PE to the neighboring $(i+1)$th PE, and finally s_i is computed as $\bar{s}_i - c_i \cdot 2^b + c_{i-1}$. In the unlikely event that one of the s_i is still $\geq 2^b$ the carry propagation is repeated until all s_i are $< 2^b$. Here we note that it requires only one fast instruction on the MasPar to check if there are still any carries to be propagated. To complete the addition modulo n, we simply subtract n from s, using a similar technique; if $s - n$ is negative then s is the final outcome, otherwise it is $s - n$.

Examples that require r carry propagation steps are easy to construct, and a depth $O(\log r)$ carry propagation tree would give a better worst case performance. Our simplistic approach, however, works on average much faster because the second carry propagation step hardly ever occurs.

Multiplication modulo n within blocks is more complicated. As in the first elliptic curve implementation we used the Montgomery representation to avoid divisions. Let $R = 2^{b \cdot r}$, the smallest power of 2^b larger than n. The Montgomery representation $\tilde{x}$ of an integer x modulo n is the integer $x \cdot R \bmod n \in \{0, 1, \ldots, n-1\}$. Addition and subtraction of numbers in Montgomery representation is not different from ordinary addition and subtraction modulo n, and is carried out as described above. Multiplication, however, becomes much simpler than ordinary multiplication modulo n. Let z be such that $z \equiv x \cdot y \bmod n$. Then $\tilde{z}$ equals $\tilde{x} \cdot \tilde{y} / R \bmod n$. This $\tilde{z}$ can be computed efficiently as follows. First compute $v = \tilde{x} \cdot \tilde{y}$. Let $v = \sum_{i=0}^{2r} v_i 2^{b \cdot i}$ and d be such that $d \cdot n \equiv -1 \bmod 2^b$, which is well-defined because n is odd. Next, for $i = 0, 1, \ldots, r - 1$ in succession replace v by $v + 2^{b \cdot i} \cdot n \cdot (v_i \cdot d \bmod 2^b)$, where the v_i for $i > 0$ are the radix 2^b digits of the v that was computed in the previous iteration. Notice that after iteration j the new v_j is zero, and that the new v is congruent to the old v modulo n. Consequently, the resulting v_0 through v_{r-1} are all zero, and the division by R can be carried out by simply shifting the resulting v to the right. The result might be $\geq n$, in which case it suffices to subtract n once to make it $< n$. Montgomery multiplication can be carried out using $2r^2 + r$ multiplications on b-bit integers.

By merging the iterations for the (ordinary) multiplication of $\tilde{x}$ and $\tilde{y}$, and the division modulo n by R, the multiplication of numbers in Montgomery representation can be done quite easily in a block of PE's. Straightforward application of the above algorithm leads to a block-wise modular multiplication that can be made to fit in blocks of only r consecutive PE's, with 3 multiplications per iteration: two on all PE's in the block (on different data), but one (the computation of $v_i \cdot d \bmod 2^b$) that operates on identical data for all PE's in the block or that could be carried out by only one PE and sent to the others. This is inefficient, because it requires time for $3r$ multiplications on r PE's in parallel, instead of the $2r + 1 = (2r^2 + 1)/r$ multiplications we hoped for. As shown in [2] the $3r$ can be improved to $2r + 2$, giving a ratio which is close to optimal. Using this method we got an acceptable speed for the modular multiplication: for a 95 digit n one modular multiplication takes about 0.003 seconds, and because $r = 11$ implies $\lceil 128/(11 + 1) \rceil = 10$ blocks per row, 1280 of these multiplications can be carried out simultaneously.

An additional advantage of the block-wise arithmetic is that all relevant values can be kept in registers, so that costly memory fetches and stores can be avoided. A detailed description can be found [2].

5. A second implementation and results

Incorporation of the modular arithmetic from the previous section into our first SIMD elliptic curve implementation led to a second version of our program that runs one curve per block instead of one curve per processing element. This implies that the number of trials depends on the size of n. For a 95 digit n we get 1280 trials, for 80 digits r becomes 9, and the number of trials goes up to 1536.

We used the following elementary method to lower the number of group operations needed to compute the kth power of the point x on the curve. According to the definition of k given in Section 3, we have $k = \prod_{i=1}^{l} q_i$, where $\{q_1, q_2, \ldots, q_l\}$ is the set of prime powers $< B_1$. The usual way of computing x^k is to first raise x to the power q_1, next raise the result to the power q_2, etc., until all q_i have been processed. Let for some integer m the *weight* $w(m)$ be defined as the number of ones in the binary representation of m. If ordinary repeated squaring and multiplication is used for the exponentiation, then the cost of the computation of $x^k \in G$ is $\sum_{i=1}^{l} \lceil \log_2 q_i \rceil$ squarings in G and $\sum_{i=1}^{l} (w(q_i) - 1)$ multiplications in G.

Let $S = S_1 \cup S_2 \cup \ldots S_s$ be a partition of $\{1, 2, \ldots, l\}$, and let $\bar{q}_j = \prod_{i \in S_j} q_i$. Clearly, x^k can also be computed by first raising x to the power $\bar{q}_1$, next the result to the power $\bar{q}_2$, etc., up to $\bar{q}_s$. The number of squarings in G needed for this computation is approximately the same as the number of squarings given above. The number of multiplications in G, however, can be made substantially smaller by choosing a partition S for which $\sum_{j=1}^r w(\bar{q}_j)$ is small. Finding the best partition with respect to this metric is in general a hard problem. In practice we will have to do with what can be found in a reasonable amount of time. Using a simple greedy algorithm (and $B_1 = 100000$) we found a partition in subsets of cardinality at most two that had approximately half the original weight; Bill Cook used this solution to derive an optimal partition under the same restrictions, but the resulting weight was not significantly lower. Next, we considered subsets of cardinality at most three. This resulted in approximately a third of the original weight. Given how much time it took us to find these triples, this is probably the best we may hope to achieve. The triples were found by means of a greedy-type algorithm on the 16K MasPar. To make the run times acceptable we processed the primes in intervals. More precisely, for each $i \in \{1, 2, \ldots, 20\}$ we determined partitions into triples of the prime powers $< 2 \cdot 10^6$ for the primes in the interval $[(i-1) \cdot 10^5, i \cdot 10^5]$. This separation into intervals is also useful because it allows a choice of bounds. The total run time of the program was reduced by 18% using this technique. To give an example, the primes 1028107, 1030639, and 1097101 have weights 10, 16, and 11, but their product has weight 8; in binary

$$11111011000000001011 \cdot 11111011100111101111 \cdot 10000101111101100011101 =$$

$$1000000100010000001000000010000000000000001000001000000000000001.$$

This example is remarkable but by no means exceptional.

For n with $r = 11$ it takes about 34 hours to complete 1280 elliptic curve trials with $B_1 = 10^6$ and $B_2 = 2 \cdot 10^7$. For n with $r = 12$ it takes approximately $34 * 12/11$ hours to complete 1152 trials with the same bounds, and other timings can be derived similarly. Consequently, this second elliptic curve program allows a much more balanced choice of the parameters for a search of factors in the 30 digit range. We have used the program to factor various numbers from the list of composite numbers from [1] and many numbers from the 'Partition List' of the 'RSA Data Security Factoring Challenge.' To date the largest factor we have found has 40 digits, which was a new elliptic curve factoring record:

$$p(11279) = 2^6 \cdot 5 \cdot 8418\,735626\,949973\,617503 \cdot$$

$$1232\,079689\,567662\,686148\,201863\,995544\,247703 \cdot$$

$$78\,507734\,924917\,342278\,622201\,969372\,653526\,213641\,483293.$$

The number factored was the 89 digit product of the last two factors, and $p(11279)$ denotes the 11279th partition number. The factorization was found by one of 1408 trials with $B_1 = 10^6$ after $q = 1208269$ in the second phase, which means that we have been quite lucky since finding the factor at this point happens with probability 0.7%*.

We found another, unexpected, application of our SIMD elliptic curve program in [6]. In this paper a new class of pseudo random generators is introduced with the remarkable property that they have provably long periods. Such generators are based on integers b, r, and s such that $m = b^r \pm b^s \pm 1$ is prime, and have a period equal to the order of b modulo m. The size of b depends on the type of pseudo random number generator that is needed; typically it varies between 16 and 64 bits. Consequently, primes m as above can fairly easily be found, if r and s are not excessively large (typically they vary between 10 and 50). Establishing the order of b, however, requires factoring $m - 1$, which might be (and usually is) hard. We found, among others, that $b = 2^{53} - 1052$ has order $(m - 1)/2$ modulo the 511 digit prime $m = b^{32} + b^{16} - 1$, where the factorization of the factor $b^8 + 1$ of $m - 1$ was found using our SIMD elliptic curve implementation:

$$b^8 + 1 = 257 \cdot 16673 \cdot 275422\,758002\,613571\,762817 \cdot$$

$$29\,464604\,796724\,055573\,394001 \cdot 65726\,748717\,597552\,856331\,429857 \cdot$$

$$18\,955214\,139747\,587789\,390113\,133069\,907813\,629489.$$

This was one of the few examples where the global inversion on the front end produced a composite factor: one block had found the 24 digit factor and simultaneously an other block had found the 26 digit factor. This b and m provide an attractive way to generate double-precision floating point pseudo random numbers using just addition.

---

* This corresponds to a 0.0006% probability of success per curve. For comparison, the 42 digit factor referred to above, was found with $B_1 = 2 \cdot 10^6$, and $B_2 \approx 10^8$, which leads to a 0.003% probability of success per curve. The number of curves used in that factorization is unknown to us.

References

1. J. Brillhart, D. H. Lehmer, J. L. Selfridge, B. Tuckerman, S. S. Wagstaff, Jr., *Factorizations of $b^n \pm 1$, $b = 2, 3, 5, 6, 7, 10, 11, 12$ up to high powers*, second edition, Contemporary Mathematics **22**, Amer. Math. Soc., Providence, 1988.

2. B. Dixon, A. K. Lenstra, *Systolic Montgomery multiplication*, in preparation.

3. A. K. Lenstra, H. W. Lenstra, Jr., *Algorithms in number theory*, Chapter 12 in: J. van Leeuwen (ed.), *Handbook of theoretical computer science*, Volume A, *Algorithms and complexity*, Elsevier, Amsterdam, 1990.

4. A. K. Lenstra, M. S. Manasse, *Factoring by electronic mail*, Advances in Cryptology, Eurocrypt '89, Lecture Notes in Comput. Sci. **434** (1990), 355–371.

5. H. W. Lenstra, Jr., *Factoring integers with elliptic curves*, Ann. of Math. **126** (1987), 649–673.

6. G. Marsaglia, A. Zaman, *A new class of random number generators*, Ann. of Appl. Prob. 1 (1991), 462–480.

7. *MasPar MP-1 principles of operation*, MasPar Computer Corporation, Sunnyvale, CA, 1989.

8. U. M. Maurer, Y. Yacobi, *Non-interactive public key cryptography*, Advances in Cryptology, Eurocrypt '91, Lecture Notes in Comput. Sci., to appear.

9. P. L. Montgomery, *Modular multiplication without trial division*, Math. Comp. **44** (1985), 519–521.

10. P. L. Montgomery, *Speeding the Pollard and elliptic curve methods of factorization*, Math. Comp. 48 (1987), 243–264.

11. R. D. Silverman, S. W. Wagstaff, Jr., *A practical analysis of the elliptic curve factoring algorithm*, manuscript.

The Eurocrypt'92 Controversial Issue
Trapdoor Primes and Moduli

Introduction

Motivated by the public controversy surrounding the draft standard for digital signatures (DSS) proposed by NIST, the Program Committee of Eurocrypt'92 decided to hold a panel discussion on the larger issue of trapdoor primes and moduli. The panel members were:

Yvo Desmedt, University of Wisconsin

Peter Landrock, Aarhus University

Arjen Lenstra, Bellcore

Kevin McCurley, Sandia National Laboratories

Andrew Odlyzko, AT&T Bell Laboratories

Rainer Rueppel, R^3 Security Engineering

Miles Smid, National Institute of Standards and Technology

Each of the panel members was given time to make a personal statement on the subject. Then an open discussion followed. For this report each of the panel members was asked to provide a summary of his own personal statement. The following contributions are ordered in the same sequence as the statements were given at Eurocrypt'92.

For people interested in the public discussion on DSS, the special section in the July 1992 issue of the Communications of the ACM provides further information on the ongoing debate.

Rainer A. Rueppel

R^3 Security Engineering

There is an increased awareness that the electronic exchange of data requires security. Security comes in two flavours: authenticity and confidentiality. Authenticity means that the receiver can verify the origin and the integrity of a received message, confidentiality means that only the intended recipient is able to read the message. Two parties wishing to communicate securely must use exactly the same cryptographic algorithms and must be in possession of the right keys. To provide security in open systems requires national and international standards. Therefore, in the US the National Institute for Standardization (NIST) has developed and proposed the Digital Signature Standard (DSS). The objective of the DSS is to provide authenticity. Why is there such an uproar over this standards proposal which after all seems to be a step in the right direction ?

R.A. Rueppel (Ed.): Advances in Cryptology - EUROCRYPT '92, LNCS 658, pp. 194-199, 1993.
© Springer-Verlag Berlin Heidelberg 1993

The public debate reveals various levels of interests. There is a level of social and political interests. Citizens and corporate users are concerned about the privacy and integrity of their communications. The government is concerned about the country's economy, about national security and law enforcement. There is some room for mutual distrust on this level.

But there is also a level of personal and business interests. There are manufacturers who want to protect their investments in security technology, there are patent holders who are concerned about royalties, there are scientists who use their reputation against or in favour of the DSS. Not many other research fields react so nervously to claims and allegations as cryptology. But after all, the security of modern computer and communication systems is a matter of trust; it cannot be proven, it can only be reassured to some degree.

It might be that the harsh attacks and criticism of the DSS will harm the larger issue of integrity and privacy protection. For non-experts it is difficult to follow the scientific discussion, the result might be a general distrust of all cryptographic techniques and standards. It might also be that the harsh attacks will improve the DSS (a first result is NIST's increase of the allowable size of the prime p). But the issue is not RSA or DSS or any other signature standard. In all likelihood there will be more than one signature standard anyway. The main issue is to establish trust in the integrity and the privacy of stored and communicated data without inhibiting the flow of information or the access to information services. Regarding the present situation where there is no security available on public telecommunication services, any step in the direction of secure systems must be considered a progress.

Arjen K. Lenstra

Bellcore

In this note, a trapdoor for a public key cryptosystem is an additional piece of information about the public key that undermines the computational infeasibility to derive the corresponding secret private key.

In RSA, the construction of a public/private key pair requires, as far as we know, the knowledge of secret information (i.e., the factorization) about the modulus. Therefore, in RSA one necessarily has to trust the vendor of the modulus. It hardly makes sense for a vendor to put trapdoors in the moduli he sells because the secret information is available to him anyway. If the modulus is not bought but generated using some software package, either all programmers involved in the production of that software package have to be trusted or the software has to be verified very carefully, which is far beyond the capabilities of the average user: it is not hard to put a trapdoor into moduli in such a way that the trap is undetectable for outsiders but easily recognizable for insiders.

In discrete logarithm based cryptosystems no further secret information about the modulus (i.e., the prime) is needed to construct a public/private key pair. Therefore, a sufficiently large prime from any source can safely be used, at least if it does not contain a trapdoor. For users who cannot be sure how much to trust the entity from which they receive their prime this may pose a threat. A well-known way to trap a prime p is to generate it such that p-1 has only small factors. This works for any size p, but is easily detectable, and therefore poses no real threat. Another trap is to generate p in such a way that p divides $f(X/Y)*(Y**d)$ for an integral polynomial f of small degree d with small coefficients, and integers X and Y close to $p**(1/d)$. For the discrete logarithm problem modulo such primes a fairly recent algorithm can be used which is based on the number field sieve, and which is faster than any of the previous methods. It is expected that this algorithm is currently only practical for trapped primes up to about 600 bits, so this trap only makes sense for primes up to that size. Furthermore, this kind of trap can be detected, although this requires more work than an average user will be able to invest. The probability that a randomly chosen prime turns out to be trapped in this way is negligibly small.

Miles E. Smid

National Institute of Standards and Technology

DSA "Trapdoor"

A claim was made that a dishonest Certification Authority could purposely select a value of p for its own users which would permit the Certification Authority to recover the private keys of the users. This property was called a "trapdoor" in the proposed NIST Digital Signature Algorithm (DSA).

Response:

No evidence of an intent to put a "trapdoor" in the DSA has been presented.

The NIST proposed Digital Signature Standard (DSS) specifies a digital signature algorithm. It does not discuss all the ways the algorithm may be used or misused. The Qualifications section of the DSS Announcement states that "The responsible authority in each agency or department shall assure that an overall implementation provides an acceptable level of security." The proposed DSS specifically states that, "Systems for certifying credentials and distributing certificates are beyond the scope of this standard." Therefore, one would not expect an algorithm specification standard to cover the case of a dishonest certification authority.

The DSS allows users to generate their own primes, p and q. The DSS also allows the user to use primes generated by a trusted party or a Certification Authority. If primes are known to be randomly generated, the user can even accept primes generated by a distrusted Certification Authority. One can construct special primes that are considered weak, and if they were used the private keys of the users might be recovered. (Note that many other algorithms have similar weak values.) However, the probability of generating a weak prime at random is infinitesimally small. (The probability of generating a weak p at random has been estimated to be less than 10^{-90}.)

Two responses to the NIST request for comments pointed out that the use of a one-way function, such as the NIST proposed Secure Hash Algorithm (SHA), in the process that generates p and q could ensure that weak values occur only randomly. By making publicly known the input to the SHA, the resulting p, the resulting q, and the process, the user would be able to verify that weak primes were not purposely constructed. A technique which makes use of the SHA in the generation of DSA primes will be proposed by NIST.

Warning! As with all systems using a Certification Authority, the Certification Authority must be trusted to correctly establish the binding between the user's identity and the user's public key.

Kevin S. McCurley

Sandia National Laboratories

In their seminal 1976 paper on public-key cryptography, Diffie and Hellman described a trap-door cipher as one that ". . . allows the designer to break the system after he has sold it to a client and yet falsely maintain his reputation as a builder of secure systems." The subject of trapdoor moduli has lately received much attention in the popular press, particularly as it applies to the U.S. draft standard for digital signatures, known as DSS. In my opinion, the situation has been wildly distorted by the press, leading to a general distrust of DSS that lacks any serious scientific justification.

So far, it has not been demonstrated that trapdoor moduli for the discrete logarithm problem can be constructed such that a) they are hard to detect, and b) knowledge of the trapdoor provides a quantifiable computational advantage for parameter sizes that could actually be computed by known methods, even with foreseeable machines.

Even if trapdoor keys can eventually be constructed, this will have few consequences for DSS. Many people skilled in the art will recognize that there are numerous ways to build trapdoor moduli for RSA, but all this means is that one needs to be careful in the method of selecting keys. For example, if a user is provided software from another party to generate keys for a cryptographic scheme, then the user needs to trust the provider of the software to produce keys that are truly random and free of any predictability.

Most of the attention devoted to the issue of trapdoor primes and moduli has been motivated by political and business influences rather than scientific concerns. Some people hold political views that cause them to distrust DSS for the simple reason that it was proposed by a US government agency. To this day, some people believe that the US government designed DES to incorporate a trap-door, making it easy for them to break it. No evidence has ever been put forward for the existence of such a trapdoor, but conspiracy theories persist. Another source of influence is the fact that cryptographic policy and standards can have serious business consequences. Care needs to be taken to distinguish conclusions that are motivated by such concerns rather than objective scientific judgements.

Yvo Desmedt

University of Wisconsin

Because NIST has not standardized a prime in the DSS proposal, the discussion on trapdoor primes is not an issue.Moreover the probability of having a weak prime is small.

It is unfortunate that many comments on the DSS proposal, e.g., claims that NIST has lost considerable credibility with the non-military cryptographic research community, have been exaggerations. It seems that these comments have no scientific grounds and that, unfortunately, other interests have overshadowed a scientific discussion. The real issue was the size of p, which now has been adjusted by NIST. Implementations which neither use p's of 1024 bits nor allow to update p to a 1024 bit should not receive a standard certification, but only be validated for a very short time. Although there is no scientific evidence today that q could be too small, the DSS proposal could stimulate a lot of research on breaking such discrete logarithms.

There is a need for future standards, such as one for privacy protection, one for very fast authentication, and standards that allow to fulfill different needs such as threshold signatures (signatures in which the secret key is shared such that a threshold of shareholders can sign, but less cannot) which can be obtained based on RSA, but it is not known how to achieve them (in a practical way) using DSS.

Andrew Odlyzko

AT&T Bell Laboratories

1. Progress in factoring and discrete logarithms

There has been substantial progress on both fronts in the last 15 years. When the RSA system was invented, the largest integers that experts could be sure of being able to factor with the algorithms and computers that were available to them at that time were on the order of 38 to 45 decimal digits. Today, integers of between 115 and 130 digits can be factored. Many of the projections that one sees of where the field is going take into account the progress in computer technology, with individual machines becoming much faster and more of them becoming available on networks. However, it is only prudent to allow for progress in algorithms. If one considers what it was that allowed the advance from factoring 38 decimal digit integers 15 years ago to factoring 115 decimal digit integers today, it seems that only about half of this was due to computer technology, with the other half coming from better algorithms. There is no reason to expect that the future will be any different. Terefore the moduli that are used should be large enough to guard against such algorithmic improvements.

2. Trapdoor primes and moduli

Trapdoor primes and moduli are not a significant issue. The only methods that have been suggested for constructing such primes and moduli yield only a slight advantage to the person who chooses them, an advantage that depends on the use of particular algorithms for factoring and discrete logarithms. If one chooses large enough moduli to guard against advances in algorithms and computers that one can prudently expect, this threat disappears.

Peter Landrock

Aarhus University

In any known public key scheme based on properties of primes, there seems be so me primes that are weaker than others. Thus it must be considered in the key generation phase first of all if these keys should be avoided explicitly by the key generation program, and - if so - secondly how they can be avoided. An important property of any key generation program is that the keys be chosen randomly from a key space, which is sufficiently large. Any key generation program not achieving this at an acceptable level, which can be estimated by its creator, is dangerous to use.

Once the properties of weak keys have been identified, the strategy to try to avoid them should depend on an estimate of the probability that the key generation program will return a prime with that property. For instance, if DES keys are generated in a random manner, there is no need to check if a weak or semi-weak key is returned. The probability that this will happen is so small that it can be completely ignored .

Consequently, if the estimation shows that the probability for the occurrence of weak primes is sufficiently small, the property can be ignored. We have seen some papers classifying all primes with a certain property. Then a warning is issued that primes with this property should be avoided by the key generation program. However, in most situations, any phenomenon that can be classified in this manner is so unlikely to occur that there is no need to worry.

If the propability in question is not insignificant, it will be necessary to avoid primes with that kind of property in the key generation. As an example, it can be estimated that the propability that a randomly chosen prime of 256 bits is strong as an RSA key is not sufficiently large, whereas a prime of 512 bits is strong with sufficiently high probability. (See [1])

An independent problem which needs attention is that of a sufficiently good random generator. However, this must be solved as a separate and equally important issue, which has nothing to do with prime generation.

[1] J. Brandt, I. Damgaard and P. Landrock, "Speeding up up prime number generation," Abstracts of ASIACRYPT'91, Fujiyoshida, Japan.

Fast Exponentiation with Precomputation [*]
(Extended Abstract)[**]

Ernest F. Brickell[1], Daniel M. Gordon[2],
Kevin S. McCurley[1], and David B. Wilson[3]

[1] Division 1423, Sandia National Laboratories, Albuquerque, NM 87185
[2] Department of Computer Science, University of Georgia, Athens, GA 30602
[3] Department of Mathematics, M.I.T., Cambridge, MA 02139

Abstract. In several cryptographic systems, a fixed element g of a group (generally $\mathbf{Z}/q\mathbf{Z}$) is repeatedly raised to many different powers. In this paper we present a practical method of speeding up such systems, using precomputed values to reduce the number of multiplications needed. In practice this provides a substantial improvement over the level of performance that can be obtained using addition chains, and allows the computation of g^n for $n < N$ in $O(\log N/\log\log N)$ group multiplications. We also show how these methods can be parallelized, to compute powers in $O(\log\log N)$ group multiplications with $O(\log N/\log\log N)$ processors.

1 Introduction

The problem of efficiently evaluating powers has been studied by many people (see [6, Sect. 4.6.4] for an extensive survey). One standard method is to define an *addition chain*. Let $l(n)$ denote the length of the shortest addition chain for an exponent n (and the smallest number of multiplications possible by this approach). Then it is known that

$$\lceil \log n \rceil \leq l(n) \leq \lfloor \log n \rfloor + \nu(n) - 1, \tag{1}$$

where logs are to base 2 and $\nu(n)$ is the number of ones in the binary representation of n.

Addition chains can be used to great advantage when the exponent n is fixed (as in the RSA cryptosystem), and the goal is to quickly compute x^n for randomly chosen bases x. We shall consider a slightly different problem in this paper. For many cryptosystems (e.g. [3],[4],[8],[1]), the dominating computation is to compute for a fixed base g the power g^n for a randomly chosen exponent n. For this problem, we achieve a substantial improvement over addition chains by storing a set of precomputed values.

[*] This research was supported in part by the U.S. Department of Energy under contract number DE-AC04-76DP00789, in part by the Defense Advanced Research Projects Agency under Grant N00014-91-J-1698, and in part by an ONR-NDSEG fellowship.

[**] PATENT CAUTION: This document may reveal patentable subject matter.

R.A. Rueppel (Ed.): Advances in Cryptology - EUROCRYPT '92, LNCS 658, pp. 200-207, 1993.
© Springer-Verlag Berlin Heidelberg 1993

We will assume that g is an element of a group such as $\mathbb{Z}/q\mathbb{Z}$, where q is a large integer (say 512 bits). We address the problem of repeatedly calculating powers of g up to g^N, where N is also large. In the Schnorr scheme [8], N is about 140 bits, the DSS scheme uses N of 160 bits, and the Brickell-McCurley scheme [4] uses N of 512 bits. Our results will apply to any group, so that the speedups work as well in an elliptic curve group as for modular exponentiation. Specialized results can also be derived from this approach for the special case of $GF(p^k)$, but we will defer these to the full paper. We will assume that operations other than multiplications in the group will use negligible time.

As an example of the practicality of the schemes that we present, consider the exponentiation required for Diffie-Hellman key exchange using a 512-bit prime and 512-bit random exponent. Using the square-and-multiply scheme (see [6], page 442), we would expect to perform $765 + 3/2^{512}$ modular multiplications on average and 1022 multiplications in the worst case, using storage of at least 64 bytes. Results in [2] report that addition chains of length around 608 can be computed, resulting in a 21% improvement over the average number for the binary method. It follows from (1) that addition chains cannot do better than 512 multiplications for a 512 bit exponent. For one of the schemes that we present here, we expect to perform fewer than 105 modular multiplications on average and 106 in the worst case, using storage of 23168 bytes. This gives better than a seven-fold speedup on average, and about a ten-fold speedup in the worst case over the square and multiply method. Even very small amounts of storage can produce dramatic speedups.

For the rest of this paper, it will be assumed that g is fixed, and n is uniformly distributed on $\{0, \ldots, N-1\}$.

2 Basic strategies

Using the square-and-multiply method, g^n may be computed using at most $2\lceil \log N \rceil - 2$ multiplications, and on average $\leq 3\lceil \log N \rceil/2$ multiplications. By storing a set of precomputed values, we want to reduce the number of multiplications to compute g^n.

One simple method (see [5]) is to precompute the set $\{g^{2^i} | i = 1, \ldots, \lceil \log N \rceil - 1\}$. Then g^n may be computed in $\nu(n) - 1$ multiplications, using $\lceil \log N \rceil$ storage, by multiplying together the powers corresponding to nonzero digits in the binary representation of n.

There is no reason that powers of 2 have to be stored. Suppose we instead precompute and store $g^{x_0}, \ldots, g^{x_{m-1}}$ for some integers $x_0, \ldots, x_{m-1}$. If we are then able to find a decomposition

$$n = \sum_{i=0}^{m-1} a_i x_i, \tag{2}$$

where $0 \le a_i \le h$ for $0 \le i < m$, then we can compute

$$g^n = \prod_{d=1}^{h} c_d^d, \tag{3}$$

where $c_d = \prod_{a_i=d} g^{x_i}$.

If (3) were computed using optimal addition chains for $1, 2, \ldots, h$, the total number of multiplications to compute g^n would be about $m + O(h \log h)$. However, (3) can be computed much more efficiently, as the following result shows.

Theorem 1. *Suppose* $n = \sum_{i=0}^{m-1} a_i x_i$, *where* $0 \le a_i \le h$. *If* g^{x_i} *is precomputed for each* $0 \le i < m$, *and if* $m + h \ge 2$, *then* g^n *can be computed with* $m + h - 2$ *multiplications.*

Proof. The following is an algorithm to compute g^n.

```
b ← 1
a ← 1
for d = h to 1 by −1
        for each i such that a_i = d
               b ← b * g^{x_i}
        a ← a * b.
return a.
```

It is easy to prove by induction that, after going through the loop i times, we have $b = c_h c_{h-1} \cdots c_{h-i+1}$ and $a = c_h^i c_{h-1}^{i-1} \cdots c_{h-i+1}$. After traversing the loop h times, it follows that $a = \prod_{d=1}^{h} c_d^d$.

It remains to count the number of multiplications performed by the algorithm. We shall count only those multiplications where both multiplicands are unequal to 1, since the others can be accomplished simply by assignments. We may assume $n \ne 0$. There are m digits, so the $b \leftarrow b * g^{x_i}$ line gets executed at most m times. The $a \leftarrow a * b$ line gets executed h times. Finally, at least two of these multiplications are free since a and b are initially 1. $\qquad \Box$

Embodied in the algorithm is a method for computing the product $\prod_{d=1}^{h} c_d^d$ in at most $2h - 2$ multiplications. We can argue that, in the absence of any relations between the c_d's, this is optimal. Notice that if we take any algorithm to compute $\prod_{d=1}^{k} c_d^d$ and remove multiplications involving c_k, we have computed $\prod_{d=1}^{k-1} c_d^d$, which takes $2k - 4$ multiplications by our induction hypothesis. There cannot be only one multiplication by c_k, since then c_k would be raised to the same power as whatever it was multiplied by. Therefore at least two extra multiplications are needed.

The most obvious use for (3) is to represent the exponent in base b, using at most $m = \lceil \log_b N \rceil$ digits to do so, and precompute g^{b^k}, for $k = 1, \ldots, \lceil \log_b N \rceil - 1$. Using this algorithm with a base b representation for n, Theorem 1 shows that g^n can be computed in at most $\lceil \log_b N \rceil + b - 3$ multiplications. For a randomly chosen exponent n, we expect that a digit will be zero about $1/b$ of the time, so that on average we expect the $b \leftarrow b * g^{x_i}$ line to be executed $\frac{b-1}{b}\lceil \log_b N \rceil$ times, giving an expected number of multiplications that is at most $\frac{b-1}{b}\lceil \log_b N \rceil + b - 3$. For a 512-bit exponent, the optimal value of b is 26. This method requires at most 127.8 multiplications on average, 132 multiplications in the worst case, and requires 109 stored values.

Note that some minimal effort may be required to convert the exponent from binary to base b, but this is probably negligible compared to the modular multiplications (certainly this is the case for exponentiation in $\mathbb{Z}/q\mathbb{Z}$). Even if this is not the case, then we can simply use base 32, which allows us to compute the digits for the exponent by extracting 5 bits at a time. Using this choice, the scheme will require at most 128.8 multiplications on average, 132 multiplications in the worst case, and 109 stored values.

3 Other number systems

The major problem with the general approach described in the previous section is that we must be able to compute a representation of the form (2). Subject to this constraint, the goal is to choose the parameters to optimize the necessary number of multiplications for a given amount of storage. In this section we shall explore some approaches to this problem.

As our first example of this approach, if $b > 1$, then every integer n such that $|n| \leq (b^m - 1)/2$ may be represented as $\sum_{i=0}^{m-1} a_i b^i$, where each $a_i \in [-\lceil (b - 1)/2 \rceil, \lceil (b-1)/2 \rceil]$ (see Theorem 2 below). If the powers $g^{\pm 1}, g^{\pm b}, \ldots, g^{\pm b^{m-1}}$ are precomputed, then we compute

$$c_d = \prod_{|a_j|=d} g^{\text{sign}(a_j)b^j} \quad .$$

In this case, $m = \lceil \log_b(2N + 1) \rceil$, $h = \lceil (b-1)/2 \rceil$, and the worst case number of multiplications required is $\lceil \log_b(2N + 1) \rceil + \lceil (b - 1)/2 \rceil - 2$. Moreover, since the probability that a digit is nonzero is again at most $(b - 1)/b$, the average number of multiplications required is bounded above by $\lceil \log_b(2N+1) \rceil (b-1)/b + \lceil (b - 1)/2 \rceil - 2$. The storage required is for $2\lceil \log_b(2N + 1) \rceil$ values. For a 512-bit exponent, a good base is 45, resulting in 111.91 multiplications on average and 114 multiplications in the worst case, using 188 stored values.

At one extreme, if we take $h = 1$, we can completely bypass the computation of $\prod_{d=1}^{h} c_d^d$ by storing instead all values g^{db^i}, $1 \leq d < b, 0 \leq i \leq \lceil \log_b N \rceil - 1$, and perform at most $\lceil \log_b N \rceil (b-1)/b - 1$ multiplications on average, and $\lceil \log_b N \rceil - 1$ multiplications in the worst case. For example, with $N = 2^{512}$ we might take $b = 256$ and $h = 1$ to derive a method that takes 62.75 multiplications on average, and 63 multiplications in the worst case. The problem with this method is that it requires $(b-1)\lceil \log_b N \rceil$ stored values. For this case that is 16320 stored values, or 1,044,480 bytes of storage.

By slightly increasing the value of h, we can reduce either the storage or time required. For instance, taking $h = 2$, let $M_2 = \{d \mid 1 \leq d < b , \omega_2(d) \equiv 0$

(mod 2)}, where $\omega_p(d)$ is the largest power of p that divides d, i.e., $k = \omega_p(d)$ if and only if $p^k \parallel d$. It suffices to store the values, $\{g^{db^i} | d \in M_2\}$. Then for $1 \leq a_i < b$, $g^{a_i b^i} = g^{db^i}$ or g^{2db^i} for some $d \in M_2$. Using the same base, this only increases the time by one multiplication, but reduces the storage substantially. For example, with $b = 256$, we achieve an average of at most 63.75 multiplications but reduce the storage to 10880 values, or 696,320 bytes.

Increasing h or decreasing b further increases the time and lowers the storage. Continuing this line of reasoning, we can take $M_3 = \{d \mid 1 \leq d < b, \omega_2(d) + \omega_3(d) \equiv 0 \pmod 2\}$. For example, if we take a base of $b = 128$ for a 512 bit exponent, then we arrive at a method that requires an average of at most 74.42 multiplications, using storage for 5624 values.

In the remainder of this section we shall describe a method that allows us to reduce the amount of computation without such a huge increase in the amount of storage. Call a set of integers D a *basic digit set* for base b if any integer can be represented in base b using digits selected from the set D.

Before we examine the problem of finding basic digit sets for our problem, we should first remark that the difficulty of finding a representation using digits from D is almost exactly the same difficulty as finding the (ordinary) base b representation. The algorithm for finding such a representation was published by Matula [7], and a particularly simple description was later given in [6, Exercise 4.1.19].

In searching for good basic digit sets, we can make use of the following result of Matula [7], which provides a very efficient algorithm for determining if a set is basic.

Theorem 2. *Suppose that D is a complete residue system modulo b. Let $d_{\min} = \min\{s | s \in D\}$ and $d_{\max} = \max\{s | s \in D\}$. Then D is a basic digit set for base b if there are representations for each i with*

$$\frac{-d_{\max}}{b-1} \leq i \leq \frac{-d_{\min}}{b-1}$$

using digits from D.

In the method that we consider now, we shall store powers g^{mb^j}, for $j \leq \lceil \log_b N \rceil$ and m in a set M of multipliers. We need to choose M and h for which

$$D(M, h) = \{km | m \in M, 0 \leq k \leq h\}$$

is a basic digit set. Given a representation $n = \sum_{i=0}^{m-1} d_i b^i$ in terms of this basic digit set, we can represent $d_i = m_i k_i$ and compute

$$g^n = \prod_{k=1}^{h} \left(\prod_{k_i = k} g^{m, b^i} \right)^k = \prod_{k=1}^{h} c_k^k . \tag{4}$$

Another class of multiplier sets is provided by the following:

Theorem 3. *If b is odd, $M = \{\pm 1, \pm 2\}$, and $h = \lfloor b/3 \rfloor$, then $D(M, h)$ is a basic digit set.*

Tables 1 and 2 summarize the effects of the various methods presented above on the storage and complexity of the parameters that might be used for the DSS and Brickell-McCurley schemes, namely 160 and 512 bit exponents respectively. The larger sets of multipliers were found by a computer search. Large sets of good multipliers become harder to find, and use increasing amounts of storage for progressively smaller reductions in computation.

Table 1. Selected parameters for a 160-bit exponent ($N = 2^{160}$). By comparison, the binary method requires about 237 multiplications on average, and 318 multiplications in the worst case.

b	M	h	expected time	worst-case time	storage
12	$\{1\}$	11	50.25	54	45
19	$\{\pm 1\}$	9	43.00	45	76
29	$\{\pm 1, \pm 2\}$	9	39.83	41	134
36	$\{\pm 1, 9, \pm 14, \pm 17\}$	7	36.11	37	219
36	M_3	3	31.14	32	620
64	M_2	2	26.58	27	1134
128	M_3	3	23.82	24	1748
256	M_2	2	20.92	21	2751

Table 2. Selected parameters for a 512-bit exponent ($N = 2^{512}$). By comparison, the binary method requires about 765 multiplications on average and 1022 in the worst case.

b	M	h	expected time	worst-case time	storage
26	$\{1\}$	25	127.81	132	109
45	$\{\pm 1\}$	22	111.91	114	188
53	$\{\pm 1, \pm 2\}$	17	104.28	106	362
67	$\{\pm 1, \pm 2, \pm 23\}$	16	98.72	100	512
64	M_3	3	85.66	87	3096
122	M_3	3	74.39	75	5402
256	M_2	2	63.75	64	10880

For some of the lines of each table, the expected times are actually upper bounds for the expected time. For the others, the expected times were calculated using the assumption that the probability of a digit being zero for any base b basic digit set is $1/b$. We have not proven this in general, but it's a reasonable heuristic that matches empirical results.

For a given N and amount of storage, it seems difficult to prove that a scheme is optimal. However, we can show that the above schemes are asymptotically optimal. Storing powers g^{rb^j} for r in a fixed set of multipliers, the optimal value of b is about $\log N/(\log\log N)^2$, for which $(1+o(1))\log N/\log\log N$ multiplications and $O(\log N/\log\log N)$ stored values are needed. The following theorem shows that we cannot do better with a reasonable amount of storage.

Theorem 4. *If the number of stored values is less than $\log^k N$ for $k \geq 1$, then the number of multiplications required is at least $(1/k + o(1))(\log N/\log\log N)$.*

4 Parallelizing the algorithm

The first method for computing a power g^n that we presented in Sect. 2 consisted of three main steps:

1. Determine a representation $n = a_0 + a_1 b + \ldots + a_{m-1} b^{m-1}$.

2. Calculate $c_d = \displaystyle\prod_{\substack{j=0 \\ a_j = d}}^{m-1} g^{b^j}$ for $d = 1, \ldots, h$.

3. Calculate $g^n = \displaystyle\prod_{d=1}^{h} c_d^d$.

As we mentioned previously, the algorithm of Matula makes the first step easy, even with a large set of multipliers. Most time is spent in the second and third steps. Both of these may be parallelized. Suppose we have h processors. Then for step 2, each processor can calculate its c_d separately. The time needed to calculate c_d depends on the number of a_j's equal to d. Thus the time for step 2 will be the d with the largest number of a's equal to it.

This is equivalent to the maximum bucket occupancy problem: given $k + 1$ balls randomly distributed in h buckets, what is the expected maximum bucket occupancy? This is discussed in [10], in connection with analysis of hashing algorithms. Taking b and h to be $O(\log N/\log\log N)$, so $(k+1)/h = \Theta(1)$, the expected value is

$$\frac{\log h}{\log\log h} = O\left(\frac{\log\log N}{\log\log\log N}\right) .$$

For step 3, each processor can compute c_d^d for one d using a standard addition chain method, taking at most $2\log h$ multiplications. Then the c_d^d's may be combined by multiplying them together in pairs repeatedly to form g^n (this is referred to as *binary fan-in multiplication* in [9]). This takes $\log h$ time.

Therefore, taking $h = O(\log N/\log\log N)$, we may calculate powers in time $O(\log\log N)$ with $O(\log N/\log\log N)$ processors. For example, storing only powers of b, we may compute powers for a 140-bit exponent in the time necessary for 13 multiplications using 15 processors, taking $b = 16$ and $M = \{1\}$. For a 512-bit exponent, we can compute powers with 27 processors in the time for 17 multiplications, using $b = 28$.

The disadvantage to this method is that each processor needs access to each of the powers g^{b^i}, so we either need a shared memory or every power stored at every processor. An alternative approach allows us to store only one power at each processor.

For this method, we will have k processors, each of which computes one $g^{a_i b^i}$ using a stored value and an addition chain for a_i. This will take at most $2 \log h$ time. Then the processors multiply together their results using binary fan-in multiplication to get g^n. The total time spent is at most $2 \log h + \log k$, which is again $O(\log \log N)$ time with $O(\log N / \log \log N)$ processors.

If the number of processors is not a concern, then the optimal choice of base is $b = 2$, for which we need $\log N$ processors and $\log \log N$ time. We could compute powers for a 512-bit exponent with 512 processors in the time for 9 multiplications, and for a 140-bit exponent with 140 processors in the time for 8 multiplications. Taking a larger base reduces the number of processors, but increases the time.

Acknowledgment. We would like to thank Professor Tsutomu Matsumoto of Yokohama National University for informing us of reference [5], and for providing a partial translation.

References

1. A Proposed Federal Information Processing Standard for Digital Signature Standard, *Federal Register*, Volume 56, No. 169, August 31, 1991, pp. 42980-42982.
2. J. Bos and M. Coster, Addition Chain Heuristics, in *Advances in Cryptology - Proceedings of Crypto '89, Lecture Notes in Computer Science*, Volume 435, Springer-Verlag, New York, 1990, pp. 400–407.
3. W. Diffie and M. Hellman, New Directions in Cryptography, *IEEE Transactions on Information Theory* 22 (1976), 472-492.
4. E.F. Brickell and K.S. McCurley, An Interactive Identification Scheme Based on Discrete Logarithms and Factoring, to appear in *Journal of Cryptology*.
5. Ryo Fuji-Hara, Cipher Algorithms and Computational Complexity, Bit 17 (1985), 954-959 (in Japanese).
6. D.E. Knuth, *The Art of Computer Programming*, Vol. 2, *Seminumerical Algorithms*, Second Edition, Addison-Wesley, Massachusetts, 1981.
7. D.W. Matula, Basic digit sets for radix representation, *Journal of the ACM*, 29 (1982), pp. 1131-1143.
8. C.P. Schnorr, Efficient signature generation by smart cards, to appear in *Journal of Cryptology*.
9. D.R. Stinson, Some observations on parallel algorithms for fast exponentiation in $GF(2^n)$, *Siam. J. Comput.*, 19, (1990), pp. 711-717.
10. J.S. Vitter and P. Flajolet, Average-case analysis of algorithms and data structures, in *Handbook of Theoretical Computer Science*, ed. J. van Leeuwen, Elsevier, Amsterdam, 1990, pp. 431-524.

Batch Diffie-Hellman Key Agreement Systems and their Application to Portable Communications

Michael J. Beller   Yacov Yacobi

Bellcore
New Jersey, U.S.A.

June 19, 1991

Abstract.
RSA (Rivest, Shamir and Adleman) is today's most popular public key encryption scheme. Batch-RSA (due to Fiat) is a method to compute many ($n/log_2^2(n)$, where n is the security parameter) RSA decryption operations at a computational cost approaching that of one normal decryption. It requires that all the operations use the same modulus, but distinct, relatively prime in pairs, short, public exponents. A star-like key agreement scheme could use such a system to slash computational complexity at the center. We show a real life example of such a system – secure portable telephony. Unfortunately, in this system Batch-RSA cannot be employed effectively, due to a delay component which arises from the nature of RSA key exchange. We show that mathematical ideas similar to Fiat's can lead to a Batch-Diffie-Hellman key agreement scheme, that does not suffer such delay and is comparable in efficiency to Batch-RSA. We prove that with some precautions, this system is as hard to break as RSA with short public exponent. In practice our method improves processing time at the center by a factor of 6 to 17 when compared to (non-batch) Diffie-Hellman schemes with full-size exponents and moduli in the practical range. Smaller improvements (on the order of 1.6 to 3) are obtainable when compared to a Diffie-Hellman scheme employing abbreviated exponents.

1   Introduction

RSA (Rivest, Shamir and Adleman) [13] is today's most popular public key encryption scheme. Batch-RSA [7] is a method to compute many ($n/log^2(n)$, where n is the security parameter, throughout logarithms are to the base 2) RSA decryption operations at a computational cost approaching that of one normal decryption. It requires that all the operations use the same modulus, but distinct, relatively prime in pairs, short, public exponents. A star-like key agreement scheme could use such a system to slash complexity at the center. We show a real life example of such a system – secure portable telephony. Unfortunately, in this system Batch-RSA can not be employed effectively,

R.A. Rueppel (Ed.): Advances in Cryptology - EUROCRYPT '92, LNCS 658, pp. 208-220, 1993.

due to a delay component which arises from the nature of RSA key exchange. We then show that mathematical ideas similar to Fiat's can lead to a Batch-Diffie-Hellman key agreement scheme, that does not suffer such delay and is comparable in efficiency to Batch-RSA. We prove that with some precautions, this system is as hard to break as RSA with short public exponent.

In practice our method improves processing time at the center by about an order of magnitude when compared to (non-batch) Diffie-Hellman schemes with full-size exponents and moduli in the practical range. Smaller improvements (on the order of 1.6 to 3) are obtainable when compared to a Diffie-Hellman scheme employing abbreviated exponents.

Section 2 describes Fiat's original Batch-RSA method, in section 3 we describe how we plan to use it with a Diffie-Hellman-like system, and in section 4 we analyze the security of our proposed system. Section 5 of this paper describes a manner in which the method of this paper can be applied to improve efficiency in a real portable communications system (PCS), and quantifies the achievable improvement. Section 6 explains why batch RSA introduces unacceptable delays, and motivates batch DH.

2 Batch-RSA

Suppose we have to compute $m_i^{1/e_i} \bmod N$ for $i = 0, 1, 2, \ldots, b-1$, where b is the batch size, and the e_i's are relatively small (and the inverses $1/e_i \bmod \lambda(N)$ are large, $\lambda(N) = (p-1) \cdot (q-1)/2$). The main idea of Batch-RSA is to compute first $c \equiv \prod_{i=0}^{b-1} m_i^{p/e_i} \bmod N$, where $p = \prod_{i=0}^{b-1} e_i$, using a special efficient binary tree structure, to be described later. The second phase is to compute $m \equiv c^{1/p} \bmod N$, which is a full size modular exponentiation (but the cost is spread over b computations). The last phase is to break $m \equiv \prod_{i=0}^{b-1} m_i^{1/e_i} \bmod N$ into its b separate components $m_i^{1/e_i} \bmod N$, $i = 0, 1, 2, \ldots, b-1$, which is the desired output of our computation. This is done using the binary tree developed in the first step in a very efficient way.

The binary tree: To simplify explanations we assume that $b = 2^k$, for some k. Create a complete binary tree where the leaves are labeled $m_0, m_1, m_2, \ldots, m_{b-1}$. A path is identified with the corresponding binary sequence $\in \Sigma^*$, $\Sigma = \{0, 1\}$. We use the symbol ϵ to denote the string of length zero. Right sons are associated with 1, and left sons with 0. We refer to any arc in the tree using the unique path leading to it, i.e. $\eta \in \Sigma^{k'}$, $k' < k$ is the arc in depth k' from the root, which is approached when traversing the tree from the root according to η. If η is of length k then it leads to a leaf labeled m_i s.t. η is the binary representation of i. Let $x \in \Sigma$. ηx denotes a sequence composed of x concatenated to the right of η, $\bar{x}$ denotes the complement of x. Each arc η has label $l(\eta)$. Arcs in the tree are labeled bottom-up according to the following rule.

- Let ηx be a path leading to a leaf associated with message m_i, then $l(\eta \bar{x}) = e_i$.

- For $\eta \in \Sigma^{k'}$, $k' < k-1$, $l(\eta x) = l(\eta \bar{x} 0) \cdot l(\eta \bar{x} 1)$.

The above labeling procedure is independent of the actual messages $\{m_i\}$. It depends on the exponents only, hence if those are fixed, this procedure may be done off line, once and for all. Nodes in the tree contain data, which depend on both the messages and the exponents. We refer to each node by the path leading to it. The data stored in node η is denoted $d(\eta)$. Initially the content of each leaf i is the corresponding message m_i. The content of each node in the tree is computed bottom-up, after its sons were computed using $d(\eta) \equiv d(\eta 0)^{l(\eta 0)} \cdot d(\eta 1)^{l(\eta 1)} \bmod N$. It follows that the content of the root is the desired $d(\epsilon) \equiv \prod_{i=0}^{b-1} m_i^{p/e_i} \bmod N$. This concludes the first phase (see Fig-1).

After computing $M \equiv d(\epsilon)^{1/p} \equiv \prod_{i=0}^{b-1} m_i^{1/e_i} \bmod N$ (second phase), we use the tree (top-down) to break M into its components (third phase). This is done recursively as follows.

Let $0, 1, \ldots, q-1$ be the leaves associated with the left son of the root (and $q, q+1, \ldots, b-1$ with the right son). Note that $l(0) = \prod_{i=0}^{q-1} e_i$, $l(1) = \prod_{i=q}^{b-1} e_i$, and $p = l(0) \cdot l(1)$. Using the Chinese Remaindering algorithm [1] compute X, s.t. $X \equiv 0 \bmod e_i$, $i = 0, \ldots, q-1$, and $X \equiv 1 \bmod e_i$, $i = q, \ldots, b-1$. Here we use the fact that the e_i's are relatively prime in pairs. From the construction of X it follows that $X \equiv 0 \bmod l(0)$ and $X \equiv 1 \bmod l(1)$, hence there exist X_1 and X_0 s.t. $X = l(1) \cdot X_1 + 1$ and $X = l(0) \cdot X_0$. Denote $M_0 \equiv \prod_{i=0}^{q-1} m_i^{1/e_i} \bmod N$, and $M_1 \equiv \prod_{i=q}^{b-1} m_i^{1/e_i} \bmod N$. For convenience we use the shorthand l_0, l_1, d_0, d_1 instead of $l(0), l(1), \ldots$, etc. Since $M_0 \equiv d_0^{l_0/p} \bmod N$, and $M_1 \equiv d_1^{l_1/p} \bmod N$, we have $d_1 \equiv M_1^{l_0} \bmod N$ and $d_0 \equiv M_0^{l_1} \bmod N$. $c \equiv d_0^{l_0} \cdot d_1^{l_1} \bmod N$, hence, $M^X \equiv c^{X/p} \equiv d_0^{l_0(l_1 X_1 + 1)/p} \cdot d_1^{l_1 l_0 X_0/p} \equiv d_0^{X_1} d_1^{X_0} d_0^{l_0/p}$. But the last multiplicand is just M_0, hence $M_0 \equiv M^X/(d_0^{X_1} \cdot d_1^{X_0}) \bmod N$, and $M_1 \equiv M/M_0 \bmod N$. The d's were computed in phase I, and are stored in the nodes. The X's do not depend on the messages and may be computed off-line. The process repeats recursively, until at the leaves the desired output is reached (see Fig-2).

As is shown in [7] the total complexity of a batch computation approaches that of one full size exponentiation.

Using Batch-RSA together with Montgomery's modular reduction

Montgomery's modular reduction [11] is very popular among public-key implementors [6]; so we must address the question "can the two methods coexist?" The answer is positive both theoretically and practically. Montgomery's method addresses additions and multiplications, which comprise the bulk of the computation in Batch-RSA (and Batch-DH, which will be discussed later in the paper). The only operations which can not be done using Montgomery's method are the two divisions that have to be done in each node when going down the tree (phase 3). Asymptotically they are negligible, and practically they are small compared to the larger exponentiations done in each node. The reason is the following: We have to do modular $a \boxminus b/cd$, $e \equiv f/a$, where each of b, c, d, f is given as a Montgomery number (i.e. a multiple of $R = 2^n \bmod N$, where n is the length in bits of the modulus). Let $\mathcal{M}(a \cdot b)$ denote a Montgomery multiplication of a and b, (i.e. $\mathcal{M}(ab) \equiv abR^{-1} \bmod N$). Here is a proposed procedure

Begin

1. $cd = \mathcal{M}(c \cdot d);$

2. regular $a \equiv b/cd \bmod N;$ (result: non-Montgomery number)

3. regular $e \equiv f/a \bmod N;$ (result: a Montgomery number)

4. $a \equiv \mathcal{M}(a \cdot (R^2 \bmod N));$ (to get back a Montgomery number).

End

So the cost of the (relatively small number of) divisions is inflated by a small factor (2) for the loss of the advantage associated with the use of Montgomery's reduction method. Also added to each division is the cost of step 4 above.

There is some increased software complexity due to the need for both regular and Montgomery numerical routines. However, this small increase in code size is not significant in the central office application.

3   Batch Diffie-Hellman

In this section we show how the ideas of Batch-RSA can be extended to support Batch Diffie-Hellman. We use the example of a portable communications system which has many portable radio telephones ("portables") accessing a central office via a matrix of fixed access points called "ports".

Let (S_i, P_i) be the secret and public keys, respectively, of portable i, $i = 1, \ldots, n$, and let (S'_j, P'_j) be the secret and public key of a port j. The central facility is trusted by all ports. (In section 5 we show that the fixed assignment of keys to ports–i.e. key j to port j–is naive, and leads to an inefficient real-time system. Nonetheless, this example is adequate for the purposes of this section.)

In the basic Diffie-Hellman [5] scheme there is some prime modulus, N, common to the whole system, and some primitive element of $GF(N)$, denoted α, s.t. $(\forall i)[P_i \equiv \alpha^{S_i} \bmod N]$, similarly, $(\forall j)[P'_j \equiv \alpha^{S'_j} \bmod N]$, and a session key between i and j is $SK_{ij} \equiv \alpha^{S_i \cdot S'_j} \bmod N$, efficiently computable exclusively by i and j (or the center on behalf of j). Many other variations exist. For example, sometimes it may be desirable to choose a composite modulus [14] [9], and most of the time we need key distribution systems that authenticate the users, and are dynamic [15].

In order for the central authority to be able to use Batch-DH we have to introduce additional constraints.

- First, the modulus should be a composite, with secret prime factorization (two large primes) known only to the central facility, and such that factorization of N is hard.

- Second, the secret key of each port j, S'_j, is not chosen at random. Rather a relatively small e_j, is chosen, and its multiplicative inverse modulo $\lambda(N)$, is computed. This is S'_j. As before, $P'_j \equiv \alpha^{S'_j} \mod N$. For modulus of size n bits we need $e_j < log_2(n)$.

- Third, the e_js must be relatively prime in pairs.

No new constraints are needed for the portable keys. They can be chosen with abbreviated secret exponents (130 bits seem currently secure) [16], or they can be identity based systems [10], or whatever. In that respect the system may be heterogeneous, i.e. users may choose inexpensive portables (with abbreviated secret exponents), or more expensive portables, with long secret exponents, and with the added convenience of being identity based [10]. This may be important if the cryptosystem is used for additional applications, such as monetary transactions.

4   Security

The following lemma is well known.

Lemma 1: Let α and β be generators of the multiplicative groups Z_p^* and Z_q^*, respectively, and let $\gamma \in Z_N^*$, $\gamma \equiv \alpha \mod p$, $\gamma \equiv \beta \mod q$, (i.e. γ is obtained from α and β using Chinese Remaindering). Then γ generates a maximal cyclic subgroup of Z_N^*, it is of size $\lambda = (p-1)(q-1)/2$ (for our choice of p and q), this cyclic subgroup, denoted M is the Cartesian product of Z_p^* and Z_q^*, and therefore is independent of the choice of α and β.

Assuming the base element, γ, for our CDH scheme is chosen according to lemma 1, we now give evidence to the security of our scheme, namely, if we assume that RSA with short public key is hard to break on the average, over the subset M, then so is CDH. This follows from Lemma 2, and the corollary which immediately follows it.

Lemma 2: Composite Diffie-Hellman key agreement scheme is at least as hard to break on the average as RSA for half the messages (here we average over M).

Proof: Assume there exists an oracle AL, s.t. for all N, β, x, y $AL(N, \beta, \beta^x, \beta^y) \equiv \beta^{xy} \mod N$, and let $e \equiv x^{-1} \mod \lambda(N)$. Let γ be the generator of the maximal cyclic subgroup M of Z_N^*, discussed in lemma 1.

Given RSA cracking problem, defined by Input: $N, e, c \equiv m^e \mod N$, $c \in M$ ($m \in M$ implies $c \in M$ by M's closure); Output: m. We use oracle AL to solve it, as follows.

1. Find $\gamma \in Z_N^*$, a generator of M (using the construction of Lemma 1).

2. Compute $\beta \equiv \gamma^e \mod N$.

3. Call oracle $AL(N, \beta, \gamma, c)$.

The oracle's answer in the last step is m, because $m \in M$ and hence $(\exists y)[m \equiv \gamma^y]$, hence $\gamma \equiv \beta^x, c \equiv \beta^y, m \equiv \beta^{xy}$, where all these congruences are modulo N.

Once γ is fixed, the above mapping is one to one, therefore, the reduction is measure preserving [2], and therefore preserves average case complexity.

Q.E.D.

Corollary: The reduction of Lemma 2 does not depend on the length of e. The case in which we are interested is CDH with short e. RSA with short public key reduces to CDH with exactly one short inverse by the same construction.

Lemma 3: Composite Diffie-Hellman key agreement scheme, in which exactly one of the exponents has a short inverse ($< O(n)$) is at most as hard to break on the average (over all messages) as RSA with short public key.

Proof: Let $AL2$ be an oracle that solves RSA problem, where the public exponent e is short ($e < O(n)$), i.e. $AL2(N, c, e) = m$, s.t. $m^e \equiv c \bmod N$. Given a CDH problem defined by Input: N, a, a^x, a^y; Output: $a^{xy} \bmod N$, where $x^{-1} \equiv e \bmod \lambda(N)$ is short, we use oracle $AL2$ to solve it as follows:

1. Find e, s.t. $(a^x)^e \equiv a \bmod N$ (since e is short exhaustive search is feasible).

2. Call oracle $AL2(N, a^y, e)$.

The answer of oracle $AL2$ is a^{xy}, as required.

In this reduction the mapping is one to one. Therefore, it is measure preserving [2], and therefore preserves average case complexity.

Q.E.D.

Lemma 4: Composite Diffie-Hellman key agreement scheme, in which both exponents have short inverses is easily breakable for $6/\pi^2$ of the instances.

Proof: If both secret exponents (x, y) have short inverses (e_i, e_j) then, as before, the adversary can find them by exhaustive search (raising the public keys to the power of $e_{...}$ until he gets α). If in addition the two short inverses are relatively prime (happens with probability $6/\pi^2$, see a theorem by Dirichlet in Knuth II, pp. 324) the adversary can break the system. First he finds a and b such that $ae_i + be_j = 1$, using extended Euclid's gcd algorithm [1]. Multiplying this equation by xy we get $ay + bx \equiv xy \bmod \lambda(N)$. It follows that the adversary can compute the session key $\alpha^{xy} \equiv (\alpha^x)^b \cdot (\alpha^y)^a \bmod N$.

Q.E.D.

Of course, the reduction of Lemma 2 does not hold when both exponents have short inverses.

5 System Considerations for Batch-Diffie-Hellman

The method of this paper can be used to improve the efficiency of network processing in a Portable Communications System key agreement and authentication subsystem. An example Portable Communications System is described in [4]. An example key agreement and authentication protocol, involving Diffie-Hellman computations, and tuned to the requirements of portable communications, is described in [3] as the last of the candidate protocols described in that paper.

The general framework of the system is that portable telephones access the network via a matrix of radio ports. Providing service involves establishing an encrypted, authenticated channel in the initial phase. The encryption is needed to protect the user's privacy from eavesdroppers using radio scanners, and the authentication is needed to protect the service provider from attempts to obtain service fraudulently (i.e. without intent to pay for the call).

During operation, the portable units make requests for service (e.g. to originate or answer a call). Each of these requests begins by employing a protocol to establish a private and authenticated channel. The protocol requires the network to perform a Diffie-Hellman exponentiation. It is necessary for the network to perform these computations quickly, so as not to unduly delay the user's access to service (and increase the holding time of the radio channel and network resources). Since there will often be many users requesting service at the same time, we can make use of the method of this paper to perform a number of Diffie-Hellman computations in parallel, and thereby increase the effectiveness of the processing power available in the network. This can ultimately decrease the cost of providing the PCS service.

The primary issue to be addressed is how much improvement can be obtained. The first question to ask is "how can we ensure that when requests come in, they can be processed in a batch?" One naive method (mentioned in section 3) is to assign a different e_j to each port. However, we can not ensure that there will always be exactly one service request available from each of the ports. Thus, this system would require an enormous (impractical) precomputation phase. This is because the method requires us to precompute a "tree" for each possible combination of exponents we wish to process simultaneously. A system with many ports will require a number of trees proportional to the number of combinations of the ports taken b at a time, where b is the optimal batch size. Any fixed assignment of e_j's to network equipment will either have enormous precomputation, or will be inefficient at obtaining the maximum batch size for processing, if it is still to provide the required response time.

Thus we propose to dynamically assign e_j's on a request-by-request basis. For example, for a given exponent length, we have an optimum batch size b. We choose b exponents e_j. When finished processing the previous batch of requests, the cryptoserver looks in its request queue. It scoops up b requests from the queue (assuming they are present–we discuss this issue later in this section) and assigns to each request one of the exponents e_j. The network immediately sends out the (precomputed) corresponding public keys α^{1/e_j}, and any other required information such as a certificate associated with the assigned public key, to each of the requesters. While the requesters do their Diffie-Hellman computations, the network will process these b computations in a batch. This "dynamic exponent" approach is not usable with Batch-RSA, as will be shown in the next section.

What is the improvement which can be obtained with this method? As shown in the complexity section above, if we assume a n-bit modulus with n-bit exponents $1/e_j$, this allows improvement by a factor of $n/log_2^2(n)$. However, currently some Diffie-Hellman systems use abbreviated exponents. It is considered that the current best algorithm for breaking Diffie-Hellman systems with abbreviated exponents is due to Pollard (the "Lambda Method of Catching Kangaroos" [12]). By examining the relationship of the complexity of "catching a Kangaroo" vs. the difficulty of factoring using the best algorithm available (the number field sieve [8]), we come up with the following formula for the required length l (in bits) of an abbreviated exponent as a function of modulus N.

$$\sqrt{2^l} = e^{(1.9+o(1)) \cdot ln(N)^{1/3} \cdot (ln ln(N))^{2/3}}$$

By algebraic manipulation, we can get the "safe" exponent size for abbreviated Diffie-Hellman exponents as a function of modulus N. The o(1) in the formula refers to a small constant, values for which have been estimated at between 0.1 and 10. In our comparisons, we use values of o(1)=0, and o(1)=1. The value 0 gives the shortest abbreviated exponents (and therefore, by comparison, casts the method of this paper in the most negative light). The value o(1)=1 represents comparison to a more typical (conservative) system, and was chosen because it leads to an abbreviated exponent size of around 190 bits for a 512-bit modulus, which is currently considered to be an acceptable abbreviated exponent size for that modulus size.

To make our comparison, we take the exponent size obtained from the above formula, and divide it by the complexity value for Batch-Diffie-Hellman computations $n/b = log^2(n)$ obtained in Section 2 (because the batch method performs b exponentiations for the cost of one full n-bit exponentiation). The results are tabulated in Table-1 for some interesting values of n. (Where n is equal to $log(N)$, i.e. the number of bits in the modulus N). The table shows gain figures, as well as the corresponding short exponent lengths for o(1)=0 and o(1)=1.

Modulus Size (bits)	Long Exponent Gain Factor	Abbreviated Exponent Gain Factor			
		o(1) = 0		o(1) = 1	
		Exp. Size	Gain	Exp. Size	Gain
512	6	126	1.6	193	2.4
768	8	151	1.6	231	2.5
1024	10	171	1.7	262	2.6
2048	17	231	1.9	353	2.9

Table 1: Gain Factors for Batch-DH Over Other DH Systems.

From the table, one can see that the method of this paper can give a factor of 6 improvement over Diffie-Hellman with full-sized exponents and 512-bit modulus. The factor increases to 17 with a 2048-bit modulus. In comparison to a Diffie-Hellman system with minimum-sized abbreviated exponents (corresponding to $o(1) = 0$), the method of this paper gives a factor of 1.6 improvement with a 512-bit modulus. As modulus sizes increase, the improvement factor also increases, reaching 2.0 near 2000 bit-modulus size. For more conservative abbreviated exponents (corresponding to $o(1) = 1$), the method of this paper gives improvements from 2.4 for 512-bit modulus, reaching 3.0 near 2000-bit modulus size. It is noted that any improvements made in attacking DH with abbreviated exponents will increase the attractiveness of the method of this paper.

It is noted that Fig-3 shows maximum possible improvement figures for this method. In order to achieve this maximum, there must be at least b requests waiting for service whenever the cryptoserver looks in its queue. In reality the requests for service are not evenly spaced in time (they're better modeled by a Poisson process). Thus, it will be difficult to assure that the server always sees b inputs in the queue, while also maintaining some reasonable performance constraint on the delay between service requests and their associated responses.

One way to mitigate this somewhat is to precompute not just batches of size b, but also $b/2$, $b/4$, etc. This will at least allow some improvement over straight n-bit exponentiation for those cases where fewer than b requests are in the queue. One caution is that, as the batch size gets below 4, it may be more efficient to use standard abbreviated-exponent DH. Given the processing scenario we've developed, we can readily switch back and forth between techniques, only using the parallel method when it is advantageous; if efficiency is not very high when there are few requests it does not matter, so long as the installed computational power is not exceeded.

6   Why not Batch-RSA?

Batch-DH and Batch-RSA are comparable in real-time computational requirements. However, in this section we show that, when used for key agreement in a system where response time is important (e.g. the Portable Communications system described above), Batch-RSA introduces unacceptable delays. As was explained in the previous section, the use of Batch-RSA or Batch-DH requires dynamic assignment of exponents to service requests in order to make the precomputation manageable while maintaining appropriate response time.

To use Batch-RSA, j has to use the same modulus N_j with different short exponents e_j's (there is no security hole here). The corresponding secret exponents d_j's are full size (In order to minimize complexity in the portable unit, we would use a medium-sized d_i).

With RSA, the network must send its public key to the portable. The portable then encrypts information using the public key, and sends results to the network. The network then computes the inverse function. Thus there is a random delay (dependent on the response time of the portable unit) between the network transmitting its public key information, and the time when the network could begin its secret computation. With DH, however, there is no such random delay. Immediately upon receipt of a public-key from the portable, the network can begin computation of the secret operation.

Thus, if batch methods are used, DH will allow the network to define a batch and compute it all at once. RSA, on the other hand, will require the network to define a batch, send out information, and wait for all the portable units involved in the batch to return responses before it can begin computation. This will reduce everyone's performance unacceptably (by making the response time for all members of the batch dependent on the slowest-responding portable in the batch). Therefore, Batch-RSA is not useful for key agreement protocols in a Portable Communications System.

7   Conclusions

We have shown how the techniques of Batch-RSA can be expanded to support Batch-Diffie-Hellman. We have also shown that Batch-Diffie-Hellman can be used to reduce processing requirements of a central public-key cryptoserver in a portable communications system by a factor of 6 to 17 over a Diffie-Hellman system with full-sized exponents. Improvements on the order of 1.6 to 3 are obtainable when compared to a Diffie-Hellman scheme employing abbreviated exponents. We have noted that the benefit of Batch Diffie-Hellman increases with new advances in attacks against Diffie-Hellman with abbreviated exponents. We have also shown that it is not feasible to employ Batch-RSA to obtain similar improvements in a key-agreement system where response time is important.

8   Acknowledgement

We wish to thank Rafi Heiman and Arjen K. Lenstra for reviewing this paper, and for their many helpful comments, and to anonymous referee for his important criticism regarding our evidence of security.

References

[1] A.V. Aho, J.E. Hopcroft, and J.D. Ullman, *The Design and Analysis of Computer Algorithms*, Addison Wesley, 1974.

[2] Ben-David, S., Chor, B., Goldreich, O., Luby, M., *On the Theory of Average Case Complexity*, Proc. STOC 1989, pp. 204-216.

[3] M. J. Beller, L. F. Chang, Y. Yacobi, *Privacy and Authentication on a Portable Communications System*, IEEE Globecom '91 Conference Proceedings, Phoenix, December 1991.

[4] D. C. Cox, *Portable Digital Radio Communications–An Approach to Tether-less Access*, IEEE Communications Magazine, Vol. 27, No. 7, July 1989.

[5] W. Diffie and M.E. Hellman, *New directions in cryptography*, IEEE Trans. on Inform. Theory, vol. IT-22, pp. 664-654, Nov. 1976.

[6] S.R. Dusse and B.S. Kaliski, *A Cryptographic Library for the Motorola DSP56000*, Advances in Cryptology: Proceedings of Eurocrypt '90, I.B. Damgard (Ed.), LNCS 473, Springer Verlag, May 1990, pp. 230-243.

[7] A. Fiat: *Batch RSA*, Proc. Crypto'89, pp 175-185.

[8] A.K. Lenstra, Private communication.

[9] K.S. McCurley, *A key distribution system equivalent to factoring*, J. Cryptology, vol. 1, no. 2, 1988.

[10] U.M. Maurer and Y. Yacobi *Non-interactive Public Key Cryptography* Proc. Eurocrypt'91.

[11] P.L. Montgomery, *Modular Multiplication Without Trial Division*, Math of Computation, Vol. 44, 1985, pp. 519-521.

[12] J.M. Pollard, *Monte Carlo Methods for Index Computation (mod P)*, Math, Comp. 32 (1978), 918-924.

[13] R.L. Rivest, A. Shamir and L. Adleman, *A method for obtaining digital signatures and public-key cryptosystems*, Communications of the ACM, vol. 21, pp. 120-126, 1978.

[14] Z. Shmuely, *Composite Diffie-Hellman public-key generating systems are hard to break*, TR 356, CS Dept., Technion, Feb. 1985.

[15] Y. Yacobi, *A key distribution "paradox"*, Proc. CRYPTO'90 Santa Barbara, CA, Aug. 11-15, 1990.

[16] Y. Yacobi, *Discrete-Log With Compressible Exponents* Proc. CRYPTO'90, Santa Barbara, CA, Aug. 11-15, 1990.

PHASE I:

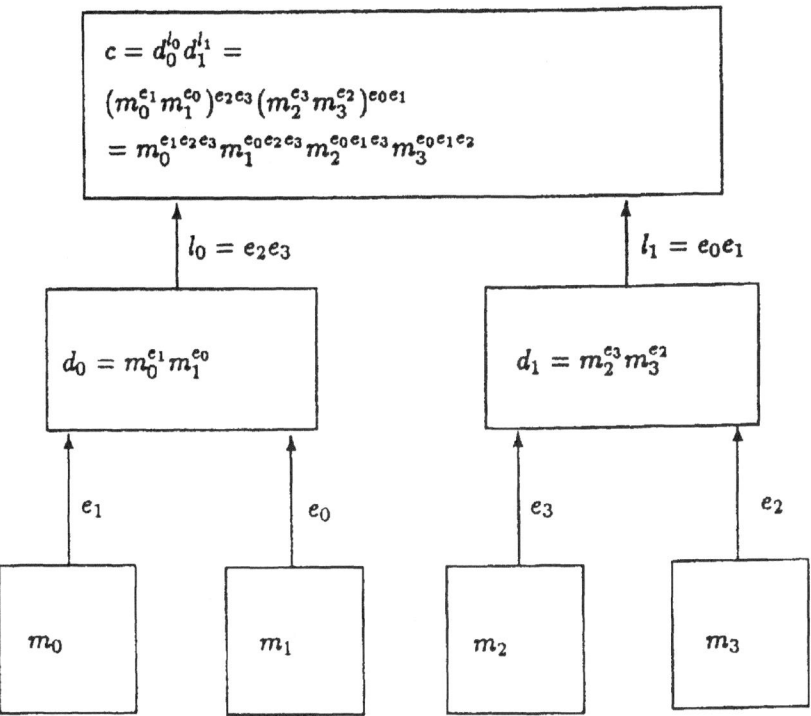

PHASE II:

$$c^{1/p} = m_0^{1/e_0} m_1^{1/e_1} m_2^{1/e_2} m_3^{1/e_3}$$

Figure 1: Phases I and II of Batch-RSA

PHASE III:

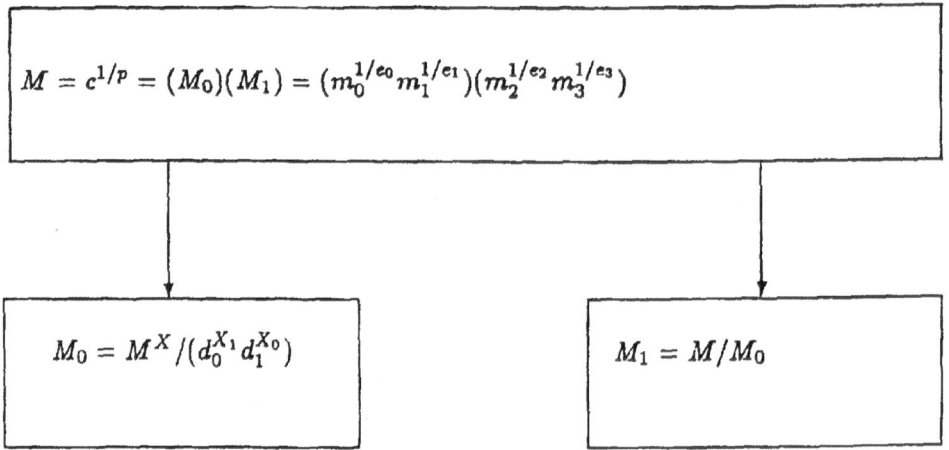

$$M = c^{1/p} = (M_0)(M_1) = (m_0^{1/e_0} m_1^{1/e_1})(m_2^{1/e_2} m_3^{1/e_3})$$

$$M_0 = M^X / (d_0^{X_1} d_1^{X_0})$$

$$M_1 = M/M_0$$

Continue recursively, with different X, X_0, X_1

in each level. d's and X's are precomputed.

Figure 2: Phase III of Batch RSA

High-Speed Implementation Methods
for RSA Scheme

Keiichi Iwamura[†], Tsutomu Matsumoto[††], and Hideki Imai[††]

† Canon Research Center, 21 Laboratory, 5-1 Wakamiya, Morinosato,
Atsugi-shi, Kanagawa 243-01, Japan

†† Yokohama National University, Division of Electrical & Computer
Engineering, 156 Tokiwadai, Hodogaya, Yokohama, 240 Japan

Abstract. This paper proposes two novel implementation methods for
the RSA cryptographic scheme. (1) The most efficient RSA implementation known to the present authors. This implementation achieves 50
Kbps at about 25 Kgates for a 512-bit exponent e and a 512-bit modulus
N. Thus the efficiency is 2.0 bps/gate. (2) A systolic architecture useful
for high-speed and efficient and flexible chip implementation of the RSA
scheme.

1 Introduction

Modular exponentiations (or powerings) are elementary operations for cryptographic transformations in public-key cryptosystems like the RSA scheme
[RSA78]: $C = M^e \bmod N$, the ElGamal schemes, and so on. These cryptosystems' security is based on the difficulty of factoring integers or that of computing discrete logarithms. To achieve enough security level, the word lengths in
the modular exponentiations should be significantly greater than those used in
conventional general-purpose computer hardwares. The required typical word
length is around 512 bits or more. A modular exponentiation can be accomplished by iterating modular multiplications. Thus, for obtaining high-speed
implementations of the RSA scheme and the like, it is quite natural to pursue
techniques for speeding up modular multiplications for integers of 512 bits or
longer.

A lot of efforts have been done in this field [Bric82]-[WQ90]. Reference
[Bric89] surveys various hardware implementations of the RSA scheme. We
add Table 1 for introducing other implementations, not mentioned in [Bric89],

Table.1: Some of the implementations for the RSA scheme so far presented

Reference	Baudrate	bits	Gates	Reference	Baudrate	bits	Gates
[Miya83]	50K	512	280K	[TAA87]	100K	256	448K
[KMT87]	70K	512	160K	[THAA88]	500K	512	640K
[Mori90]	80K	512	50K	[IWDS91]	64K	512	50K

R.A. Rueppel (Ed.): Advances in Cryptology - EUROCRYPT '92, LNCS 658, pp. 221-238, 1993.
© Springer-Verlag Berlin Heidelberg 1993

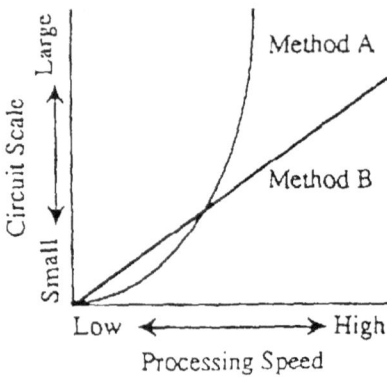

Fig.1: Efficiency and Flexibility

with known circuit scale (Gates) and known processing speed (Baudrate). It should be noted that in Table 1 [Mori90] shows an implementation only for modular multiplication and [TAA87] presents an RSA implementation for a 256-bit exponent and a 256-bit modulus. If we define the efficiency of implementation as

(efficiency of implementation) = (processing speed)/(circuit scale)

then many of hardware implementations so far reported are not so efficient. This measure may be appropriate from the practical view point of achieving high-speed and smaller circuits. The better efficiency is realized by the implementation with the smaller circuit scale and the higher processing speed.

We define another measure called the flexibility, which represents the degree of linearity between the processing speed and the circuit scale. To illustrate this notion, we examine two virtual implementations named Method A and Method B, each of which relationship between the circuit scale and the processing speed is given in Fig.1. We consider that Method A is not flexible because it is not efficient in the wide range of processing speed. In contrast, Method B is flexible since it realizes implementations with the same efficiency in the wide range of processing speed. In this paper we propose two implementation methods for modular multiplication from these points of view.

Section 2 shows a modular multiplication algorithm and circuit with the best efficiency rating of all algorithms and circuits introduced to date and known to the present authors. Based on this Section 2 described an RSA implementation achieving 50 Kbps at about 25 Kgates for a 512-bit exponent e and a 512-bit modulus N. Thus the efficiency is 2.0 bps/gate.

Section 3 presents a flexible modular multiplication algorithm based on systolic array, a way of parallel processing. This high speed processing method utilizes a pipeline processing by means of several types of processing elements (henceforth abbreviated as PEs) each of which is controlled locally and regularly. Also the algorithm is suitable for chip implementations because a chip can be constructed simply and regularly with favorite circuit scale and because the RSA processing speed increases in proportion to the number of chips. This implementation can allow an increase in speed based on the Chinese remainder theorem in the same chip implementation because it permits variable bit number for the encryption keys.

Lastly in Section 4, the results of this paper are compared with previously obtained results and are shown to be better from the practical view point. See Fig.12. Parallel processing applied to the RSA scheme gives an architecture that allows for the increasing of speed for processing and the reduction in size of the circuit scale systematically.

2 An Efficient Implementation for RSA Scheme

2.1 An Efficient Modular Multiplication Circuit

Consider the modular multiplication $R = A \cdot B \bmod N$ (with N, A, B, R being $k = m \cdot n$ bit integers). We express A in binary and B and R in radix $X = 2^m$ as follows,

$$
\begin{align}
A &= A_{k-1} \cdot 2^{k-1} + A_{k-2} \cdot 2^{k-2} + \cdots + A_1 \cdot 2 + A_0 \tag{1} \\
B &= B_{n-1} \cdot X^{n-1} + B_{n-2} \cdot X^{n-2} + \cdots + B_1 \cdot X + B_0 \tag{2} \\
R &= R_{n-1} \cdot X^{n-1} + R_{n-2} \cdot X^{n-2} + \cdots + R_1 \cdot X + R_0 \tag{3}
\end{align}
$$

where $A_{k-j} \in \{0,1\}$ $(j = 1, \cdots, k)$ and B_i, $R_i \in \{0,1\}^m$ $(i = 0, \cdots, n-1)$. Then the modular multiplication can be calculated by the consecutive execution of the following two algorithms:

ALGORITHM 1A
(Input: $A_0, \cdots, A_{k-1}$; $B_0, \cdots, B_{n-1}$; N)
(Output: $R_{k,0}, \cdots, R_{k,n-1}$)

$\quad D_{0,i} = 0$; $C_{0,i} = 0$
$\quad$**FOR** $j = 1$ **TO** k
$\quad\quad E_{j-1} = E_{j-1,n-1} \cdot X^{n-1} + \cdots + E_{j-1,0} = (C_{j-1,n-1} \cdot X^n) \bmod N$ $\quad$(4)
$\quad\quad$**FOR** $i = 0$ **TO** $n-1$
$\quad\quad\quad R_{j,i} = D_{j-1,i} \cdot 2 + C_{j-1,i-1} + A_{k-j} \cdot B_i + E_{j-1,i}$
$\quad\quad\quad D_{j,i} = dw_{m-1}(R_{j,i})$
$\quad\quad\quad C_{j,i} = up_{m-1}(R_{j,i})$
$\quad\quad$**NEXT**
$\quad$**NEXT**

where $dw_a(Z) = Z \bmod 2^a$ and $up_a(Z) = (Z - dw_a(Z))/2^a$.

ALGORITHM 1B
(Input: $R_{k,0}, \cdots, R_{k,n-1}$; N)
(Output: $R_0, \cdots, R_{n-1}$)

$$
\begin{align}
R &= R_{n-1} \cdot X^{n-1} + R_{n-2} \cdot X^{n-2} + \cdots + R_1 \cdot X + R_0 \\
&= (R_{k,n-1} \cdot X^{n-1} + R_{k,n-2} \cdot X^{n-2} + \cdots + R_{k,1} \cdot X + R_{k,0}) \bmod N \tag{5}
\end{align}
$$

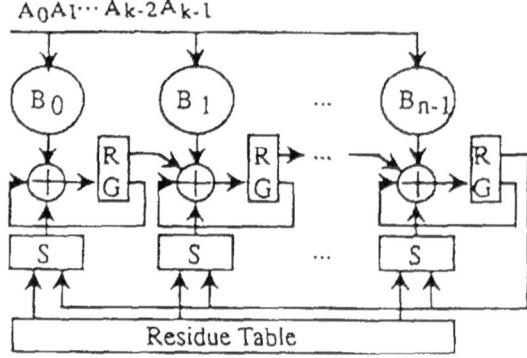

A₀A₁⋯Aₖ₋₂Aₖ₋₁

Fig.2: An efficient modular multiplication circuit

Fig.2 illustrates the principal part of our modular multiplication circuit, namely the circuit implementing ALGORITHM 1A. The index i goes from left ($i = 0$) to right ($i = n - 1$) and corresponds to the position of the register, multiplier, adder and the modular-reduction circuit, whereas the value j corresponds to the clock. Each circuit except registers is realized as below.

1) Multiplication: the multiplier is easily realized by AND gates which output B_i only when A_{k-j} is 1. Therefore the circuit scale of the multipliers is small.

2) Addition: the adder is realized by 4-input addition functions of which inputs are $A_{k-j} \cdot B_i$ from the multiplier and $E_{j-1,i}$ from the modular-reduction circuit and $D_{j-1,i} \cdot 2$ and $C_{j-1,i-1}$. Since $A_{k-j} \cdot B_i$ and $E_{j-1,i}$ are m-bit and $D_{j-1,i}$ is $(m-1)$-bit, the fact that $C_{j-1,i-1}$ is less than an m-bit value leads that the value enters the register from the adder is $(m+2)$-bit and thus, $C_{j,i}$ is 3-bit. Therefore in the case where $m \geq 3$ the circuit scale for the carry-bit registers is small compared to the whole circuit scale for registers, because there are 2 carry bits for every $m-$bit added result.

3) Modular Reduction: it is easy to see that $C_{j,i}$, the 3-bit carry, doesn't take the value [111] (See Appendix). The residue $E_{j-1,i}$ ($i = 0, \cdots, n-1$) is derived as Expression (4) from $C_{j-1,n-1}$. Since $C_{j-1,n-1}$ takes 3-bit values [000], [001], $\cdots$, [110], except for [111], each $E_{j-1,i}$ takes only 7 patterns including $E_{j-1,0} = 0$. These residue patterns can be selected by the selectors S in Fig.2. Therefore the delay time of modular reduction is short. With this circuit, the modular reduction processing is simplified because of the no need to compare divisors with dividends and operate negative numbers. Also, since $E_{j-1,n-i}$ is a true value, these values can be input into the addition circuit as they are.

The outputs $R_{k,0}, \cdots, R_{k,n-1} \in \{0,1\}^{m+2}$ from the circuit should be modified by an auxiliary small circuit for ALGORITHM 1B, which compensates the carry bits in $R_{k,i}$ so that they become all zero after the final calculation and corrects the final value of R to be less than N. Such a small circuit will have little effect on the overall processing speed. This is the reason why it is omitted from Fig.2.

2.2 Circuit Scale and Processing Speed

Assuming that the proposed circuit is fabricated by C-MOS technology, let us evaluate the circuit scale according to those of the Fujitsu Standard Cells [FujG88], [FujM88].

When k equals 512 and m equals 4, the modular multiplication circuit of roughly 25 Kgates is possible. (The residues $E_0, E_1, \cdots, E_{n-1}$ are calculated and set by another circuit from the value of N in relation of Expression (4).) Since the processing time required for 1 clock is realized at the order of $20 \sim 30$ ns using C-MOS gates, the circuit in Fig.2 needs approximately $13\mu s$ to execute one modular multiplication $R = A \cdot B \bmod N$. Subsequently, when the keys e and N are each at the 512 bit, RSA encryption using this modular multiplication circuit can be achieved at a processing speed of 50 Kbps. Also, if the 6 non-zero values for $E_{j,i}$ derived from N are used as the public-key instead of the modulus N, then the preprocessing required for $E_{j,i}$ can be eliminated.

Table 1 and the reference [Bric89] show various implementations for the RSA scheme up to date. Even though an unconditional comparison cannot be made, when efficiency is defined as the processing speed per one gate (*processing speed /circuit scale*), we attempt an evaluation of several examples of the RSA scheme. Among the RSA scheme implementations developed until now, that of [THAA88] achieves the highest processing speed of 500 Kbps, however it needs a scale greater than or equal to 640 Kgates. It yields an efficiency of 0.78 bps/gate. A modular multiplication circuit for the RSA scheme [Mori90] can achieve a processing speed of 80 Kbps using a 50 Kgates circuit scale. The modular multiplication circuit yields an efficiency of 1.6 bps/gate, and can be considered to be the most efficient before this paper.

While, the circuit proposed in 2.1 yields an efficiency of 2.0 bps/gate, and can thus be described as more efficient than that of [Mori90]. Further, since the modular multiplication circuit in [Mori90] requires the circuit for Booth algorithm, both the control and calculation algorithm of the circuit proposed herein can be considered simpler than that of [Mori90]. Consequently, the modular multiplication circuit in Fig.2 allows the most efficient implementation of the RSA scheme to date.

3 Systolic Arrays for RSA Scheme

3.1 An Approach to Design Systolic Arrays for Modular Multiplication

Modular multiplication: $R = A \cdot B \bmod N$ can be calculated by the recursive execution of

$$R_j = 2 \cdot R_{j-1} + A_{k-j} \cdot B + E_{j-1} \tag{6}$$

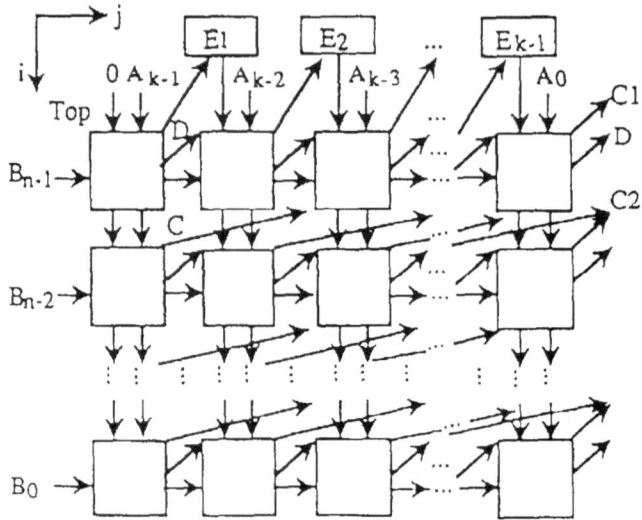

Fig.3: Data dependence graph for modular multiplication

from $j = 1$ to $j = k$, where R_0 is 0, and where E_{j-1} is the residue shown in Expression (4).

A data dependence graph can represent a version of the recursion obtained by using m-bitwise slice. We propose the graph given in Fig.3 where each column of n cells performs Expression (6) and the row of k such columns illustrate the recursive execution. In the rest of this subsection, we give an intuitive explanation of the key idea for our systolic arrays.

The inputs B flow from left to right along the horizontal arrows, and the inputs A flow downward along the vertical arrows. The result of the computation done in each cell is divided into the least significant $m - 1$ bits D and the rest of bits C (the carry bits). The 1-bit shifted D (double of D) is input to the cell in the same row and in the right next column. The 1-bit shifted C (double of C) is input to the cell in the upper next row and in the column two positions to the right. For each cell in the top row the carry bits C is translated into a residue E and E affects the right next column of cells. Such a trick is to avoid the increasing of the size of data treated in each cell and to ensure a regular and efficient pipeline structure.

Observe that between each pair of consecutive columns there are three types D, $C1$, and $C2$ of data flows as well as the flow of B. For each column, $C1$ denotes the type of C made in the column and $C2$ denotes the type of C passing through the column. Each column can be realized by a single processing element (PE) defined as Fig.5 and the row of k such columns can be implemented as the connection of k PEs. See Fig.4. The j th PE holds the value of A_{k-j} at one of its internal registers. The carry bits C produced at the top cell in a column in Fig.3 is hold at a register as S_{out} in the corresponding PE in Fig.5. The ROM

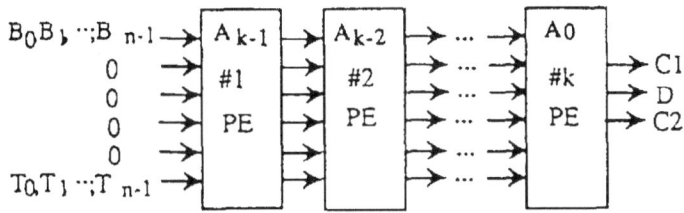

Fig.4: Systolic array for modular multiplication

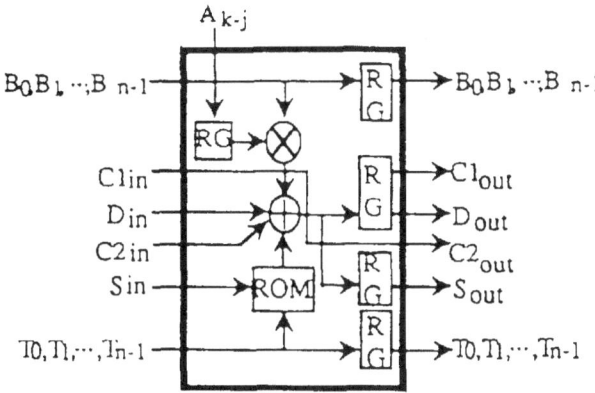

Fig.5: Processing element in Fig.4

in each PE accepts the S_{out} coming from the left next PE as
S_{in} and outputs a residue E specified by S_{in} in m-bit wise according to an input sequence of timing signals $T_{n-1}, T_{n-2}, \cdots, T_0$.

Since the form of the data D, $C1$, and $C2$ appears in the array differs from that of the input data, the modular product of two integers, output from the array cannot be directly used as another input to the same array. Thus, using only the array we cannot execute modular exponentiation. We describe in the next subsection a method to eliminate such a drawback and propose an effective systolic array.

3.2 A Systolic Array for Modular Multiplication and Compensating Carry Bits

We consider the modular multiplication $R_c = A_c \cdot B_c \mod N$, where $A_c = A + a \cdot X$, $B_c = B + b \cdot X$ and $R_c = R + r \cdot X$. A, B and R are shown in Expressions (1) - (3), respectively and a, b and r are expressed as follows,

$$a = a_{k-m} \cdot X^{n-2} + \cdots + a_{2 \cdot m} \cdot X + a_m$$
$$b = b_{n-1} \cdot X^{n-2} + \cdots + b_2 \cdot X + b_1$$
$$r = r_{n-1} \cdot X^{n-2} + \cdots + r_2 \cdot X + r_1$$

where, a_{k-j}, b_{n-i}, $r_{n-i} \in \{0, 1\}$ $(j = m, 2 \cdot m, \cdots, (n-1) \cdot m)$ $(i = 1, \cdots, n-1)$, and a_{k-j}, b_{n-i}, r_{n-i} are 1-bit carry bits by every m-bit group. The modular multiplication $R_c = A_c \cdot B_c \mod N$ is calculated as shown in ALGORITHM 2.

Fig.9 is the systolic array for ALGORITHM 2 and PEA, PEB, and PEC's processing is executed by the PE shown in Fig.6, Fig.7, and Fig.8, respectively. The neighboring PEs is connected between X_{out} and X_{in} ($X = b, B, S, D, C$ and T) In ALGORITHM 2, the value i refers to the clock, and j refers to the number for the PEs, so that from right to left $j = 1$ to $j = k + n$ refers to each position of the PEs. PEA calls 1 PEB for every m turns, and for the final calculation calls PEC instead of PEB in Fig.9.

In the first PE($j = 1$) of Fig.9, B_{n-i} and $b_{n-i}(i = 1, \cdots, n)$ are input from B_{in} and b_{in} simultaneously in order from the upper row. Also, the timing signal $T_{n-i} = n - i$ $(i = 1, \cdots, n)$ referring to the modular-reduction circuit is input from T_{in} synchronized with B_{n-i}. B_{n-i}, b_{n-i} and T_{n-i} are held back by one clock in internal registers of the PEs, then are output to the next PE from B_{out}, b_{out} and T_{out}. In the first PE the values input for D_{in}, S_{in} and C_{in} are set to 0. Each PE's composition elements and actions can be explained in the following:

PEA : The value A_{k-j+s} is preset in the internal 1-bit register of j-th PEA. The multiplier is constructed with $m + 1$ AND gates and outputs $A_{k-j+s} \cdot B_{n-i}$ and $A_{k-j+s} \cdot b_{n-i}$. The adder is based on the 5-input from the multiplier outputs $A_{k-j+s} \cdot B_{n-i}$ and $A_{k-j+s} \cdot b_{n-i}$, the modular-reduction circuit's output $E_{j-1,n-i}$, and the previous PE's output $2 \cdot dw_{m-1}(R_{j-1,n-i})$ and the output $2 \cdot C_{j-2,n-i-1}$ of the two PEs before. Since $A_{k-j+s} \cdot B_{n-i}$ and $E_{j-1,n-i}$ are m-bit and $dw_{m-1}(R_{j-1,n-i})$ is $m - 1$-bit and $A_{k-j+s} \cdot b_{n-i}$ is 1-bit, the fact that $2 \cdot C_{j-2,n-i-1}$ is less than an m-bit value leads that the value enters the register from the adder is $m + 2$-bit. The residue $E_{j-1,n-i}$ is derived from Expression (7) in relation of $up_{m-1}(R_{j-1,n-1}) = C_{j-1,n-1}$. Since B_{n-i} and $E_{j-1,n-i}$ are in the same digit, $E_{j-1,n-i}$ ($i = 1, \cdots, n$) can be gradually output according to the timing signal T_{n-i} which is synchronized with B_{n-i}. Since $C_{j-1,n-1}$ is 3-bit and T_{n-i} is $\log n$-bit, the output circuit for $E_{j-1,n-i}$ can be realized by ROM (read-only-memory) holding input of $3 + \log n$ bits and output of m bits and the selector and the register for the 3 bits to save the value of $C_{j-1,n-1}$.

PEB : In the internal register of this PE the value $a_{k-j+s+1}$ is preset. Since $a_{k-j+s+1}$ and A_{k-j+s} in the previous PE correspond to the same bit in the binary expression, $dw_m(R_{j-1,n-i})$ and $C_{j-2,n-i-1}$ are not doubled and $C_{j-1,n-i}$ is $up_m(R_{j-1,n-i})$.

PEC : This PE aims at to make the output r_{n-i} ($i = 1, \cdots, n$) one bit. This PE adds the outputs D_{out}, S_{out}, and C_{out} from the previous PE and creates the single value $R_{k+n,n-i}$. Next the PE delays the value $dw_m(R_{k+n,n-i})$ in the register and adds it to the value $up_m(R_{k+n,n-i})$, and produces $R_{k+n+1,n-i} \in \{0, 1\}^{m+1}$ ($i = 1, \cdots, n$). The most significant bits $up_m(R_{k+n,n-1})$ is saved in a separate register. Since the residue E_{k+n+1}

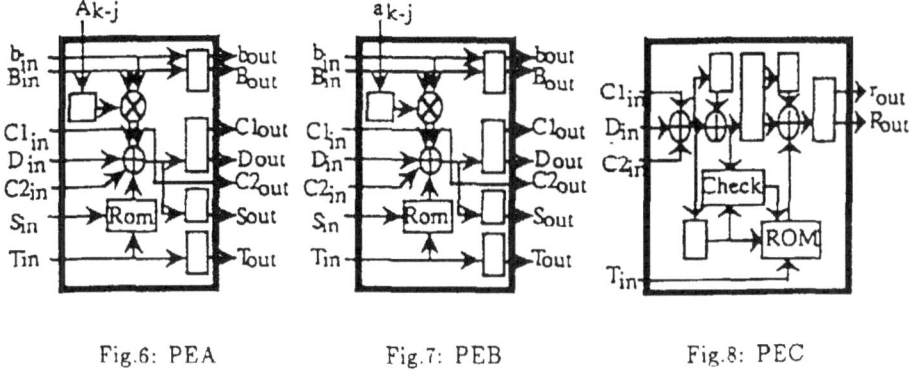

Fig.6: PEA   Fig.7: PEB   Fig.8: PEC

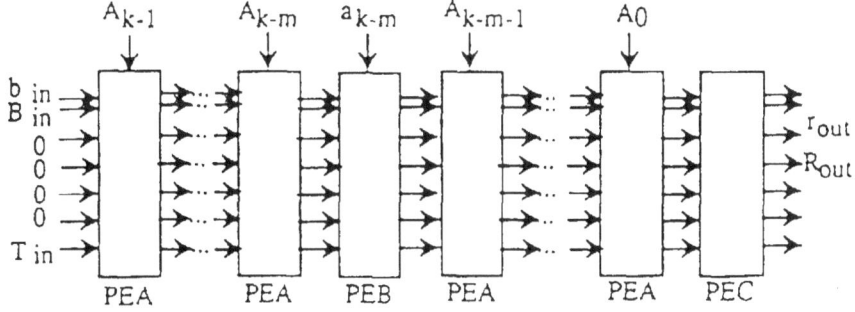

Fig.9: Systolic-array for modular multiplication using PEA, PEB and PEC

is calculated by $(CC \cdot X^n) \bmod N$, $R_{k+n+2,n-i} = dw_m(R_{k+n+1,n-i}) + up_m(R_{k+n+1,n-i-1}) + E_{k+n+1,n-i}$ and $R_{k+n+2,n-i} \in \{0,1\}^{m+1}$ $(i = 1, \cdots, n)$. However, pre-modular multiplication is performed in the check circuit shown as C in Fig.8 to set the most significant bit $r_{n-1} = 0$. Since the most significant bits of residue E_{ad} from the value CC is calculated in advance, the pre-modular multiplication $dw_m(R_{k+n+1,n-1}) + up_m(R_{k+n+1,n-2}) + E_{ad}$ can be calculate. If the most significant bit of the pre-modular multiplication is 1, the fact that E_{k+n+1} takes $((CC + 1) \cdot X^n) \bmod N$ leads $r_{n-1} = 0$. This checking circuit is executed with 3 bits of ROM and an adder.

Consequently, the repeating modular multiplication such as the RSA enciphering and deciphering can be achieved with a systolic array as shown in Fig.9.

ALGORITHM 2

(Input: $A_0, \cdots, A_{k-1}$; $a_m, \cdots, a_{k-m}$; $B_0, \cdots, B_{n-1}$; $b_1, \cdots, b_{n-1}$; N)
(Output: $R_0, \cdots, R_{n-1}$; $r_0, \cdots, r_{n-1}$)

$\quad R_{0,n-i=0}$; $C_{0,n-i=0}$
$\quad$ **FOR** $s = 0$ **TO** $k/m - 1$
$\qquad$ **FOR** $c = 1$ **TO** m
$\qquad\quad j = s \cdot (m+1) + c$
$\qquad\quad E_{j-1} = E_{j-1,n-1} \cdot X^{n-1} + \cdots + E_{j-1,0}$
$\qquad\qquad\quad = (up_{m-1}(R_{j-1,n-1}) \cdot X^n) \bmod N \qquad\qquad (7)$
$\qquad\quad$ **FOR** $i = 1$ **TO** n

P
E $\qquad\qquad R_{j,n-i} \quad = 2 \cdot dw_{m-1}(R_{j-1,n-i}) + 2 \cdot C_{j-2,n-i-1}$
A $\qquad\qquad\qquad\qquad + A_{k-j+s} \cdot (B_{n-i} + b_{n-i}) + E_{j-1,n-i}$
$\qquad\qquad C_{j-1,n-i} = up_{m-1}(R_{j-1,n-i})$
$\qquad\quad$ **NEXT**
$\qquad$ **NEXT**
$\qquad$ **IF** $s = k/m - 1$ **THEN** $\quad$ **GOTO PEC**
$\qquad j = j + 1$
$\qquad E_{j-1} = E_{j-1,n-1} \cdot X^{n-1} + \cdots + E_{j-1,0}$
$\qquad\qquad\quad = (up_m(R_{j-1,n-1}) \cdot X^n) \bmod N \qquad\qquad (8)$
$\qquad$ **FOR** $i = 1$ **TO** n

P
E $\qquad\qquad R_{j,n-i} \quad = dw_m(R_{j-1,n-i}) + C_{j-2,n-i-1}$
B $\qquad\qquad\qquad\qquad + a_{k-j+s+1} \cdot (B_{n-i} + b_{n-i}) + E_{j-1,n-i}$
$\qquad\qquad C_{j-1,n-i} = up_m(R_{j-1,n-i})$
$\qquad$ **NEXT**
$\quad$ **NEXT**
$\qquad$ **FOR** $i = 1$ **TO** n

P
E $\qquad\qquad R_{k+n,n-i} \quad = R_{k+n-1,n-i} + C_{k+n-2,n-i-1}$
C $\qquad\qquad R_{k+n+1,n-i} = dw_m(R_{k+n,n-i}) + up_m(R_{k+n,n-i-1})$
$\qquad\qquad CC \qquad\qquad = (up_m(R_{k+n,n-1}) + up_m(R_{k+n+1,n-1}))$
$\qquad\qquad E_{ad} \qquad\quad = ((CC \cdot X^n) \bmod N)/X^{n-1}$
$\qquad\qquad$ **IF** $dw_m(R_{k+n+1,n-1}) + up_m(R_{k+n+1,n-2}) + E_{ad} \geq X$
$\qquad\qquad$ **THEN** $CC = CC + 1$
$\qquad\qquad E_{k+n+1} \quad = E_{k+n+1,n-1} \cdot X^{n-1} + \cdots + E_{k+n+1,0}$
$\qquad\qquad\qquad\qquad\quad = (CC \cdot X^n) \bmod N$
$\qquad\qquad R_{k+n+2,n-i} = dw_m(R_{k+n+1,n-i}) + up_m(R_{k+n+1,n-i-1})$
$\qquad\qquad\qquad\qquad\quad + E_{k+n+1,n-i}$
$\qquad\qquad R_{n-i} \qquad = dw_m(R_{k+n+2,n-i})$
$\qquad\qquad$ **IF** $i = 1$ **THEN** $r_{n-1} = 0$ **ELSE** $r_{n-i} = up_m(R_{k+n+2,n-i})$
$\qquad$ **NEXT**

where $dw_a(Z) = Z \bmod 2^a$ and $up_a(Z) = (Z - dw_a(Z))/2^a$.

3.3 Circuit Scale and Processing Speed

At first we evaluate the circuit scales of PEA, PEB and PEC, and show them in Table.2 for $m = 8$.

Table.2: Circuit scale of PEs for $m = 8$

	PEA	PEB	PEC
multiplication	20	20	0
addition	294	294	464
modular reduction	404	306	404
delay	94	94	38
total	812	714	906

Table.3: Efficiency of PEA for m

m	16	32	64	128
multiplication	31	55	103	199
addition	451	891	1771	3531
modular reduction	410	410	349	318
delay	138	232	425	819
PEA's circuit scale	1030	1603	2648	4867
number of PEs	544	528	520	516
total circuit scale	561K	847K	1377K	2512K
processing speed	400K	800K	1600K	3200K
efficiency	0.71	0.94	1.16	1.27

In order to construct a systolic array shown in Fig.9, 512 PEAs, 63 PEBs and 1 PEC are necessary. For simplicity's sake, if the systolic array in Fig.9 consists of 576 PEAs, its circuit scale becomes 468 Kgates.

Since the processing time needed for 1 clock is the time to go from the selector to pass the ROM and the adder, if a bi-polar ROM is used a processing time of about 50 ns can be achieved. When the RSA scheme keys e and N are 512 bits, a processing speed for the RSA scheme can achieve at about 200Kbps. This means that even if data is entered at the speed of 200Kbps successively, real-time processing by means of the pipeline processing of the systolic array without buffer overflow is possible.

Table.3 shows PEA's circuit scale and processing speed for selected m's. Since the delay time for the adder increases as the value of m increases even when m is doubled it can be considered that the processing speed will not double, but, rather, somewhat decrease. However, if the processing speed as m increases is compared to the heap of the delay time, the influence of the delay time is ignorably small. Consequently, we realize that as m increases the efficiency increases.

Also, since the functions of PEA, PEB and PEC resemble each other, a common PE can be constructed only increasing a small circuit scale as compared with that of PEA. The common PE calculates a partial modular multiplication such as $R_j = 2 \cdot R_{j-1} + A_{k-j} \cdot B_c \bmod N$ and/or $R_j = R_{j-1} + a_{k-j} \cdot B_c \bmod N$

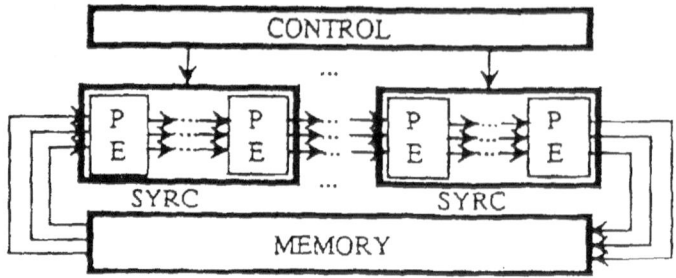

Fig.10 : Chip implementation using Systolic RSA Chip

, and the pipeline structure by q PEs realizes partial modular multiplications q times. Consequently, processing speed of a modular multiplication can be increased in proportion to the increase of the number of the PEs. Further, the pipeline structure with the PEs can keep the same efficiency for an increase in speed. For example, if we take $m = 8$, a circuit scale of $q*812$ gates (with q being an integer representing the number of PEs, $131072 \geq q \geq 1$) and a processing speed of $q \cdot 347$ bps can thus be achieved for the RSA scheme implementation. If we take $m = 128$, the RSA scheme implementation at a circuit scale of $q \cdot 4867$ gates and a processing speed of $q \cdot 6.2$ Kbps can be achieved. Consequently, it is shown that this method can realize an efficient RSA scheme implementation for the increase in speed.

3.4 Chip Implementation Using the Systolic-Array Approach

As shown in 3.2, since the repeating modular multiplication circuit can be constructed using optional q PEs ($131072 \geq q \geq 1$), we construct a modular multiplication circuit with favorite circuit scale. At that point the creation of chips (hereafter referred to as SYRCs, Systolic RSA Chips) in units with favorite number of PEs, in combination with RAM, both of which can be controlled by external programming thus affecting a way to achieve the RSA scheme implementation. The external programming can be constructed in a flexible way using the ROM. Also, if a greater processing speed is required, many SYRCs can be connected as shown in Fig.10. Fig.11 shows the difference of processing speed between using 2 SYRCs and using 3 SYRCs in the implementation of Fig.10. It is easy to understand that the implementation using 3 SYRCs is 1.5 times faster than that of using 2 SYRCs. Like these examples, this implementation can achieve a speed-up proportional to the number of SYRCs and can realize high-speed real-time processing. The same characteristic can be also easily accomplished by changing the number of times the SYRC processing occurs with a control circuit.

While, since this method is based on a systolic array, this method is suitable for the VLSI (Very Large Scale Integration) implementation. If VLSI is utilized,

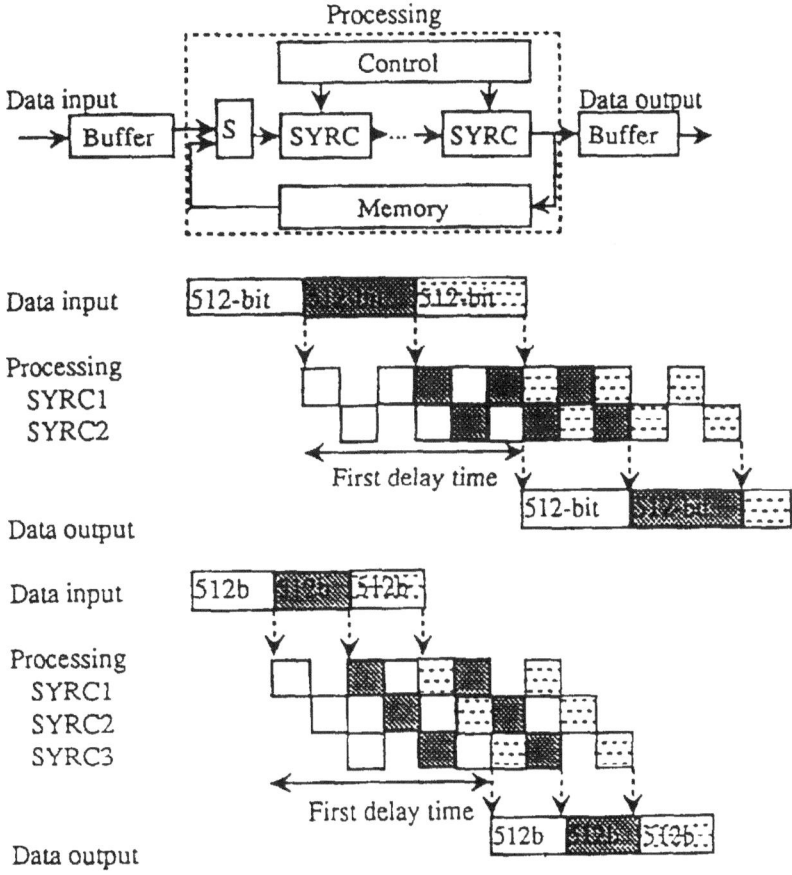

Fig.11 : Difference of processing speed between using 2 SYRCs and using 3 SYRCs

high-speed RSA scheme implementation in one chip is easily constructed because of the regularity and simplicity of the structure. Also, since the processing that occurs in one PE is simple integer calculations, even without putting the PEs onto a separate chip the modular multiplication algorithm shown in **3.1** can be realized using the normal DSP or CPU etc.. As shown above, the modular multiplication method demonstrated in this section is useful because it is extraordinarily easy to circuitize and improvements in speed are easily attainable.

3.5 Another Utility of the Systolic-Array Approach

It is known that the processing can be sped up with a same circuit scale using the CRT (Chinese Remainder Theorem) [QC82]. When an RSA implementation using the CRT is attempted, the problem of creating different bit numbers for multipliers and divisors makes use of the same modular multiplication circuit for both enciphering and deciphering difficult to realize. However, since the modular multiplication circuit in this section enables an easy trade-off between circuit scale and processing events, enciphering and deciphering can be realized on the same circuit if we change the processing events for the difference in multipliers and divisors. This can be easily affected by simply changing the control of the number of feedback inputs to the SYRC for enciphering and deciphering with a control circuit. Using the CRT to increase processing speed can yield at most a quadrupled rate on the same circuit scale. Consequently, when the method outlined in this section uses the CRT, the processing speed demonstrated in the previous section can be increased by approximately 4 times, thus making the construction of an extremely efficient RSA scheme implementation possible. Also, since the RSA deciphering using the CRT can be basically achieved in parallel processing, when modular multiplication is achieved with multiple chips, this paper's method is suitable from this point of view.

This also means that effective support for changes in the bits of keys e and N can be achieved with the method shown in this section. In order to increase the security of the cryptosystem, even when N is made larger, if the number of feedbacks to the SYRC and the size of ROM for the modular reduction is increased then support is possible. Although this lowers the processing speed, if the number of SYRCs is increased, it is possible to maintain the same processing speed.

4 Conclusion

Each $+$ in Fig.12 shows the RSA implementations with known processing speeds and circuit scales. Except for [THAA88], a trend of pursuing smaller circuit scales within 100Kbps and of getting higher efficiency can be observed. If the horizontal co-ordinate is the same, then the lower the vertical co-ordinate the more efficient the system is. In particular, the propose in Section 2 exhibits the best efficiency, i.e., 2.0 bps/gate. For processing speeds more than 100Kbps, however, like the implementation described in [THAA88], no efficient way has been reported for constructing appropriate scaled circuits for the RSA implementation. Any conventional method, even one in Section 2, would require a circuit scale which rapidly grows as the processing speed increases.

Section 3 resolved this difficulty by introducing the systolic array architecture and attained RSA implementations of constant efficiency as shown in lines a and b in Fig.12, where $m = 64$ is adopted. In these systolic implementations the circuit scale increases linearly in the required processing speed. And the efficiency is improved if the value of m is increased.

The systolic approach gives a systematic way of designing RSA hardwares in a wide range of circuit scales and processing speeds. Its characteristics are listed below:

1) Since the number of PEs on a chip is optional, we can make a chip to favorite scale. To increase the processing speed, one would merely have to increase the number of the chips.

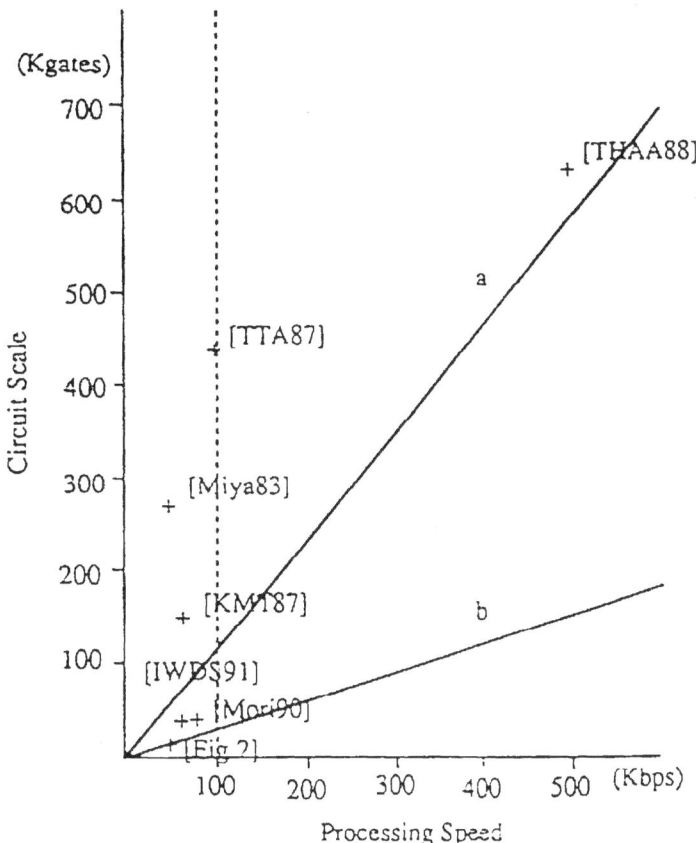

Fig.12 : Comparison of implementations proposed in this paper with those proposed so far

2) This method allows the same chip to implement the deciphering or signing (the secret transformation) with the aid of the CRT and the enciphering or signature-verification (the public transformation). Using C-MOS gates, which can be produced cheaply in large quantity due to the present-day semi-conductor technology, a one-chip RSA implementation achieving 64Kbps but using at most 20Kgates can be realized readily by adopting CRT-based secret transformations and moderate-sized public exponents e.

3) A systolic array architecture is simple and regular, and thus, suitable for VL-SIs. An RSA implementation using VLSI enables a single chip to achieve Mega bps.

4) Since the calculations processed within the PEs are simple integer calculations, an efficient RSA implementation can be constructed even by the use

of CPUs or DSPs.

In conclusion, the methods outlined in this paper lead to the constructive and effective ways for implementing the RSA scheme.

On the other hand it is known that the Montgomery method [Mont85] is an effective algorithm for modular multiplication. We already have a short paper [IMI92] proposing systolic arrays for the Montgomery method which can realize more efficient and more flexible implementation in the wide range of processing speed.

References

[RSA78] R.Rivest, A.Shamir, and L.Adleman, "A method of obtaining digital signatures and public key cryptosystems," Comm. of ACM, Vol.21, No.2, pp.120-126, Feb. 1978.

[Brick89] E.F.Brickell, "A Survey of hardware implementation of RSA," Advances in Cryptology — CRYPTO'89, pp.368-370, Springer-Verlag.

[Bric82] E.F.Brickell, "A fast modular multiplication algorithm with applications to two key cryptography," Advances in Cryptology, Proc. of CRYPTO'82, pp.51-60, Plenum.

[DK90] S.R.Dusse and B.S.Kaliski Jr., "A cryptographic library for the Motorola DSP56000," Advances in cryptology — EUROCRYPT'90, pp.230-244, Springer-Verlag.

[Even90] S.Even, "Systolic modular multiplication," Advances in Cryptology — CRYPTO'90 pp.619-624, Springer-Verlag.

[HDVG87] F.Hoornaert, M.Decroos, J.Vandewalle, and R.Govaerts, "Fast RSA-hardware: Dream or reality ?" Advances in Cryptology — CRYPTO'88, pp.257-264, Springer-Verlag.

[HTAA90] T.Hasebe, N.Torii, M.Azuma, and R.Akiyama, "Implementation of high speed modular exponentiation calculation," Proc. IEICE Spring Conference, A-284, 1990.

[IWDS91] P.A.Ivey, S.N.Walker, S.Davidson, and J.M.Stern, "A VLSI architecture for RSA encryption," Secure Design and Test of Crypto-Chips, IFIP WG10.5 Workshop, Oct. 1991.

[Kawa88] S.Kawamura, "A modulo multiplication algorithm using a small size residue table," Proc. Sympo. on Cryptography and Information Security, Session J, 1988.

[KMT87] Y.Kano, N.Matsuzaki, and M.Tatebayashi, "A modulo exponentiation LSI using high-order modified Booth's algorithm," Proc. Workshop on Cryptography and Information Security, WCIS87-11, 1987.

[Koch85] M.Kochanski, "Developing an RSA chip," Advances in Cryptology — CRYPTO'85, pp.350-357, Springer-Verlag.

[Miya83] S.Miyaguchi, "A fast computing scheme for RSA public-key cryptosystem and its VLSI organization," Trans. Info. Processing Soc. Japan, Vol.24, No.6, pp.764-771, Nov. 1983.

[Mori90] H.Morita, "A fast modular multiplication algorithm based on a radix 4 and its application," Trans. IEICE, Vol.73, No.7, July 1990.

[Omur90] J.K.Omura, "A public key cell design for smart card chips," Proc. 1990 Int. Symp. Info. Theory and Its Applications, 65-1, 1990.

[ORSPT86] G.Orton, M.Roy, P.Scott, L.Peppard, and S.Tavares, "VLSI implementation of public-key encryption algorithms," Advances in Cryptology — CRYPTO'86, pp.277-301, Springer-Verlag.

[OSA90] H.Orup, E.Svendsen, and E.Andreasen, "VICTOR an efficient RSA hardware implementation," Advances in Cryptology — EUROCRYPT'90, pp.245-252, Springer-Verlag.

[OT88] E.Okamoto and K.Tanaka, "A realization of RSA cryptosystem using digital signal processor, " Proc. Sympo. on Cryptography and Information Security, Session J, 1988.

[QC82] J.J.Quisquater and C.Couvreuer, "Fast decipherment algorithm for RSA public-key cryptosystem," Electron. Letters Vol.18, No.21, pp.905-907, Oct. 1982.

[Rive84] R.L.Rivest, "RSA chips (past/present/future)," Advances in Cryptology — EUROCRYPT'84, pp.159-168, Springer-Verlag.

[Taka88] K.Takaragi, "Hardware of RSA encryption," Proc. Sympo. on Cryptography and Information Security, Session J, 1988.

[TAA87] N.Torii, M.Azuma, and R.Akiyama, "A study on high speed RSA encryption LSI using parallel processing," Proc. IEICE National Convention, p.1388, 1987.

[THAA88] N.Torii, T.Hasebe, M.Azuma, and R.Akiyama, "The hardware technologies for RSA encryption systems," Proc. Sympo. on Cryptography and Information Security, Session J, 1988.

[TKS90] K.Takabayashi, S.Kawamura and A.Shinbo, "A modular exponentiation method using a fast constant multiplication algorithm," Proc. IEICE Spring Conference, A-285, 1990.

[Walt91] C.D.Walter, "Fast modular multiplication by operand scaling," Abstracts of CRYPTO'91, pp.8-1-8-6.

[WQ90] S.Waleffe and J.J.Quisquater, "CORSAIR: a smart card for public key cryptosystems," Advances in Cryptology — CRYPTO'90, pp.475-487, Springer-Verlag.

[FujG88] Fujitsu Limited, "ASIC technical information," GATI0172C, 1988.

[FujM88] Fujitsu Limited, "CMOS standard cell design manual," MATI10702, 1988.

[Mont85] P.L.Montgomery, "Modular multiplication without trial division," Math.of Computation, Vol.44, pp.519-521, 1985.

[IMI92] K.Iwamura, T.Matsumoto and H.Imai, "Systolic-arrays for modular exponentiation using Montgomery method," presented in Rumpsession of Eurocrypt'92, May 24-28, 1992.

Appendix

Theorem

In ALGORITHM 1A when $A_{k-j} \cdot B_i$, $E_{j-1,i}$, $D_{j-1,i}$ and $C_{j-1,n-i}$ are respectively m-bit, m-bit, $(m-1)$-bit and 3-bit values, the most significant 3 bits of

$$R_{j,i} = A_{k-j} \cdot B_i + E_{j-1,i} + 2 \cdot D_{j-1,i} + C_{j-1,i-1}$$

is less than [111] for $m \geq 3$.

Proof

Because $A_{k-j} \cdot B_i \leq 2^m - 1$, $E_{j-1,i} \leq 2^m - 1$, $2 \cdot D_{j-1,i} \leq 2^m - 2$ and $C_{j-1,i-1} \leq 7$, we have

$$R_{j,i} \leq (2^2 + 2^1) \cdot 2^{m-1} + 2^1 + 2^0.$$

Thus, if $m - 1 \geq 2$, that is, $m \geq 3$, then the most significant 3 bits of the right-hand side is equal to [110].

A Simplified and Generalized Treatment of Luby-Rackoff Pseudorandom Permutation Generators

Ueli M. Maurer

Institute for Theoretical Computer Science
ETH Zürich
CH-8092 Zürich, Switzerland
Email address: maurer@inf.ethz.ch

Abstract. A paper by Luby and Rackoff on the construction of pseudorandom permutations from pseudorandom functions based on a design principle of the DES has recently initiated a burst of research activities on applications and generalizations of these results. This paper presents a strongly simplified treatment of these results and generalizes them by pointing out the relation to locally random functions, thereby providing new insight into the relation between probability-theoretic and complexity-theoretic results in cryptography. The first asymptotically-optimal construction of a locally random function is presented and new design strategies for block ciphers based on these results are proposed.

Keywords. Locally random function, Pseudorandom function, Pseudorandom permutation, Luby-Rackoff permutation generator.

R.A. Rueppel (Ed.): Advances in Cryptology - EUROCRYPT '92, LNCS 658, pp. 239-255, 1993.
© Springer-Verlag Berlin Heidelberg 1993

1. Introduction

In a celebrated complexity-theoretic paper [9], Luby and Rackoff described a construction of a pseudorandom permutation generator from any pseudorandom function generator that was motivated by a study of the Data Encryption Standard (DES, cf. [4]). Much research has recently been based on this paper (e.g., [11], [12], [13], [14]). The main goal of the present paper is to give a simplified and generalized treatment of the results of [9] by suggesting an information-theoretic rather than complexity-theoretic interpretation based on the concept of locally random functions whose treatment is an independent goal of this paper. It is shown that the proof of the Main Lemma of [9], which originally required three pages of highly technical definitions, claims and arguments, can be strongly simplified and interpreted essentially as an application of the birthday paradox, thus providing much more insight. Moreover, the central proposition of [9], which was used to establish the relation between probability-theoretic arguments and complexity-theoretic results and which was unfortunately stated without proof, is shown to be unnecessary and somewhat misleading.

Local randomness is an important concept in theoretical computer science with several applications. Intuitively, a family of functions is locally random of degree k if for every set of at most k arguments, the function values for these arguments for a randomly (from the family) chosen function are independent and uniformly distributed. In other words, a randomly (from the family) chosen function behaves precisely like a truly random function as long as it is evaluated for at most k arguments. Similarly, a sequence generator is locally random of degree k [10] if for a randomly selected seed, every subset of k (or less) digits is completely random. Clearly, a locally random sequence generator can be obtained from a locally random function by "reading out" the function values for a given enumeration of the arguments, but the converse is not true in general because a sequence generator need not have the property that arbitrary digits can be accessed efficiently (only consecutive digits must be efficiently computable).

The usefulness of local randomness has previously been observed (e.g., [1], [3], [6], [7]) and was referred to as k-wise independence. However, our treatment is more general in that (1) families of functions that are only "almost" locally random of degree k and (2) polynomial-time computable functions with super-polynomial degree of local randomization are considered, allowing applications in complexity theory as well as for the design of practical block ciphers.

The results of Luby and Rackoff are discussed in Section 3. Locally random functions are introduced in Section 4, and an alternative interpretation and generalization of the results of Luby and Rackoff based on this new concept are described Section 5. Some further applications of locally random functions and a new design strategy for block ciphers are discussed in Section 6.

2. Terminology

Our terminology is similar to that of [9]. Let $\{0,1\}^n$ denote the set of binary strings of length n, let F^n denote the set of all 2^{n2^n} functions $\{0,1\}^n \to \{0,1\}^n$, and let P^n denote the subset of functions of F^n that are permutations of $\{0,1\}^n$, i.e., invertible or one-to-one. For $f_1 \in F^n$ and $f_2 \in F^n$, $f_1 \circ f_2$ denotes the composition of f_1 and f_2, i.e., $f_1 \circ f_2(x) = f_2(f_1(x))$.

For two binary strings a and b, $a \bullet b$ denotes their concatenation and when a and b have the same length, $a \oplus b$ denotes their bit-by-bit *exclusive or*. The string consisting of the t rightmost bits of a string a is denoted by $[a]_t$. In particular, $[i]_t$ for a non-negative integer $i < 2^t$ denotes the representation of i by t bits (with possible leading zeroes).

When an argument of a function is replaced by a set of arguments this will denote the multiset of resulting function values. In all cases, a random choice of an object x from a set or multiset S of objects (denoted by $x \in_R S$) will be such that each object is equally likely to be chosen, taking into account multiple occurrencies in multisets. We refer to Section 4 of [9] for definitions of a pseudorandom number (or bit) generator (PRNG), of a pseudorandom function generator (PRFG) and of a pseudorandom permutation generator (PRPG). A function $f : \mathbb{N} \to \mathbb{N}$ is called superpolynomial if for every polynomial Q, $f(n) > Q(n)$ for all sufficiently large n. Finally, $\#S$ denotes the cardinality of the set or multiset S, and all logarithms in this paper are to the base 2.

3. Luby-Rackoff Pseudorandom Permutation Generators

Levin [8] gave a construction of a PRNG from any one-way function, and Goldreich, Goldwasser and Micali [5] devised a method for constructing a PRFG from any PRNG and hence, by Levin's result, also from any one-way function. (As a by-product of this research a simpler construction of a PRFG from any PRNG will be described in Section 4.) A PRNG can be used for encryption in a so-called additive stream cipher but a PRFG cannot directly be used for (block) encryption because pseudorandom functions are not invertible in general. Luby and Rackoff considered the problem of constructing a secure block encryption algorithm, i.e., a (secure) pseudorandom permutation generator, from any (secure) PRFG, and hence from any PRNG or from any one-way function. We refer to [9] for definitions.

Motivated by the round structure of the Data Encryption Standard DES (cf. [4]), Luby and Rackoff defined a mapping $H : F^n \times F^n \times F^n \to P^{2n}$ assigning every triple of functions in F^n a permutation in P^{2n}. Let L and R denote the left and right half of a $2n$-bit string $L \bullet R$ and let for $f \in F^n$ the permutation $\bar{f} \in P^{2n}$ be defined as

$$\bar{f}(L \bullet R) = R \bullet [L \oplus f(R)],$$

i.e., the right half of the argument appears unchanged as the left half of the result and the right half of the result is equal to $L \oplus f(R)$. This corresponds to one round of DES. For a list of functions, $f_1, \ldots, f_s \in F^n$, let the function (actually a permutation) $\psi(f_1, \ldots, f_s) : \{0,1\}^{2n} \rightarrow \{0,1\}^{2n}$ be defined by

$$\psi(f_1, \ldots, f_s) = \bar{f}_1 \circ \cdots \circ \bar{f}_s,$$

i.e., $\psi(f_1, \ldots, f_s)(L \bullet R) = \bar{f}_s(\bar{f}_{s-1}(\cdots \bar{f}_1(L \bullet R) \cdots))$. Note that H can now be defined by $H(f_1, f_2, f_3) = \psi(f_1, f_2, f_3)$ (cf. Figure 1), where

$$\psi(f_1, f_2, f_3)(L \bullet R) = [R \oplus f_2(L \oplus f_1(R)] \bullet [L \oplus f_1(R) \oplus f_3(R \oplus f_2(L \oplus f_1(R)))].$$

Luby and Rackoff considered the problem of distinguishing, by use of an oracle circuit, a function randomly chosen from F^{2n} from a function randomly chosen from the much smaller set $\psi(F^n, F^n, F^n)$. An oracle circuit C_{2n} is a circuit with oracle gates, i.e., gates with a $2n$-bit input and a $2n$-bit output, where all oracle gates in a circuit evaluate the same fixed function in F^{2n} (for details see [9]). Let

$$P[C_{2n}(f){=}1 : f \in_R \psi(F^n, F^n, F^n)]$$

and

$$P[C_{2n}(f){=}1 : f \in_R F^{2n}]$$

denote the probabilities that C_{2n} outputs 1 if the oracle gates are evaluated for a function chosen randomly from $\psi(F^n, F^n, F^n)$ and from F^{2n}, respectively. We hope that this notation, which differs slightly from that of [9], is more intuitive. The Main Lemma of [9] is as follows.

Main Lemma of [9]. *Let C_{2n} be an oracle circuit with k oracle gates such that no input value is repeated to an oracle gate. Then*

$$\left| P[C_{2n}(f){=}1 : f \in_R \psi(F^n, F^n, F^n)] - P[C_{2n}(f){=}1 : f \in_R F^{2n}] \right| \leq k^2 / 2^n.$$

From the following discussion it will become clear that the restriction to circuits whose oracle gates must have different inputs, and hence also the proposition stated (unfortunately without proof) in [9] above the Main Lemma, are unnecessary and somewhat misleading. The result can be stated as a purely probability-theoretic result having no direct relation to complexity theory, and will in Section 5 be interpreted as a result on locally random functions.

Let $g : (\{0,1\}^{2n})^k \rightarrow \{0,1\}$ be a function taking as input k $2n$-bit strings. For a given set of k arguments $x_1, \ldots, x_k$, let in analogy to the above definitions

$$P[g(f(x_1), \ldots, f(x_k)) = 1 : f \in_R \psi(F^n, F^n, F^n)]$$

and

$$P_g \triangleq P[g(f(x_1), \ldots, f(x_k)) = 1 : f \in_R F^{2n}] \tag{1}$$

be defined as the probabilities that $g(f(x_1), \ldots, f(x_k)) = 1$ when f is chosen randomly from $\psi(F^n, F^n, F^n)$ and from F^{2n}, respectively. Note that P_g can alternatively be defined as

$$P_g = P[g(r_1, \ldots, r_k) = 1]$$

where $r_1, \ldots, r_k$ are independent and randomly selected from $\{0,1\}^{2n}$. Again equivalently, P_g can also be defined as

$$P_g = \frac{\#\{(r_1, \ldots, r_k) \in (\{0,1\}^{2n})^k : g(r_1, \ldots, r_k) = 1\}}{2^{2nk}}.$$

Lemma 1. *For every function* $g : (\{0,1\}^{2n})^k \rightarrow \{0.1\}$ *and for every set of* k *arguments* $x_1, \ldots, x_k$,

$$\left| P[g(f(x_1), \ldots, f(x_k)) = 1 : f \in_R \psi(F^n, F^n, F^n)] - P_g \right| \leq k^2/2^n.$$

Clearly, Lemma 1 is also true for every function $g : (\{0,1\}^{2n})^{k'} \rightarrow \{0,1\}$ with $k' < k$. It demonstrates that there exists no set of k arguments, whether adaptively chosen or not, and whether distinct or not, that would allow an oracle circuit with these arguments as the inputs to the oracle gates to achieve

$$\left| P[C_{2n}(f) = 1 : f \in_R \psi(F^n, F^n, F^n)] - P[C_{2n}(f) = 1 : f \in_R F^{2n}] \right| > k^2/2^n.$$

The Main Lemma of [9] is hence an immediate consequence of Lemma 1. (It is easy to see that the converse is also true.) Moreover, it is obvious that probabilistic strategies cannot be better than deterministic ones for distinguishing a function from a random function since the deterministic function g could be defined as that resulting for the optimal choice of the randomizer.

Proof of Lemma 1. Let f_1, f_2 and f_3 be functions randomly chosen from F^n, and let $f = \psi(f_1, f_2, f_3)$. Let $x_i = L_i \bullet R_i$ for $1 \leq i \leq k$ be the k arguments of f, and define S_i, T_i and V_i for $1 \leq i \leq k$ as follows (cf. Figure 1):

$$S_i = L_i \oplus f_1(R_i),$$

$$T_i = f_2(S_i) \oplus R_i$$

and

$$V_i = f_3(T_i) \oplus S_i.$$

Note that when the evaluation of f for the argument x_i is viewed as a three-round process (similar to three rounds of DES), the outputs of the first, second and third round are $R_i \bullet S_i$, $S_i \bullet T_i$ and $T_i \bullet V_i = f(L_i \bullet R_i)$, respectively. We may for the rest of the proof assume without loss of generality that the x_i, $1 \le i \le k$, are distinct. Choosing identical arguments provides no new information and can thus certainly not help.

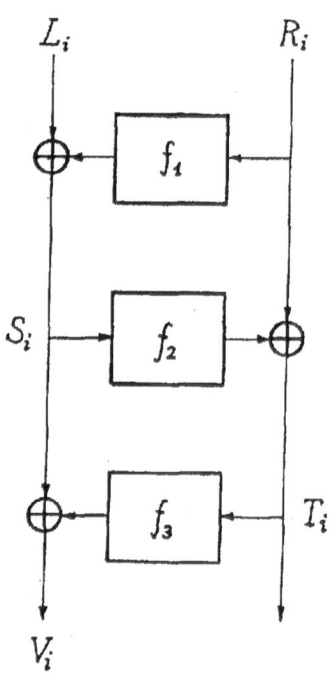

Figure 1.   Computation of $\psi(f_1, f_2, f_3)(L \bullet R)$ as $T_i \bullet V_i$. For the sake of simplicity, the swaps of left and right half between rounds are not shown explicitly.

Let $\mathcal{E}_S$ and $\mathcal{E}_T$ denote the events that $S_1, \ldots, S_k$ are distinct and that $T_1, \ldots, T_k$ are distinct, respectively, and let $\mathcal{E}$ be the event that both $\mathcal{E}_S$ and $\mathcal{E}_T$ occur. If $\mathcal{E}_S$ occurs, then $T_1 = R_1 \oplus f_2(S_1), \ldots, T_k = R_k \oplus f_2(S_k)$ are completely random because f_2 is a random function and hence $f_2(S_1), \ldots, f_2(S_k)$ are completely random. Similarly, if $\mathcal{E}_T$ occurs, then $V_1 = S_1 \oplus f_3(T_1), \ldots, V_k = S_k \oplus f_3(T_k)$ are completely random because f_3 is a random function. Thus if both $\mathcal{E}_S$ and

$\mathcal{E}_T$ occur, $f(x_1) = T_1 \bullet V_1, \ldots, f(x_k) = T_k \bullet V_k$ are completely random and thus $f = \psi(f_1, f_2, f_3)$ behaves precisely like a function chosen randomly from F^{2n}. Therefore the distinguishing probability is upper bounded by

$$\left| P[g(f(x_1), \ldots, f(x_k)) = 1 : f \in_R \psi(F^n, F^n, F^n)] - P_g \right| \leq 1 - P[\mathcal{E}].$$

We now derive an upper bound on $1 - P[\mathcal{E}] = P[\overline{\mathcal{E}}]$, where $\overline{\mathcal{E}}$ denotes the complementary event of $\mathcal{E}$. $\overline{\mathcal{E}}$ is the union of the $\binom{k}{2}$ events $\{S_i = S_j\}$ for $1 \leq i < j \leq k$ and the $\binom{k}{2}$ events $\{T_i = T_j\}$ for $1 \leq i < j \leq k$. The probability of the union of several events is upper bounded by the sum of the probabilities, and hence

$$1 - P[\mathcal{E}] = P[\overline{\mathcal{E}}] \leq \sum_{1 \leq i < j \leq k} P[S_i = S_j] + \sum_{1 \leq i < j \leq k} P[T_i = T_j]. \qquad (2)$$

For $i \neq j$ we have

$$P[S_i = S_j] = \begin{cases} 2^{-n} & \text{if } R_i \neq R_j \\ 0 & \text{if } R_i = R_j. \end{cases} \qquad (3)$$

Note that when $R_i = R_j$, then $P[S_i = S_j] = 0$ since by assumption $L_i \bullet R_i \neq L_j \bullet R_j$ and hence $L_i \neq L_j$. Equation (3) shows that

$$P[S_i = S_j] \leq 2^{-n}$$

for $i \neq j$. By a similar argument we obtain

$$P[T_i = T_j] \leq 2^{-n}$$

for $i \neq j$. The total number of terms on the right side of (2) is $2\binom{k}{2} = k(k-1) < k^2$. Lemma 1 follows. $\square$

An interpretation and generalization of this result based on locally random functions, which are introduced in the following section, will be presented in Section 5.

4. Random Functions and Locally Random Functions

A random function $r : \{0, 1\}^n \to \{0, 1\}^n$ is a function that assigns to all arguments $x \in \{0, 1\}^n$ independent and completely random values $r(x) \in \{0, 1\}^n$. The trivial implementation of a random function as a table requires the generation of $n2^n$ random bits during a precomputation phase and $n2^n$ bits of memory to store the table.

A random function can alternatively be implemented as a device (or procedure) that, when given as input an argument x that was never given before, generates a random output $r(x)$ and stores the pair $(x, r(x))$ in a table ordered according to x, and when given as input an argument x for which r was previously evaluated, outputs $r(x)$ stored in the table. An advantage of the latter implementation is that when r needs to be evaluated for at most t arguments, $2tn$ bits of memory are required and at most $2tn$ random bits need to be generated. However, the computation time for each argument is $O(\log t)$ compared to $O(1)$ for an implementation based on a pregenerated table (of exponential size), and hence depends on the number t of arguments.

When the computation time of an algorithm accessing a random function, implemented as described above, is polynomially (in n) bounded, so are the total computation time and memory requirements of the resulting algorithm, including the random function. In other words, although a random function seems at first to require an exponential amount of memory, any polynomial-time algorithm using random functions can be implemented in polynomial time and polynomial space. This observation is the key argument of the proof of Theorem 1 of [9]. We would like to point out (without making further use of this result) that the same observation can be used to present a construction of a PRFG from a PRNG that is much simpler (albeit less practical) than that proposed by Goldreich, Goldwasser and Micali [5] for proving the following proposition.

Proposition 1 [5]. *Pseudorandom function generators exist if and only if pseudorandom number generators exist.*

Randomness is often an expensive and limited resource. Moreover, a dependence of the function evaluation time and memory requirement on the number t of arguments for which a function is evaluated is most often intolerable. Therefore, an important concept is that of a locally random function, i.e., a function that behaves like a random function as long as it is evaluated for at most k arguments for some parameter k.

Definition 1. A family $\mathcal{F}_{\mathcal{Z}} = \{f_z : z \in \mathcal{Z}\}$ of functions $f_z : \{0,1\}^n \to \{0,1\}^m$ is an (n, m, k) *locally random function (LRF)* with key space $\mathcal{Z}$ if for every subset $\{x_1, \ldots, x_k\}$ of $\{0,1\}^n$, $f_z(x_1), \ldots, f_z(x_k)$ are uniformly distributed over $\{0,1\}^m$ and jointly statistically independent, when z is randomly selected from $\mathcal{Z}$.

$\mathcal{F}_{\mathcal{Z}}$ could alternatively be viewed as a single function $\mathcal{Z} \times \{0,1\}^n \to \{0,1\}^n$. A random function $\{0,1\}^n \to \{0,1\}^n$ is an $(n, n, 2^n)$ LRF. The restriction to binary digits is made without essential loss of generality. The above definition is purely combinatorial, i.e., no restriction on the computation time is made. LRFs will be generalized below to take into account both minor deviations from complete randomness of any k function values and efficient (i.e., polynomial-time) computability.

An important question is for which choices of parameters n, m, k and $|\mathcal{Z}|$ there exist LRFs. Because it is impossible to expand deterministically a sequence of random bits into a longer sequence of (independent) random bits, it is obvious that for an (n, m, k) LRF with key space $\mathcal{Z}$,

$$|\mathcal{Z}| \geq 2^{km} \tag{4}$$

must hold. It may appear to be somewhat surprising that, for any n, m and k with $m = n$ or m a multiple of n, equality in (4) can be achieved. This follows from the following well-known proposition, which can be proved by observing that the $d+1$ coefficients of a polynomial of degree d over a field can be interpolated from any set of $d+1$ arguments and the corresponding polynomial values. When $m < n$, equality in (4) cannot be achieved.

Proposition 2. *Let* $p_0, \ldots, p_{k-1}$ *be randomly selected n-bit strings. The function*

$$p : \{0, 1\}^n \longrightarrow \{0, 1\}^n : \quad x \mapsto p(x) = p_{k-1} x^{k-1} + \cdots + p_1 x + p_0,$$

where all quantities are considered as representations of elements of the finite field $GF(2^n)$, *is a* (n, n, k) *LRF with minimal key space* $\mathcal{Z} = \{0, 1\}^{kn}$.

For the sake of completeness we state the following proposition, which is an immediate consequence of Theorem 1 in [10]. Let $\mathcal{Z} = \{0, 1\}^v$, i.e., the key consists of v binary digits.

Proposition 3. *There exists a* (n, m, k) *LRF if*

$$mk \leq \frac{v}{n + log_2 m}$$

and there exists no $(n, 1, k)$ *LRF if*

$$k > \frac{2(v + n + 1)}{n - \log_2 v + 1}.$$

In order to state our results on PRFGs and PRPGs in terms of LFRs, we need to generalize the concept of LRFs in two different ways. As a first generalization, the condition of true randomness of any k function values must be somewhat relaxed. Instead of introducing the new concept of "almost" locally random functions we generalize LRFs by introducing a fourth parameter, ϵ, believing that this generalization will be intuitive rather than ambiguous. (Note that a $(n, m, k, 0)$ LRF will be the same as a (n, m, k) LRF.)

Definition 1'. A family $\mathcal{F}_{\mathcal{Z}} = \{f_z : z \in \mathcal{Z}\}$ of functions $f_z : \{0, 1\}^n \rightarrow \{0, 1\}^m$ is an (n, m, k, ϵ) *locally random function* with key space $\mathcal{Z}$ if for all functions g :

$(\{0,1\}^m)^k \to \{0,1\}$ and for every subset $\{x_1,\ldots,x_k\}$ of $\{0,1\}^n$, for z randomly selected from $\mathcal{Z}$,

$$\left| P[g(f_z(x_1),\ldots,f_z(x_k)) = 1] - P[g(r_1,\ldots,r_k) = 1] \right| \le \epsilon,$$

where $r_1,\ldots,r_k$ are independent and randomly selected from $\{0,1\}^m$.

Note that

$$P[g(f_z(x_1),\ldots,f_z(x_k)) = 1] = \frac{\#\{z \in \mathcal{Z} : g(f_z(x_1),\ldots,f_z(x_k)) = 1\}}{\#\mathcal{Z}}.$$

Clearly, an (n,m,k,ϵ) LRF is also an (n,m,k',ϵ') LRF for every $k' \le k$ and $\epsilon' \ge \epsilon$. Moreover, a (n,m,k,ϵ) LRF can easily be modified by deleting some output bits to yield a (n,m',k,ϵ) LRF for any $m' < m$. Conversely, a (n,cm,k,ϵ) LRF with key space $\mathcal{Z}^c$ can for $c > 1$ be obtained from a (n,m,k,ϵ) LRF $\mathcal{F}_{\mathcal{Z}}$ with key space $\mathcal{Z}$ by a simple concatenation of c copies of $\mathcal{F}_{\mathcal{Z}}$ with independent keys.

A second generalization of LRFs is necessary in order to be consistent with other asymptotic definitions in complexity theory, in particular those used in [9].

Definition 2. A *locally random function generator (LRFG)* with key length function $l(n)$ and degree of local randomization $k(n)$ is a family $\mathcal{F} = \{\mathcal{F}^n_{\{0,1\}^{l(n)}} : n \in \mathbb{N}\}$, where $\mathcal{F}^n_{\{0,1\}^{l(n)}}$ is a $(n,n,k(n),\epsilon(n))$ LRF with key space $\{0,1\}^{l(n)}$ that is (for every given argument and key) computable in time polynomial in n, independent of the number of previous evaluations, where $\epsilon(n)$ vanishes faster than $1/Q(n)$ for every polynomial $Q(n)$ (i.e., $1/\epsilon(n)$ is superpolynomial in n).

5. Complexity-theoretic Applications of Locally Random Functions

The construction of [9] for a PRPG is based on the following observation which can be formalized. Let $\mathcal{F}$ be a LRFG (or LRPG) with key length function $l(n)$ whose degree of local randomization $k(n)$ is superpolynomial in n (which implies that $l(n)$ is superpolynomial in n). Let $\mathcal{F}'$ be the PRFG (or PRPG) resulting from $\mathcal{F}$ when the $n \cdot l(n)$ random bits are substituted by a function generator $\mathcal{G}$ generating the corresponding amount of (pseudorandom) bits. Then $\mathcal{F}'$ is a PRFG (or PRPG) under the assumption that $\mathcal{G}$ is a PRFG.

For PRFGs, this construction seems to be of little value because a PRFG is required for constructing another PRFG. For the case of a PRPG, however, this observation allows to relax the problem of constructing a PRPG to the two problems of constructing a PRFG and a LRFG with superpolynomial degree of local randomization that is also a permutation generator.

For the construction of a LRF described in Proposition 2, the evaluation time is proportional to the degree k of local randomization because for every x, $f(x)$ depends on every internal random bit, i.e., on all the coefficients of the polynomial $p(x)$. A different construction for which every function value depends only on an negligible fraction of the key bits must hence be used for obtaining a superpolynomial degree k of local randomization while retaining the polynomial evaluation time, as is required for a generalized interpretation of the Luby-Rackoff results. Before presenting such constructions we point out that the mapping $H : F^n \times F^n \times F^n \rightarrow P^{2n}$ gives a construction of LRFGs from other LRFGs.

Theorem 1. *Let $\mathcal{F}_i$ for $i = 1, 2, 3$ be three independent (n, n, k, ϵ_i) LRFs. $\psi(\mathcal{F}_1, \mathcal{F}_2, \mathcal{F}_3)$ is a $(2n, 2n, k, \epsilon)$ LRF for all k, where $\epsilon = k^2 2^{-n} + \epsilon_1 + \epsilon_2 + \epsilon_3$.*

Proof. The fact that $\mathcal{F}_i$ is a (n, n, k, ϵ_i) LRF can be expressed as

$$\left| P[g(f(x_1), \ldots, f(x_k)) = 1 : f \in_R \mathcal{F}_i] - P_g \right| \leq \epsilon_i \tag{5}$$

where P_g is defined in (1) with F^{2n} replaced by F^n. Since using a randomized strategy for distinguishing $\mathcal{F}_i$ from a random function cannot be better than using the best deterministic function g, as mentioned before, inequality (5) implies that

$$\left| P[g(f(x_1), \ldots, f(x_k)) = 1 : f \in_R \psi(F^n, F^n, F^n)] \right.$$
$$\left. - P[g(f(x_1), \ldots, f(x_k)) = 1 : f \in_R \psi(\mathcal{F}_1, F^n, F^n)] \right| \leq \epsilon_1$$

and

$$\left| P[g(f(x_1), \ldots, f(x_k)) = 1 : f \in_R \psi(\mathcal{F}_1, F^n, F^n)] \right.$$
$$\left. - P[g(f(x_1), \ldots, f(x_k)) = 1 : f \in_R \psi(\mathcal{F}_1, \mathcal{F}_2, F^n)] \right| \leq \epsilon_2$$

and

$$\left| P[g(f(x_1), \ldots, f(x_k)) = 1 : f \in_R \psi(\mathcal{F}_1, \mathcal{F}_2, F^n)] \right.$$
$$\left. - P[g(f(x_1), \ldots, f(x_k)) = 1 : f \in_R \psi(\mathcal{F}_1, \mathcal{F}_2, \mathcal{F}_3)] \right| \leq \epsilon_3.$$

The proof is completed by combining these three inequalities with Lemma 1 and observing that these four probability differences define four adjacent (but possibly overlapping) subintervals of $[0, 1]$ whose total span can be at most the sum of the four interval lengths. □

Note that the functions of $\psi(\mathcal{F}_1, \mathcal{F}_2, \mathcal{F}_3)$ are actually permutations. Lemma 1 follows immediately as a corollary of this theorem since F^n with key space $\{0,1\}^{n2^n}$ is an $(n, n, 2^n, 0)$ LRF.

Instead of implementing the functions f_1, f_2, f_3 in the Luby-Rackoff construction $H(f_1, f_2, f_3)$ directly as some pseudorandom functions, the construction H can be applied iteratively. For instance, a pseudorandom permutation $f : \{0,1\}^{4n} \to \{0,1\}^{4n}$ can be implemented as $f = H(f_1, f_2, f_3)$ where $f_1 = H(f_{11}, f_{12}, f_{13})$, $f_2 = H(f_{21}, f_{22}, f_{23})$ and $f_3 = H(f_{31}, f_{32}, f_{33})$ and where f_{ij} are pseudorandom functions $\{0,1\}^n \to \{0,1\}^n$. Let

$$H^{(s)} : (F^n)^{3^s} \to P^{2^s n}$$

be the s-fold iterative application of the Luby-Rackoff construction H which requires 3^s functions $\{0,1\}^n \to \{0,1\}^n$ as inputs. The following Corollary to Theorem 1 gives a characterization of this iterative construction as a result on locally random functions.

Corollary 1. *When $H^{(s)}$ is applied to 3^s independent (n, n, k, ϵ) LRFs then the resulting function is a $(2^s n, 2^s n, k, \epsilon')$ LRF where $\epsilon' = k^2 \sum_{i=1}^{s} 3^{i-1} 2^{-2^{s-i} n} + 3^s \epsilon$.*

Although the mapping H serves the originally intended purpose of proving an important complexity-theoretic result, randomness is used wastefully: The degree of local randomization is only on the order of the square root of the number of key bits. The best (in terms of efficient use of key bits) previously known asymptotic construction is the LRFG $\{\psi(F^n, F^n, F^n, F^n, F^n) : n \in \mathbb{N}\}$ with superpolynomial degree of local randomization and with key length function $l(n) = 5n2^n$, which was proved in [11] to have degree $k(n) = \Omega(2^{2n/3}) = \Omega(l(n)^{2/3})$ of local randomization.

In the following we present an alternative construction of LRFGs that achieves a local randomization of degree $k(n) = \Omega(l(n)^\alpha)$ for any $\alpha < 1$. Such LRFGs lead to alternative constructions of PRFGs and PRPGs based on PRFGs.

Let d be a parameter of the following construction, let $r_i : \{0,1\}^t \to \{0,1\}^n$ for $1 \le i \le d$ be random functions, let $c = \lceil \log_2 d \rceil$ and let P be a $(n + c, t, 2d)$ LRF. For example, P could be implemented as $P(\xi) = [p(\xi)]_t$ (the t least significant bits of $p(\xi)$), where p is a polynomial $p(u) = p_{2d-1}u^{2d-1} + \cdots + p_1 u + p_0$ of degree $2d - 1$ over $GF(2^{n+c})$ and the key of P consists of the $2d$ coefficients $p_{2d-1}, \ldots, p_0$. The total number of random bits required for implementing r and P is hence $l(n) = 2^t dn + 2d(n + c)$.

Theorem 2. *The family of functions $\mathcal{F}^{(n)} = \{f_z : z \in \{0,1\}^{l(n)}\}$ defined by*

$$f_z(x) = \sum_{i=0}^{d-1} r_i(P(x \bullet [i]_c)),$$

where the sum is bit-wise modulo 2 andthe $l(n) = 2^t dn + 2d(n + c)$ bits of z are used in some manner to implement the functions r_i and as the key of P, is a $(n, n, k, \gamma(k))$ LRF for all k, where $\gamma(k) = k^{d+1} 2^{-dt}$.

Proof. Let a_{ij}, $1 \leq i \leq d$, $1 \leq j \leq k$ be the input to function r_i when $\mathcal{F}^{(n)}$ is evaluated for the jth argument x_j. Let $\mathcal{E}$ be the event that for every x_j, $1 \leq j \leq k$, there exists an i_j, $1 \leq i_j \leq d$, such that $a_{i,j} \neq a_{i,m}$ for all $m \neq j$. If $\mathcal{E}$ occurs, then for $1 \leq j \leq k$ at least one of the terms in the sum forming $f_z(x_j)$ is a random variable that is completely random and independent of all the other terms occurring in the evaluations of $f_z(x_1), \ldots, f_z(x_k)$, and hence $f_z(x_1), \ldots, f_z(x_k)$ are independent and completely random. The complementary event $\overline{\mathcal{E}}$ is the union over $1 \leq j \leq k$ and over $m_i \in \{1, \ldots, j-1, j+1, \ldots, k\}$ for $1 \leq i \leq d$ of the $k(k-1)^d$ events

$$\{a_{ij} = a_{im_i} \text{ for } 1 \leq i \leq d\}.$$

Because the a_{ij} are $2d$-wise independent, each of these events has probability 2^{-dt} and hence

$$P[\overline{\mathcal{E}}] \leq k^{d+1} 2^{-dt}. \quad \square$$

The following argument demonstrates that the construction of Theorem 2 yields an asymptotically optimal locally random function. Let

$$\mathcal{G} = \{\mathcal{F}^{(n)} : n \in \mathbb{N}\}, \tag{6}$$

where $\mathcal{F}^{(n)}$ is the LRF from Theorem 2 with $d = t$ and t is any function of n such that $2^t/t$ is superpolynomial in n. For instance, $t(n) = \lceil (\log n)^{1+\delta} \rceil$ for some fixed $\delta > 0$. The key length function of $\mathcal{G}$ is

$$l(n) = 2t(n + \lceil \log_2 t \rceil) + 2^t tn = \Theta(2^t tn).$$

For

$$k(n) = 2^{t(t-1)/(t+1)}$$

we have $\gamma(k) = 2^{-t}$. It is straight-forward to prove that

$$\lim_{n \to \infty} \frac{l(n)^\alpha}{k(n)} = 0$$

for all $\alpha < 1$ and thus we have the following result.

Corollary 2. $\mathcal{G}$ as defined in (6) is a LRFG with degree of local randomization $k(n) = \Omega(l(n)^\alpha)$ for any $\alpha < 1$.

We suggest as an open problem to devise a LRFG with superpolynomial key length function $l(n)$ and local randomization of degree $k(n) = \Omega(l(n))$.

One can prove that even when $r_1, \ldots, r_d$ are taken to be the same random function $r : \{0, 1\}^t \rightarrow \{0, 1\}^n$ rather than independent random functions, the resulting family of functions is a LRF also satisfying $k(n) = \Omega(l(n)^\alpha)$ for any $\alpha < 1$.

Theorem 3. *The family of functions $\mathcal{F}^{(n)} = \{f_z : z \in \{0, 1\}^{l(n)}\}$ defined by*

$$f_z(x) = \sum_{i=0}^{d-1} r(P(x \bullet [i]_c)),$$

where the $l(n) = 2^t n + 2d(n + c)$ bits of z are used in some manner to implement the function r and as the key of P, is a $(n, n, k, \gamma(k))$ LRF, where $\gamma(k) = k^{d+1}(2d2^{-t})^d$, for all $k \geq \sqrt{2^t/d}$.

6. Concluding Discussion

An important application of locally random functions is in the area of probabilistic algorithms, where randomness is often an expensive resource and therefore simulated by a pseudorandom generator with a random seed. In the analysis of a probabilistic algorithm that uses blocks of random bits at various stages it is sometimes sufficient to require that the random blocks be only k-wise independent rather than jointly independent. For instance, the "birthday paradox" holds not only for independent random birthdays but also when the birthdays are only pairwise independent. Furthermore, pairwise independence of a set of random variables is sufficient for proving that the variance of the sum of the random variables equals the sum of the variances (e.g., see [3]).

Local randomness has also several applications in cryptography. A first (in the author's opinion misinterpreted) cryptographic application is the design of cipher systems "provably secure" against enemies with unlimited computational resources. Schnorr [13] suggested to simulate the random keystream of the one-time pad by a keystream that is only locally random. If an eavesdropper can examine at most k (arbitrarily chosen) bits of the keystream, where k is the degree of local randomization, such a system offers the same perfect security as the one-time pad, even if the eavesdropper has infinite computing power. Of course, as is pointed out in [10] where Schnorr's idea is generalized, the drawback of such a system is that clearly k cannot be greater than the length of the secret key (the seed) and thus the assumption that an eavesdropper cannot obtain more than k keystream bits is generally completely unrealistic. Another example is the "provably secure" block cipher described in [14] which suffers from an even stronger weakness because the number of plaintext-cryptogram pairs an eavesdropper is allowed to obtain is upper bounded by only the square root of the key size. Loosely speaking, an enemy is guaranteed to spend at least

100 years breaking the cipher if the user of the system is willing to spend 10'000 years for only loading the secret key into the system. Clearly, if such a long secret key were available, the users would be better off using a one-time pad to begin with.

A related but much more important cryptographic application of local randomness is the design of conventional cryptographic algorithms using a secret key of only moderate size. The basic idea, which could be further formalized, is to design a system that uses an (impractically) large amount of secret random bits and to prove it secure against enemies with unlimited computational resources for a suitable definition of security. If the random bits are replaced by pseudorandom bits generated by a pseudorandom number generator or a pseudorandom function generator with only a short secret key, the system can clearly not retain its unconditional security. However, failure of this modified system to be *computationally* secure for the same definition of security implies failure of the pseudorandom (number or function) generator to be *computationally indistinguishable* from a random generator since any breaking algorithm for the cryptosystem would yield a distiguishing algorithm for the pseudorandom generator. Therefore, the modified system can be proved secure under the assumption that the component pseudorandom generators are secure. Although no pseudorandom generator has been proved secure, to rely on such an unproven assumption may be worth-while as it allows to clarify the principles on which a cipher's security is based.

The most trivial and widely used application of the described idea are conventional additive stream ciphers which can trivially be "proved" secure under the assumption that the keystream generator is a pseudorandom number generator according to [2]. Another less trivial application is for the design of block ciphers. A block cipher can be obtained from an efficiently computable locally random function by replacing the random bits by one or several pseudorandom function.

A further cryptographic application of local randomness may be for the key scheduling in secret-key ciphers where a relatively short key must be stretched to a sequence of subkeys (e.g. round keys of a block cipher).

Acknowledgment

Jacques Patarin has independently found a different construction of a LRFG with superpolynomial key length function $l(n)$ and local randomization of degree $k(n) = \Omega(l(n)^\alpha)$ for any $\alpha < 1$.

References

[1] N. Alon, O. Goldreich, J. Hastad and R. Peralta, Simple constructions of almost k-wise independent random variables, *Proceedings of the 31st IEEE Symposium on Foundations of Computer Science*, pp. 544-553, 1990.

[2] M. Blum and S. Micali, How to generate cryptographically strong sequences of pseudo-random bits, *SIAM Journal on Computing*, Vol. 10, pp. 96-113, 1981.

[3] B. Chor and O. Goldreich, On the power of two-point based sampling, *Journal of Complexity*, Vol. 5, No. 1, pp. 96-106, 1989.

[4] D.E. Denning, *Cryptography and Data Security*, Addison-Wesley, Reading, MA, 1983.

[5] O. Goldreich, S. Goldwasser and S. Micali, How to construct random functions, *Journal of the Association for Computing Machinery*, Vol. 33, pp. 792-807, 1986.

[6] A. Joffe, On a set of almost deterministic k-independent random variables, *The Annals of Probability*, Vol. 2, No. 1, pp. 161-162, 1974.

[7] H.O. Lancaster, Pairwise statistical independence, *Ann. Math. Statist.*, Vol. 36, pp. 1313-1317, 1965.

[8] L.A. Levin, One-way functions and pseudorandom generators, *Proc. 17th ACM Symposium on Theory of Computing*, pp. 363-364, 1985.

[9] M. Luby and C. Rackoff, How to construct pseudorandom permutations from pseudorandom functions, *SIAM Journal on Computing*, Vol. 17, No. 2, pp. 373-386, 1988.

[10] U.M. Maurer and J.L. Massey, Local randomness in pseudo-random sequences, *Journal of Cryptology*, Vol. 4, No. 2, pp. 135-149, 1991.

[11] J. Patarin, Etude des générateurs de permutations basés sur le Schéma du D.E.S., Ph. D. Thesis, INRIA, Domaine de Voluceau, Le Chesnay, France, 1991. An extract appeared in: J. Patarin, New results on pseudorandom permutation generators based on the DES scheme, *Advances in Cryptology – CRYPTO'91*, J. Feigenbaum (Ed.), Lecture Notes in Computer Science, Vol. 576, Springer-Verlag, pp. 301-312, 1992.

[12] J. Pieprzyk, How to construct pseudorandom permutations from single pseudorandom functions, *Advances in Cryptology – EUROCRYPT'90*, I.B. Damgård (Ed.), Lecture Notes in Computer Science, Vol. 473, Springer-Verlag, pp. 140-150, 1991.

[13] C.P. Schnorr, On the construction of random number generators and random function generators, *Advances in Cryptology – EUROCRYPT'88*, C.G. Günther (Ed.), Lecture Notes in Computer Science, Vol. 330, Springer-Verlag, pp. 225-232, 1988.

[14] Y. Zheng, T. Matsumoto and H. Imai, Impossibility and optimality results on constructing pseudorandom permutations, *Advances in Cryptology - EUROCRYPT'89*, J.-J. Quisquater et al. (Eds.), Lecture Notes in Computer Science, Vol. 434, Springer-Verlag, pp. 412-421, 1990.

HOW TO CONSTRUCT PSEUDORANDOM AND SUPER PSEUDORANDOM PERMUTATIONS FROM ONE SINGLE PSEUDORANDOM FUNCTION

Jacques PATARIN

Bull CP8, 68 route de Versailles - B.P.45 - 78430 Louveciennes - France

Abstract

In this paper we will solve two open problems concerning pseudorandom permutations generators.

1. We will see that it is possible to obtain a pseudorandom permutation generator with only three rounds of DES - like permutation and a single pseudorandom function. This will solve an open problem of [6].

2. We will see that it is possible to obtain a super pseudorandom permutation generator with a single pseudorandom function. This will solve an open problem of [5]. For this we will use only four rounds of DES - like permutation.

For example, we will see that if ζ denotes the rotation of one bit, $\psi(f, f, f \circ \zeta \circ f)$ is a pseudorandom function generator. And $\psi(f, f, f, f \circ \zeta \circ f)$ is a super pseudorandom function generator.

Here the number of rounds used is optimal. It should be noted that here we introduce an important new idea in that we do not use a composition of f, i times, but $f \circ \zeta \circ f$ for the last round, where ζ is a fixed and public function.

1 Introduction

In their important paper [1], M. Luby and C. Rackoff showed how to construct a pseudorandom permutation generator from a pseudorandom function generator and the application of three rounds of DES - like permutations. (This structure is notated as $\psi(f, g, h)$). Later, Zheng, Matsumoto and Imai [6] showed that it is impossible to make a pseudorandom permutation with some composition of a single pseudorandom function f and three rounds of DES - like permutations, i.e. $\psi(f^i, f^j, f^k)$ is not pseudorandom for any i, j, k. At the end of their article, they raised two open problems :

R.A. Rueppel (Ed.): Advances in Cryptology - EUROCRYPT '92, LNCS 658, pp. 256-266, 1993.

a) Whether it is possible to use one single function f to construct a pseudorandom permutation, by more than three applications of DES - like transformation.

b) Whether it is possible that $\psi(f, f, \hat{f})$ be a pseudorandom permutation, where $\hat{f}$ is constructed from f with $\hat{f} \neq f^m$ for any $m \in \mathbb{N}$.

The problem a) was solve by J. Pieprzyk in [4] where he explains that $\psi(f, f, f, f^2)$ is a pseudorandom permutation (but not a super pseudorandom permutation as we will see).

But problem b) remained open. Recently, we have solved this problem b) and will explain our results in this paper.

Another open problem was :

c) How to construct super pseudorandom permutations from one single pseudorandom function.
This notion of super pseudorandomness was introduced by M. Luby and C. Rackoff. It means that the block cryptosystem is secure against a chosen plaintext / ciphertext attack. They said that $\psi(f, g, h, e)$ is super pseudorandom, where $f, g, h,$ and e are four independent pseudorandom functions. (A proof of this property is given in [3]).
If only one pseudorandom function is used, we (in [3]), and J. Pieprzyk and S. Sadeghiyan (in [5]), have independently found that $\psi(f, f, f, f^2)$ is not super pseudorandom. So an open problem is how to construct a super pseudorandom permutation from a single pseudorandom function. This problem c) is the second problem that we have solved.

2 Notations

The notation we use, is similar to [2] and [4].

- $I_n = \{0, 1\}^n$ is the set of all 2^n binary strings of length n.

- For $a, b \in I_n$, $[a, b]$ will be the string of length $2n$ of I_{2n} which is the concatenation of a and b.

- For $a, b \in I_n$, $a \oplus b$ stands for bit-by-bit exclusive- or of a and b.

- o is the composition of function.

- f^2 is $f \circ f$.

- The set of all functions from I_n to I_n is F_n. It consists of $|F_n| = 2^{n.2^n}$ elements.

- The set of all permutations from I_n to I_n is B_n, so $B_n \subset F_n$. And $|B_n| = (2^n)!$.

- Let f be a function of F_n. Let $L, R, S,$ and T be elements of I_n. Then by definition $\psi(f_1)$ is the permutation from I_{2n} to I_{2n} such that :

$$\forall (L, R) \in I_n^2, \psi(f_1)[L, R] = [S, T] \Leftrightarrow \begin{cases} S &= R \\ T &= L \oplus f_1(R). \end{cases}$$

- Let $f_1, f_2, \ldots, f_k$, be k functions of F_n. $\psi(f_1, \ldots, f_k)$ is the permutation from I_{2n} to I_{2n} defined by :

$$\psi(f_1, \ldots, f_k) = \psi(f_k) \circ \ldots \circ \psi(f_2) \circ \psi(f_1).$$

For example, we have :

$$\psi(f_1, f_2, f_3)[L, R] = [S, T] \quad \Leftrightarrow \quad \begin{cases} S &= R \oplus f_2(L \oplus f_1(R)) \\ T &= L \oplus f_1(R) \oplus f_3(S) \end{cases}$$

$$\psi(f_1, f_2, f_3, f_4)[L, R] = [S, T] \quad \Leftrightarrow \quad \begin{cases} S &= L \oplus f_1(R) \oplus f_3(R \oplus f_2(L \oplus f_1(R))) \\ T &= R \oplus f_2(L \oplus f_1(R)) \oplus f_4(S) \end{cases}$$

Remark :

$\psi(f_1, \ldots, f_k)$ is in fact a k iteration DES Scheme where the S-boxes are replaced by the functions $f_1, \ldots, f_k$.

We will assume that the definitions of permutation generator, distinguishing circuit, normal and inverse oracle gates, pseudorandom permutation generator and super pseudorandom permutation generator are known. These definitions, may be found in [1] for example.
In all this article the number of computations that can perform a distinguishing circuit is not bounded. But the number of oracle gates of a distinguishing circuit is a fixed integer m.

3 Definition of the "spreading" of a permutation ζ

To explain the results that we obtained, we will need to define the notion of the "spreading" of a permutation. This notion will be useful to formulate our results in general.

Definition 1 *Let ζ be a permutation of I_n. (Then, $\zeta \in B_n$). By definition, the "spreading" of ζ is the smallest integer λ such that :*
$\forall L \in I_n$, *the equation $x \oplus \zeta(x) = L$ has at most λ solutions in I_n.*

Examples

1. If ζ is the identity function, then $\lambda = 2^n$. This is because for $L = 0$ the equation $x \oplus x = L$ has 2^n solutions.

2. Let ζ be the rotation of one bit, such that the first bit becomes the second bit, the second bit the third, ..., and the last bit becomes the first bit.

So if $x = x_1 x_2 \ldots x_n$, where $x \in I_n$, $\zeta(x) = x_n x_1 x_2 \ldots x_{n-1}$.

Then we will see that $\lambda = 2$.
Let $L = \ell_1 \ell_2 \ldots \ell_n$.

Then $x \oplus \zeta(x) = L$ if and only if :
$$\begin{cases} x_1 & \oplus & x_n = \ell_1 \\ x_2 & \oplus & x_1 = \ell_2 \\ & \vdots & \\ x_n & \oplus & x_{n-1} = \ell_n \end{cases} \qquad (I)$$

where $\forall i, 1 \leq i \leq n, \ell_i = 0$ or 1, and $x_i = 0$ or 1.

First case : $x_n = 0$

Then the $(n-1)$ first equations of (I) give us $x_1, x_2, \ldots, x_{n-1}$. And then if $\ell_n = x_n \oplus x_{n-1}$ the last equation is true. If not, $x_n = 0$ is not possible.

Second case : $x_n = 1$

In the same way, the system (I) then has zero or one solution if $x_n = 1$.

Conclusion :

Let ζ be the rotation of one bit. Then we have seen that for all $L \in I_n$, the equation $x \oplus \zeta(x) = L$ has at most two solutions. (And for $L = 0$ it has exactly two solutions). So the "spreading" of the function rotation of one bit is $\lambda = 2$, as claimed.

Remark :

It is possible to find some permutations with $\lambda = 1$. As pointed out by Mr Lothrop Mittenthal after my presentation of this paper, in this case a permutation ζ is call "Complete Mappings". Complete Mappings have been studied for differents raisons before, and more details about them are given in [7].

In this paper, we will present two theorems that will use our notion of "spreading". These theorems are :

Theorem 3.1 *Let $\zeta = (\zeta_n)_{n \in N}$ be a sequence of permutations, $\zeta_n \in B_n$.*
Let λ_n be the spreading of ζ_n.
If for all polynomial $P(n)$ we have :

$$\frac{\lambda_n P(n)}{2^n} \quad n \xrightarrow{} +\infty \quad 0 \quad then :$$

$\psi(f, f, f \circ \zeta \circ f)$ *is a pseudo-random function generator, where f is a single pseudo-random function.*
(When $f \in F_n$, the notation $f \circ \zeta \circ f$ means $f \circ \zeta_n \circ f$).

Theorem 3.2 *With the same notations, and the same condition on ζ, we have :*
$\psi(f, f, f, f \circ \zeta \circ f)$ *is a super pseudo-random function generator.*

These two theorems solve the two open problems b) and c) discussed in paragraph 1. Theorem 3.1. solves problem b) and Theorem 3.2 solves problem c).

For example, if ζ is the function rotation of one bit, $\lambda_n = 2$ and by Theorem 3.2 we have found a super pseudorandom permutation generator by using only one single pseudorandom function f. Moreover we have found many of these super pseudorandom permutation generators :

each ζ such that $\dfrac{\lambda_n P(n)}{2^n}$ $n \xrightarrow{} +\infty$ 0 (for all polynomial $P(n)$) will give us such a generator.

We will now see the ideas that we have used to prove these two new theorems. Paragraphs 4 and 5 discuss about theorem 3.1, and paragraph 6 and 7 will deal with theorem 3.2. And in paragraph 8 we will compare our results with one single pseudorandom function with what can be obtained with two pseudorandom functions.

4 A "basic property" of $\psi(f, f, f \circ \zeta \circ f)$

To prove Theorem 3.1, we have first proved a "basic property" of $\psi(f, f, f \circ \zeta \circ f)$.

Theorem 4.1 *Basic property of* $\psi(f, f, f \circ \zeta \circ f)$.

Let ζ be a permutation of I_n. Let λ be the spreading of ζ.
Let $[L_i, R_i], 1 \leq i \leq m$, be a sequence of m distinct elements of I_{2n}. (Distinct means that $i \neq j \Rightarrow R_i \neq R_j$ or $L_i \neq L_j$).
And let $[S_i, T_i], 1 \leq i \leq m$, be a sequence of m distinct elements of I_{2n} such that if $i \neq j$, then $S_i \neq S_j$.
Then the number H of functions f of F_n such that

$$\forall i, 1 \leq i \leq m, \psi(f, f, f \circ \zeta \circ f)[L_i, R_i] = [S_i, T_i]$$

is $H \geq \dfrac{|F_n|}{2^{2nm}} \left(1 - \dfrac{6m^2}{2^n} - \dfrac{\lambda m}{2^n}\right)$.

The proof of this theorem 4.1 is not very difficult : it is just a combinatorial evaluation. In [3] a complete proof of this Theorem is given.

5 $\psi(f, f, f \circ \zeta \circ f)$ is pseudorandom, if ζ is well chosen

Let ϕ be a distinguishing circuit.

- We will denote by $\phi(F)$ its output (1 or 0) when its oracle gates are given the values of a function F.
  Let ζ be a fixed permutation.

- We will denote by P_1 the probability that $\phi(F) = 1$ when f is a function randomly chosen in F_n, and $F = \psi(f, f, f \circ \zeta \circ f)$.

  So $P_1 = \dfrac{\text{Number of } f \in F_n \text{ such that } \phi(\psi(f, f, f \circ \zeta \circ f)) = 1}{|F_n|}$.

- We will denote by P_1^* the probability that $\phi(F) = 1$ when F is randomly chosen in F_{2n}.

So $P_1^* = \dfrac{\text{Number of function } F \in F_{2n} \text{ such that } \phi(F) = 1}{|F_{2n}|}$.

From Theorem 4.1, it is possible to prove this theorem :

Theorem 5.1 *For every distinguishing circuit ϕ with m oracle gates, we have :*

$$|P_1 - P_1^*| \leq \frac{6m^2}{2^n} + \frac{\lambda m}{2^n} + \frac{m(m-1)}{2.2^n},$$

where λ is the spreading of ζ.
And then : $|P_1 - P_1^*| \leq \dfrac{6.5m^2}{2^n} + \dfrac{\lambda m}{2^n}$.

(See [3] for the complete proof). In the APPENDIX we will give a general result (Theorem A1) which shows that Theorem 5.1 is just a consequence of Theorem 4.1.
And theorem 3.1 is an easy consequence of this theorem 5.1.

Remarks

1. Let Q_1 be the probability that $\phi(F) = 1$ when f, g, h are three independant functions randomly chosen in F_n and $F = \psi(f, g, h)$. Then, in [1] M. Luby et C. Rackoff have proved that : for every distinguishing circuit ϕ whith m oracle gates, we have :
$|Q_1 - P_1^*| \leq \dfrac{m^2}{2^n}$.
It is useful to compare this property with our property of theorem 5.1 :
$|P_1 - P_1^*| \leq \dfrac{6.5m^2}{2^n} + \dfrac{\lambda m}{2^n}$.
In theorem 5.1 the inequality is a little worse, but the important thing is that we use only one function f randomly chosen in F_n. When these theorems are used in order to obtain a cryptosystem, the functions f, g, h will generally be generated by a pseudorandom functions generator. So by using $\psi(f, f, f \circ \zeta \circ f)$ instead of $\psi(f, g, h)$ the lenght of the secret key will generaly be divided by three (because we need to generate only one pseudorandom function instead of three).

2. Theorem 5.1 is true because it concerns a distinguishing circuit. These circuits have only normal oracle gate. If we use super distinguishing circuit, with normal and inverse oracle gates, then the property of theorem 5.1 will not be true. This is because here we use only three rounds of DES - like permutations. And in [1] M. Luby and C. Rackoff have proved that in this case a generator is never super pseudorandom.

6   A basic "property" of $\psi(f, f, f, f \circ \zeta \circ f)$

We will now present the ideas that we have used to prove Theorem 3.2. The proof is very similar to the proof of theorem 3.1. First, we proved this theorem :

Theorem 6.1 *basic property of $\psi(f, f, f, f \circ \zeta \circ f)$.*

Let ζ be a permutation of I_n. Let λ be the spreading of ζ.
Let $[L_i, R_i], 1 \leq i \leq m$, be a sequence of m distinct elements of I_{2n}. (Distinct means that $i \neq j \Rightarrow R_i \neq R_j$ or $L_i \neq L_j$).
Let $[S_i, T_i], 1 \leq i \leq m$, be also a sequence of m distinct elements of I_{2n}. (Distinct means that $i \neq j \Rightarrow S_i \neq S_j$ or $T_i \neq T_j$).
Then the number H of functions f of F_n such that

$$\forall i, 1 \leq i \leq m, \psi(f, f, f, f \circ \zeta \circ f)[L_i, R_i] = [S_i, T_i]$$

is $H \geq \dfrac{|F_n|}{2^{2nm}} \left(1 - \dfrac{10.5 m^2}{2^n} - \dfrac{\lambda m}{2^n} \right)$.

See [3] for the complete proof.

7 $\psi(f, f, f, f \circ \zeta \circ f)$ is super pseudorandom, if ζ is well chosen

Let ζ be a fixed and public permutation. (For example a rotation of one bit).
Let ϕ be a super distinguishing circuit. (This is a circuit with normal and inverse oracle gates. See [1] for precise definitions).

- We will denote by $\phi(F)$ its output (1 or 0) when its normal oracle gates are given the values of a permutation F, and its inverse oracle gates are given the values of F^{-1}.

- We will denote by P_1 the probability that $\phi(F) = 1$ when f is a function randomly chosen in F_n, and $F = \psi(f, f, f, f \circ \zeta \circ f)$.

- We will denote by P_1^{**} the probability that $\phi(F) = 1$ when F is randomly chosen in B_{2n}.

From theorem 6.1, it is possible to prove this theorem :

Theorem 7.1 *For every super distinguishing circuit ϕ with m oracle gates, we have :*

$$|P_1 - P_1^{**}| \leq \frac{10.5 m^2}{2^n} + \frac{\lambda m}{2^n} + \frac{m(m-1)}{2 \cdot 2^{2n}}.$$

And then :

$$|P_1 - P_1^{**}| \leq \frac{11 m^2}{2^n} + \frac{\lambda m}{2^n}.$$

(See [3] for the proof). In the APPENDIX we will give a general result (Theorem A2) which shows that Theorem 7.1 is just a consequence of Theorem 6.1.
And theorem 3.2 is an easy consequence of this theorem 7.1.

Remark

If for ζ we take the identity function, then $\lambda = 2^n$ (see paragraph 3).
Then $\dfrac{\lambda P(n)}{2^n}$ does not tend to 0 for any polynomial P.
So the theorem 3.2 can not conclude in this case that $\psi(f, f, f, f^2)$ is super pseudorandom.
And in fact it is possible to show that $\psi(f, f, f, f^2)$ is not super pseudorandom.
(See [5] or [3]).

8   Generators with two functions

Although that it is not exactly the thema of this article, we will now briefly survey the properties of the generators with two independant functions f and g randomly chosen in F_n. All these properties can be proved with the "coefficient H technique" which is given in Appendix. Most of these properties have been find before this article. With these properties we will then comment a result claimed in [5], and we will explain why we think that this result is wrong.

The properties

1. $\psi(f, f, g)$ and $\psi(f, g, g)$ are pseudorandom, but not super pseudo-random. They are not super pseudo-random beacause here there is only three rounds.

2. $\psi(f, g, f)$ is not pseudorandom. This is easy to see because this permutation is its own inverse if left and right halves of inputs and outputs are swapped.

   Remark : These results on 1. and 2. have first been find by Ohnishi (see [6]).

3. $\psi(f, f, f, g)$ and $\psi(f, g, g, g)$ are super pseudorandom.

4. $\psi(f, f, g, g)$ is also super pseudorandom.

5. $\psi(f, g, g, f)$ is not pseudorandom. This is easy to see because this permutation is its own inverse if left and right halves are swapped.

6. $\psi(f, g, f, g)$ is super pseudorandom.

7. $\psi(f, f, g, f)$ and $\psi(f, g, f, f)$ are super pseudorandom.

Because of these results, we think that the "Corollary 1" of [5] is wrong.
This "Corollary 1" claim :

Let $(f_1, f_2, \ldots, f_i)$ be i functions of F_n such that $G_1 = \psi(f_i, \ldots, f_1)$ be a pseudorandom permutation.
Then G_1 is super pseudorandom if and only if $G_2 = \psi(f_i, \ldots, f_2)$ and $G_3 = \psi(f_1, \ldots, f_{i-1})$ are pseudorandom permutations.
But we have just seen that :

1. $\psi(f, f, f, g)$ is super pseudorandom and $\psi(f, f, f)$ is not pseudorandom.

2. $\psi(f, g, f, g)$ is super pseudorandom, but $\psi(f, g, f)$ and $\psi(g, f, g)$ are not pseudorandom. Notice here that f_2 and f_{i-1} are independant.

So this is why we think that this "Corollary 1" of [5] is wrong.

9 Conclusion

In this paper we have explained the ideas that we have used to solve two open problems about pseudorandom permutation generators. We have presented a notion of "spreading" for a permutation, and this notion is very useful for our results. Another new idea was to use functions $f \circ \zeta \circ f$, where ζ is a fixed and public permutation with small "spreading", and f a pseudorandom function.
Using this sort of function give us very different results from those obtained when using only compositions of f.
Our main result is that it is possible to construct super pseudorandom permutations from a single pseudorandom function. Such permutations give block cryptosystems secure against chosen plaintext / ciphertext attack. Finally, we have explained that there are many such permutations : any $\psi(f, f, f, f \circ \zeta \circ f)$, where ζ is a fixed and public permutation with small "spreading" will be a super pseudorandom permutation constructed from the pseudorandom function f. And here the number of rounds used (four) is also optimal.

References

[1] M. Luby and C. Rackoff, *How to construct pseudorandom permutations from pseudorandom functions*, SIAM Journal and Computing, 17(2) : 373-386, April 1988.

[2] J. Patarin, *New results on pseudorandom permutation generators based on the DES Scheme*, Abstracts of Crypto'91, p. 7-2, 7-7.

[3] J. Patarin, *Etude des générateurs de permutations basés sut le schéma du DES*, Thèse, November 1991, INRIA, Domaine de Voluceau, Le Chesnay, France.

[4] J. Pieprzyk, *How to construct pseudorandom permutations from Single Pseudorandom Functions*, EUROCRYPT'90, Århus, Denmark, May 1990.

[5] B. Sadeghiyan and J. Pieprzyk, *On necessary and sufficient conditions for the construction of super pseudorandom permutations*, Abstracts of Asiacrypt'91, November 1991, p. 117-123.

[6] Y. Zheng, T. Matsumoto and H. Imai, *Impossiblility and optimality results on constructing pseudorandom permutations*, Abstract of EUROCRYPT'89, Houthalen, Belgium, April 1989, p. 412-421.

[7] L. J. Paige, *Complete Mappings of finite groups*, J. Math. 1, 111-116, 1951.

APPENDIX

The "coefficients H technique" for proving pseudo-randomness and super-pseudorandomness

We will formulate here the two main Theorems that we generally used in order to prove some pseudorandom or super-pseudorandom properties.

These theorems show a technique (we call it the "coefficient H technique") for proving such properties. More details and variants of these technique (some generalisations exist) are given in [3] with the proof of these theorems.

Theorem A1 ("coefficient H technique for pseudorandomness")

Let Λ be a pseudorandom permutation generator such that if $(f_1, \ldots, f_p)$ are p functions of F_n, $\Lambda(f_1, \ldots, f_p) \in B_{2n}$.

Let α be a real, $\alpha > 0$.

If :

(1) For all sequence $[L_i, R_i], 1 \leq i \leq m$, of m distinct elements of I_{2n} (distinct means that $i \neq j \Rightarrow R_i \neq R_j$ or $L_i \neq L_j$) and for all sequences $[S_i, T_i], 1 \leq i \leq m$, of m elements of I_{2n} such that if $i \neq j$ then $S_i \neq S_j$, we have :
the number H of p-tuple of functions $(f_1, \ldots, f_p)$ such that

$$\forall i, 1 \leq i \leq m, \Lambda(f_1, \ldots, f_p)[L_i, R_i] = [S_i, T_i]$$

is $H \geq \dfrac{|F_n|^p}{2^{2nm}}(1 - \alpha)$.

Then :

(2) For all distinguishing circuit ϕ with m oracle gates, we have :

$$|P_1 - P_1^*| \leq \alpha + \frac{m(m-1)}{2 \cdot 2^n}.$$

Where P_1 is the probability that $\phi(F) = 1$ when $F = \Lambda(f_1, \ldots, f_p)$ and $(f_1, \ldots, f_p)$ are p functions randomly (and independantly) chosen in F_n.

And P_1^* is the probability that $\phi(F) = 1$ when F is randomly chosen in F_{2n}.

Notice that here there is no limitation in the number of computations that can perform the distinguishing circuits in order to analyse the m values given by its oracle gates.

Example

With this Theorem A1 we can obtain a new proof of the result of M. Luby and C. Rackoff about $\psi^3(f_1, f_2, f_3)$, for example.

First, it is possible to prove this property :

Property of $\psi(f_1, f_2, f_3)$

For all sequence $[L_i, R_i], 1 \leq i \leq m$, of m distinct elements of I_{2n} and for all sequence $[S_i, T_i], 1 \leq i \leq m$, of m elements of I_{2n} such that if $i \neq j$ then $S_i \neq S_j$, we have : the number H of 3-tuple of functions (f_1, f_2, f_3) such that

$$\forall i, 1 \leq i \leq m, \psi^3(f_1, f_2, f_3)[L_i, R_i] = [S_i, T_i]$$

is $H \geq \dfrac{|F_n|^3}{2^{2nm}}\left(1 - \dfrac{m(m-1)}{2.2^n}\right)$.

Then, from Theorem A1 we obtain :

$$|P_1 - P_1^*| \leq \frac{m(m-1)}{2^n} \quad \text{as claimed.}$$

Theorem A2 ("coefficient H technique for super-pseudorandomness")

If :

(1) For all sequence $[L_i, R_i], 1 \leq i \leq m$, of m distinct elements of I_{2n}, and for all sequences $[S_i, T_i], 1 \leq i \leq m$, of m distinct elements of I_{2n} the number H of p-tuple of functions $(f_1, \ldots, f_p)$ such that

$$\forall i, 1 \leq i \leq m, \Lambda(f_1, \ldots, f_p)[L_i, R_i] = [S_i, T_i]$$

is $H \geq \dfrac{|F_n|^p}{2^{2nm}}(1 - \alpha)$.

Then :

(2) For all super distinguishing circuit ϕ with m super oracle gates (normals or inverses), we have :

$$|P_1 - P_1^{**}| \leq \alpha + \frac{m(m-1)}{2.2^{2n}}.$$

Where P_1 is the probability that $\phi(F) = 1$ when $F = \Lambda(f_1, \ldots, f_p)$ and $(f_1, \ldots, f_p)$ are randomly (and independantly) chosen in F_n.
And P_1^{**} is the probability that $\phi(F) = 1$ when F is randomly chosen in B_{2n}.

Notice that here there is also no limitation in the number of computations that can perform the super distinguishing circuits in order to analyse the m values given by its super oracle gates.

A Construction for Super Pseudorandom Permutations from A Single Pseudorandom Function

Babak Sadeghiyan *   Josef Pieprzyk †

Abstract

In this paper, we show how to construct a super pseudorandom permutation generator from a single pseudorandom function generator, based on DES-like permutations. First, we show $\psi(g, 1, f, g, 1, f)$, which consists of six rounds of DES-like permutations with two different pseudorandom functions and a fixed permutation, is super psuedorandom. Then, we show that with replacing a two-fold composition of one of the pseudorandom functions instead of the other one it is possible to construct a super pseudorandom permutation from a single pseudorandom function, where we need six rounds of DES-like permutations and six references to the pseudorandom function.

1 Introduction

Pseudorandom bit generators have many cryptographic applications. Classical pseudorandom bit generators are deterministic algorithms which generate binary strings that look like a truly random one. A pseudorandom bit generator passes a given test if the results of the test are similar for both the pseudorandom and a truly random one. If the statistics of pseudorandom generated strings are different from the statistics of truly random ones, the bit generator fails some statistical test. However, even if a generator passes all known statistical tests, it is sometimes possible to predict the next bit knowing some previous ones. The notion of unpredictability of the next bit was first introduced by Yao [6]. He showed that a generated string is indistinguishable from a random one, if and only if, knowing the first s bits of the string, predicting the $s + 1$ bit is difficult, in other words

---

*Department of Computer Science, University College, University of New South Wales, Australian Defence Force Academy, Canberra, A.C.T. 2600, Australia.

†Department of Computer Science, University of Wollongong, Wollongong, N.S.W. 2500, Australia.

R.A. Rueppel (Ed.): Advances in Cryptology - EUROCRYPT '92, LNCS 658, pp. 267-284, 1993.
© Springer-Verlag Berlin Heidelberg 1993

the next bit is unpredictable. Goldreich, Goldwasser and Micali introduced the notion of pseudorandom function generators [1] and showed how to construct a pseudorandom function generator from a pseudorandom bit generator adopting Yao's unpredictability property of the next bit.

Luby and Rackoff showed how to construct a pseudorandom invertible permutation generator with three pseudorandom function generators and application of three rounds of DES-like permutations [2] (this structure is notated as $\psi(h, g, f)$). A practical implication of their work is that a private key block cipher which can be proven to be secure against chosen plaintext attack can be constructed. Their result is quite astonishing since it is not based on any unproven hypothesis.

Later Pieprzyk [4] showed that it is possible to construct a pseudorandom permutation generator based on a single pseudorandom function and four rounds of DES-like permutations, i.e., $\psi(f^2, f, f, f)$ is pseudorandom and it is secure against a chosen plaintext attack. Luby and Rackoff also introduced the notion of super pseudorandomness, where the block cryptosystem is secure against a chosen plaintext/ciphertext attack. They suggested that $\psi(h, g, f, e)$ is super pseudorandom, which is a structure with four pseudorandom functions in four DES-like permutations. Recently, in [5] the necessary and sufficient conditions for the construction of super pseudorandom permutation generators based on DES-like permutations have been investigated and it has been suggested that $\psi(g, g, f, f)$ is super pseudorandom, which is a structure with two pseudorandom functions and four DES-like permutations.

It is a question of how to construct super pseudorandom permutations from a single pseudorandom function.

In this paper, we answer the above question and present a construction based on a single pseudorandom function which is super pseudorandom. We take advantage of the structure of the optimal perfect randomiser $\psi(g, 1, f, g, 1, f)$ presented in [3], where not only the output is independent from the input but also the two branches of the output are independent of each other. First, we show that the perfect randomiser is super pseudorandom. Then, we present a construction based on a single pseudorandom function, with replacing a two-fold composition of one of the pseudorandom functions instead of the other one, i.e., $\psi(f^2, 1, f, f^2, 1, f)$, which is indistinguishable from the randomiser. Finally, we show that the presented construction is super pseudorandom. Hence, it is possible to construct a super pseudorandom permutation from a single pseudorandom function, where we need six rounds of DES-like permutations and six references to the pseudorandom function.

2 Notations

The notation we use, is similar to [4]. The set of all integers is denoted by N. Let $\Sigma = \{0,1\}$ be the alphabet we consider. For $n \in N$, Σ^n is the set of all 2^n binary strings of length n. The concatenation of two binary strings x, y is denoted by $x \parallel y$. The bit by bit exclusive-OR of x and y is denoted by $x \oplus y$. By $x \in_r S$ we mean that x is chosen from a set S uniformly at random. By f a function we mean a trasformation from Σ^n to Σ^n. The set of all functions on Σ^n is denoted by H_n, i.e., $H_n = \{f \mid f : \Sigma^n \to \Sigma^n\}$ and it consists of $2^{n 2^n}$ elements. The composition of two functions f and g is defined as $f \circ g(x) = f(g(x))$. The i-fold composition of f is denoted by f^i. A function f is a permutation if it is a 1 to 1 and onto function. The set of all permutations on Σ^n is defined by P_n and it consists of $2^n!$ elements.

3 Preliminaries

This section provides the preliminary definitions and notions which are used throughout the paper.

Definition 1 *We associate with a function $f \in H_n$ the DES-like permutation $D_{2n,f} \in P_{2n}$ as,*

$$D_{2n,f}(L \parallel R) = (R \oplus f(L) \parallel L)$$

where R and L are n-bit strings, i.e., $R, L \in \Sigma^n$.

Definition 2 *Having a sequence of functions $f_1, f_2, \ldots, f_i \in H_n$, we define the composition of their DES-like permutations as $\psi \in P_{2n}$, where*

$$\psi(f_i, \ldots, f_2, f_1) = D_{2n,f_i} \circ D_{2n,f_{i-1}} \circ \ldots \circ D_{2n,f_1}$$

Definition 3 *Let $l(n)$ be a polynomial in n, a function generator $F = \{F_n : n \in N\}$ is a collection of functions with the following properties:*

- *Indexing: Each F_n specifies for each k of length $l(n)$ a function $f_{n,k} \in H_n$.*

- *Poly-time evaluation: Given a key $k \in \Sigma^{l(n)}$, and a string $x \in \Sigma^n$, $f_{n,k}(x)$ can be computed in polynomial time in n.*

Definition 4 *An oracle circuit C_n is an acyclic circuit which contains Boolean gates of type AND, OR and NOT, and constant gates of type zero and one, and a particular kind of gates named oracle gates. Each oracle gate has an n-bit input and an n-bit output and it is evaluated using some function from H_n. The oracle circuit C_n has a single bit output.*

Definition 5 *The size of an oracle circuit C_n is the total number of connections between gates, Boolean gates, constant gates and oracle gates.*

Definition 6 *A distinguishing circuit family for a function generator F is an infinite family of circuits $\{C_{n_1}, C_{n_2}, \ldots\}$, where $n_1 < n_2 < \ldots$, such that for some pair of constants c_1 and c_2 and for each n there exist a circuit C_n such that:*

- *The size of C_n is less than or equal to n^{c_1}.*

- *Let $\mathrm{Prob}\{C_n[H_n] = 1\}$ be the probability that the output bit of C_n is one when a function is randomly selected from H_n and used to evaluate oracle gates. Let $\mathrm{Prob}\{C_n[F_n] = 1\}$ be the probability that the output bit of C_n is one when a key k of length $l(n)$ is randomly chosen and $f_{n,k}$ is used to evaluate the oracle gates.*

The distinguishing probability for C_n is greater than or equal to $\frac{1}{n^{c_2}}$, that is,

$$| \mathrm{Prob}\{C_n[H_n] = 1\} - \mathrm{Prob}\{C_n[F_n] = 1\} | \geq \frac{1}{n^{c_2}}$$

Definition 7 *A function generator F is pseudorandom if there is no distinguishing circuit family for F.*

How to construct pseudorandom permutation from pseudorandom functions was first presented by Luby and Rackoff applying a DES-like structure. The following lemma is due to Luby and Rackoff and has been stated in [2] as the main lemma.

Lemma 1 *Let $f_1, f_2, f_3 \in_r H_n$ be indepedent random functions and C_{2n} be an oracle circuit with $m < 2^n$ oracle gates, then*

$$| \mathrm{Prob}\{C_{2n}[P_{2n}] = 1\} - \mathrm{Prob}\{C_{2n}[\psi(f_3, f_2, f_1)] = 1\} | \leq \frac{m^2}{2^n}$$

In other words, a block cryptosystem with three rounds of DES-like permutation and three different random functions is secure against chosen plaintext attack, when a cryptanalyst can ask for only a polynomial number of plaintexts. Later, Pieprzyk et al. considered the composition of Luby - Rackoff modules in [3]. They showed that the composition of two Luby - Rackoff modules where the intermediate layers are replaced with random permutations, would produce a pseudorandom permutation where not only the output is independent of the input but the two branches of output are also independent of each other. In [3], it was also suggested that the independency of output branches would still be held even if the intermediate layers are replaced with some fixed permutation. The following theorem is from [3], but is stated in a different word.

Theorem 1 *Let $f_1, f_2, h_1, h_2 \in_r H_n$ be four different random functions. There is no distinguishing circuit for $\Psi_2 = \psi(h_2, 1, f_2) \circ \psi(h_1, 1, f_1)$, and it is also a perfect randomiser, i.e., for all oracle gates used by the distinguishing circuit, the outputs are independent from the inputs and are independent from each other.*

When the block cryptosystem is secure against chosen plaintext / ciphertext attack, it is called super pseudorandom. This notion only applies for invertible permutations and is stated formally in three following definitions.

Definition 8 *A permutation generator F is a function generator such that each function $f_{n,k}$ is 1 to 1 and onto. Let $\overline{F} = \{\overline{F_n} : n \in N\}$, where $\overline{F_n} = \{\overline{f}_{n,k} : k \in \Sigma^{l(n)}\}$, where $\overline{f}_{n,k}$ is the inverse function of $f_{n,k}$. F is called invertible if $\overline{F}$ is also a permutation generator.*

Definition 9 *A super distinguishing family of circuits for an invertible permutation generator F is an infinite family of circuits $\{SC_{n_1}, SC_{n_2}, \ldots\}$, where $n_1 < n_2 < \ldots$, where each circuit is an oracle circuit containing two types of oracle gates, normal and inverse, such that for some pair of constants c_1 and c_2 and for each n there exist a circuit SC_n such that:*

- *The size of SC_n is less than or equal to n^{c_1}.*

- *Let $\mathrm{Prob}\{SC_n[P_n] = 1\}$ be the probability that the output bit of SC_n is one when a permutation p is randomly selected from P_n and p and $\overline{p}$ are used to evaluate normal and inverse oracle gates. Let $\mathrm{Prob}\{SC_n[F_n] = 1\}$ be the probability that the output bit of SC_n is one when a key k of length $l(n)$ is randomly chosen and $f_{n,k}$ and $\overline{f}_{n,k}$ is used to evaluate the normal and inverse oracle gates respectively.*

The distinguishing probability for SC_n is greater than or equal to $\frac{1}{n^{c_2}}$, that is,

$$| \mathrm{Prob}\{SC_n[P_n] = 1\} - \mathrm{Prob}\{SC_n[F_n] = 1\} | \geq \frac{1}{n^{c_2}}$$

Definition 10 *A permutation generator F is super pseudorandom if there is no super distinguishing circuit family for F.*

If F is a super pseudorandom permutation generator, it is secure against a chosen plaintext/ciphertext attack where a cryptanalyst can interactively choose plain blocks and see their encryptions and choose encryptions and see their corresponding plaintext blocks.

4 Construction of Super Pseudorandom Permutations

Luby and Rackoff [2] suggested that, it is possible to make a super pseudorandom permutation with four independent random functions, i.e., if $f_1, f_2, f_3, f_4 \in H_n$ are independent random functions then $\psi(f_4, f_3, f_2, f_1)$ is a super pseudorandom permutation. It is a matter of question whether it is possible to build a super pseudorandom permutation with a single random functions.

First a definition for independent permutations is given, which would be used in the proof of Theorem 2.

Definition 11 *A D-distinguishing family of circuits for two invertible pseudorandom permutation generators* (Π_1, Π_2) *is an infinite family of circuits* $\{DC_{n_1}, DC_{n_2}, \ldots\}$, *where* $n_1 < n_2 < \ldots$, *where each circuit is an oracle circuit containing two types of oracle gates, such that for some pair of constants* c_1 *and* c_2 *and for each* n *there exist a circuit* DC_n *such that:*

- *The size of* DC_n *is less than or equal to* n^{c_1}.

- *Let* $\mathrm{Prob}\{DC_n[P_n, P_n] = 1\}$ *be the probability that the output bit of* DC_n *is 1, when two permutations* p_1 *and* p_2 *are chosen independently and randomly from* P_n *and are used to evaluate the two types of oracle gates of* DC_n *respectively. Let* $\mathrm{Prob}\{DC_n[\Pi_1, \Pi_2] = 1\}$ *be the probability that the output bit of* DC_n *is one when a key* k *of length* $l(n)$ *is randomly chosen and* $p_{1,k}$ *and* $p_{2,k}$ *are used to evaluate the two types of oracle gates respectively.*

The distinguishing probability for DC_n *is greater than or equal to* $\frac{1}{n^{c_2}}$, *that is,*

$$| \mathrm{Prob}\{DC_n[P_n, P_n] = 1\} - \mathrm{Prob}\{DC_n[\Pi_1, \Pi_2] = 1\} | \geq \frac{1}{n^{c_2}}$$

Definition 12 *Two pseudorandom permutation generators* Π_1 *and* Π_2, *are two independent permutation generators, if there is no D-distinguishing oracle circuit family for* (Π_1, Π_2).

Note that, the D-distinguishing oracle circuits are generalizations of the distinguishing circuits and the super distinguishing circuits, if two simple circuits be excluded. If there is no D-distinguishing circuit for a permutation generator (Π_1 and Π_1), then there is no distinguishing circuit family for Π_1, provided that the D-distiguishing circuit is not an identity test, (e.g., giving an input to two types of oracles and comparing the outputs). Moreover, if there is no D-distinguishing circuit for a permutation generator Π_1 and its inverse $\overline{\Pi}_1$, then there is no super distinguishing circuit family for Π_1 provided that the D-distiguishing circuit is not an inversion test, (e.g., giving an input to one type of oracle and feeding the other type of oracle with this result and comparing the output with the original input).

4.1 Super Pseudorandomness of $\psi(h, 1, f, h, 1, f)$

To construct a super pseudorandom permutation generator based on a single pseudorandom function, we first show that $G_1 = \psi(h, 1, f) \circ \psi(h, 1, f)$ is a super pseudorandom permutation generator, then, we show that if h is substituted with f^2, $G_1 = \psi(f^2, 1, f, f^2, 1, f)$ would remain super pseudorandom. To show that G_1 is a super pseudorandom permutation generator, we first show that $G_2 = \psi(h, 1, f, h, 1)$ and $G_3 = \psi(f, 1, h, f, 1)$ are not only pseudorandom but also two independent permutations. Then we show that G_1 is super pseudorandom.

Lemma 2 *Let $h, f \in_r H_n$ be independent random functions and $G_2 = \psi(h, 1, f, h, 1)$ then*

$$| \operatorname{Prob}\{C_{2n}[G_2] = 1\} - \operatorname{Prob}\{C_{2n}[P_{2n}] = 1\} | \leq \frac{m^2}{2^n} + \frac{m^2}{2^{2n}}$$

where C_{2n} is any polynomial size distinguishing circuit with $m < 2^n$ oracle gates.

Proof :

When the distinguisher examines an oracle, the input is a $2n$ bit string $(L \parallel R)$ and the output would also be a $2n$ bit string $(S \parallel T)$ where

$$
\begin{aligned}
S &= L \oplus R \oplus f(L \oplus h(L \oplus R)) \oplus h(R \oplus h(L \oplus R) \oplus f(L \oplus h(L \oplus R))) \\
T &= R \oplus h(L \oplus R) \oplus f(L \oplus h(L \oplus R))
\end{aligned}
$$

For two different experiments, $(L_i \parallel R_i)$ should be different from $(L_j \parallel R_j)$, so either $L_i \neq L_j$ or $R_i \neq R_j$ or both are different. If there is no leakage of information from the input to the output, the distinguisher cannot make any sensible decision on the generator used to evaluate the oracle gates. The leakage of information happens when there are at least one pair of oracle gates such that their outputs are related to their inputs. Let X be a random variable notating the ouput of h, the random function in the second round of the DES-like structure of G_2, and let Y be a random variable notating the output of f, the random function in the third round of DES-like structure of G_2 (see Figure 1).

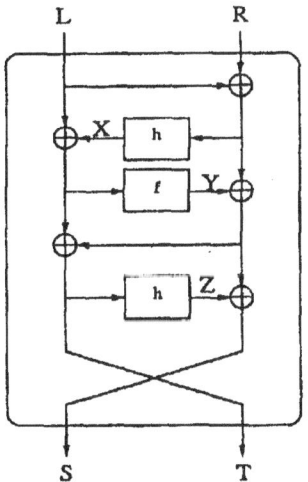

Figure 1: Random variables X and Y in G_2

The leakage of information happens in two cases:

1. X is bad: it happens when there are a pair of oracle gates O_i, O_j with the input random variables $R_i \neq R_j$ and $L_i = L_j = L$ such that the random function h

collides. Hence, for this case we assume that $x_i = x_j = x$ where $x_i = h(L \oplus R_i)$ and $x_j = h(L \oplus R_j)$. It is obvious that if X is bad then always $T_i \oplus T_j = R_i \oplus R_j$. The probability that h collides in a pair of oracle gates among m oracles would be equal to $\frac{m(m-1)}{2} \frac{1}{2^n}$.

2. Y is bad: it happens when there are a pair of oracle gates O_i, O_j with the input random variables $R_i = R_j = R$ and $L_i \neq L_j$ such that the random function f collides. Hence, for this case we assume that $y_i = y_j = y$ where $y_i = f(L_i \oplus h(L_i \oplus R))$ and $y_j = f(L_j \oplus h(L_j \oplus R))$. It is obvious that if Y is bad and also X is bad then always $T_i \oplus T_j = L_i \oplus L_j$. The probability that f collides and also h collides in a pair of oracle gates among m oracles would be equal to $\frac{m(m-1)}{2} \frac{1}{2^{2n}}$.

The probability that a distinguishing circuit for G_2 could be constructed would be equal to $\frac{m(m-1)}{2^{n+1}} + \frac{m(m-1)}{2^{2n+1}}$. On the other hand, when a permutation p is chosen randomly, the probability that p can satisfy the distinguishing circuit relation is $\frac{m}{2^n} + \frac{m}{2^{2n}}$. So, an upper bound on the probability of distinguishing would be

$$| \operatorname{Prob}\{C_{2n}[P_{2n}] = 1\} - \operatorname{Prob}\{C_{2n}[\psi(h,1,f,h,1)] = 1\} | \leq \frac{m^2}{2^n} + \frac{m^2}{2^{2n}}$$

$\square$

Lemma 3 Let $h, f \in_r H_n$ be independent random functions and $G_3 = \psi(f,1,h,f,1)$, then

$$| \operatorname{Prob}\{C_{2n}[G_3] = 1\} - \operatorname{Prob}\{C_{2n}[P_{2n}] = 1\} | \leq \frac{m^2}{2^n} + \frac{m^2}{2^{2n}}$$

where C_{2n} is a distinguishing circuit with $m < 2^n$ oracle gates.

Proof : Since structure of G_3 is similar to the structure of G_2 except that the locations for g and f are replaced, a proof similar to the proof of Lemma 2 can be suggested for the probability of distinguishing of G_3 from a random permutation, and is omitted here. $\square$

Note that when m is a polynomial in n, the probability of distinguishing G_2 or G_3 from a random permutation becomes less than 1 over any polynomial in n.

Lemma 4 Let $h, f \in_r H_n$ be independent random functions, and $G_2 = \psi(h,1,f,h,1)$ and $G_3 = \psi(f,1,h,f,1)$, then

$$| \operatorname{Prob}\{DC_{2n}[G_2, G_3] = 1\} - \operatorname{Prob}\{DC_{2n}[P_{2n}, P_{2n}] = 1\} | \leq \frac{m^2}{2^{2n}}$$

where DC_{2n} is a D-distinguishing circuit with two types of oracle gates and $m < 2^n$ the number of oracle gates.

Proof : As the distinguisher has two types of oracle gates, one probability is calculated when two permutations are chosen independently and randomly from P_{2n} and are used to evaluate the oracle gates. The other probability is calculated when the distinguisher chooses f, h independently and randomly from H_n and uses them in the G_2 and G_3 structures, which are applied for the evaluation of the oracle gates. When the distinguisher examines an oracle, the input is a $2n$ bit string $(L \parallel R)$ and the output would also be a $2n$ bit string $(S \parallel T)$.

When G_2 is examined the output would be

$$S = L \oplus R \oplus f(L \oplus h(L \oplus R)) \oplus h(R \oplus h(L \oplus R) \oplus f(L \oplus h(L \oplus R)))$$
$$T = R \oplus h(L \oplus R) \oplus f(L \oplus h(L \oplus R))$$

and when G_3 is examined the output would be

$$S = L \oplus R \oplus h(L \oplus f(L \oplus R)) \oplus f(R \oplus f(L \oplus R) \oplus h(L \oplus f(L \oplus R)))$$
$$T = R \oplus f(L \oplus R) \oplus h(L \oplus f(L \oplus R))$$

As both G_2 and G_3 are pseudorandom, there is no leakage of information from their input to their output. The D-distinguisher could only make a decision if there were at least one pair of oracle gates such that their outputs were related to each other.

Let X_2 be a random variable notating the ouput of h, the random function in the second round of the DES-like structure of G_2, and X_3 be a random variable notating the ouput of f, the random function in the second round of the DES-like structure of G_3. Let Y_2 be a random variable notating the output of f, the random function in the third round of DES-like structure of G_2, and Y_3 be a random variable notating the output of h, the random function in the third round of DES-like structure of G_3 (see Figure 2).

The distinguisher can make a sensible decision if any of the two following cases happens:

1. When there are a pair of oracle gates O_i, O_j such that the random functions f and h collide: Hence, for this case we assume that $x_{i,2} = x_{j,3} = x$ where $x_{i,2} = h(L_i \oplus R_i)$ and $x_{j,3} = f(L_j \oplus R_j)$ and $y_{i,2} = y_{j,3} = y$ where $y_{i,2} = f(L_i \oplus h(L_i \oplus R_i))$ and $y_{j,3} = h(L_j \oplus f(L_j \oplus R_j))$.

   In this case, when the input random variables $R_i \neq R_j$ and $L_i = L_j = L$ then always $T_i \oplus T_j = R_i \oplus R_j$, and when the input random variables $R_i = R_j = R$ and $L_i \neq L_j$ then always $S_i \oplus S_j = L_i \oplus L_j$. The probability that f and h collide in a pair of oracle gates among m oracles would be equal to $\frac{m(m-1)}{2} \frac{1}{2^{2n}}$.

2. When there are a pair of oracle gates O_i, O_j such that the random function f collides and the random function h collides: For this case we assume that $x_{i,2} = y_{j,3}$ where $x_{i,2} = h(L_i \oplus R_i)$ and $y_{j,3} = h(L_j \oplus f(L_j \oplus R_j))$, and $y_{i,2} = x_{j,3}$ where $y_{i,2} = f(L_i \oplus h(L_i \oplus R_i))$ and $x_{j,3} = f(L_j \oplus R_j)$

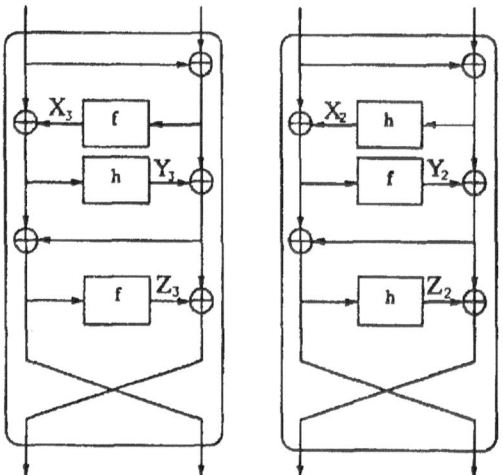

Figure 2: X_3, Y_3 random variables in G_3 and X_2, Y_2 in G_2

In this case, when the input random variables $R_i \neq R_j$ and $L_i = L_j = L$ then always $T_i \oplus T_j = R_i \oplus R_j$. The probability that f collides and also h collides in a pair of oracle gates among m oracles would be equal to $\frac{m(m-1)}{2} \frac{1}{2^{2n}}$.

The probability that a D-distinguishing circuit could be constructed for (G_2, G_3) would be equal to $\frac{m(m-1)}{2^{2n+1}} + \frac{m(m-1)}{2^{2n+1}}$. On the other hand, when two permutations p_1 and p_2 are chosen independently and randomly the probability that they can satisfy the distinguishing circuit relation would be $\frac{2m}{2^{2n}}$. So, an upper bound on the probability of distinguishing would be

$$| \operatorname{Prob}\{DC_{2n}[P_{2n}, P_{2n}] = 1\} - \operatorname{Prob}\{DC_{2n}[G_2, G_3] = 1\} | \leq \frac{m^2}{2^{2n}}$$

$\square$

Theorem 2 Let $f_1, f_2 \in_r F_n$ be independently chosen pseudorandom functions, and $G_2 = \psi(f_2, 1, f_1, f_2, 1)$ and $G_3 = \psi(f_1, 1, f_2, f_1, 1)$. When

$$| \operatorname{Prob}\{DC_{2n}[G_2, G_3] = 1\} - \operatorname{Prob}\{DC_{2n}[P_{2n}, P_{2n}] = 1\} | < \frac{1}{n^{c_2}}$$

for any polynomial size D-distinguishing circuit and for any constant c_2, then $G_1 = \psi(f_2, 1, f_1, f_2, 1, f_1)$ is a super pseudorandom permutation generator.

Proof: First, it is necessary to show that $G_3 = \psi(f_1, 1, f_2, f_1, 1)$ and $\overline{G}_1 = \psi(f_1, 1, f_2, f_1, 1, f_2)$ are independent of each other. In order to prove this claim, by contradiction assume that they are not independent, and there is a D-distinguishing circuit such that for a constant c_2

$$| \operatorname{Prob}\{DC_{2n}[G_3, \overline{G}_1] = 1\} - \operatorname{Prob}\{DC_{2n}[P_{2n}, P_{2n}] = 1\} | > \frac{1}{n^{c_2}}$$

Without changing the inequality relation, we have,

$$| \operatorname{Prob}\{DC_{2n}[G_3, \overline{G}_1] = 1\} - \operatorname{Prob}\{DC_{2n}[G_3, G_3] = 1\} +$$
$$\operatorname{Prob}\{DC_{2n}[G_3, G_3] = 1\} - \operatorname{Prob}\{DC_{2n}[P_{2n}, P_{2n}] = 1\} | > \frac{1}{n^{c_2}}$$

then,

$$| \operatorname{Prob}\{DC_{2n}[G_3, \overline{G}_1] = 1\} - \operatorname{Prob}\{DC_{2n}[G_3, G_3] = 1\} | +$$
$$| \operatorname{Prob}\{DC_{2n}[G_3, G_3] = 1\} - \operatorname{Prob}\{DC_{2n}[P_{2n}, P_{2n}] = 1\} | > \frac{1}{n^{c_2}}$$

If $| \operatorname{Prob}\{DC_{2n}[G_3, G_3] = 1\} - \operatorname{Prob}\{DC_{2n}[P_{2n}, P_{2n}] = 1\} | > \frac{1}{n^{c_2}}$, G_3 would not be pseudorandom, as the D-distinguishing circuit is not a test for identity. This contradicts Lemma 3. Furthermore, if $| \operatorname{Prob}\{DC_{2n}[G_3, \overline{G}_1] = 1\} - \operatorname{Prob}\{DC_{2n}[G_3, G_3] = 1\} | > \frac{1}{n^{c_2}}$, then the oracle circuit virtually distinguishes f_1 from a randomly chosen function. This also contradicts our assumption that f_1 is a pseudorandom function. Since both cases conclude to contradictions, then G_3 and $\overline{G}_1$ should be independent of each other, and there would be no D-distinguishing circuit for $(\overline{G}_1, G_3)$.

Considering the independency of G_2 and G_3, it is given that

$$| \operatorname{Prob}\{DC_{2n}[G_2, G_3] = 1\} - \operatorname{Prob}\{DC_{2n}[P_{2n}, P_{2n}] = 1\} | < \frac{1}{n^{c_2}}$$

Note that G_2 and G_3 are not inverse of each other, so the D-distinguishing circuit cannot be a test for inversion. Without changing the sign of inequality, the above relation can be expanded as,

$$| \operatorname{Prob}\{DC_{2n}[G_1, \overline{G}_1] = 1\} - \operatorname{Prob}\{DC_{2n}[G_1, \overline{G}_1] = 1\} +$$
$$\operatorname{Prob}\{DC_{2n}[G_3, \overline{G}_1] = 1\} - \operatorname{Prob}\{DC_{2n}[G_3, \overline{G}_1] = 1\} +$$
$$\operatorname{Prob}\{DC_{2n}[G_2, G_3] = 1\} - \operatorname{Prob}\{DC_{2n}[P_{2n}, P_{2n}] = 1\} | < \frac{1}{n^{c_2}}$$

With reordering and separation of absolute values, we get:

$$|| \operatorname{Prob}\{DC_{2n}[G_2, G_3] = 1\} - \operatorname{Prob}\{DC_{2n}[G_3, \overline{G}_1] = 1\} | -$$
$$| \operatorname{Prob}\{DC_{2n}[G_3, \overline{G}_1] = 1\} - \operatorname{Prob}\{DC_{2n}[G_1, \overline{G}_1] = 1\} | -$$
$$| \operatorname{Prob}\{DC_{2n}[G_1, \overline{G}_1] = 1\} - \operatorname{Prob}\{DC_{2n}[P_{2n}, P_{2n}] = 1\} || < \frac{1}{n^{c_2}}$$

Since, it was assumed that G_2 and G_3 are two independent permutations, and so are G_3 and $\overline{G}_1$, then $| \operatorname{Prob}\{DC_{2n}[G_2, G_3] = 1\} - \operatorname{Prob}\{DC_{2n}[G_3, \overline{G}_1] = 1\} |$ would be less than $\frac{1}{n^{c_2}}$, since

$$|| \operatorname{Prob}\{DC_{2n}[G_2, G_3] = 1\} - \operatorname{Prob}\{DC_{2n}[G_3, \overline{G}_1] = 1\} | \le$$
$$| \operatorname{Prob}\{DC_{2n}[G_2, G_3] = 1\} - \operatorname{Prob}\{DC_{2n}[P_{2n}, P_{2n}] = 1\} | +$$
$$| \operatorname{Prob}\{DC_{2n}[G_3, \overline{G}_1] = 1\} - \operatorname{Prob}\{DC_{2n}[P_{2n}, P_{2n}] = 1\} | < \frac{1}{n^{c_2}}$$

Hence

$$| \operatorname{Prob}\{DC_{2n}[G_1, \overline{G}_1] = 1\} - \operatorname{Prob}\{DC_{2n}[P_n, P_n] = 1\} | \ +$$
$$| \operatorname{Prob}\{DC_{2n}[G_3, \overline{G}_1] = 1\} - \operatorname{Prob}\{DC_{2n}[G_1, \overline{G}_1] = 1\} | \ < \ \frac{1}{n^{c_2}}$$

So, each of the above absolute values would be less than $\frac{1}{n^{c_2}}$. In other words

$$| \operatorname{Prob}\{DC_{2n}[G_1, \overline{G}_1] = 1\} - \operatorname{Prob}\{DC_{2n}[P_{2n}, P_{2n}] = 1\} | < \frac{1}{n^{c_2}}$$

Hence G_1 and $\overline{G}_1$ are independent of each other, and G_1 is a super pseudorandom permutation. $\qquad \square$

The above lemmas and theorem show that when the number of oracle gates m is a polynomial of n then the probability of making a circuit which distinguishes $G_2 = \psi(f_2, 1, f_1, f_2, 1)$ from $G_3 = \psi(f_1, 1, f_2, f_1, 1)$ is less than 1 over any polynomial in n, and the probability of making a super distinguishing circuit for $G_1 = \psi(f_2, 1, f_1) \circ \psi(f_2, 1, f_1)$ is also less than 1 over any polynomial in n. In other words, G_1 is secure against chosen plaintext / ciphertext attack if the cryptanalyst is permitted to access to only a polynomial number of queries.

4.2 Super Pseudorandomness of $\psi(f^2, 1, f, f^2, 1, f)$

In the above relations, an upper bound on the probabilities has been defined, where f_1 and f_2 are two independently chosen pseudorandom functions. In the following lemmas and theorem, we show that if f_2 is substituted with $f_1{}^2$, there would be an increase of the upper bound of the above probabilities of distinguishing. Nevertheless, if the number of oracles would be limited to some polynomial in n, the corresponding probabilities of distinguishing would remain less than 1 over any polynomial in n.

Lemma 5 *Let $f, h \in_r H_n$ and C_{2n} be a distinguishing circuit with $m < 2^n$ oracle gates, then*

$$| \operatorname{Prob}\{C_{2n}[\psi(f^2, 1, f, f^2, 1)] = 1\} - \operatorname{Prob}\{C_{2n}[\psi(h, 1, f, h, 1)] = 1\} | \leq \frac{2m^2}{2^n}$$

Proof : Since both f and h can be considered as two sequences of 2^n independent and uniformly distributed n-bit random variables, for an argument $a \in \Sigma^n$, $f(a)$ and $h(a)$ are two independent n-bit strings. When the input to an oracle is $(L \parallel R)$ and the oracle is evaluated with $\psi(h, 1, f, h, 1)$, each branches of the outputs, i.e., S and T, is always a sum of two random variables generated by the functions f and h. That is

$$S = L \oplus R \oplus f(L \oplus h(L \oplus R)) \oplus h(R \oplus h(L \oplus R) \oplus f(L \oplus h(L \oplus R)))$$
$$T = R \oplus h(L \oplus R) \oplus f(L \oplus h(L \oplus R))$$

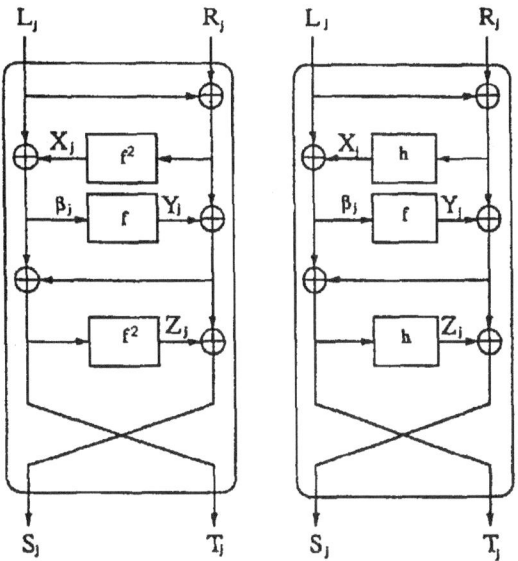

Figure 3: X, Y and Z random variables in $\mathcal{G}_2$ and G_2

where all the random variables $Y_i = f(L_i \oplus h(L_i \oplus R_i))$ are independent from random variables

$$
\begin{aligned}
X_j &= h(L_j \oplus R_j) \\
Z_j &= h(R_j \oplus h(L_j \oplus R_j) \oplus f(L_j \oplus h(L_j \oplus R_j)))
\end{aligned}
$$

When the input to an oracle is $(L \parallel R)$ and the oracle is evaluated with $\psi(f^2, 1, f, f^2, 1)$, each branch of the output, i.e., S and T, is always a sum of two random variables generated by the function f. That is

$$
\begin{aligned}
S &= L \oplus R \oplus f(L \oplus f^2(L \oplus R)) \oplus f^2(R \oplus f^2(L \oplus R) \oplus f(L \oplus f^2(L \oplus R))) \\
T &= R \oplus f^2(L \oplus R) \oplus f(L \oplus f^2(L \oplus R))
\end{aligned}
$$

If $R = 0$

$$
\begin{aligned}
S &= L \oplus f(L \oplus f^2(L)) \oplus f^2(f^2(L) \oplus f(L \oplus f^2(L))) \\
T &= f^2(L) \oplus f(L \oplus f^2(L))
\end{aligned}
$$

If $L = 0$

$$
\begin{aligned}
S &= R \oplus f^3(R)) \oplus f^2(R \oplus f^2(R) \oplus f^3(R)) \\
T &= R \oplus f^2(L \oplus R) \oplus f(L \oplus f^2(L \oplus R))
\end{aligned}
$$

If $L = 0$ and $R = 0$

$$
\begin{aligned}
S &= f^3(0) \oplus f^2(f^2(0) \oplus f^3(0)) \\
T &= f^2(0) \oplus f^3(0)
\end{aligned}
$$

When all the oracle gates are evaluted with $\psi(f^2, 1, f, f^2, 1)$ or all the oracle gates are evaluated with $\psi(h, 1, f, h, 1)$, a distinguisher generates "1" on its output with the same probability if all the random variables $Y_i = f(L_i \oplus f^2(L_i \oplus R_i))$ are independent from random variables

$$X_j = f^2(L_j \oplus R_j)$$
$$Z_j = f^2(R_j \oplus f^2(L_j \oplus R_j) \oplus f(L_j \oplus f^2(L_j \oplus R_j)))$$

In other words, a distinguisher with m oracle gates generates "1" on its output when there is one oracle gate O_i such that β_i the input value to the random function f is equal to either of

$$\beta_i = \begin{cases} f(L_j \oplus R_j) \\ f(R_j \oplus f^2(L_j \oplus R_j) \oplus f(L_j \oplus f^2(L_j \oplus R_j))) \end{cases}$$

for some $j = 1, \ldots, m$. The probability that in a given oracle the input to the f function takes a value equal to any of $2m$ internal random values in m oracle gates with different inputs is $\frac{2m}{2^n}$. The probability that a circuit distinguishes $\psi(f^2, 1, f, f^2, 1)$ from $\psi(h, 1, f, h, 1)$ is equal to the probability that two of the oracles generate dependant random variables. Hence

$$\mid \mathrm{Prob}\{C_{2n}[\psi(f^2, 1, f, f^2, 1)] = 1\} - \mathrm{Prob}\{C_{2n}[\psi(h, 1, f, h, 1)] = 1\} \mid \leq \frac{2m^2}{2^n}$$

The probability of distinguishing $\psi(h, 1, f, h, 1)$ from a random permutation was given in Lemma 2, as a result an upper bound on the overall probability of distinguishing $\mathcal{G}_2 = \psi(f^2, 1, f, f^2, 1)$ from a random permutation is

$$\mid \mathrm{Prob}\{C_{2n}[\mathcal{G}_2] = 1\} - \mathrm{Prob}\{C_{2n}[P_{2n}] = 1\} \mid \leq \frac{3m^2}{2^n} + \frac{m^2}{2^{2n}}$$

where $m < 2^n$ is the number of oracle gates. Hence, when the distinguisher is bounded by a polynomial number of oracle gates the probability of distinguishing $\psi(f^2, 1, f, f^2, 1)$ from a random permutation is less than 1 over any polynomial in n. $\qquad\square$

Lemma 6 *Let $f, h \in_r H_n$ and C_{2n} be a distinguishing circuit with $m < 2^n$ oracle gates, then*

$$\mid \mathrm{Prob}\{C_{2n}[\psi(f, 1, f^2, f, 1)] = 1\} - \mathrm{Prob}\{C_{2n}[\psi(f, 1, h, f, 1)] = 1\} \mid \leq \frac{2m^2}{2^n}$$

Proof : A proof similar to the proof of Lemma 5 can be given here. When the input to an oracle is $(L \parallel R)$ and the oracle is evaluated with $\psi(f, 1, f^2, f, 1)$, each branch of the output, i.e., S and T, is always a sum of two random variables generated by the function f. That is

$$S = L \oplus R \oplus f^2(L \oplus f(L \oplus R)) \oplus f(R \oplus f(L \oplus R) \oplus f^2(L \oplus f(L \oplus R)))$$
$$T = R \oplus f(L \oplus R) \oplus f^2(L \oplus f(L \oplus R))$$

If $R = 0$

$$
\begin{aligned}
S &= L \oplus f^2(L \oplus f(L)) \oplus f(f(L) \oplus f^2(L \oplus f(L))) \\
T &= f(L) \oplus f^2(L \oplus f(L))
\end{aligned}
$$

If $L = 0$

$$
\begin{aligned}
S &= R \oplus f^3(R)) \oplus f(R \oplus f(R) \oplus f^3(R)) \\
T &= R \oplus f(R) \oplus f^3(R)
\end{aligned}
$$

If $L = 0$ and $R = 0$

$$
\begin{aligned}
S &= f^3(0) \oplus f(f(0) \oplus f^3(0)) \\
T &= f(0) \oplus f^3(0)
\end{aligned}
$$

When all the oracle gates are evaluted with $\psi(f, 1, f^2, f, 1)$ or all the oracle gates are evaluated with $\psi(f, 1, h, f, 1)$, a distinguisher generates "1" on its output with the same probability if all the random variables $Y_i = f^2(L_i \oplus f(L_i \oplus R_i))$ are independent from random variables

$$
\begin{aligned}
X_j &= f(L_j \oplus R_j) \\
Z_j &= f(R_j \oplus f(L_j \oplus R_j) \oplus f^2(L_j \oplus f(L_j \oplus R_j)))
\end{aligned}
$$

In other words, a distinguisher with m oracle gates generates "1" on its output when there is one oracle gate O_i such that β_i, the input value to the random function f, is equal to $f(L_j \oplus f(L_j \oplus R_j))$ for some $j = 1, \ldots, m$. As in each oracle, f have been used in two different layers, the probability that in a given oracle, β_i the input to the f functions takes a value equal to any of m internal random values Y_i, i.e., output of f^2 layer in m oracle gates with different inputs, is $\frac{2m}{2^n}$. The probability that a circuit distinguishes $\psi(f, 1, f^2, f, 1)$ from $\psi(f, 1, h, f, 1)$ is equal to the probability that two of the oracles generate dependant random variables. Hence

$$
\mid \mathrm{Prob}\{C_{2n}[\psi(f, 1, f^2, f, 1)] = 1\} - \mathrm{Prob}\{C_{2n}[\psi(f, 1, h, f, 1)] = 1\} \mid \leq \frac{2m^2}{2^n}
$$

The probability of distinguishing $\psi(f, 1, h, f, 1)$ from a random permutation was given in Lemma 3, as a result an upper bound on the overall probability of distinguishing $\mathcal{G}_3 = \psi(f, 1, f^2, f, 1)$ from a random permutation is

$$
\mid \mathrm{Prob}\{C_{2n}[\mathcal{G}_3] = 1 - \mathrm{Prob}\{C_{2n}[P_{2n}] = 1\} \mid \leq \frac{3m^2}{2^n} + \frac{m^2}{2^{2n}}
$$

where $m \leq 2^n$ is the number of oracle gates. Hence, when the distinguisher is bounded by a polynomial number of oracle gates the probability of distinguishing $\psi(f, 1, f^2, f, 1)$ from a random permutation is less than 1 over any polynomial in n. $\qquad \square$

Theorem 3 *Let $f \in_r F_n$ be a pseudorandom function, then $\mathcal{G}_1 = \psi(f^2, 1, f, f^2, 1, f)$ is a super pseudorandom permutation.*

Proof Sketch: To prove that $\mathcal{G}_1$ is a super pseudorandom permutation generator, we first show that $\mathcal{G}_2 = \psi(f^2, 1, f, f^2, 1)$ and $\mathcal{G}_3 = (f, 1, f^2, f, 1)$ are independent of each other. As it was shown in Lemma 5 and Lemma 6, in a D-distinguishing circuit when the input to an oracle is $(L \parallel R)$ and the oracle is evaluated with $\psi(f^2, 1, f, f^2, 1)$, the output is

$$
\begin{aligned}
S &= L \oplus R \oplus f(L \oplus f^2(L \oplus R)) \oplus f^2(R \oplus f^2(L \oplus R) \oplus f(L \oplus f^2(L \oplus R))) \\
T &= R \oplus f^2(L \oplus R) \oplus f(L \oplus f^2(L \oplus R))
\end{aligned}
$$

and when the oracle is evaluated with $\psi(f, 1, f^2, f, 1)$, the output is

$$
\begin{aligned}
S' &= L \oplus R \oplus f^2(L \oplus f(L \oplus R)) \oplus f(R \oplus f(L \oplus R) \oplus f^2(L \oplus f(L \oplus R))) \\
T' &= R \oplus f(L \oplus R) \oplus f^2(L \oplus f(L \oplus R))
\end{aligned}
$$

The six random variables involved are:

$$
\begin{aligned}
X_2 &= f^2(L \oplus R) \\
Y_2 &= f(L \oplus f^2(L \oplus R)) \\
Z_2 &= f^2(R \oplus f^2(L \oplus R) \oplus f(L \oplus f^2(L \oplus R)))
\end{aligned}
$$

and

$$
\begin{aligned}
X_3 &= f(L \oplus R) \\
Y_3 &= f^2(L \oplus f(L \oplus R)) \\
Z_3 &= f(R \oplus f(L \oplus R) \oplus f^2(L \oplus f(L \oplus R)))
\end{aligned}
$$

S, T, S' and T' are always a sum of two random variables generated by the function f. If a random variable in a output branch of an oracle becomes dependent on a random variable in any output branch of another oracle, the other random variables are always independent of each other. For example, if X_2 be equal to X_3, the probability that Y_2 be equal to Y_3 is $\frac{1}{2^n}$. Likewise, if $Y_2 = Y_3$, the probability that $Z_2 = Z_3$ is $\frac{1}{2^n}$. Hence, the probability of dependency of two branches is equal to $\frac{1}{2^n}$, which is equal to the dependency of two output branches in two different oracle gates when instead of f^2 an independent random function such as h was applied. Here, we calculate an upper bound on the independency probability. As it was shown earlier in Lemma 5 that $\psi(f^2, 1, f, f^2, 1)$ and $\psi(h, 1, f, h, 1)$ are indistinguishable from each other, and it was shown in Theorem 2 that $\psi(h, 1, f, h, 1)$ and $\psi(f, 1, h, f, 1)$ are independent of each other, and it was shown in Lemma 6 that $\psi(f, 1, h, f, 1)$ and $\psi(f, 1, f^2, f, 1)$ are indistinguishable from each other. The probability that there would be a D-distinguishing circuit for $\mathcal{G}_2$ and $\mathcal{G}_3$ can be written as:

$$
\mid \mathrm{Prob}\{DC_{2n}[\mathcal{G}_2, \mathcal{G}_3] = 1\} - \mathrm{Prob}\{DC_{2n}[P_{2n}, P_{2n}] = 1\} \mid < \frac{1}{n^{c_2}}
$$

The above probability statement can be expanded as,

$$
\begin{aligned}
& | \quad \mathrm{Prob}\{DC_{2n}[\mathcal{G}_2, \mathcal{G}_3] = 1\} - \mathrm{Prob}\{DC_{2n}[\mathcal{G}_2, G_3] = 1\} \\
& + \quad \mathrm{Prob}\{DC_{2n}[\mathcal{G}_2, G_3] = 1\} - \mathrm{Prob}\{DC_{2n}[G_2, G_3] = 1\} \\
& + \quad \mathrm{Prob}\{DC_{2n}[G_2, G_3] = 1\} - \mathrm{Prob}\{DC_{2n}[P_{2n}, P_{2n}] = 1\} |
\end{aligned}
$$

where $G_2 = \psi(h, 1, f, h, 1)$ and $G_3 = \psi(f, 1, h, f, 1)$. With reordering and separation of absolute values, we get:

$$
\begin{aligned}
& | \quad | \mathrm{Prob}\{DC_{2n}[\mathcal{G}_2, \mathcal{G}_3] = 1\} - \mathrm{Prob}\{DC_{2n}[\mathcal{G}_2, G_3] = 1\} | \\
& + \quad | \mathrm{Prob}\{DC_{2n}[\mathcal{G}_2, G_3] = 1\} - \mathrm{Prob}\{DC_{2n}[G_2, G_3] = 1\} | \\
& + \quad | \mathrm{Prob}\{DC_{2n}[G_3, G_3] = 1\} - \mathrm{Prob}\{DC_{2n}[P_{2n}, P_{2n}] = 1\} ||
\end{aligned}
$$

If the sum of the above probabilities is less than 1 over any polynomial in n, then $\mathcal{G}_2$ and $\mathcal{G}_3$ would essentially be independent of each other according to the definition. When $f, h \in_r H_n$, with applying procedures similar to the proofs of Lemma 5, Lemma 6 and Lemma 3, it can be shown that the first term is less than $\frac{2m^2}{2^n}$, and the second term is less than $\frac{2m^2}{2^n}$, and the third term is less than $\frac{m^2}{2^n} + \frac{m^2}{2^{2n}}$, respectively. So, the sum of these three probabilities, makes a bound on the probability for making a D-distinguishing circuit with m oracle gates for $\mathcal{G}_2$ and $\mathcal{G}_3$, which is $\frac{5m^2}{2^n} + \frac{m^2}{2^{2n}}$. When there are a polynomial number of oracle gates, i.e., m is a polynomial in n, then

$$
| \mathrm{Prob}\{DC_{2n}[\mathcal{G}_3, \mathcal{G}_2] = 1\} - \mathrm{Prob}\{DC_{2n}[P_{2n}, P_{2n}] = 1\} | < \frac{1}{n^{c_2}}
$$

With applying a proof similar to the proof of Theorem 2, it can be shown that when $\mathcal{G}_2$ and $\mathcal{G}_3$ are independent, $\mathcal{G}_1$ is super pseudorandom. $\qquad\square$

5 Conclusions and Open Problems

We have shown that it is possible to construct a super pseudorandom permutation generators applying a single pseudorandom function. We took advantage of the stucture of optimal perfect randomizer, presented in [3]. We first showed that $\psi(h, 1, f, h, 1, f)$ is a super pseudorandom permtation. Then, we showed that by substituting f^2 instead of h, the probability of making a distinguisher still would remain less than 1 over any polynomial in n. Although, such a block cryptosystem is less than practical, it can be seen as an attempt towards a construction of practical ones which are provably secure against chosen plaintext / ciphertext attack without relying on any unproven hypothesis. If a cryptosystem is super pseudorandom it can be applied as a hashing algorithm, where it is assured that a cryptanalyst cannot find a collision without trying at least 2^n different messages. Two open problems can be posed from the result of this paper. The structure applies 6 rounds of DES-like permutations. The first open problem is whether the proposed structure is an optimal structure, and whether any other structure can be suggested which needs less number of rounds. It can be seen that $\psi(f, f^2, f, f, f)$ is a good candidate, but we have not got a justification for this claim yet. The second open problem is whether the proposed structure can be adopted to improve the quality of existing cryptosystems such as DES or LOKI against differential cryptanalysis without a need for redesigning their S boxes.

ACKNOWLEDGMENT

We would like to thank Professor Jennifer Seberry and Mr. Andy Quaine for their attention and support.

References

[1] O. Goldreich, S. Goldwasser, and S. Micali. How to construct random functions. *Journal of the ACM*, 33(4):792–807, 1986.

[2] Michael Luby and Charles Rackoff. How to Construct Pseudorandom Permutations from Pseudorandom Functions. *SIAM Journal on Computing*, 17(2):373–386, 1988.

[3] J. Pieprzyk and B. Sadeghiyan. Optimal perfect randomizers. In *Abstracts of ASIACRYPT '91*, pages 130–135, 1991.

[4] Josef Pieprzyk. How to Construct Pseudorandom Permutations from Single Pseudorandom Functions. In *Advances in Cryptology - EUROCRYPT '90*, volume 473 of *Lecture Notes in Computer Science*, pages 140–150. Springer-Verlag, 1991.

[5] B. Sadeghiyan and J. Pieprzyk. On necessary and sufficient conditions for the construction of super pseudorandom permutations. In *Abstracts of ASIACRYPT '91*, pages 117–123, 1991.

[6] A. C. Yao. Theory and applications of trapdoor functions. In *the 23rd IEEE Symposium on the Foundations of Computer Science*, pages 80–91, 1982.

How to Break a "Secure"
Oblivious Transfer Protocol

Donald Beaver

313 Whitmore Lab, Penn State University, State College, PA 16802, USA,
(814) 863-0147, beaver@cs.psu.edu.

Abstract.

We show how to break a protocol for Oblivious Transfer presented at Eurocrypt 90 [11]. Armed with a new set of definitions for proving the security of interactive computations, we found difficulties in proving the protocol secure. These difficulties led us to a simple attack that breaks the OT protocol in a subtle but fundamental way. The error that we found may be present in a wide variety of secure protocols. It reveals a fundamental flaw in the traditional definition of Oblivious Transfer itself.

1 Introduction

Solid proofs are a lacking but essential requirement for cryptography. Whereas a failed claim in complexity theory might mean an algorithm gives errors sometimes, a failed claim to security might provide a huge potential for malicious destruction of data, resources, and dependability.

Proofs are often lacking because clear and simple definitions are hard to come by. The value of good definitions goes beyond the confidence they inspire in proven results. When good definitions are present, a concise proof can usually be found; the lack of one, or the difficulty of finding one, often indicates that a theorem is incorrect. In fact, truly clear definitions and proof techniques often turn up counterexamples when applied to an incorrect conjecture.

In [1, 3], Beaver proposed a concise set of definitions that provide not only a clear, intuitive understanding of security in interactive protocols but that support direct and modular proofs. In this paper, we show how these definitions revealed a subtle flaw in a protocol claimed to be "provably" secure.

We consider the problem of Oblivious Transfer (OT), a fundamental cryptographic problem introduced by Rabin [17]. OT is a two-party protocol in which Alice transmits a bit b to Bob, and Bob receives the bit with probability $\frac{1}{2}$. Alice must not learn whether Bob received the bit. OT forms the basis for a wide variety of cryptographic protocols [18, 15, 6, 8].

Based on assuming that determining quadratic residuosity is hard, Rabin suggested an elegant but inefficient protocol for OT, requiring the generation of a large integer $n = pq$ (with p, q prime) for each bit to be sent [17].

Den Boer suggested an efficient protocol for OT in Eurocrypt '91 [11], relying on the same assumption but allowing the re-use of n and ensuring unconditional security of b. We show that this protocol is insecure.

R.A. Rueppel (Ed.): Advances in Cryptology - EUROCRYPT '92, LNCS 658, pp. 285-296, 1993.

The flaw is that Alice can send a bit b without knowing what b is — potentially giving her information about the quadratic residuosity of numbers of her choice, without her knowing the factors of n. In fact, after a single transfer she may be able to break all subsequent transfers.

Interestingly, the protocol *is* provably secure according to the traditional list of required security properties, if one confirms "Alice does not learn whether Bob received b" using the approach of of zero-knowledge [14]. As we shall see, zero-knowledge is insufficient as as a measure of the security of OT (and of other protocols). Rabin's protocol (apparently serendipitously) does not suffer from the same problem, but our analysis shows how to break Rabin's protocol in a network of three or more players.

Instead of proposing a new requirement for OT, we define the security of OT according to a unifying property called *resilience*. Resilience captures all known security properties and, we submit, it implies all security properties *a priori*. Unforeseen properties such as transparency (the inability to bluff) arise as newly-observed implications of our unified definition, as this paper exemplifies.

Using resilience, we were able to expose the subtle flaw in the OT protocol within minutes. We fix the flaw and sketch a proof of security, thus salvaging den Boer's brilliant idea and providing an efficient protocol for Oblivious Transfer.

2   Oblivious Transfer: the Traditional Approach

Traditionally, the goal of Oblivious Transfer is to find a protocol satisfying certain properties:

1. Alice sends bit b but does not know whether Bob receives it;
2. Bob receives $(1, b)$ or $(0, 0)$ [*resp.* "received b" or "got nothing"] with equal probability, but receives no additional information;
3. Both players can abort the protocol by deviating from it in a syntactic sense (*eg.* Alice does not send 0/1 or Bob sends "quit").

The usual formalization of "no additional information" or "does not know" uses a zero-knowledge approach: a simulator must demonstrate that Alice (or Bob) could generate an accurate view of the interaction based *only* on given, limited information. But zero-knowledge is not enough to guarantee that Alice learns nothing.

We show that a previously-overlooked property is essential, namely that Alice not be able to bluff her way through the protocol:

4. Alice must know the effective bit b she sends.

Rather than adding to a potentially incomplete list of properties, we examine a single property, *resilience*, which implies (1)-(4) above and, by virtue of its clarity and wide applicability, seems to capture properties as yet unobserved.

A simple and direct fix for protocols lacking property (4) is to require Alice to give a zero-knowledge proof that she knows her effective input b. This turns out to be sufficient to construct a protocol secure under our unified definition. First, however, let us examine why the ability to bluff should be restricted and how it can be used to break den Boer's protocol.

3 Oblivious Transfer in an Ideal World

In the ideal case, Alice would give b to a trusted host (noisy channel) that would then send $(1, b)$ or $(0, 0)$ to Bob with equal probability, without informing Alice which one it sent. In a very real sense, the goal of Oblivious Transfer is to implement this ideal protocol without having a trusted host available.

Imagine for the moment that a trusted host is available – perhaps in the form of a quantum channel [9]. Consider the following scenario: Alice sends b_1 to Bob, and later says, "I think I got some static on the line; can we test it?" Bob agrees, and Alice sends a second bit b_2, and Bob reports his second result $((0, 0)$ or $(1, b_2))$.

Certainly, in the ideal case, Alice would not learn anything about the results of the first execution, eg. whether Bob received b_1 or not. Any implementation of OT should not allow later executions to compromise earlier ones, even when Bob reveals the later results.

4 Oblivious Transfer: Background

4.1 Notes on Cryptography

A **Blum integer** is a product $n = pq$ of two equally-sized primes of the form $p = 4k + 3$, $q = 4l + 3$; let BLUM_k be the set of such numbers of size k. A number x is a **quadratic residue** (mod n) iff it has a square root (mod n). The integers modulo Blum integer n having Jacobi symbol $+1$ form a multiplicative group, Z_n^+, of which half are residues. Define $Q_n(x) = 0$ iff $x \in Z_n^+$ is a residue, or 1 if not; note $Q_n(ab) = Q_n(a) \oplus Q_n(b)$. For Blum integers, $Q_n(-1) = 1$.

The notation $x \leftarrow X$ indicates x is sampled uniformly at random from set X. An **ensemble** is a function mapping a pair (z, k), with $z \in \Sigma^*$ and $k \in \mathbf{N}$, to a distribution on strings of size at most $(|z|k)^c$ for some c. If P and Q are ensembles, and if M is a TM or function, define the **distinguishing power** of M to be

$$\delta_M(z, k) \stackrel{\text{def}}{=} |\Pr[M(P(z, k)) = 1] - \Pr[M(Q(z, k) = 1)]|.$$

Ensembles P and Q are **statistically indistinguishable**, written $P \approx Q$, if for all functions M, and for all z and c, $\delta_M(z, k) = 0(k^{-c})$. Ensembles P and Q are **computationally indistinguishable**, written $P \tilde{\approx} Q$, if for all probabilistic poly-time TM's M, and for all z and c, $\delta_M(z, k) = 0(k^{-c})$.

The **Quadratic Residuosity Assumption (QRA)** states that random residues in BLUM_k are computationally indistinguishable from random non-residues. A more general version states the same for all products of two primes.

RABIN-OT(b, k)

1. Alice chooses $n = pq \leftarrow$ BLUM$_k$, remembers (p, q), selects $t \leftarrow \mathbf{Z}_n^*$, and sends $(n, (-1)^b t^2)$ to Bob.
2. Bob chooses $r \leftarrow \mathbf{Z}_n^*$ and sends $x \equiv r^2 \pmod{n}$ to Alice.
3. Alice chooses s randomly from the four square roots of x and sends s to Bob.
4. If $s \boxminus \pm r$ then Bob concludes $(0, 0)$; otherwise Bob factors n using $\gcd(r + s, n)$, computes $b = Q_n((-1)^b r^2)$, and concludes $(1, b)$.

Fig. 1. Rabin's Oblivious Transfer protocol: Alice sends b to Bob; k is a security parameter.

4.2 An Early Implementation

Figure 1 describes Rabin's protocol for OT, for security parameter k. A flaw noted long ago is that Bob might choose x in a different manner, obtaining illegal information from the protocol: for example, a square root of a number whose roots he does not know. The simple correction is to require Bob to prove in zero-knowledge that he already knows a square root of x. A second flaw is that Alice may cheat by using an n that is not a product of two primes; thus, she too should prove in zero-knowledge that n is the product of two primes. The corrected version satisfies a definition of OT using the property list given above, assuming on the QRA; in fact, it satisfies our definitions given below, if Alice must also prove she *knows* the two factors of n.

Unfortunately, Rabin's protocol is inefficient: it requires Alice to generate a new, large Blum integer for every bit to be transferred. Furthermore, the secrecy of b is not unconditional: it depends on Bob's inability to determine quadratic residuosity.

5 Breaking an Oblivious Transfer Protocol

At Eurocrypt '91, den Boer [11] presented a protocol (see Fig. 2) that requires only *one* generation of a large product of two primes and that ensures the secrecy of bit b unconditionally. In contrast to Rabin's protocol, den Boer's protocol assumes that *Bob* knows the factors of n while Alice does not. Together, Alice and Bob generate a number $J^c a$ (where J is a nonresidue) whose residuosity is random and unknown to Alice. Alice "encodes" b as $x = J^b r^2$; she sends x and $y = J^c a x^{-1}$ to Bob. If $Q_n(J^c a) = 0$, implying that $Q_n(x) = Q_n(y) = b$, then Bob can compute b, else he cannot.

The fundamental flaw in this protocol is that Alice can cheat by selecting x without knowing its residuosity, effectively transmitting some bit b without knowing its value. This apparently innocuous flaw ("Who cares if Alice knows less?") is far more significant than it seems. One main advantage to den Boer's protocol is that it does not require repeated generation of large Blum integers

---

Boer-OT(b, k)

1. Bob chooses $n \leftarrow pq$, remembers (p, q), selects a nonresidue J, selects $a \leftarrow \mathbf{Z}_n^+$, and sends (n, J, a) to Alice. Bob proves in zero-knowledge that n is the product of two primes and J is a nonresidue.
2. Alice selects $c \leftarrow \{0, 1\}$ and $r \leftarrow \mathbf{Z}_n^*$ and sets $z = J^c u$. Alice computes $x = J^b r^2 \pmod{n}$ and sets $y = zx^{-1} \pmod{n}$. Alice sets $(u, v) = (x, y)$ if $x < y$, else $(u, v) = (y, x)$. (This is equivalent to taking (x, y) in random order). She sends (u, v) to Bob.
3. Bob checks that $uv \equiv a$ or $uv \equiv Ja$. Bob computes $Q_n(uv)$; if 0, he outputs $(1, Q_n(u))$ ("received $Q_n(u)$"), else he outputs $(0, 0)$ ("received nothing").

Fig. 2. Den Boer's Oblivious Transfer protocol.

---

for every bit to be sent; it can also be used directly for string-transfer (by using the same a for many bits). But Alice can break the protocol exactly in those situations where more than one bit must be transferred. Even though the protocol may seem secure when used only once, subtle flaws such as these preclude its use as a black-box subroutine — an essential property for cryptographic protocols.

To be concrete, let us consider the simple scenario described in §3. Alice sends b_1 and later sends b_2. (We use subscripts 1 and 2 to denote the two executions.) By directly asking Bob as described in §3 (or perhaps more realistically, by observing Bob's later behavior) Alice learns whether Bob received b_2 or not — without being told anything directly about the results of sending b_1. Because the sending of b_2 and b_1 should be independent, this should not really be a problem.

But a clever Alice misbehaves during the second execution, setting $x_2 \leftarrow a_1 t_2^2$ (where a_1 was the value used for b_1). If Bob reports he received b_2, then Alice now knows that $Q_n(a_1) = b_2$, so she can calculate precisely whether Bob received bit b_1!

Knowing that Alice can send bits without knowing their value, the interested reader is invited to consider other more and less subtle ways to break the protocol or at least to gain unfair advantage. Alice's ability to bluff her way through is essential to her attack.

The devil's advocate may complain that Bob should never go along with Alice's later requests, to prevent Alice from deducing anything. In this case, Bob's actions must always be completely independent of the results he obtains — this includes cases where he detects cheating, since Alice can derive subtle and compromising information even from an accusation of cheating. This is an extremely stringent handicap to put on a protocol, hardly applicable to any realistic situation. Protocols should be secure enough to be treated as black-boxes, called at will and independently, without interdependencies that compromise security.

6 Why OT Has Always Been Defined Incorrectly

Until now, the requirement that Alice *know* the bit b that she sends has not been made explicit (to the best of our knowledge). Protocol BOER-OT fails exactly for this reason. Rabin's OT protocol survives because Alice knows the factors of n and hence could calculate the effective bit b that she sent, even if she generated numbers illegally. (Let us point out that even this assurance is subject to the assumption that Alice's proof of $n \in \text{BLUM}_k$ demonstrates *knowledge* of p and q).

Intuitively, bluffing allows Alice to observe effects that she could not predict by herself. If we design a protocol that requires Alice's attack to be *transparent* (namely, Alice's effective bit b should be predictable from her *view*), then Alice cannot play subtle games without knowing in advance what their effects will be. Thus she does not gain information to which she is not entitled. This is a situation quite different from zero-knowledge.

We could add "Alice knows b" to the list of properties required by OT, but we would have no guarantee that our human powers of observation have still not overlooked some essential property. Instead, we turn to a single property, *relative resilience*, that defines exactly what it means for a real protocol to implement an *ideal* specification, without having to list separate properties.

7 Defining and Proving Security

We define a general security reduction among protocols that states precisely how one protocol implements another securely and fault-tolerantly. We sketch results developed in [3] for the information-theoretic setting and in [2] for resource-bounded computation.

Information-Theoretic Security. Intuitively, protocol α is as secure as protocol β if the attack of any allowed adversary $\mathcal{A}$ against α wreaks as much havoc on β as on α. Essentially, $\mathcal{A}$ gains the same *information* and wields the same *influence* over correct outputs in β as in α. Of course, $\mathcal{A}$ might not understand β (*eg.*, β might have a different communication format), so we give it an *interface* $\mathcal{I}$. The interface provides a convincing α-environment to $\mathcal{A}$, attempting to bring $\mathcal{A}$ to a final state as though $\mathcal{A}$ had really attacked α. At the same time, $\mathcal{I}$ attacks β, getting the information it needs and attempting to induce the same results in honest processors as when $\mathcal{A}$ attacks α. (These two goals are inseparable; $\mathcal{I}$ must achieve them simultaneously.) Thus the view of $\mathcal{A}$ (its *information*) and the outputs of nonfaulty players (reflecting $\mathcal{A}$'s *influence* on correctness) are the same in both protocols. If ADV_α is the set of allowable adversaries against α and ADV_β that against β, $\mathcal{I}(\mathcal{A})$ should of course be in ADV_β.

An execution of a protocol α with n players on inputs $x_1, \ldots, x_n$ (and auxiliary inputs $a_1, \ldots, a_n$ for the players, a_A for $\mathcal{A}$) induces some distribution on outputs $y_1, \ldots, y_n$ and views $v_1, \ldots, v_n$ of good players and the output/view y_A of $\mathcal{A}$. We let $[\mathcal{A}, \alpha](\mathbf{x} \circ \mathbf{a} \circ a_A, k)$ denote the distribution on $(v_A, y_1, y_2, \ldots, y_n)$, namely on adversary-view and honest-outputs. Ranging over all possible inputs and security parameters, this collection of distributions is an ensemble, $[\mathcal{A}, \alpha]$. When $\mathcal{A}$ is allowed to attack another protocol β through an interface $\mathcal{I}$, the corresponding ensemble is denoted $[\mathcal{I}(\mathcal{A}), \beta]$. The restriction to player outputs (resp., adversary view) is denoted by subscript Y (resp., Y_A).

[Aside: for technical reasons, we consider a slight modification allowing the adversary to perform "post-protocol corruption," namely to elect to compute and corrupt even after the protocol is finished. In this case (and only in this case), $\mathcal{A}$ receives *all player outputs* (but not views) and $\mathcal{I}$ must continue to provide $\mathcal{A}$ with accurate α-views when $\mathcal{A}$ corrupts new players. This tests the ability of $\mathcal{I}$ to create accurate α-views — which, technically, it is otherwise not required to do after the protocol is finished. Without further explanation (but see [3]), such strong interfaces permit us to prove that sequential protocol concatenation is secure. The ensembles $[\mathcal{A}, \alpha]$ and $[\mathcal{I}(\mathcal{A}), \beta]$ are defined to include the cases when $\mathcal{A}$ elects post-protocol corruption.]

An interface is called **parsimonious** if it corrupts the same pattern of players as that listed in the adversary-view v_A it induces.

The following definition is a preliminary formalization of the notion that, if effects of attacks on α match those of interface-assisted attacks on β, then α is as secure as β.

Definition 1. Let $\mathtt{ADV}_\alpha$ denote a class of adversaries allowed to attack α. Protocol α is **(info-theoretically) as resilient as** β if there exists a parsimonious interface $\mathcal{I}$ such that, for all $\mathcal{A} \in \mathtt{ADV}_\alpha$, we have $\mathcal{I}(\mathcal{A}) \in \mathtt{ADV}_\beta$ and

$$[\mathcal{A}, \alpha] \approx [\mathcal{I}(\mathcal{A}), \beta].$$

Ideal Protocols. An **ideal protocol** contains one or more *trusted hosts* that are incorruptible. All desirable security properties are, by definition, observations about an ideal situation. The **ideal protocol** $\mathtt{ID}(F)$ **for function** F consists of a trusted host that accepts inputs, computes F, and returns the outputs. The **ideal OT protocol** contains a trusted host that accepts b from Alice and sends $(1, b)$ or $(0, 0)$ to Bob with equal probability; either player can send a quit message to the host to abort the protocol. We shall declare a protocol secure if it achieves what an ideal protocol achieves; but first, we consider some computational issues.

Computational Issues. In the computational setting, we are worried about obtaining information that is not efficiently computable, so we require that $\mathcal{I}$ be poly-time regardless of how it accesses $\mathcal{A}$ (*eg.* as a black-box, resettable black-box, *etc.*). This restricts $\mathcal{A}$'s information to be a feasible function of the information $\mathcal{I}$ gains in β. A subtle but crucial point to note is that in β, $\mathcal{I}$ *knows* explicitly the messages it sends; it cannot learn its effective inputs, because it must send them itself. (We define "effective input" as the collection of messages sent and received by $\mathcal{I}$, generalizing the intuitive but restricted notion.) To make our mapping from α to β accurate, we must require that $\mathcal{A}$ cannot discover its effective inputs; it must know them.

To this end, we require a *translator* that maps progressive stages of $\mathcal{A}$'s view to the messages sent and received by $\mathcal{I}$ in β. This approach to specifying effective inputs from views generalizes the input-committal function introduced in 1989 by Beaver [2]. We consider stages because we must ensure that $\mathcal{A}$ knows its input before receiving a response, *ie.* that the execution of α corresponds temporally to the execution of β. To avoid notational inconvenience and save space, we now restrict our attention to OT, where in the ideal protocol $\beta = \text{ID}(OT)$, Alice merely sends one message, and Bob merely receives one message. Thus the issues of timing are simplified, and we require merely that a translator map faulty Alice's view (in α, the implementation) to the bit b that $\mathcal{I}$ sends on her behalf (likewise, the translator maps faulty Bob's α-view to the pair $(0,0)$ or $(1,b)$, a much simpler task).

Let $\mathcal{T}$ be a translator, namely a machine that (synchronously) maps v_A to the messages sent and received by $\mathcal{I}$. The ensemble $[\mathcal{A}, \alpha, \mathcal{T}]$ is induced by running α attacked by $\mathcal{A}$, and including the outputs of $\mathcal{T}$ run on the views of the participants ($\mathcal{A}$ and honest players). The ensemble $[\mathcal{A}, \beta]$ is now taken to include the conversations (transcripts of messages sent and received) by $\mathcal{A}$ and the honest players. The output by $\mathcal{T}$ should match the conversations.

Definition 2. A **transparent interface** is an interface-translator pair $(\mathcal{I}, \mathcal{T})$. Protocol α is **computationally as resilient as** β, written $\alpha \succeq \beta$, if there exists a transparent interface $(\mathcal{I}, \mathcal{T})$ such that:

(1.) $\mathcal{I}(\mathcal{A}) \in \text{ADV}_\beta$;
(2.) $\mathcal{I}(\mathcal{A})$ is probabilistic poly-time;
(3.) $\mathcal{I}$ is parsimonious;
(4.) $\mathcal{T}$ is probabilistic poly-time;
(5.) The effects of attacks on α match those for β:[1]

$$[\mathcal{A}, \alpha, \mathcal{T}] \approx [\mathcal{I}(\mathcal{A}), \beta] \text{ and } [\mathcal{A}, \alpha, \mathcal{T}]_Y \approx [\mathcal{I}(\mathcal{A}), \beta]_Y.$$

We remark specifically that the *entire view* of $\mathcal{A}$ is considered for transparency. An alternate definition based on just the messages that $\mathcal{A}$ sends and receives but not its internal bits (*cf.* "traffic" in [16]) would imply that there exists absolutely no protocol for OT — nor could there be any secure encryption, for that matter.

Definition 3. We say α **implements** β if $\alpha \succeq \beta$ and $\beta \succeq \alpha$.

Definition 4. Protocol α is a **resilient protocol** for F if it implements the ideal protocol for F. Protocol α is a **resilient OT protocol** if it implements the ideal protocol for OT.

---

[1] The requirement that $[\mathcal{A}, \alpha, \mathcal{T}]_Y \approx [\mathcal{I}(\mathcal{A}), \beta]_Y$ (*ie.* statistical indistinguishability of nonfaulty outputs) addresses an ongoing philosophical debate: should an answer be mathematically correct or just indistinguishable from a correct answer? There are pros and cons, but the important point is that this decision is independent of the rest of the definition. Compare Beaver, Micali, and Rogaway in [8].

We remark for the interested reader that these definitions support *proofs* that resilience is transitive and that sequential (black-box) compositions preserve resilience (see [1, 3]).

Let us also remark that the inclusion of a translator provides a notion of security more stringent than otherwise necessary: the theory stands on its own if the translator is omitted, as long as we continue to require that $\mathcal{I}$ be poly-time. Even if we do not require the translator, the arguments of §8 show that a proof of security — *ie.* a satisfactory interface — cannot be found. But the translator helps make it explicit that the adversary must know the effective inputs it uses, and it provides a useful tool to detect vulnerability to bluffing attacks.

8 Finding and Fixing Security Holes

In an attempt to find a transparent interface that maps attacks on protocol BOER-OT to attacks on the ideal OT protocol, two problems arose. We consider primarily the harder situation, when $\mathcal{A}$ chooses to corrupt Alice.

The first is fixable: how would $\mathcal{I}$ come up with bit b to send to the trusted host? Interface $\mathcal{I}$ could create an "environment" for an adversary $\mathcal{A}$ that manipulates Alice, by playing the role of Bob. The problem is that $\mathcal{I}$ would then obtain bit b from $\mathcal{A}$ only half the time, so it might have to "reset" $\mathcal{A}$ until $\mathcal{I}$ gets the bit b that $\mathcal{A}$ sends. Then $\mathcal{I}$ can send this bit to the trusted host, and in the ideal protocol, Bob receives it with probability $\frac{1}{2}$. Clearly, $\mathcal{I}$ is poly-time.

The second problem is inescapable. Even though $\mathcal{I}$ can send bits to the trusted host with the same probability as $\mathcal{A}$ (hence inducing a correct distribution on final outputs in the ideal scenario), it does not make $\mathcal{A}$'s attack transparent: $\mathcal{A}$ might not be able to compute the bit b it effectively sent. Provably, no polynomial-time machine $\mathcal{M}$ can determine whether $\mathcal{I}$ sends 0 or 1 to the trusted host, based on $\mathcal{A}$'s view. Assume otherwise; consider an adversary $\mathcal{A}$ that makes Alice generate x at random. Then the output of $\mathcal{M}$ must be $Q_n(x)$, which is what the interface passes on to the trusted host. Intuitively, the output of $\mathcal{M}$ is the *effective* bit that corrupt Alice sends. But this would mean that Quadratic Residuosity is not hard. Thus:

- Either the QRA is untrue and protocol BOER-OT is therefore invalid, or
- the QRA is true and the protocol is insecure because Alice's view is not translatable, *ie.* because it fails to have a transparent interface, *ie.* because it permits a "bluffing" attack in which Alice does not know b.

8.1 How to Fix den Boer's Protocol

Clearly, we must require Alice to prove that she knows $Q_n(u)$ or $Q_n(v)$. It is a fairly straightforward exercise to come up with a direct number-theoretic method for Alice to demonstrate such knowledge without revealing whether u is x or y (we may adapt [13] to these purposes, which is especially suitable because the verifier has already generated a suitable n). Note, however, that this significantly increases the round complexity and message complexity of the protocol. Fortunately, many such demonstrations can be done in parallel. Fortunately as well, it seems that the added computations seem less expensive than the alternative: generating many large Blum integers. Our main theorem becomes:

BB-OT(b, k)

1. Bob chooses $n = pq \leftarrow$ BLUM$_k$, remembers (p, q), selects $a \leftarrow Z_n^+$, and sends (n, a) to Alice. Bob proves in zero-knowledge that $n \in$ BLUM$_k$.
2. Alice selects $c \leftarrow \{0, 1\}$ and $r \leftarrow Z_n^*$ and sets $z = (-1)^c a$. Alice computes $x = (-1)^b r^2 (\bmod\ n)$ and sets $y = zx^{-1} (\bmod\ n)$. Alice sets $(u, v) = (x, y)$ if $x < y$, else $(u, v) = (y, x)$. She sends (u, v) to Bob and proves in zero-knowledge that she knows the residuosity of one of u or v.
3. Bob checks that $uv \equiv \pm a (\bmod\ n)$. Bob computes $Q_n(uv)$; if 0, he outputs $(1, Q_n(u))$ ("received $Q_n(u)$"), else he outputs $(0, 0)$ ("received nothing").

Fig. 3. Our corrections to den Boer's protocol, along with some modifications. ("BB" represents "Beaver/Boer" or "Boer/Beaver.")

Theorem 5. *There exists an efficient protocol for OT that is computationally resilient.*

Proof Sketch. Figure 3 outlines the corrected protocol. We can now find a transparent interface and prove the modified version secure. If $\mathcal{A}$ corrupts Alice, $\mathcal{I}$ simulates Bob internally and corrupts ideal-Alice. Whenever Alice fails to prove that she knows how (u, v) was constructed (*ie.* that she knows b), $\mathcal{I}$ sends quit to the trusted host, causing ideal-Bob to output ABORT, just as the "real" Bob does. If the proof succeeds, then the interface can in fact derive the bit b from Alice's view (briefly: by resetting a copy of $\mathcal{A}$ and making different challenges, thereby extracting b), and it sends b to the trusted host. Because the proof of knowledge is such that a successful proof indicates b can be efficiently extracted from $\mathcal{A}$'s view, not only $\mathcal{I}$ but $\mathcal{T}$ as well can extract $\mathcal{I}$'s message b in poly-time.

If $\mathcal{A}$ corrupts Bob, $\mathcal{I}$ corrupts ideal-Bob in ID(OT) and obtains $(0, 0)$ or $(1, b)$ from the trusted host. $\mathcal{I}$ plays the part of Alice in BB-OT. If $\mathcal{I}$ got $(0, 0)$, it plays using $b = 0$ and resets $\mathcal{A}$ until it fails to receive the bit. If $\mathcal{I}$ got $(1, b)$, it plays using b and resets $\mathcal{A}$ until it does receive a bit. Ignoring detectable cheating, these require one expected reset. If $\mathcal{A}$ tries to cheat, then with equal probability, $\mathcal{I}$ resets $\mathcal{A}$ or accepts the cheating and sends quit to the trusted host, so that honest-ideal-Alice outputs abort with the same probability as honest-real-Alice. □

Some Remarks. One must be extremely careful in formalizing the notion of "proof of knowledge." It is quite easy to come up with a notion that seems fine in isolation but which fails when protocols are executed in parallel, unless some identification scheme is available.

An interesting flaw in Rabin's protocol comes to light when one applies our definitions to parallel executions. Although provably secure individually or in sequential composition, Rabin's protocol (even with corrections) is insecure when used in parallel, unless some identification scheme is available. One serious problem is that a proof that n is the product of two primes is not exactly a proof that Alice knows those two primes. A full description of the attack and of necessary conditions to ensure the resilience of *parallel* composition exceeds our space bounds here but is forthcoming [4].

It should be clear that deriving b from Alice's *view* is possible when she gives a satisfying proof of knowledge. We emphasize that deriving b from Alice's *conversation* should never be easy, or else Bob could do it himself. In this case, one must define input-committal/transparency with respect to views.

8.2 Yet Another Protocol

We mention an OT protocol developed by the author with Nicol So, which led the author to den Boer's protocol and inspired this paper. Like den Boer's protocol, this OT protocol does not require repeated generation of large Blum integers.[2]

1. Bob chooses $n = pq \leftarrow \text{BLUM}_k$, remembers (p, q), selects $a \leftarrow \mathbf{Z}_n^+$, and sends (n, a) to Alice. Bob proves in zero-knowledge that $n \in \text{BLUM}_k$.
2. Alice chooses $r \leftarrow \mathbf{Z}_n^*$ and random bit d, computes $z = [(-1)^d a]^b r^2 \pmod{n}$, and sends (d, z) to Bob. She proves that z was computed properly and that she knows (b, r).
3. Bob concludes $(d \oplus Q_n(a), [d \oplus Q_n(a)] \cdot Q_n(z))$, meaning: if $d = Q_n(a)$, Bob received nothing; else he received $Q_n(z) = b$.

Like the corrected BOER-OT protocol, this protocol requires that Alice demonstrate that she behaved and that she knows b.

9 Conclusion

We have found and fixed a flaw in a recently published protocol for Oblivious Transfer [11]. The flaw was found by applying a new, robust definition for security and fault-tolerance, which we call *resilience*. Resilience expresses the idea that one protocol is as secure as another if the results of attacks on the first are the same as those on the second. Using our definitions, we were able to identify the flaw quickly and even to give a direct fix for it.

A correction for the flawed step introduces a significant amount of communication, but in practical and computational terms it seems less costly than

---

[2] References to any other appearances of this or similar protocols would be greatly appreciated.

repeatedly generating large products of two primes. We have recently developed a non-generic proof of knowledge optimized specifically for this OT protocol [5], performing much better than using a generic proof of knowledge to correct the protocol. Thus, we believe that the computational advantages of den Boer's elegant OT protocol can be salvaged, and we can provide provably-secure OT at low cost.

Acknowledgements. Thanks to Nicol So for many discussions of OT. Thanks to Claude Crépeau for pointing out den Boer's paper.

References

1. D. Beaver. *Security, Fault Tolerance, and Communication Complexity in Distributed Systems.* PhD Thesis, Harvard University, Cambridge, 1990.

2. D. Beaver. "Formal Definitions for Secure Distributed Protocols." *Proceedings of the DIMACS Workshop on Distributed Computing and Cryptography,* Princeton, NJ, October, 1989, J. Feigenbaum, M. Merritt (eds.).

3. D. Beaver. "Foundations of Secure Interactive Computing." *Proceedings of Crypto 1991,* 377–391.

4. D. Beaver. "The Security of Protocols Executed in Parallel." In preparation.

5. D. Beaver. "Efficient and Provably Secure Oblivious Transfer." Manuscript, 1992.

6. D. Beaver, S. Goldwasser. "Multiparty Computation with Faulty Majority." *Proceedings of the 30th FOCS,* IEEE, 1989, 468–473.

7. D. Beaver, S. Haber. "Cryptographic Protocols Provably Secure Against Dynamic Adversaries." Eurocrypt 1992.

8. D. Beaver, S. Micali, P. Rogaway. "The Round Complexity of Secure Protocols." *Proceedings of the 22st STOC,* ACM, 1990, 503–513.

9. C. Bennett, G. Brassard, C. Crépeau, M. Skubiszewska. "Practical Quantum Oblivious Transfer." *Proceedings of Crypto 1991,* 351–366.

10. M. Blum. "How to Exchange (Secret) Keys." *ACM Trans. Comput. Sys.* 1:2, May, 1983, 175–193.

11. B. den Boer. "Oblivious Transfer Protecting Secrecy." *Proc. of Eurocrypt 1991,* 31–45.

12. G. Brassard, D. Chaum, C. Crépeau. "Minimum Disclosure Proofs of Knowledge." *J. Comput. System Sci.* **37** (1988), 156–189.

13. G. Brassard, C. Crèpeau. "Non-Transitive Transfer of Confidence: A Perfect Zero-Knowledge Interactive Protocol for SAT and Beyond." *Proceedings of the 27th FOCS,* IEEE, 1986, 188–195.

14. S. Goldwasser, S. Micali, C. Rackoff. "The Knowledge Complexity of Interactive Proof Systems." *SIAM J. Comput.* 18:1 (1989), 186–208.

15. J. Kilian. "Founding Cryptography on Oblivious Transfer." *Proceedings of the 20th STOC,* ACM, 1988, 20–29.

16. S. Micali, P. Rogaway. "Secure Computation." *Proc. of Crypto 1991,* page 9.8 [sic], and incomplete preliminary version distributed at conference.

17. M. Rabin. "How to Exchange Secrets by Oblivious Transfer." TR-81, Harvard, 1981.

18. A. Yao. "Protocols for Secure Computations." *Proceedings of the 23rd FOCS,* IEEE, 1982, 160–164.

UNIFORM RESULTS IN
POLYNOMIAL–TIME SECURITY

Paul Barbaroux

L.R.I. bât.490
Université Paris XI
91405 Orsay
France

Abstract

Most security results can be established both in the non–uniform and the uniform model of computation. Nonetheless, non–uniform results are often much easier to obtain than their uniform version. In this paper we initiate a general framework in which the classical sampling technique can be applied to obtain uniform results. Our main theorem gives sufficient conditions under which a non–uniform result can be extended to a uniform one. As a consequence, we derive the uniform version of Schrift and Shamir's generalization of Yao's theorem on the universality of the next–bit test.

1 Introduction

A (perfect) pseudorandom source of bits is a probability distribution which cannot be distinguished in probabilistic polynomial time from a truly random source.

This notion was defined by Yao [8], who also showed that it is equivalent to the following simpler one: no bit of the source can be predicted from the previous ones in probabilistic polynomial time with probability significantly greater than 1/2. In other words, if all the bits "resist" to be predicted by some probabilistic polynomial time adversary, then the source can be used by probabilistic polynomial time algorithms as a truly random source of coin tosses. In short, this result states the universality of the next–bit test.

More generally, two distributions are indistinguishable if they cannot be distinguished by a probabilistic polynomial–time algorithm.

Schrift and Shamir [7] generalized Yao's result by finding a suitable version of the next–bit test (which they called the " comparative next–bit test ") which is universal for two arbitrary distributions.

In fact, their proof is established in the non–uniform model of security, where adversary algorithms are non–uniform Turing machines or sequences of boolean circuits, without any assumption on the computability of the sequence.

R.A. Rueppel (Ed.): Advances in Cryptology - EUROCRYPT '92, LNCS 658, pp. 297-306, 1993.
© Springer-Verlag Berlin Heidelberg 1993

In general, in computational security, most results have a non–uniform and a uniform version (depending on which model of security is chosen). Historically, non-uniform results were obtained first, and extending them to the uniform case is often a hard task.

The main technique for obtaining uniform results is now classical. It can be found in Levin [5]: it consists in replacing true probabilities by approximations computed from random samplings.

The purpose of this paper is twofold. We prove that Schrift and Shamir's next-bit test remains universal in the uniform model of security. More importantly, we initiate a general framework in which the sampling technique can always be applied to obtain uniform results. Our main theorem specifies some conditions which are sufficient to extend a non–uniform result into a uniform one. In fact, the universality of the next–bit test becomes a consequence of our main theorem.

The theory of one–way functions and pseudorandom generators involves different notions of resistance of some object against an adversary. Two distributions are indistinguishable if they resist to be distinguished by a probabilistic polynomial-time adversary. A one–way function is a function which resists to be inverted. A bit is hard–core on a function f if it resists to be predicted using the knowledge of $f(x)$.

Several well known results can be restated in terms of reduction of one notion of resistance to another. For example, the universality of the next–bit test can be expressed as follows: the resistance of two distributions against an arbitrary adversary reduces to their resistance against the particular one who tries to distinguish them by the next–bit prediction. Let us quote a few other:

— Goldreich–Levin [2]: "if there exists a one–way function, then there exists a function with a hard–core bit". This result has been proved in [2] in both models of security.

— Impagliazzo–Levin–Luby: "if there exists a one–way function, then there exists a pseudorandom generator". This theorem was first proved under different types of restrictive assumptions ([1],[5]), then in the general case in [4], in the non-uniform model of security. It was then proved in the uniform model by Hastad [3].

— Rompel [6]: "if there exists a one–way function, then there exists a secure signature scheme".

In section 2, we give a few notations and some basic definitions about uniform and non–uniform computation. In section 3, we formalize the very general notion of resistance by defining a *security scheme* and the *resistance* of a security scheme. Basically, a security scheme is a predicate saying that some algorithm significantly succeeds in attacking some object. Then we define the *reduction* between two security schemes (reduction of a scheme to another means that the problem of proving the resistance of the former reduces to the same problem for the latter). Finally we illustrate our definitions by restating several theorems in terms of reducibility between security schemes.

In section 4, we introduce the notion of a *virtual algorithm* which will be crucial for extending non–uniform results to uniform ones. Intuitively, a virtual algorithm can compute in a single (virtual) step the accepting probability of a probabilistic polynomial–time algorithm. Then we define approximations of virtual algorithms, where the virtual steps are replaced by approximations with polynomially small error. Our main theorem states that under some technical assumptions, if a reduction is achievable using some approximations of a virtual algorithm, then it is in fact achievable in the uniform model. We show then that Schrift and Shamir's reduction [7] can be extended to satisfy the assumptions of our main theorem. This will prove our second theorem about the uniform universality of the next–bit test.

2   Uniform vs. non–uniform algorithms

Notations
We write Σ for the set $\{0,1\}$; Σ^* denotes the set of all finite strings of bits, and Σ^N the set of all infinite strings of bits. For a string $x \in \Sigma^*$, $|x|$ denotes the length of x, x_i the i-th bit of x, and x_i^j the substring of x from bit i to bit j. The concatenation of x and y is denoted by xy. Following the usual definition, a *distribution ensemble* is a sequence $(D_n)_{n \in N}$, where D_n is a probability distribution on Σ^n. We write $x \in_D E$ when x is picked up at random in the set E according to distribution D. U_n denotes the uniform distribution on Σ^n. U denotes the usual distribution on Σ^N (infinite sequence of independent and unbiased coin tosses).

When not specified otherwise, all the algorithms we consider are supposed to *run in polynomial time*. As usual, a uniform algorithm is a Turing Machine (or any equivalent model of computation). A non uniform algorithm is a Turing machine provided with a sequence $(a_n)_{n \in N}$ of advices in Σ^* (it can also be viewed as a sequence (C_n) of boolean circuits of polynomial size). When not specified, these algorithms are supposed to be *probabilistic*. For simplicity of the notations, we represent the sequence of random coin tosses by a random input which is an *infinite* string $\omega \in_U \Sigma^N$. In the probabilistic case the algorithms run in time bounded by a polynomial function of the input length, independently from the random string ω.

Definition 1 *A distribution ensemble* (D_n) *is **uniformly samplable** if there exists a uniform algorithm* $A(n, \omega)$ *such that, for every n, D_n is the distribution generated by taking $\omega \in_U \Sigma^N$ and applying $A(n, .)$.*

In other words, for every n and every event E we have:

$$\Pr[y \in E] = \Pr[A(n, \omega) \in E], \text{ where } y \in_{D_n} E \text{ and } \omega \in_U \Sigma^N.$$

3   Security schemes

Definition 2 *A **security scheme** is a predicate* $P(\phi, n, c)$, *whose free variables are supposed to be some function ϕ, an integer n, and a positive constant c.*

In the following definition and the rest of the paper, we will often identify an algorithm with the function it computes.

Definition 3 *We say that an algorithm A breaks the security scheme P (or is a breaker of P) if there exists a constant c such that, for an infinite number of values n, $P(A, n, c)$ is true.*

Definition 4 *A security scheme P is resistant in the non–uniform (resp. uniform) model of security if no non–uniform (resp. uniform) algorithm can break P.*

Definition 5 *Given two security schemes P and Q, we say that P is reducible to Q if the resistance of Q implies the resistance of P.*

The notion of reducibility depends on the model of security (uniform or non–uniform). We will always take the same model for both P and Q (that is: the adversary algorithms A and B are of the same nature: both uniform or not).

Example 1

Let (D_n) and (D'_n) be two distribution ensembles. Let $P_1(A, n, c)$ be

" $|\Pr[A(x, \omega) = 1] - \Pr[A(y, \omega) = 1]| > \frac{1}{n^c}$
    where $x \in_{D_n} \Sigma^n$, $y \in_{D'_n} \Sigma^n$, $\omega \in_U \Sigma^N$ "

We call P_1 the *distinguishing scheme* of (D_n) and (D'_n). Resistance of P_1 means that (D_n) and (D'_n) are indistinguishable.

Example 2

Let (D_n) and (D'_n) be two distribution ensembles. Let $P_2(A, n, c)$ be

" A on input n computes some integer i such that $1 \le i \le n-1$, then acts on Σ^i and satisfies $\Pr[A(x_1^i, \omega) = x_{i+1}] - \Pr[A(y_1^i, \omega) = y_{i+1}] > \frac{1}{n^c}$, where $x \in_{D_n} \Sigma^n$, $y \in_{D'_n} \Sigma^n$, $\omega \in_U \Sigma^N$ ".

P_2 is the *next–bit distinguishing scheme* of (D_n) and (D'_n). Reducibility of P_1 to P_2 means that the next–bit test is universal. Schrift and Shamir [7] proved that indeed, in the non–uniform model, P_1 reduces to P_2. Our notion of resistance of P_1 corresponds to their notion of "passing the comparative next–bit test". Note that they took the difference of the two probabilities in absolute value, but the absolute value can obviously be dropped in the non–uniform model by considering the complementary event if necessary.

The reason for the non–uniformity of their result is that the integer i is obtained by the pigeonhole principle, and therefore not *computed* from n. Here we derive the uniform version as a corollary of our main theorem. In the uniform version, we obtain a stronger result if we drop the absolute value, since the sign, a priori, cannot be computationally decided.

Example 3

Let (f_n) be a polynomial–time computable sequence of functions : $\Sigma^n \to \Sigma^n$. Let $P_3(A, n, c)$ be

" $\Pr[A(f_n(x), \omega) \in f_n^{-1}(f_n(x))] > \frac{1}{n^c}$ where $x \in_{U_n} \Sigma^n, \omega \in_U \Sigma^N$ ".

P_3 is the *inversion scheme* of (f_n). Resistance of P_3 means that (f_n) is a one–way function.

Example 4

Let $(f_n) : \Sigma^n \to \Sigma^n$, and $(b_n) : \Sigma^n \to \Sigma$, and let $P_4(A, n, c)$ be

"$\Pr[A(f_n(x), \omega) = b_n(x)] > \frac{1}{2}(1 + \frac{1}{n^c})$ where $x \in_{U_n} \Sigma^n, \omega \in_U \Sigma^{Nn}$".

P_4 is the *prediction scheme* of the bit family (b_n) from (f_n). Resistance of P_4 means that (b_n) is hard–core on (f_n).(cf. [2])

Example 5

Let $(f_n) : \Sigma^n \to \Sigma^n$, and $P_4(A, n, c)$ be

"$\Pr[A(r, f_n(x), \omega) = r \odot x] > \frac{1}{2}(1 + \frac{1}{n^c})$ where $x, r \in_{U_n} \Sigma^n, \omega \in_U \Sigma^{Nn}$".

where $r \odot x$ denotes the boolean inner product of r and x. P_5 is the *inner product prediction scheme* of (f_n). The theorem of Goldreich and Levin [2] asserts that P_5 is reducible to P_3, in both models of security.

4 Obtaining uniform results

Looking carefully to the proof of the universality of the next–bit test in [7], one can see that the problem of achieving the uniform universality of the next–bit test comes from the non–computability (in polynomial time) of $\Pr_\omega[A(x, \omega) = 1]$, when given an algorithm $A(x, \omega)$. In fact, this is a rather general phenomenon: many non–uniform results can be trivially extended to the uniform case under the condition that we can compute such probabilities. Unfortunately, this is in general not the case since the straightforward computation of this probability takes exponential time in n.

Levin [5] showed that the exact probability computation is not really necessary. In fact, it can be replaced by a suitable approximation, such as the median of average values of random samplings. In this section, we formalize this general phenomenon and specify when this sampling technique can be applied. Then we can derive easily the uniform version of Schrift and Shamir's result.

Definition 6 *A virtual algorithm is a deterministic uniform algorithm for which the evaluation of exact probabilities is allowed and regarded as an elementary operation.*

In other words, a virtual algorithm M can use a "black box" which receives as input some subroutine $A : \Sigma^N \to \Sigma$ of M, and outputs $\Pr_\omega[A(\omega)] = 1]$.

Given a real number r, we call an ϵ-approximation of r any real number r' such that $|r - r'| \leq \epsilon$.

Definition 7 *Given a virtual algorithm $M(x)$, an integer n, and some positive constant c, an (n, c)-approximation of M is a virtual algorithm which runs like M and for which, on every input $|x|$ of length n, the black box gives, instead of its normal output, an n^{-c}-approximation.*

For all other inputs, we do not require any specific behaviour from the algorithm. We do not require either that an (n, c)-approximation run in polynomial time. (In fact, it is easy to construct a virtual algorithm M such that a suitable (n, c)-approximation of M might run forever, even on inputs of length n).

Definition 8 *A security scheme P is accessible if for every algorithm $A(x, \omega_1, \omega_2)$ which gets two random inputs ω_1 and ω_2, and for any positive constants c_1, c_2 such that $c_1 > c_2$, we have for n large enough:*

$$\Pr[P(A(.,.,\omega_2), n, c_2) \text{ is false }] \leq 2^{-n} \Rightarrow P(a, n, c_1) \text{ is true.}$$

It is easy to see that schemes P_2, P_3, P_4, P_5 in the examples of section 3 are accessible. More generally, any security scheme which says that some "suitable" expectation (depending on A, and taken over inputs x of length n and $\omega \in \Sigma^N$) is greater than $\frac{1}{n^c}$, is accessible. Note that this is not the case for P_1, since this scheme corresponds to an expectation which is taken in absolute value.

Theorem 1 *(Main) Let P and Q be two security schemes, such that Q is accessible. Assume that for every positive constant c, there exist positive constants c_1, c_2, c_3 and a virtual algorithm $M(A)$ (receiving A as a subroutine) such that for n large enough, every (n, c_2)-approximation $\tilde{M}(A)$ of $M(A)$ satisfies the following two conditions:*

1. $\tilde{M}(A)$ has running time bounded by n^{c_3} on inputs of length n

2. $P(A, n, c) \Rightarrow Q(\tilde{M}(A), n, c_1)$

Then P is uniformly reducible to Q.

We begin the proof with some lemmas which will be our tools for the sampling technique.

Lemma 1 *For every positive constant c, there exists a polynomial–time probabilistic algorithm $M(A, x, \omega')$ which, when given some algorithm $A(\omega)$ as a subroutine and $x = 1^n$ as input, computes with probability at least $\frac{7}{8}$ (on ω') an n^{-c}-approximation $\tilde{p}$ of $p = \Pr[A(\omega) = 1]$.*

Proof
The algorithm M is defined as follows:

compute $N = \lceil 2n^{2c} \rceil$
pick up independently at random $\omega_1, \cdots, \omega_N$
(the collection $< \omega_1, \cdots, \omega_N >$ will form ω')
compute $\tilde{p} = \frac{1}{N}($ number of i s.t. $A(\omega_i) = 1)$

Let $X_i(\omega_1, \cdots, \omega_N) = A(\omega_i)$. Then the X_i's are independent Bernouilli random variables with expectation $E(X_i) = p$ and variance $V(X_i) = p(1-p) \leq 1/4$. Chebyshev's inequality gives

$$\Pr[|\tilde{p} - p| \geq n^{-c}] \leq \frac{V(X_i)n^{2c}}{N} \leq \frac{n^{2c}}{4N} \leq \frac{1}{8}$$

Lemma 2 *Let* $(X_i)_{1 \leq i \leq 2S+1}$ *be independent random variables and* p, ϵ *such that* $\Pr[|X_i - p| \geq \epsilon] \leq 1/8$. *Let* Y *denote the median of the* X_i *'s. Then*

$$\Pr[|Y - p| \geq \epsilon] \leq \frac{1}{S}$$

Proof

We have $\Pr[|Y - p| \geq \epsilon] = \Pr[|X_i - p| \geq \epsilon$ for at least $S + 1$ subscripts i]
$$= \sum_{i=S+1}^{2S+1} \binom{2S+1}{i} \left(\frac{1}{8}\right)^i \left(1 - \frac{1}{8}\right)^{2S+1-i}$$
$$\leq \sum_{i=S+1}^{2S+1} \binom{2S+1}{i} \left(\frac{1}{8}\right)^{S+1}$$
$$\leq 2^{2S+1} \left(\frac{1}{8}\right)^{S+1}$$
$$= \frac{1}{2^{S+2}} \leq \frac{1}{2^S}$$

Corollary 1 *For every positive constant* c, *there exists a probabilistic polynomial-time algorithm* $M(A, x, y, \omega')$ *which, on inputs* $x = 1^n$ *and* $y = 1^S$, *computes a* n^{-c}-*approximation of* $p = \Pr_\omega[A(\omega) = 1]$ *with probability (on* ω'*) at least* $1 - 2^{-S}$

Proof
Just repeat $2S + 1$ times the algorithm of lemma 1 with independent random samplings, then take the median, and apply lemma 2.□

Proof of theorem 1
   Let A be a uniform breaker of P. Then there exists c such that, for an infinite number of values n, $P(A, n, c)$ is true. Let $c_1, c_2, c_3, M(A)$ be such that they satisfy the two conditions of the theorem.
   We define an algorithm B which will be a uniform breaker of Q: B, on input x, computes $S = n + \lceil c_3 \log_2 n \rceil$, where $n = |x|$. Then it simulates $M(A)$ while the number of computational steps of $M(A)$ does not exceed n^{c_3}. Whenever this number of steps is not sufficient to finish the simulation, B stops (and fails). Moreover, each time $M(A)$ makes a call to the black box for computing some probability p, B replaces this call by computing itself an n^{-c_2}-approximation $\tilde{p}$ of p with probability

of failure 2^{-S} (cf. Corollary 1), using independent random samplings for all these computations.

The total number of calls of $M(A)$ to the black box is at most n^{c_3}. Moreover, for each call we have $\Pr[|\tilde{p}-p| > n^{-c_2}] \leq 2^{-S}$, so with probability $\geq 1-n^{c_3}2^{-S} \geq 1-2^{-n}$, B is a (n, c_2)-approximation of $M(A)$. Whenever this happens, the forced halt does not occur before the end of the execution.

Let n be such that $P(A, n, c)$ is true, and large enough for the conditions of the theorem 1 to be satisfied. Then with probability $\geq 1 - 2^{-n}$ $Q(B, n, c_1)$ is true (using cond. 2). Finally B is a probabilistic uniform algorithm using 2 kinds of random inputs (those of A, and the random samplings, denoted by ω'), such that, for an infinite number of values of n, with probability on ω' greater than $1 - 2^{-n}$, $P(B(., \omega'), n, c_1)$ is true (cond. 2). Hence using the accessibility of Q, B is a uniform breaker of Q. $\square$

From this theorem we derive the following:

Theorem 2 (Uniform universality of the next–bit text)
Given two uniformly samplable distribution ensembles, their distinguishing scheme is uniformly reducible to their next–bit distinguishing scheme.

Proof
We have seen that the next–bit distinguishing scheme P_2 is accessible.

We will show that Schrift and Shamir's proof for the non–uniform universality yields in fact constants c_1, c_2, c_3 and a virtual algorithm $M(A)$ satisfying the assumptions of the main theorem.

The virtual algorithm M is defined as follows (there are two parts in the algorithm: first M chooses an integer i in $\{1, \cdots, n-1\}$ and then predicts x_{i+1} from x_1^i).

for $i = 1$ to n
    compute $p_i = \Pr[A(y_1^i z_{i+1}^n, \omega) = 1]$ and $p_i' = \Pr[A(\tilde{y}_1^i z_{i+1}^n, \omega) = 1]$,
        where $y \in_{D_n} \Sigma^n$, $\tilde{y} \in_{D_n'} \Sigma^n$, $\omega \in_U \Sigma^N$. The computation is possible
        using the "black box" operation allowed in virtual algorithms:
        since (D_n) and (D_n') are both uniformly samplable,
        one can generate in polynomial time D_n and D_n' from the integer n,
        using the usual distribution on Σ^N.
    compute $q_i = p_i - p_i'$
choose an i (for instance, the first) such that $q_{i+1} - q_i > \frac{2}{3}n^{-(c+1)}$.
    (if there does not exist such an i, then the algorithm fails)
        (Here ends the first part of the algorithm)
choose at random $z_{i+1}^n \in_{U_{n-i}} \Sigma^{n-i}$ and $\omega \in_U \Sigma^N$ if $q_{i+1} - q_i > 0$ then
    if $A(x_1^i z_{i+1}^n, \omega) = 1$ then output z_{i+1} else output $1 - z_{i+1}$
else
    if $A(x_1^i z_{i+1}^n, \omega) = 1$ then output $1 - z_{i+1}$ else output z_{i+1}

Let us verify that the assumptions of the theorem 1 are satisfied. The existence of c_3 is obvious, since every approximation of $M(A)$ has the same running time as $M(A)$. Let us take $c_1, c_2 > c + 1$. Then for n large enough we have

$$n^{-c_2} \le \frac{1}{12} n^{-(c+1)} \tag{1}$$

and

$$n^{-c_1} \le \frac{1}{3} n^{-(c+1)} \tag{2}$$

Let $\tilde{M}$ be a (n, c_2)–approximation of $M(A)$, that is: $\tilde{M}$ replaces p_i by $\tilde{p}_i$ s.t. $|\tilde{p}_i - p_i| \le n^{-c_2}$ (and the same for p'_i) and therefore q_i by $\tilde{q}_i$ s.t. $|\tilde{q}_i - q_i| \le 2n^{-c_2}$ and $q_{i+1} - q_i$ by $\tilde{q}_{i+1} - \tilde{q}_i$ s.t.

$$|(\tilde{q}_{i+1} - \tilde{q}_i) - (q_{i+1} - q_i)| \le 4n^{-c_2} \tag{3}$$

Suppose now $P_1(A, n, c)$ is true, that is:

$$|\Pr[A(y, \omega) = 1] - \Pr[A(\tilde{y}, \omega) = 1]| > n^{-c}$$

where $y \in_{D_n} \Sigma^n$, $\tilde{y} \in_{D'_n} \Sigma^n$, $\omega \in_U \Sigma^N$. Then by the pigeonhole principle (cf. [7]) there exist an i s.t. $|q_{i+1} - q_i| > n^{-(c+1)}$, that is, using (1),

$$|\tilde{q}_{i+1} - \tilde{q}_i| \ge |q_{i+1} - q_i| - 4n^{-c_2} > n^{-c+1} - 4n^{-c_2} \ge \frac{2}{3} n^{-(c+1)} \tag{4}$$

so $\tilde{M}$ does not fail and finds such an i. Then for this i we have

$$\frac{2}{3} n^{-(c+1)} < |\tilde{q}_{i+1} - \tilde{q}_i| < |q_{i+1} - q_i| + 4n^{-c_2} < |q_{i+1} - q_i| + \frac{1}{3} n^{-(c+1)}$$

so

$$\frac{1}{3} n^{-(c+1)} < |q_{i+1} - q_i| \tag{5}$$

Now we have to prove that $\tilde{M}$ can decide the sign of $q_{i+1} - q_i$ by looking at the sign of $\tilde{q}_{i+1} - \tilde{q}_i$. But if these two quantities were of opposite signs, this would imply, using (4) and (5):

$$|(\tilde{q}_{i+1} - \tilde{q}_i) - (q_{i+1} - q_i)| > n^{-(c+1)}$$

and therefore, using (3):

$$4n^{-c_2} > n^{-(c+1)}$$

which contradicts the choice of c_2.

Now the proof ends as in [7]: it is easy to see that

$$\Pr[\tilde{M}(y_1^i, \omega') = y_{i+1}] = \frac{1}{2} + \epsilon(p_{i+1} - p_i)$$

and

$$\Pr[\tilde{M}(\tilde{y}_1^i, \omega') = \tilde{y}_{i+1}] = \frac{1}{2} + \epsilon(p'_{i+1} - p'_i),$$

where ϵ denotes the sign of $q_{i+1} - q_i$. Therefore $\Pr[\tilde{M}(y_1^i, \omega') = y_{i+1}] - \Pr[\tilde{M}(\bar{y}_1^i, \omega') = \tilde{y}_{i+1}]$ is positive, and equals $|q_{i+1} - q_i| > \frac{1}{3}n^{-(c+1)} \geq n^{-c_1}$. Hence $P_2(\tilde{M}, n, c_1)$ is true.

Bibliography

[1] O. Goldreich, H. Krawcyzk, M. Luby, "On the Existence of Pseudorandom Generators,", Proc. FOCS 1988, pp. 12–24.

[2] O. Goldreich, L.A Levin, "A Hard–Core Predicate For All One–Way Function", Proc. STOC 1989, pp. 25–32.

[3] J. Hastad, "Pseudorandom Generators Under Uniform Assumptions", Proc. STOC 1990, pp. 395–404.

[4] R. Impagliazzo, L.A. Levin, M. Luby, "Pseudo–Random Generation from One–Way Functions", Proc. STOC 1989, pp. 12–24.

[5] L.A. Levin, "One–Way Functions and Pseudorandom Generators", Combinatorica 7(4), pp. 357–363, 1987.

[6] J. Rompel, "One–Way Functions are Necessary and sufficient for Secure Signatures", Proc. STOC 1990, pp. 387–394.

[7] A.W. Schrift, A.Shamir, "On the Universality of the Next Bit Test", Proc. CRYPTO 1990, pp. 394–408.

[8] A.C. Yao, "Theory and Applications of Trapdoor Functions", Proc. FOCS 1982, pp. 80–91.

Cryptographic Protocols Provably Secure Against Dynamic Adversaries

Donald Beaver[1] and Stuart Haber[2]

[1] 313 Whitmore Lab, Penn State University, University Park, PA 16802, U.S.A.
beaver@cs.psu.edu
[2] Bellcore, 445 South St., Morristown, NJ 07960-1910, U.S.A.
stuart@bellcore.com

Abstract. We introduce new techniques for generating and reasoning about protocols. These techniques are based on protocol transformations that depend on the nature of the adversaries under consideration. We propose a set of definitions that captures and unifies the intuitive notions of correctness, privacy, and robustness, and enables us to give concise and modular proofs that our protocols possess these desirable properties.

Using these techniques, whose major purpose is to greatly simplify the design and verification of cryptographic protocols, we show how to construct a multiparty cryptographic protocol to compute any given feasible function of the parties' inputs. We prove that our protocol is secure against the malicious actions of any adversary, limited to feasible computation, but with the power to eavesdrop on all messages and to corrupt any *dynamically chosen* minority of the parties. This is the first proof of security against dynamic adversaries in the "cryptographic" model of multiparty protocols. We assume the existence of a one-way function and allow the participants to erase small portions of memory. Our result combines the superior resilience of the cryptographic setting of [GMW87] with the stronger (dynamic) fault pattern of the "non-cryptographic" setting of [BGW88,CCD88].

1 Introduction

A large body of recent work in distributed computing has addressed the problem of constructing protocols whereby n parties can cooperatively compute a function of n arguments, each secretly held by one party. Much of this work can be classified either as the "cryptographic" approach or as the "non-cryptographic" approach. In the non-cryptographic scenario, one posits the existence of a complete network of private communications channels connecting all pairs of users, something that may not exist in practice. In this optimistic situation, [10, 12, 28, 3] have shown that one can tolerate misbehavior by any dynamically chosen colluding minority of faulty players. In the cryptographic scenario, on the other hand, we rely on unproven assumptions about the complexity of certain computational problems instead of on "physical" assumptions about the network. Here again, [23, 24] have shown that one can tolerate faulty behavior by a minority of the players, but the proof techniques in the literature seem to require that the

R.A. Rueppel (Ed.): Advances in Cryptology - EUROCRYPT '92, LNCS 658, pp. 307-323, 1993.

subset of misbehaving parties be chosen "statically," that is at the beginning of the protocol execution.

Our contribution. We introduce a new technique for enhancing the security of multiparty protocols. Beginning with an "ideal protocol" for a computational problem, we proceed by a sequence of reductions in order to generate a maximally secure protocol. One result we can achieve by this technique is to combine the virtues of the best known constructions for both the *cryptographic* and the *non-cryptographic* scenarios, so that we are able to achieve all desired properties of a multiparty protocol for the evaluation of any feasible probabilistic function on private inputs: our protocol does not require private channels, it tolerates a fully dynamic, fully Byzantine adversary, and it is very simple to describe. With novel, concise, and penetrating definitions, our techniques support direct, modular proofs that provide a deep understanding not only of the bounds of cryptographic protocols but of the nature of security itself.

Defining security; desirable properties; proving security. In the past few years many different multiparty protocols have been proposed [23, 24, 17, 10, 12, 2, 28, 3, 7, 16] for various purposes, and authors have given separate arguments for the correctness, privacy, resilience, independence, fairness, robustness, simultaneity, etc. of the computations performed by these protocols. It can be very difficult to compare different attempts to define the same property, let alone to compare the (more or less formal) definitions of these different properties. Continuing the work of [3, 5, 6], our new definitions enable us to write surprisingly simple and clear proofs of the security of many of the protocol transformations in the literature—including, of course, the new ones we introduce in this abstract. Following Chor and Rabin [13], we argue that "in the slippery business of distributed cryptographic protocols, simpler proofs are important."

The introduction of zero-knowledge proofs [21] made two important contributions to the field of cryptography: first, the new idea of proving an assertion to be true without giving away any additional information; and second, the technical insight that feasible simulatability is a good way to talk about and reason about "not giving away information." The idea of zero-knowledge proofs inspired much fruitful research, resulting in many new procedures. In trying to state the properties of these procedures precisely, researchers have had trouble. Distracted, as it were, by the technical difficulties of simulation, they have been drawn away from the main point, which is to achieve, in the absence of trusted parties, various goals that would be easy to achieve in a world where trusted hosts of various sorts were available.

Here, we get back to the point. We describe a way to formally compare two different protocols, and then we compare any proposed protocol to the ideal protocol—usually ridiculously easy to design—that one can write for an ideal world that (by *fiat*) includes trusted hosts. Zero-knowledge uses Yao's notion of indistinguishability of ensembles [31] in reasoning about feasible simulation. Our definitions and proof techniques use indistinguishability to reason not just about the *information* an adversary can glean from a protocol execution but also about the *influence* the adversary can have on the course of an execution.

Static *vs.* dynamic adversaries. The difference between tolerating static and tolerating dynamic adversaries is crucial to distributed system security. It is inconceivable that in exactly the scenario where all the transmitted information is exposed to the adversary (even though in an encrypted form), the adversary is not allowed to take advantage of this in its corruption policy. The major difficulty in the case of cryptographic protocols is that, having access to on-line communication, a dynamic adversary's behavior can depend on this communication in an unpredictable manner. Standard simulation arguments fail, preventing even a proof of *privacy* against a dynamic adversary from having appeared in the literature (despite claims of dynamic security (*cf.* [24, 16])).

It has recently come to the authors' attention that cryptographic techniques similar to ours have been used previously to address the question of simulating private-channels protocols in the cryptographic model [15]. However, this work was not primarily concerned with the question of properly defining the desired security, and the proof sketches are stated in terms of rather incomplete definitions.

With clear and simple definitions and proof techniques, we are able to describe a simple protocol transformation demonstrating that if processors can erase very small portions of memory, then *any result developed for the non-cryptographic model holds also for the cryptographic model, even in the presence of dynamic adversaries.*

Efficiency. Our protocol transformation is remarkably efficient in terms of overhead. The simplifying but unrealistic assumption that private channels exist provides efficient protocols for the non-cryptographic model; our transformation preserves the simplicity of non-cryptographic protocols while deftly handling the difficulties encountered in proof techniques for the cryptographic scenario. At little additional cost we supply the guarantee of *provable security* with the ease of simple protocol design.

Remarks. From a practical standpoint, our assumption that processors may erase memory is far less unreasonable than absolutely private channels, given physical circuits with volatile memory that is destroyed upon tampering. Our transformations require each processor to erase a very small portion of memory, on the order of the size of the keys needed to encrypt or decrypt messages.

Our purpose in this abstract is, first, to demonstrate the power of our new technique to generate and reason about protocols; second, to apply the technique in order to show that it is *possible* to achieve security against a fully dynamic adversary in the cryptographic scenario; and, third, to *prove* that we have achieved this level of security.

Contents. Section 2.2 defines relative resilience, our main new technical tool; §2.3 proves some supporting theorems; §3 describes modular protocol transformations and proves that they produce a secure protocol. Details of our proofs are contained in the appendix.

2 Definitions and proof techniques

2.1 Background mechanics

Background. Let $\Sigma = \{0,1\}$ with symbols such as delimiters (#) encoded in a natural way. Let $[n] = \{1, \ldots, n\}$ and let $\mathbf{x} = (x_1, \ldots, x_n)$. Let $\text{dist}(X)$ be the set of distributions on a set X (normally finite), and let $uniform(X)$ be the uniform distribution. Let PPTM be the set of probabilistic polynomial-time TM's. Let PFF be the set of functions mapping $(\Sigma^m)^n \rightarrow \text{dist}((\Sigma^m)^n)$ that are described by poly-size circuit families $\{C_F(n, m)\}$ where each circuit has a certain number of distinguished, "random" inputs.

The difference $|P - Q|$ between distributions $P, Q \in \text{dist}(X)$ is $\sum_X |\Pr_P[x] - \Pr_Q[x]|$. A **probabilistic function** is a function whose range contains distributions, namely $f : X \rightarrow \text{dist}(Y)$. The **composition** of probabilistic functions g and f is given by $\Pr_{g \circ f(x)}[z] = \sum_y \Pr_{f(x)}[y] \Pr_{g(y)}[z]$. An **ensemble** is a probabilistic function $P : \Sigma^* \times \mathbf{N} \rightarrow \text{dist}(\Sigma^*)$ such that $\Pr_{P(z,k)}[x] = 0$ for $|x| > k^c$ and some fixed c. Two ensembles P and Q are computationally indistinguishable ($P \approx Q$) if

$$(\forall T \in \text{PPTM})(\forall c > 0)(\exists k_0)(\forall k \geq k_0)(\forall z) \quad |T(P(z, k)) - T(Q(z, k))| < k^{-c}$$

where $T(D)$ indicates the probability T outputs 1 on input distribution D. Two *families* of ensembles $\{P_i\}$ and $\{Q_i\}$ are **uniformly**[3] **computationally indistinguishable** if

$$(\forall T \in \text{PPTM})(\forall c > 0)(\exists k_0)(\forall k \geq k_0)(\forall z) \quad (\forall i) \; |T(P_i(z, k)) - T(Q_i(z, k))| < k^{-c}.$$

Cryptographic tools. When we take a protocol designed for a network with private channels and try to implement it without such channels, the only method we have available to send private messages is to use cryptographic means. Therefore, in order to prove the security of our protocols, we shall make the complexity-theoretic assumption that **one-way trapdoor permutations** exist. Let **pkc** denote a key generator for a polynomially secure **probabilistic public-key encryption scheme** [20]; this is a probabilistic algorithm that, on input 1^K and $\kappa(K)$ random bits, produces an (encryption key, decryption key) pair that we write (E, D). For an encryption key E, we write $c = E(m, \rho)$ for an encryption of message m using random bits ρ. Let **prg** denote a **cryptographically strong pseudorandom number generator**, an algorithm mapping K bits to $B(K)$ bits [27, 26]. We will also make use of **zero-knowledge proofs of knowledge**; in our application, these will be proofs of knowledge of m performed by the sender of a ciphertext $c = E(m, \rho)$ [14, 30, 19, 9].

Networks and protocols. A **player** is an interactive PPTM having a random tape, input tape, output tape, and work tape. The I/O tapes may encode several (n) different tapes for communication with several machines. The superstate S_i of player M_i is a string describing its finite control, current state, and contents of

---

[3] "Uniform" refers to uniform convergence, not TM-computability.

all tape squares read or written so far. With messages $\mu(1, i), \ldots, \mu(n, i)$ written on its input tape, M_i induces a transition $\delta(S_i, \mu(1, i) \cdots \mu(n, i)) = (S_i', \boldsymbol{\mu})$, where $\boldsymbol{\mu}$ is a list of outgoing messages to be sent on specific communication channels.

A **channel** is a probabilistic function $C : \Sigma^* \to \text{dist}(2^{\text{N} \times \text{N} \times \Sigma^*})$. For example, a **private channel** from i to j satisfies $\text{Pr}_{C(m)}[\{(i, j, m)\}] = 1$, while a **broadcast channel** from i is $\text{Pr}_{C(m)}[\{(i, 1, m), \ldots, (i, n, m)\}] = 1$. Other channels, such as oblivious transfer (noisy) channels, are equally easily described. A **network** is a set of channel functions.

A (synchronous) **protocol** is a collection of sets of players $\Pi = \{(M_1, \ldots, M_n)\}_{n \in \text{N}}$. It is implementable on a network if the messages output by players specify channels in that network. For each n (number of players), m (size of inputs), k (protocol security parameter), $\mathbf{x} \in (\Sigma^m)^n$ (inputs), and $\mathbf{a} \in (\Sigma^*)^n$ (auxiliary inputs), a protocol Π induces a distribution $\Pi(n, m, k, \mathbf{x} \circ \mathbf{a})$ on outputs by running player M_i on input tape $X_i = 1^n \# 1^m \# 1^k \# x_i \# a_i$. Letting $z = 1^n \# 1^m \# 1^k \# \mathbf{x} \# \mathbf{a}$, this defines an ensemble. The mechanics of protocol execution are straightforward but tedious (see *e.g.* [3, 5]): a sketch with loose notation follows. The notation $\mu(A, B, r)$ denotes the messages sent from players in set A to those in B during round r. Let $R(n, m, k)$ be the number of rounds taken by a protocol (alternatively, the protocol can run until all nonfaulty players decide to halt).

Synchronous Protocol Execution
For $i = 1..n$ do
    $\mu^{in}(i, i, 0) = X_i$
For $r = 1..R(n, m, k)$ do
    For $i = 1..n$ do **in parallel**
        /* Compute locally (function Local): */
            $(S_i^r, \mu^{out}(i, [n], r)) \leftarrow \delta(S_i^{r-1}, \mu^{in}([n], i, r-1))$
        /* Apply channel functions (function Channel): */
            $\mu^{del}([n], [n], r) \leftarrow C(\mu^{out}([n], [n], r))$
        /* Deliver messages output from channels */
                /* (function Deliver): */
        $\mu^{in}([n], [n], r+1) \leftarrow \mu^{del}([n], [n], r)$

The main point is to consider the induced probabilistic functions. Let $S_g = [(S_1, \mu([n], 1)), \ldots, (S_n, \mu([n], n))]$ represent a global state of the system. A round and an execution of a synchronous protocol are described by

$$\text{Round}(S_g) = \text{Deliver}(\text{Channel}(\text{Local}(S_g)))$$
$$\text{Exec}(S_g) = \text{Round}^{R(n, m, k)}(S_g).$$

Let $Y_i : \Sigma^* \to \Sigma^*$ be an **output function**; the output of M_i is $y_i = Y_i(S_i^{R(n, m, k)})$. The **view** v_i of M_i is the list of states and messages it has seen. In the **memoryless model**, however, the view consists only of the final state and output; intermediate states and tape contents are *not* recorded, and information can thus be *erased* by a player. The view of a player during an intermediate round of the protocol includes all its states and messages up to that point; in the memoryless model, an intermediate view contains only the current state, tape contents, and incoming and outgoing messages.

Adversaries. An **adversary** $\mathcal{A}$ is an interactive TM with *one* communication line, on which it makes *corruption requests*. In this paper, all adversaries are probabilistic polynomial-time TM's. A **Byzantine** adversary requests either the view of a player i or requests to replace outgoing messages from corrupted players. Note that an adversary may change input and random tapes before the protocol starts (*i.e.* before any messages are generated). A **passive** adversary cannot change messages. If the adversary superstate is S_A, let $T = T(S_A)$ denote the set of players it has corrupted. A t-**adversary** satisfies $|T(S_A)| \leq t$. A **static** adversary satisfies $T(S_A) = T_0$ for some fixed T_0. A **rushing, dynamic** adversary sees $\mu^{del}(\overline{T}, T, r)$ (step 2.2.2) before choosing whom to corrupt and how to corrupt them. We remark that our adversary model is quite robust; we allow players to be interrupted and corrupted even in the middle of a computation. The function **Fault** provides $\mathcal{A}$ with requested information and allows it to compute a new request. The function **Replace** allows $\mathcal{A}$ to change outgoing messages $\mu^{out}(T, [n], r)$, which are then passed through channels, changing some of the messages in $\mu^{del}(T, [n], r)$ in step 2.2.2. An execution of a protocol with adversary is:

$$\mathbf{RoundA}(S_g, S_A) = \mathbf{Deliver}(\mathbf{Replace}(\mathbf{Fault}^t(\mathbf{Channel}(\mathbf{Local}(S_g, S_A)))))$$

$$\mathbf{ExecA}(S_g, S_A) = \mathbf{RoundA}^{R(n,m,k)}(S_g, S_A).$$

Let a_A be an auxiliary input for $\mathcal{A}$ and let y_A denote its output $y_a = Y_A(S_A)$. An execution thus maps $\mathbf{x} \cdot a \cdot a_A$ to $\mathbf{y} \cdot \mathbf{v} \cdot y_A$, where $\mathbf{v}$ is the list of player views (including final states) and $\mathbf{y} = (Y_1(S_1), \ldots, Y_n(S_n))$ is the list of outputs. We define the induced ensemble that describe adversary and player outputs as

$$[\mathcal{A}, \Pi]' = (Y_1(S_1), \ldots, Y_n(S_n), Y_A(S_A)),$$

and we use a subscript A (resp. $1..n$) to refer to the restriction to $Y_A(S_A)$ (resp. $(Y_1(S_1), \ldots, Y_n(S_n))$).

2.2 Security

We come to the important definitions: *relative resilience* and *absolute resilience*.

The principle behind zero-knowledge [21] is that a simulator that produces accurate verifier (adversary) views, given only information "$x \in L$," shows that a real execution leaks no additional information. This idea covers only half the picture of interaction, however. In addition to *information*, there is the *influence* an adversary has on the outputs of nonfaulty players. (This is overlooked by ZK because a faulty verifier doesn't "influence" the final output of the prover, which is irrelevant in ZK proof systems—only the verifier's decision is considered.) In a protocol, the influence is reflected in the distributions on final outputs of nonfaulty players. Previous approaches to security attempted to deal with desired properties (*e.g.* correctness, independence) separately; we unify them all by choosing to consider the ensemble $[\mathcal{A}, \Pi]$.

We present a tool called **relative** resilience, introduced by the first author in [3, 5], that allows us to compare the security of one protocol to another. Resilience is a combination of several properties, including privacy and correctness. In order to say that α is as resilient as β, we should like that an adversary who attacks α is also allowed to attack β. If its *information* and *influence* on α are the same as on β, then α is intuitively as secure—or *resilient*—as β. But $\mathcal{A}$ may be incompatible with β for many reasons: the network may differ from protocol α, the communication format may differ, and so on. We give $\mathcal{A}$ an *interface* $\mathcal{I}$ to allow it to attack β. Because the interface should not give $\mathcal{A}$ undeserved extra power, the combination $\mathcal{I}(\mathcal{A})$ (interface together with adversary, regarded as a single machine) should not exceed the power allowed to adversaries attacking β. For example, when adversaries are poly-time bounded, the interface itself must be poly-time as well.

Definition. *An* **interface** *is an interactive TM $\mathcal{I}$ with two tapes, an "environment simulation" tape and an "adversarial" tape. On the former, $\mathcal{I}$ receives and responds to messages from an adversary; on the latter, $\mathcal{I}$ passes on requests and receives responses as an adversary in its own right. An* **interface from α to β** *is such that for all $\mathcal{A} \in A_\alpha$ we have $\mathcal{I}(\mathcal{A}) \in A_\beta$, where A_α (resp., A_β) is the class of allowed adversaries for α (resp., β).*

The interface is the proper generalization of a simulator to an interactive setting where all properties (not just privacy) are important. Unless otherwise specified, we shall be concerned with t-bounded polynomial-time message-rushing, dynamic, Byzantine adversaries.

A preliminary definition for resilience captures the essential intuition: protocol α is *weakly as resilient as* protocol β if there exists an interface satisfying $[\mathcal{A}, \alpha]' \approx [\mathcal{I}(\mathcal{A}), \beta]'$. While this would suffice for a single protocol execution, we require an additional property in order to show that concatenating protocols preserves resilience.

Post-Protocol Corruption. After a protocol is finished, the views become auxiliary inputs to later protocols. Therefore, an interface for an early protocol α should be able to generate views of players that had not been corrupted *during* α, so that an interface for a later protocol can include these views in the auxiliary information captured by the later adversary. An interface satisfying our weak definition might not even be well-defined for corruption requests occurring *after* the protocol is finished. We therefore consider an interface $\mathcal{I}$ designed to respond to post-protocol requests. The adversary is allowed to send request ppc, in which case it receives *all* outputs $(y_1, \ldots, y_n)$, but *not* the views. The interface receives nothing. Adversary $\mathcal{A}$ can then request new corruptions j (up to its t-limit) and $\mathcal{I}$ must respond with outputs and newly-synthesized views. $\mathcal{I}$ is allowed to continue requesting corruptions in completed protocol β (up to its t-limit), obtaining accurate output y_j, input x_j, and view v_j^β of β—from which it must synthesize a convincing view v_j^α of α. Thus, $\mathcal{A}$ is able to test—in a strong fashion—the ability of the interface to answer continual requests, even after the given protocols are finished.

The variable Y_A^{ppc} refers to $\mathcal{A}$'s output when it is allowed to elect post-protocol corruption. The induced protocol ensemble is denoted:

$$[\mathcal{A}, \Pi] = (Y_1(S_1), \ldots, Y_n(S_n), Y_A^{ppc}(S_A)).$$

We can now concisely define relative resilience:

Definition. (Relative Resilience) *Protocol α is as **resilient** as protocol β, written $\alpha \succeq \beta$, if there exists an interface $\mathcal{I}$ from α to β such that for all $\mathcal{A} \in \mathbf{A}_\alpha$,*

$$[\mathcal{A}, \alpha] \approx [\mathcal{I}(\mathcal{A}), \beta].$$

Protocol α is **as private as** β if $[\mathcal{A}, \alpha]_A \approx [\mathcal{I}(\mathcal{A}), \beta]_A$ and it is **as correct as** β if $[\mathcal{A}, \alpha]_{1..n} \approx [\mathcal{I}(\mathcal{A}), \beta]_{1..n}$.

Absolute Resilience. The measure of absolute resilience is given by the standard of an *ideal* protocol. A **real** protocol has a t-adversary class and n players; no player is above corruption. An **ideal** protocol contains one or more *trusted hosts* (players $(n+1), (n+2), \ldots$) who cannot be corrupted. The **ideal t-adversary class** includes all t-bounded poly-time adversaries that corrupt players in the range $1..n$ (excluding trusted hosts $n+1$, $n+2$, *etc.*). If $F \in \mathrm{PFF}$ is a probabilistic finite function, the **ideal protocol** for F, $\mathrm{ID}(F)$, has two rounds: all players send their inputs to trusted host $(n+1)$, who then computes F and returns the values.

Definition. (Resilience) Π *is a t-resilient protocol for F if $\Pi \succeq \mathrm{ID}(F)$.*

Computational issues. Although our definitions are presented with resource-bounded adversaries in mind, our approach works equally well when the notions of *statistical* or *perfect* indistinguishability is employed, and adversary classes include unbounded TM's (or even nonrecursive functions).

Comparison to other definitions. Several authors [18, 8, 4] have considered a *fault-oracle* approach, an extension of ZK in which a simulator constructs an adversary's view based on a single request for a computation of F, which thereby induces player outputs as well. Though the fault-oracle approach can be regarded as a step in the right direction, it is limited to comparing a protocol to a function computation, and it is rather inflexible. It does not support modular proofs: it is not clear how the concatenation of two secure protocols can be proven secure, since one must convert two oracle calls to a single one. Other approaches [7, 25] also suffer from the limitation that arbitrary, different protocols cannot easily be compared.

Zero-Knowledge at One Blow. In the ideal zero-knowledge proof system, denoted $\mathrm{ID}(L)$, player P sends "$x \in L$" to trusted host TH, who calculates whether $x \in L$ and sends "$x \in L$" to V if so. The host otherwise sends "?" to indicate a failed proof. One of the more attractive uses of *relative resilience* is an equivalent definition of the classical notion of zero-knowledge proof system [21] in one sentence: A two-party protocol $\Pi = \langle P, V \rangle$ is a **zero-knowledge proof system** for L iff it is as resilient as $\mathrm{ID}(L)$, against static 1-adversaries.

2.3 Proof techniques and now-provable folk theorems

Many intuitively justified approaches become provable using relative resilience. Pitfalls in the naive application of intuition are brought out by our formal definitions. The contribution of this section is a formal statement of provable theorems and the necessary conditions under which they hold. Concise proofs of the non-cryptographic versions of these theorems appear in the dissertation of the first author [5]. Full proofs for the cryptographic setting are short and direct but would require more space than permitted here.

The concatenation $\alpha_2 \circ \alpha_1$ is defined by setting the inputs and auxiliary inputs $\mathbf{x}(2) \cdot \mathbf{a}(2) \cdot a_A(2)$ to protocol α_2 to be the outputs and views $\mathbf{y}(1) \cdot \mathbf{v}(1) \cdot y_A(1)$ generated by running α_1 on the original inputs $\mathbf{x}(1) \cdot \mathbf{a}(1) \cdot a_A(1) = \mathbf{x} \cdot \mathbf{a} \cdot a_A$. We say that protocol family $\{\alpha_i\}$ is **uniformly as t-resilient as** family $\{\beta_i\}$ if there exists a family $\{\mathcal{I}_i\}$ of interfaces so that for all adversaries $\mathcal{A} \in \mathbf{A}_{\alpha}$, the ensemble families $\{[\mathcal{A}, \alpha_i]\}$ and $\{[\mathcal{I}_i(\mathcal{A}), \beta_i]\}$ are uniformly indistinguishable. Recall that the term "uniform" denotes not the Turing-machine notion but the mathematical notion of uniform convergence. If the requirement of uniform convergence is relaxed, counterexamples to the results stated below are easy to find [3], demonstrating unforeseen pitfalls in the blind use of folk-theorems.

Define the **composition** of $f(n, m, k)$-many ensembles from $\{P_i\}$ as $P^f(z, k) = P_{f(n,m,k)}(P_{f(n,m,k)-1}(\cdots P_1(z, k) \cdots). k)$. The **concatenation** of $f(n, m, k)$-many protocols from $\{\alpha_i\}$ is the protocol α^f such that $\alpha^f(n, m, k, \mathbf{x} \cdot \mathbf{a} \cdot a_A) = \alpha_{f(n,m,k)} \circ \cdots \circ \alpha_1(n, m, k, \mathbf{x} \cdot \mathbf{a} \cdot a_A)$.

Lemma 1. (Composing poly-many ensembles) *If $\{P_i\}$ is uniformly computationally indistinguishable from $\{Q_i\}$ and if $f(n, m. k)$ is polynomially bounded, then $P^f \approx Q^f$.*

Proof. A proof for perfect and statistical indistinguishability appears in [3, 5]; the proof for computational indistinguishability is direct and follows the same lines. □

Theorem 2. (Concatenating poly-many protocols) *If $\{\alpha_i\}$ is uniformly as t-resilient as $\{\beta_i\}$ and if $f(n, m, k)$ is polynomially bounded, then $\alpha^f \succeq \beta^f$.*

Proof. The proof uses Lemma 1 and follows a proof for statistical resilience in [3]. □

Theorem 3. (Transitivity of $\succeq$) *$\alpha \succeq \beta$ and $\beta \succeq \gamma$ imply $\alpha \succeq \gamma$. Polynomially many applications work as well: If $\{\alpha_{i+1}\}$ is uniformly as t-resilient as $\{\alpha_i\}$,[4] and if $f(n, m, k)$ is polynomially bounded. then $\alpha_f \succeq \alpha_0$.*

Proof. See [3, 5]; the interface $\mathcal{I}$ from α_f to α_0 internally runs $f(n, m, k)$ nested interfaces $\mathcal{I}_{f,f-1} \circ \cdots \circ \mathcal{I}_{2,1} \circ \mathcal{I}_{1,0}$. □

---

[4] Roughly speaking, $(\forall i)\alpha_{i+1} \succeq \alpha_i$.

The **ideal vacuous protocol**, ID(0), returns no result. A t-**threshold scheme** is a pair of protocols (SHA, REC) computing probabilistic functions (sha, rec) such that

1. **rec** is t-robust (*i.e.*, insensitive to $\leq t$ changes in inputs);
2. **sha** is t-private (*i.e.*, ID(sha) $\succeq$ ID(0));
3. **rec** $\circ$ **sha**$(x) = x$.

Define **hide**(H) = **sha** $\circ$ H $\circ$ **rec**, the probabilistic function that reconstructs a shared value, computes H, and shares it again. The standard share/compute/reveal paradigm for multiparty protocols is to express F as $F^f = \circ_1^f F_i$ and compute **rec** $\circ$ $[\circ_1^f \text{hide}(F_i)]$ $\circ$ **sha**. That is, inputs are secretly shared; intermediate values are secretly computed but not revealed; then the final output is reconstructed. The "folk theorem" claiming that this method is secure, but whose intuitive statement is sometimes false, is formalized and proven as:

Theorem 4. ("Completeness" paradigm) *Let* (SHA, REC) *be a t-threshold scheme, and let* $F = F^f = \circ_1^f F_i$ *for some polynomially bounded* $f(n, m, k)$. *If* $\{\alpha_i\}$ *is uniformly as t-resilient as* $\{\text{ID}(\text{hide}(F_i))\}$, *then*

$$\text{REC} \circ [\circ_1^f \alpha_i] \circ \text{SHA} \succeq \text{ID}(F^f).$$

Proof. *Uniformity* is essential. The proof uses Theorems 2 and 3 and the robustness and privacy of (rec, sha) but is omitted (*cf.* [3, 5], however). □

3 Achieving security against dynamic adversaries

In this section we describe our construction of a cryptographic multiparty protocol that is t-resilient against *dynamic* adversaries for $t < n/2$, for any $F \in \text{PFF}$. Recall that n is the number of players and m is the size of the inputs. Let $K = k + n + m$. A **feasible** protocol has polynomial-size messages: assume that each message from i to j has fixed length $L(K)$, for a total of $B(K) = O(K^c)$ bits over $B(K)/L(K)$ rounds, for some c. If p_{ij} is a string of $B(K)$ bits, let $p_{ij}(r)$ denote the rth block of $L(K)$ bits. Let **pkc** be a public-key generator requiring $\kappa(K)$ random bits, and let **prg** be a cryptographically strong pseudorandom generator from K to $B(K)$ bits. Figure 1 describes the transformation and subprotocols used in the transformation and proof of security for our main results.

Theorem 5. (Main Result I) *If one-way trapdoor functions exist, then in the memoryless model, private channels can be replaced by public (broadcast) channels at a cost of 2 extra rounds: for all feasible Π,* TRANSFORM$(\Pi) \succeq \Pi$.

TRANSFORM(Π) = PAD(Π) ∘ SENDSEED.

PAD(Π)   /* Use pads, not channels */
- Input for player i is $X_i = (x_i, p_{i1}, \ldots, p_{in}, p_{1i}, \ldots, p_{ni})$
- Run Π on x_i, except replace steps "(r) $i \to j : m_{ij}[\to m'_{ij}]$" by:
  - (r) $c_{ij} \leftarrow p_{ij}(r) \oplus m_{ij}$
  - (r) i broadcasts: $c_{ij}[\to c'_{ij}]$
  - $(r+1)$ receive c'_{ji}; $m'_{ji} \leftarrow p_{ji}(r) \oplus c'_{ji}$.

SENDPAD   /* Send random pad over private channel */
1    $p_{ij} \leftarrow uniform(\{0,1\}^{B(K)})$
     $i \to j : p_{ij}[\to p'_{ij}]$
2   receive p'_{ji}; output $(x_i, p_{i1}, \ldots, p_{in}, p'_{1i}, \ldots, p'_{ni})$

SENDPSPAD   /* Send pseudorandom pad over private channel */
1    $s_{ij} \leftarrow uniform(\{0,1\}^K)$; $p_{ij} \leftarrow \mathrm{prg}(s_{ij})$
     $i \to j : s_{ij}[\to s'_{ij}]$
2   receive s'_{ji}; if not K bits then $s'_{ji} \leftarrow 0^K$
     $p'_{ji} \leftarrow \mathrm{prg}(s'_{ji})$
     erase all except output $(x_i, p_{i1}, \ldots, p_{in}, p'_{1i}, \ldots, p'_{ni})$

SENDSEED   /* Send pseudorandom pad using encryption */
1    $r_i \leftarrow uniform(\{0,1\}^{\kappa(K)})$; $(E_i, D_i) \leftarrow \mathrm{pkc}(1^K, r_i)$
     i broadcasts: $E_i[\to E'_i]$
2   receive E'_j; if bad then $E'_j \leftarrow$ identity function
     $s_{ij} \leftarrow uniform(\{0,1\}^K)$; $p_{ij} \leftarrow \mathrm{prg}(s_{ij})$
     $r_{ij} \leftarrow uniform(\{0,1\}^{\epsilon(K)})$; $\sigma_{ij} \leftarrow E'_j(s_{ij}, r_{ij})$
     i broadcasts: $\sigma_{ij}[\to \sigma'_{ij}]$;
     i and j execute a zero-knowledge proof of i's knowledge of s_{ij}
3   receive σ'_{ji}; if proof fails or $s'_{ji} \leftarrow D_i(\sigma'_{ji})$ fails then $s'_{ji} \leftarrow 0^K$
     $p'_{ji} \leftarrow \mathrm{prg}(s'_{ji})$
     erase all except output $(x_i, p_{i1}, \ldots, p_{in}, p'_{1i}, \ldots, p'_{ni})$

Fig. 1. Subprotocols for our transformation. Code for player i; take $1 \leq i, j \leq n, j \neq i$. "$(r)$ $i \to j : m[\to m']$" means i should send m to j at round r over a private channel, and m' denotes the message actually sent. See text for $\mathrm{pkc}, \mathrm{prg}, \kappa(K), \epsilon(K), B(K), p_{ij}(r)$.

Proof. TRANSFORM(Π) assumes only broadcast channels whereas Π may use private channels. The proof is modular: first, replace private channels by uniform one-time pads sent in a preprocessing stage over private channels; then replace uniform one-time pads by pseudorandom pads also sent initially over private channels; finally, replace initial private channels by public-key encryptions of the seeds for the pads. Keys and seeds are erased before the body of the protocol starts. The proof follows from Lemmas 7, 8, 9 (listed below), using transitivity (Theorem 3):

$$\text{TRANSFORM}(\Pi) = \text{PAD}(\Pi) \circ \text{SENDSEED} \succeq \text{PAD}(\Pi) \circ \text{SENDPsPAD}$$
$$\succeq \text{PAD}(\Pi) \circ \text{SENDPAD} \succeq \Pi. \qquad \Box$$

Theorem 6. (Main Result II) *If one-way trapdoor functions exist, then in the memoryless model, for $t(n) < n/2$ and any probabilistic finite function $F \in \text{PFF}$, there is a computationally t-resilient protocol for F secure against dynamic, message-rushing, Byzantine adversaries.*

Proof. Rabin, Ben-Or [28] and Beaver [3] demonstrate the existence of a protocol $\Pi(F)$ using private and broadcast channels that is statistically (hence computationally) t-resilient against dynamic, rushing, Byzantine t-adversaries, for any $t(n) < n/2$ and $F \in \text{PFF}$. (Beaver's proof uses "resilience" as defined here.) Theorem 5 suffices.[5] $\Box$

Proofs of the following lemmas are given in the Appendix.

Lemma 7. *One-time pads sent initially over private channels are as secure as private channels: for all feasible Π, $\text{PAD}(\Pi) \circ \text{SENDPAD} \succeq \Pi$.*

Lemma 8. *A pseudorandomly generated pad (with erasing) is as good as a uniformly random pad: for all feasible Π, $\text{PAD}(\Pi) \circ \text{SENDPsPAD} \succeq \text{PAD}(\Pi) \circ \text{SENDPAD}$.*

Lemma 9. *Generating a pad by encrypting and sending a seed (with erasing) is as good as sending the pad over a private channel: for all feasible Π, $\text{PAD}(\Pi) \circ \text{SENDSEED} \succeq \text{PAD}(\Pi) \circ \text{SENDPsPAD}$.*

Remark. The overhead of the transformation is small enough to be practical. Rather than store a large pad at the outset, a provably secure and practical modification of the protocol permits every player to extend its pad each round (or every few rounds) using the last, unused K bits of the previous pad as a seed which it then erases. No additional messages need be sent. Furthermore, the use of an underlying noncryptographic protocol with a simple algorithm appears more efficient and implementable than the use of, say, n^2 two-party zero-knowledge proofs of behavior after each round.

4 Open Questions

These definitions and proofs suggest a number of stimulating lines of research:

- Zero knowledge and interactive proofs are defined by [21, 1] against *static* adversaries; *i.e.*, the cases of faulty P and faulty V are considered separately. Do there exist "dynamic" zero-knowledge proof systems, which are secure against dynamic adversaries, who choose P or V based on some preliminary, and thus committed, portion of the interaction?

---

[5] We remark that if the protocols of [10, 12, 8] are proven secure using our definitions, then a *constant-round* protocol for $t < n/3$ exists.

- Our protocol employs the *statistical* zero-knowledge proofs used by Rabin, Ben-Or and Beaver [28, 3] in an underlying private-channel layer to protect against Byzantine adversaries. The standard, widely-adopted approach introduced in [23], using *computational* zero-knowledge proofs on the top broadcast-level to guarantee behavior, seems to fail in a dynamic setting. Are computational two-party zero-knowledge proofs useful for interactive computation with dynamic security?

- Is it possible to prove that erasing is *necessary* in order to achieve security against dynamic adversaries in cryptographic protocols? If this is true, we would have a nice characterization of the difference between "cryptographic" and "non-cryptographic" multi-party computing: either private channels or tape-erasing is necessary in order to withstand a dynamic adversary.

- The use of public-key encryptions to replace private channels seems intuitive. Does the apparent failure of this approach without erasing keys suggest a subtle way that public-key encryptions fail when used in combination, or does it suggest that zero-knowledge based definitions of protocol security (virtually the only sort ever considered in computational models) are not suitable for a general approach to security?

References

1. L. Babai, S. Moran. "Arthur-Merlin Games: A Randomized Proof System, and a Hierarchy of Complexity Classes." *J. Comput. System Sci.* **36** (1988), 254–276.

2. J. Bar-Ilan, D. Beaver. "Non-Cryptographic Fault-Tolerant Computing in a Constant Expected Number of Rounds of Interaction." *Proceedings of PODC,* ACM, 1989, 201–209.

3. D. Beaver. "Secure Multiparty Protocols and Zero Knowledge Proof Systems Tolerating a Faulty Minority." *J. Cryptology,* 4:2, 1991, 75–122. An earlier version appeared as "Secure Multiparty Protocols Tolerating Half Faulty Processors" in *CRYPTO '89,* G. Brassard, ed., Springer–Verlag LNCS **435**, 1990.

4. D. Beaver. "Formal Definitions for Secure Distributed Protocols." *Proceedings of the DIMACS Workshop on Distributed Computing and Cryptography,* Princeton, NJ, October, 1989, J. Feigenbaum, M. Merritt (eds.).

5. D. Beaver. *Security, Fault Tolerance, and Communication Complexity in Distributed Systems.* Ph.D. Thesis, Harvard University, Cambridge, 1990.

6. D. Beaver. "Foundations of Secure Interactive Computation." *Proceedings of Crypto '91* (to appear).

7. D. Beaver, S. Goldwasser. "Multiparty Computation with Faulty Majority." *Proceedings of the 30th FOCS,* IEEE, 1989, 468–473.

8. D. Beaver, S. Micali, P. Rogaway. "The Round Complexity of Secure Protocols." *Proceedings of the 22nd STOC,* ACM, 1990, 503–513.

9. M. Bellare and O. Goldreich. "On Defining Proofs of Knowledge." *Proceedings of Crypto '92* (to appear).

10. M. Ben-Or, S. Goldwasser, A. Wigderson. "Completeness Theorems for Non-Cryptographic Fault-Tolerant Distributed Computation." *Proceedings of the 20th STOC,* ACM, 1988, 1–10.

11. G. Brassard, D. Chaum, C. Crépeau. "Minimum Disclosure Proofs of Knowledge." *J. Comput. System Sci.* **37** (1988), 156–189.

12. D. Chaum, C. Crépeau, I. Damgård. "Multiparty Unconditionally Secure Protocols." *Proceedings of the 20th STOC*, ACM, 1988, 11–19.

13. B. Chor, M. Rabin. "Achieving Independence in a Logarithmic Number of Rounds." *Proceedings of the 6th PODC*, ACM, 1987.

14. U. Feige, A. Fiat, and A. Shamir. "Zero knowledge proofs of identity." *J. of Cryptology*, 1:2, 1988, 77–94.

15. P. Feldman. "One Can Always Assume Private Channels." Unpublished manuscript, 1988.

16. P. Feldman, S. Micali. "Optimal Algorithms for Byzantine Agreement." *Proceedings of the 20th STOC*, ACM, 1988, 148–161. (The reader of this paper is referred for the relevant result to Feldman's Ph.D. Thesis, *Optimal Algorithms for Byzantine Agreement* (MIT, 1988), where it apparently does not appear; but see [15].)

17. Z. Galil, S. Haber, M. Yung. "Cryptographic Computation: Secure Fault-Tolerant Protocols and the Public-Key Model." *Proceedings of Crypto 1987*, Springer–Verlag, 1988, 135–155.

18. Z. Galil, S. Haber, and M. Yung. "Minimum-Knowledge Interactive Proofs for Decision Problems." *SIAM J. Comput.* 18:4 (1989), 711–739.

19. Z. Galil, S. Haber, and M. Yung. "Interactive public-key cryptosystems." Submitted for publication, 1991.

20. S. Goldwasser, S. Micali. "Probabilistic Encryption." *J. Comput. System Sci.* **28** (1984), 270–299.

21. S. Goldwasser, S. Micali, C. Rackoff. "The Knowledge Complexity of Interactive Proof Systems." *SIAM J. Comput.* 18:1 (1989), 186–208.

22. S. Goldwasser, M. Sipser. "Private Coins vs. Public Coins in Interactive Proof Systems." *Proceedings of the 18th STOC*, ACM, 1986, 59–68.

23. O. Goldreich, S. Micali, A. Wigderson. "Proofs that Yield Nothing but Their Validity and a Methodology of Cryptographic Protocol Design." *Proceedings of the 27th FOCS*, IEEE, 1986, 174–187.

24. O. Goldreich, S. Micali, A. Wigderson. "How to Play Any Mental Game, or A Completeness Theorem for Protocols with Honest Majority." *Proceedings of the 19th STOC*, ACM, 1987, 218–229.

25. S. Goldwasser, L. Levin. "Fair Computation of General Functions in Presence of Immoral Majority." *Proceedings of Crypto 1990*.

26. J. Håstad. "Pseudo-Random Generators under Uniform Assumptions." *Proceedings of the 22nd STOC*, ACM, 1990, 395–404.

27. R. Impagliazzo, L. Levin, and M. Luby. "Pseudorandom Generation from One-Way Functions." *Proceedings of the 21st STOC*, ACM, 1989, 12–24.

28. T. Rabin, M. Ben-Or. "Verifiable Secret Sharing and Multiparty Protocols with Honest Majority." *Proceedings of the 21st STOC*, ACM, 1989, 73–85.

29. A. Shamir. "How to Share a Secret." *Communications of the ACM*, **22** (1979), 612–613.

30. M. Tompa and H. Woll. "Random self-reducibility and zero knowledge interactive proofs of possession of information." *Proceedings of the 28th FOCS*, IEEE, 1987, 472–482.

31. A. Yao, "Theory and Applications of Trapdoor Functions." *Proceedings of the 23rd FOCS*, IEEE, 1982, 80–91.

Appendix

Conventions for conciseness. An interface $\mathcal{I}$ from α to β is **canonical** if it runs internal copies of nonfaulty α-players i on inputs $x_i = 0$ (or some other default value); when $\mathcal{A}$ requests to corrupt i in β, $\mathcal{I}$ corrupts i to obtain the real (x_i, a_i) and overwrites the internal player's state/tape to be "consistent" with (x_i, a_i) and the α-messages already sent between $\mathcal{I}$ and $\mathcal{A}$. The set of faulty players chosen by $\mathcal{A}$ through round r is written $T(r)$. The symbol i^b indicates a faulty player $(i \in T)$; i^g indicates a good player $(i \notin T)$. Bars over letters indicate internally-simulated values; lowercase letters refer to protocol α whereas capital letters (except E, D) refer to protocol β; apostrophes indicate messages from an adversary.

Even though the lemmas in this appendix hold for any feasible protocol Π, for concreteness the reader may wish to imagine Π as a verifiable secret-sharing protocol (*e.g.* [10]).

Lemma 7. *For all feasible* Π, $\text{PAD}(\Pi) \circ \text{SENDPAD} \succeq \Pi$.

Proof sketch. We remark that this lemma holds for computationally-unbounded adversary classes as well as polynomial-time adversary classes (even when erasing is not allowed). Let $\alpha = \text{PAD}(\Pi) \circ \text{SENDPAD}$, $\beta = \Pi$. Run a canonical $\mathcal{I}$ that creates "fake" pads $\bar{p}_{ij} - uniform(\{0,1\}^{B(K)})$ for all initially nonfaulty $i \notin T(0)$. Supply faulty j^b with $\bar{p}_{ij}$ from internally simulated, nonfaulty i^g; supply nonfaulty, simulated j^g with whatever pads p'_{ij} faulty i^b sends. If $\mathcal{A}$ corrupts i, get (x_i, a_i) from β and use fake $\bar{p}_{ij}$ values to supply view. Now $\mathcal{A}$ enters $\text{PAD}(\Pi)$; consider round r. If $i^g \to j^g : M_{ij}(r)$ in β (a private message not seen by $\mathcal{I}$), $\mathcal{I}$ uses the internal message $\bar{m}_{ij}(r)$ (from fake player i on input $x_i = 0$) with fake pad $\bar{p}_{ij}$ to create broadcast message $c_{ij}(r) \leftarrow \bar{m}_{ij}(r) \oplus \bar{p}_{ij}(r)$ for $\mathcal{A}$ to see. If $i^g \to j^b : M_{ij}(r)$ in β, $\mathcal{I}$ sends $M_{ij}(r) \ominus \bar{p}_{ij}(r)$ in α. If $i^b \to j^g : c_{ij}(r)$ in α, $\mathcal{I}$ sends $m_{ij}(r) \leftarrow c_{ij}(r) \oplus p'_{ij}(r)$ in β. If $\mathcal{A}$ newly corrupts i, $\mathcal{I}$ gets (x_i, a_i) and private messages $\{m_{ij}(\rho) \mid \forall \rho \le r, \forall j\}$ from β; $\mathcal{I}$ constructs consistent pads p_{ij} by, first, overwriting used portions $[\forall \rho \le r, \forall j]$ $p_{ij}(\rho) \leftarrow c_{ij}(\rho) \oplus m_{ij}(\rho)$, which effects no change on pads already given to $\mathcal{A}$ $(p_{ij} = \bar{p}_{ij}$ for $j \in T)$, and, second, keeping unused portions $[\forall \rho > r, \forall j]$ $p_{ij}(\rho) - \bar{p}_{ij}(\rho)$. Because the pads that $\mathcal{A}$ captures are uniformly random whether obtained from α or from $\mathcal{I}$, and because $\mathcal{I}$ constructs consistent sets of messages, it is not difficult to show the collection of all outputs is *identical*: $[\mathcal{A}, \alpha] = [\mathcal{I}(\mathcal{A}), \beta]$. $\square$

Lemma 8. *For all feasible* Π, $\text{PAD}(\Pi) \circ \text{SENDPSPAD} \succeq \text{PAD}(\Pi) \circ \text{SENDPAD}$.

Proof sketch. Let $\alpha = \text{PAD}(\Pi) \circ \text{SENDPSPAD}$, $\beta = \text{PAD}(\Pi) \circ \text{SENDPAD}$. Run a canonical $\mathcal{I}$ that creates "fake" seeds $\bar{s}_{ij} - uniform(\{0,1\}^K)$ and pads $\bar{p}_{ij} \leftarrow \text{prg}(\bar{s}_{ij})$ for all initially nonfaulty $i \notin T(0)$. In Step 1, do $i^g \to j^b : \bar{s}_{ij}$ in SENDPSPAD; record uniform pads $i^g \to j^b : P_{ij}$ sent in SENDPAD. Record $i^b \to$

$j^g : s'_{ij}$ in SENDPsPAD, compute $P'_{ij} \leftarrow p'_{ij} \leftarrow \mathrm{prg}(s'_{ij})$, and send $i^b \rightarrow j^g : P'_{ij}$ in SENDPAD. If $\mathcal{A}$ corrupts i, then get (x_i, a_i) from β and supply $\bar{s}_{ij}, \bar{p}_{ij}$ in the view. In Step 2, all seeds are erased; from here on, only $\bar{p}_{ij}$ need be supplied in new corruptions. In fact, $\mathcal{I}$ may later overwrite portions of $\bar{p}_{ij}$ to create consistent views; if $\bar{p}_{ij}$ were uniformly random this would make no difference; because $\bar{p}_{ij}$ is pseudorandom the resulting distribution is false, but indistinguishable (without the erased seed).

Now $\mathcal{A}$ enters PAD(Π); consider round r. If $i^g \rightarrow j^g : C_{ij}(r)$ in β, $\mathcal{I}$ sends $c_{ij}(r) \leftarrow uniform(\{0,1\}^L)$ in α; this is false but indistinguishable. If $i^g \rightarrow j^b : C_{ij}(r)$ in β, $\mathcal{I}$ knows the pads P_{ij} in β and $\bar{p}_{ij}$ in α and calculates an α-message $c_{ij}(r) \leftarrow C_{ij}(r) \oplus P_{ij}(r) \oplus \bar{p}_{ij}(r)$. If $i^b \rightarrow j^g : c_{ij}(r)$ in α, $\mathcal{I}$ sends $C_{ij}(r) \leftarrow c_{ij}(r)$ in β; player j^g in β uses the same pad $P'_{ij}(r)$ to decrypt this message as the one $(p'_{ij}(r))$ that the simulated player j^g would use in α. If $\mathcal{A}$ corrupts player i, $\mathcal{I}$ recreates pads p_{ij} as in Lemma 7. For $j \in T$ this changes nothing, but for $j \notin T$, this replaces a portion of a pseudorandom pad by a uniform pad (because c_{ij} was artificially and uniformly created).

To prove that $[\mathcal{A}, \alpha] \approx [\mathcal{I}(\mathcal{A}), \beta]$, we assume there is a machine that K^{-c}-distinguishes $[\mathcal{A}, \alpha]$ from $[\mathcal{I}(\mathcal{A}), \beta]$, in particular for some $z = x \cdot a \cdot a_A$. Say the protocols run for $R = R(n, m, k)$ rounds. Define $(Rn^2 + 1)$ hybrid distributions ξ_{IJr}, with $(I, J, r) \in \{1..n\} \times \{1..n\} \times \{1..R\} \cup \{(0,0,0)\}$, in which certain pseudorandom pads are gradually replaced by uniform pads. We say $(i, j, \rho) \leq (I, J, r)$ if $i < I$, or if $i = I$ and $j < J$, or if $i = I$ and $j = J$ and $\rho \leq r$. In ensemble ξ_{IJr} the replacement is performed for those pads $p_{ij}(\rho)$ for which $(i, j \notin T(\rho))$ and $(i, j, \rho) \leq (I, J, r)$. Choose d so that $K^{-d} < 1/K^c Rn^2$. Since $\xi_{000} = [\mathcal{A}, \alpha](x \cdot a \cdot a_A)$ and $\xi_{nnR} = [\mathcal{I}(\mathcal{A}), \beta](x \cdot a \cdot a_A)$, there is a machine that K^{-d}-distinguishes $\xi_{ij\rho}$ from $\xi_{i,j,\rho+1}$ (or from $\xi_{i,j+1,1}$, if $\rho = R$; or from $\xi_{i+1,1,1}$, if $\rho = R$ and $j = n$) for some i, j, and ρ, and hence one that K^{-d}-distinguishes $\mathrm{prg}(uniform(\{0,1\}^K))$ and $uniform(\{0,1\}^{B(K)})$, implying prg is not a cryptographically strong pseudorandom generator. $\square$

Lemma 9. *For all feasible Π,* PAD(Π) $\circ$ SENDSEED $\succeq$ PAD(Π) $\circ$ SENDPsPAD.

Proof sketch. Let $\alpha = $ PAD(Π) $\circ$ SENDSEED, $\beta = $ PAD(Π) $\circ$ SENDPsPAD. Run a canonical $\mathcal{I}$ that, through the end of SENDSEED, creates "fake" keys (E_i, D_i), seeds $\bar{s}_{ij}$, pads $\bar{p}_{ij}$, and that when given a request to corrupt i, uses the "fake" state with only x_i and a_i overwritten with the values obtained from corrupting i in SENDPsPAD. $\mathcal{I}$ simulates the proofs of knowledge of SENDSEED in a straightforward manner. $\mathcal{I}$ records all seeds S_{ij} and pads P_{ij} obtained from SENDPsPAD in corruptions of players i or j. When $i^b \rightarrow j^g : \sigma'_{ij}$ in step 2 of SENDSEED, then $\mathcal{I}$ sends $S'_{ij} \leftarrow s'_{ij} \leftarrow D_j(\sigma'_{ij})$ in step 1 of SENDPsPAD and records $P'_{ij} \leftarrow p'_{ij} \leftarrow \mathrm{prg}(s'_{ij})$. As in Lemma 8, $\mathcal{I}$ does not supply $\mathcal{A}$ with these recorded values. Rather, as before, $\mathcal{I}$ uses P_{ij} to translate messages in β to messages in α as follows. When $i^g \rightarrow j^b : C_{ij}(r)$ in β, $\mathcal{I}$ converts this to $c_{ij}(r) \leftarrow C_{ij}(r) \oplus P_{ij}(r) \oplus \bar{p}_{ij}(r)$. When $i^b \rightarrow j^g : c_{ij}(r)$ in α, $\mathcal{I}$ sends $C_{ij}(r) \leftarrow c_{ij}(r)$ in β; player j^g in β uses the same pad $P'_{ij}(r)$ to decrypt this message as the one ($p'_{ij}(r)$) that the simulated player j^g would use in α. When messages are passed between i^g and j^g, $\mathcal{I}$ uses random strings. When $\mathcal{A}$ newly corrupts i, $\mathcal{I}$ constructs pads $p_{ij}(\rho)$ for j^g and [$\forall \rho \leq r$] by obtaining $m_{ij}(\rho)$ from β and calculating $p_{ij}(\rho) \leftarrow c_{ij}(\rho) \oplus m_{ij}(\rho)$. As in Lemma 8. all distributions are exact except for the calculation of p_{ij} pads (with $j \notin T$) upon corruption, in which case the resulting distribution is uniform (because c_{ij} has uniform distribution) rather than pseudorandom.

Consider ($Rn^2 + 1$) distributions ξ_{IJr}, as in the proof of Lemma 8, in which pseudorandom pads are progressively replaced by random pads. As before, a K^{-c}-distinguisher between [$\mathcal{A}, \alpha$] and [$\mathcal{I}(\mathcal{A}), \beta$] gives a K^{-d}-distinguisher between, not $\mathbf{prg}(\textit{uniform}(\{0,1\}^K))$ and $\textit{uniform}(\{0,1\}^{B(K)})$ as in Lemma 8, but distributions $\mathcal{P}_1$ and $\mathcal{P}_4$, where $\mathcal{P}_1 = (E_j(\bar{s}_{ij}, r_{ij}), \mathbf{prg}(\bar{s}_{ij}))$ with uniformly random $E_j, \bar{s}_{ij}$, and r_{ij}, and where $\mathcal{P}_4 = (E_j(\bar{s}_{ij}, r_{ij}), p_{ij})$ with uniformly random $E_j, \bar{s}_{ij}, r_{ij}$, and p_{ij}. This does not directly contradict the strength of $\mathbf{prg}$ because an encryption of the seed is present. The seed itself, the bits used to encrypt it, and the decryption key, however, are not present; they were erased. Define $\mathcal{P}_2 = (E_j(0, r_{ij}), \mathbf{prg}(\bar{s}_{ij}))$ and $\mathcal{P}_3 = (E_j(0, r_{ij}), p_{ij})$, with $E_j, r_{ij}, \bar{s}_{ij}$, and p_{ij} uniformly random. Let $\delta = \frac{1}{3}K^{-d}$. A δ-distinguisher between $\mathcal{P}_1$ and $\mathcal{P}_2$ provides a means to break the cryptosystem pkc: a δ-distinguisher between $\mathcal{P}_2$ and $\mathcal{P}_3$ provides a means to break the pseudorandom generator $\mathbf{prg}$; and a δ-distinguisher between $\mathcal{P}_3$ and $\mathcal{P}_4$ provides a means to break the cryptosystem pkc. (The reduction from a $\mathcal{P}_1$-$\mathcal{P}_2$ distinguisher to an algorithm that successfully attacks pkc makes crucial use of the knowledge extractor corresponding to the proof of knowledge of step 2 of SENDSEED: similarly for the case of a distinguisher between the distributions $\mathcal{P}_3$ and $\mathcal{P}_4$.) Therefore there is no K^{-d}-distinguisher between $\mathcal{P}_1$ and $\mathcal{P}_4$, and hence no K^{-c}-distinguisher between [$\mathcal{A}, \alpha$] and [$\mathcal{I}(\mathcal{A}), \beta$]. $\square$

Secure Bit Commitment Function against Divertibility

Kazuo Ohta   Tatsuaki Okamoto   Atsushi Fujioka

NTT Laboratories
Nippon Telegraph and Telephone Corporation
1-2356, Take, Yokosuka-shi, Kanagawa-ken, 238-03 Japan

Abstract

Some zero-knowledge interactive proofs (*ZKIP*s) have divertibility, that is, evidence of proof issued by a genuine prover, A, can be transferred to plural verifiers, B and then C, where the intermediate verifier, B, acts as A, with A's help, to confound the other verifier C without revealing the relation between the A-B interaction and the B-C interaction. This property is a serious problem in practice, e.g. the mafia fraud attack on identification scheme and the multi-verifier attack against undeniable signatures.

This paper proposes a new concept, *security against divertibility*, and proves that Naor's bit commitment function based on pseudo-random generators is secure against divertibility under the reasonable assumption. Usage of this bit commitment in *ZKIP* can convert a *divertible ZKIP* to a *divertible-free-ZKIP* which is secure against the mafia fraud attack and the multi-verifier attack.

1 Introduction

Zero-knowledge interactive proofs [GMR] are an attractive concept in theory and in practice [GMW, FS, C]. In zero-knowledge proofs, coin flips of the prover are essential for zero-knowledgeness of the proof, while coin flips of the verifier are essential for soundness of the proof.

The usages of randomness has two sides: one positive and the other negative. On the positive side, the prover's randomness can be used to transfer some information in order to achieve positive applications, for example, identity based key distribution, digital signatures, etc [C, OO90]. On the negative side, it can be used to create a subliminal channel [DGB], where a prover can send an authenticated message, which contains a hidden message.

The verifier's randomness can be also used maliciously to realize the *mafia fraud attack* on the Fiat-Shamir scheme [DGB] and the *multi-verifier attack*

R.A. Rueppel (Ed.): Advances in Cryptology - EUROCRYPT '92, LNCS 658, pp. 324-340, 1993.

against Chaum's undeniable signature [DY], where an intermediate B can pass himself off as the genuine prover A to another verifier C, when A proves her identity or her signature to B, and B conceals any evidence that he used A's help. This concept was expressed in the *divertible zero knowledge interactive proof* and it has been proven that the commutative random-self reducible (CRSR) problem satisfies this property [OO89]. It has also been proven that wide classes of language, *NP*, have *divertible ZKIP* under some assumption [BD]. [1]

Recently, it has been proven that no undeniable signature scheme is secure against a multi-verifier attack provided that half of the verifiers are "honest (for their conspiracy)" in a more general setting than the divertible ZKIP, where the verifiers collaborate with a sub-protocol hidden from the prover, using the concept of secure function evaluations known as "mental games"[DY]. Though this negative result is theoretically exciting, there are two problems in practice (from the attacker's side). First, their collaboration can be detected by anyone who observes the transmitted data among users, since the relationship between the mental game protocol and the undeniable signature protocol can be traced. Second, if the majority of the verifiers are dishonest, the minority can believe a false proof by the majority. Therefore, in their protocol based on the mental game, a malicious verifier cannot be convinced of the correctness of the proof without believing that the majority of the verifiers are honest.

These problems imply the possibility of constructing a secure undeniable signature against a multi-verifier attack under reasonable constraints such as the non-detectivity and the dishonest majority of the verifiers. So, there was an open problem whether such a secure undeniable signature against a multi-verifier attack under reasonable constraints exists or not.

In this paper, we solve this problem: we propose a secure undeniable signature scheme against a multi-verifier attack under a reasonable scenario satisfying the non-detectivity and the dishonest majority of the verifiers. In this scenario, the interface between verifiers are based on the basic protocol between the prover and the verifier, and the relationship among the interactions cannot be traced. Hereafter we call it the *divertible scenario*. Note that the detection of abuses is difficult in this scenario and that this scenario assumes that no verifier is trusted to be honest by the others (or the majority of the verifiers can be dishonest). Therefore, this scenario satisfies the non-detectivity and the dishonest majority of the verifiers.

In order to construct this secure undeniable signature scheme against a multi-verifier attack, we propose a new concept, *secure bit commitment function against divertibility*, and prove that Naor's bit commitment function [N] based on pseudo-random generators is secure against divertibility under the reasonable assumption. Implementation of *divertible ZKIP* using the *secure bit commitment function against divertibility* ensures invulnerability against multi-verifier attacks.

---

[1] Though [ISS] tried to construct the *divertible zero knowledge interactive proof* for *IP*, their definition of divertibility was wrong. So the construction of *divertible ZKIP* for *IP* is still open problem.

In other words, we show the way to convert a *divertible ZKIP* to a *divertible-free-ZKIP* by using a secure bit commitment function against divertibility. Therefore, a divertible ZKIP can be converted to a divertible-free-ZKIP assuming the existence of a one-way function [N, ILL, H]. Then, any negative side of a divertible ZKIP, such as the mafia fraud attack and the multi-verifier attack, can be protected by the usage of a secure bit commitment function against divertibility.

Note that a "non-malleable" bit commitment scheme proposed by Dolev, Dwork and Naor [DDN] can also solve this divertible problem. Although their scheme has broader applications than ours, since their scenario does not require the non-detectivity, their scheme is, however, much less efficient than our solution. Thus there is a tradeoff between the applicable scenario and the efficiency.

2 Problems in Divertible Scenario

First we will explain problems in the divertible scenario in this section. We will define a new security concept in the divertible scenario, and clarify some properties in the next section.

Recently, new digital signature schemes having the following properties were proposed based on *ZKIP* [C, OO90]:

(1) Only signer A can prove the validity of a message to any verifier B by using A's public key or A's identity.

(2) Verifier B cannot prove the validity of the message to another verifier C (non-transitivity).

Though it was hoped that the non-transitivity property was useful in many applications, for example, undeniable signature is suitable to software distribution [C], where only paying customers are able to verify the signature of software supplier with undeniable signature procedure, its weakness was pointed out in [DY, OOF1].

We will explain the problems in the divertible scenario using Chaum's undeniable signature in more detail.

2.1 Undeniable Signature

Although an undeniable signature is similar to a digital signature in that it is a number issued by a signer that is related to the signer's public key and his message, it cannot be verified without the signer's cooperation. In order to check the validity or invalidity of the signature, this scheme consists of two parts, confirmation and disavowal protocols. Hereafter, we will explain only the confirmation protocol.

Center generates a large prime number p and selects a primitive element g of field $GF(p)$ as common information in the system. Signer A generates his secret key x, and computes y ($= g^x \pmod{p}$). He publishes y as his public key.

Signer A generates signature s $(= m^x \pmod{p})$ corresponding to a message m from p and his secret key x, and sends (m, s) to verifier B.

Verifier B verifies the validity of signature s for a message m by cooperating with signer A using the following procedures.

Protocol 1 (Confirmation Protocol)

Step 1: Verifier B generates two random integers a and b, calculates

$$X = m^a \cdot g^b \pmod{p}$$

and sends X with m to signer A.

Step 2: Signer A generates a random integer q, computes

$$
\begin{aligned}
Y &= X \cdot g^q \pmod{p} \\
Z &= Y^x \pmod{p}
\end{aligned}
$$

and sends (Y, Z) to verifier B.

Step 3: Verifier B sends a and b to signer A.

Step 4: Signer A checks the following equation

$$X \overset{?}{=} m^a \cdot g^b \pmod{p}.$$

If the check succeeds, A sends q to verifier B. If the check fails, the procedure halts.

Step 5: Verifier B checks the following equations

$$
\begin{aligned}
Y &\overset{?}{=} m^a \cdot g^{b+q} \pmod{p} \\
Z &\overset{?}{=} s^a \cdot y^{b+q} \pmod{p}.
\end{aligned}
$$

If both checks succeed, B accepts the validity of (m, s). Otherwise, B does not accept the validity of (m, s).

This protocol satisfies non-transitivity, because it is *ZKIP* and its view of communication between a prover and a verifier can be simulated easily. Thus the view is not regarded as evidence of a signature.

2.2 Abuse of Undeniable Signature

We will describe an attack that allows plural verifiers to check the validity of a signature simultaneously, in which if a malicious person takes part as one verifier, the non-transitivity of a signature is suspect.

Suppose software supplier A believes that there is only paying verifier B_1, but unfortunately B_1 is malicious. Since B_1 can convince another verifier B_2 of the software's validity using the following attack, B_1 can be paid by B_2. As a result, B_1 can use the genuine software without paying his own money. Note that supplier A does not know that there are plural verifiers in this case.

Hereafter, we consider the simple case, where only two verifiers, B_1 and B_2, use the protocol. Two verifiers, B_1 and B_2, can verify the validity of (m, s) in cooperation with signer A using the following procedures in the confirmation protocol. Since A and B_2 act in the same way as **Protocol 1**, we will describe only the procedure of B_1.

Protocol 2

Step 1: Verifier B_1 generates two random integers a and b, calculates

$$X_1 = X_2 \cdot m^a \cdot g^b \pmod{p}$$

and sends X_1 with m to signer A, where X_2 is calculated by B_2 using **step 1 of Protocol 1**.

Step 2: Verifier B_1 calculates

$$
\begin{aligned}
Y_2 &= Y_1/(m^a \cdot g^b) \pmod{p} \\
Z_2 &= Z_1/(s^a \cdot y^b) \pmod{p}
\end{aligned}
$$

and sends (Y_2, Z_2) to verifier B_2, where (Y_1, Z_1) is calculated by A using **step 2 of Protocol 1**.

Step 3: Verifier B_1 checks the following equation

$$X_2 \stackrel{?}{=} m^{a_2} \cdot g^{b_2} \pmod{p}.$$

If the check succeeds, B_1 sends $a_1 = a_2 + a \pmod{p - 1}$, $b_1 = b_2 + b$ $\pmod{p - 1}$ to signer A, where (a_2, b_2) are sent by B_2 at **step 3** of **Protocol 1**. If the check fails, the procedure halts.

Step 4: Verifier B_1 checks the following equations receiving q from A

$$
\begin{aligned}
Y_1 &\stackrel{?}{=} m^{a_1} \cdot g^{b_1+q} \pmod{p} \\
Z_1 &\stackrel{?}{=} s^{a_1} \cdot y^{b_1+q} \pmod{p}.
\end{aligned}
$$

If both checks succeed, B_1 accepts the validity of (m, s) and sends q to verifier B_2. Otherwise, B_1 does not accept the validity of (m, s) and sends q to B_2.

Remark

Strictly speaking, the above attack is detectable, since the same value, q, is transferred between A-B_1 and B_1-B_2. The typical attacks proposed in [OO89, BD, OOF2] are based on the divertible property, which is not detectable.

2.3 Discussion

Why does the above attack succeed? Because plural verifiers can verify the validity of a signature using the same challenges, X, and the same responses, Y and Z, simultaneously.

Though we will explain the challenge case, a similar discussion is true for in the response case. Denote the generation of X as $X = f(a,b) = m^a \cdot g^b$ (mod p). Then $X_1 = X_2 \cdot f(a,b) = f(a_2,b_2) \cdot f(a,b) = f(a_2 + a, b_2 + b)$ holds. Therefore, B_1 calculates the value of $X_1 = X_2 \cdot f(a,b) \, (= f(a_1,b_1))$, where X_2 is issued by B_2 and (a,b) are generated by B_1, without knowing the values of a_1 and b_1 at **Step 1** of **Protocol 2**, and calculates the values of $a_1 \, (= a_2 + a)$ and $b_1 \, (= b_2 + b)$ using (a_2, b_2) issued by B_2 at **Step 3**.

The above attack depends on the homomorphism of functions which are used to generate a challenge, X, by a verifier B, and responses, Y and Z, by a prover A. Recently, it was proven that if probabilistic encryption homomorphism exists, then all language in NP have divertible $ZKIP$ based on 'swapping techniques' [BD]. Since a *public* coin type $ZKIP$ is used in their construction, where the committed bit by a verifier is sent publicly. the homomorphism of the bit commitment from a verifier also holds.

Therefore it is important for a protocol designer to provably overcome the homomorphism of either the challenge generating function or the response generating function.

Remark

This attack is applicable even if verifiers cannot trust each other. Since B_1 engages A with a protocol similar to **protocol 1**, B_1 can trust the result of **Step 4** of **protocol 2**. Note that the usage of randomness, a and b, by verifier B_1 is necessary in order to prevent the conspiracy of false A and B_2, because the communication between A and B_2 is simulated easily if B_1 doesn't use randomness (zero-knowledgeness of **protocol 1**).

2.4 Divertible Scenario and Divertible-freeness

Hereafter, we discuss a $ZKIP$ protocol against a multi-verifier attack under a reasonable scenario, which we call the *divertible scenario*, where the interface between verifiers are based on the basic protocol between the prover and the verifier and the relationship among the interactions cannot be traced. Note that the detection of abuses is difficult in this scenario and that this scenario assumes that no verifier is trusted to be honest by the others (or the majority of the verifiers can be dishonest). Therefore, this scenario satisfies the non-detectivity and the dishonest majority of the verifiers.

Moreover, we call a ZKIP *divertible-free* if the ZKIP is secure against multi-verifier attacks in the *divertible scenario*.

3 Secure Bit Commitment in Divertible Scenario

3.1 Definition of Security

We can counter the attack based on homomorphism described in the previous section by using the following concept, *secure bit commitment function against divertibility*, to generate the challenges. Intuitively it satisfies the following property: A committer who doesn't know the value of a and b cannot calculate the value of $f(a,b)$. Remember that B_1 can calculate the value of $f(a_1,b_1)$ without knowing the value of a_1 and b_1 at **Step 1** of **protocol 2** in the above example.

Definition 3.1 (Secure Bit Commitment Function)
Let $\mathcal{Y}(x,b)$ *be a bit commitment function, where* $b \in \{0,1\}$ *is a committed bit and* $x \in \{0,1\}^*$ *is a random string* [N]. *We say that* $\mathcal{Y}$ *is secure against divertibility, if in the case where* $t = O(poly(|\mathcal{Y}(x,b)|)) > 1$, *there is no triple of expected polynomial time probabilistic Turing machines* $(\mathcal{M}, \mathcal{B}, \mathcal{X})$ *such that,* $\mathcal{M}(y_1,\ldots,y_t) = \mathcal{Y}(x^*,b^*)$ *holds with non-negligible probability, where* $y_i = \mathcal{Y}(x_i,b_i)$ $(i = 1,\ldots,t)$, $b^* = \mathcal{B}(b_1,\ldots,b_t,y_1,\ldots,y_t)$ *and* $x^* = \mathcal{X}(x_1,\ldots,x_t,y_1,\ldots,y_t)$, *and in the case where* $t = 1$, *for any* x $(|x| \le O(poly(|\mathcal{Y}(x,b)|)))$ *there is no triple* $(\mathcal{M}, \mathcal{B}, \mathcal{X})$ *such that* $\mathcal{M}(y_1,x) = \mathcal{Y}(x^*,b^*)$ *holds with non-negligible probability, where* $y_1 = \mathcal{Y}(x_1,b_1)$, $b^* = \mathcal{B}(b_1,x,y_1)$ *and* $x^* = \mathcal{X}(x_1,x,y_1)$. *Here* $\mathcal{B}$ *and* $\mathcal{X}$ *satisfy the following properties in subsection 3.3.*

Remark
The reason why the function $\mathcal{Y}$ is applied to all (x_i, b_i) is that verifier B_i engages prover A or verifier B_{i-1} with a basic protocol. Note that verifiers, $(B_1,\ldots,B_t)$, cannot trust each other in the divertible scenario.

Definition 3.2 (Secure Bit Commitment Function for multiple bits)
Let $\vec{b}$ *be a k-bit string,* $b_1\|b_2\|\ldots\|b_k$, *where* $\|$ *means concatenation. We call* $\vec{\mathcal{Y}}$ *a secure bit commitment function for multiple bits if* $\vec{\mathcal{Y}}(\vec{b},\vec{x}) = \mathcal{Y}(b_1,x_1)\| \cdots \|\mathcal{Y}(b_k,x_k)$, *where* $\mathcal{Y}$ *is a secure bit commitment function, and* $\vec{x} = x_1\|\ldots\|x_k$.

3.2 An Application of Secure Bit Commitment Function

Chaum's scheme becomes secure against multi-verifier attack when verifier B sends $w(= \vec{\mathcal{Y}}(a\|b,x))$ with (X,m) to signer A at **Step 1**, B sends x with (a,b) to A at **Step 3** and A checks $w \overset{?}{=} \vec{\mathcal{Y}}(a\|b,x)$ at **Step 4** during the confirmation protocol (**Protocol 1**).

Why is this modification secure in the divertible scenario? Let us consider the following situation: signer A convinces verifier B of the validity of A's signature in the confirmation protocol using the above protocol, and plural verifiers,

B_i $(i = 1, \ldots, t)$, try to share the validity of A's signature through B issuing $y_i = \vec{\mathcal{Y}}(a_i \| b_i, x_i)$ to B.

Assume that verifier B, which is a polynomial time probabilistic Turing machine, succeeds the multi-verifier attack, that is, B could commit the value w calculating from $y_1, \ldots, y_t$ at **Step 1** of **Protocol 1**, where we denote this function as $\mathcal{M}$, i.e., $w = \mathcal{M}(y_1, \ldots, y_t)$, and could open the values $a \| b$ and x calculating from $a_1 \| b_1, \ldots, a_t \| b_t, x_1, \ldots, x_t$ such that $w = \vec{\mathcal{Y}}(a \| b, x)$ at **Step 3 of protocol 1**, where we denote these functions as $\vec{\mathcal{B}}$ and $\vec{\mathcal{X}}$, i.e., $a \| b = \vec{\mathcal{B}}(a_1 \| b_1, \ldots, a_t \| b_t, y_1, \ldots, y_t)$, $x = \vec{\mathcal{X}}(x_1, \ldots, x_t, y_1, \ldots, y_t)$.

Since B is a polynomial time probabilistic Turing machine, the triple $(\mathcal{M}, \vec{\mathcal{B}}, \vec{\mathcal{X}})$ is so. This is a contradiction of the definition of $\vec{\mathcal{Y}}$. Thus the multi-verifier attack fails if we use a secure bit commitment function, $\vec{\mathcal{Y}}$, in this modification.

More generally, we can prove the following theorem.

Theorem 3.3 (Conversion of divertible ZKIP to divertible-free ZKIP)
Let (A, B) be a divertible ZKIP repeating the following procedure:
step 1: *A sends a message x' to B.*
step 2: *B selects a random bit $b \in \{0, 1\}$ and sends it to A.*
step 3: *A sends a message z to B.*
step 4: *B checks whether (x', b, z) satisfies a relation.*

Case a) If we construct $(\widetilde{A}, \widetilde{B})$ using a secure bit commitment function $\mathcal{Y}$ as follows:
step 1a: *$\widetilde{B}$ selects a random bit $b \in \{0, 1\}$ and a random string $x \in \{0, 1\}^*$, calculates $y = \mathcal{Y}(x, b)$ and sends it to $\widetilde{A}$.*
step 2a: *$\widetilde{A}$ sends a message x' to $\widetilde{B}$.*
step 3a: *$\widetilde{B}$ sends the bit b and the string x to $\widetilde{A}$.*
step 4a: *$\widetilde{A}$ checks whether $y = \mathcal{Y}(x, b)$ holds. If the check succeeds, $\widetilde{A}$ sends a message z to $\widetilde{B}$.*
step 5a: *$\widetilde{B}$ checks whether (x', b, z) satisfies a relation.*
Then $(\widetilde{A}, \widetilde{B})$ is a divertible-free ZKIP.

Case b) If we construct $(\hat{A}, \hat{B})$ using a secure bit commitment function $\vec{\mathcal{Y}}$ as follows:
step 1b: *$\hat{A}$ calculates two random values $\vec{z_0}, \vec{z_1}$, where $\vec{z_b}$ corresponds to the message z at step 3 when it receives a bit b at step2, selects two random strings $\vec{x_0}, \vec{x_1} \in \{0, 1\}^*$, calculates $\vec{y_i} = \vec{\mathcal{Y}}(\vec{x_i}, \vec{z_i})(i = 0, 1)$ and sends them with x' to $\hat{B}$.*
step 2b: *$\hat{B}$ sends a bit b to $\hat{A}$.*
step 3b: *$\hat{A}$ sends the message $\vec{z_b}$ and the string $\vec{x_b}$ to $\hat{B}$.*
step 4b: *$\hat{B}$ checks whether $\vec{y_b} = \vec{\mathcal{Y}}(\vec{x_b}, \vec{z_b})$ holds. If the check succeeds, $\hat{B}$ checks whether $(x', b, \vec{z_b})$ satisfies a relation.*
Then $(\hat{A}, \hat{B})$ is a divertible-free ZKIP.

Sketch of Proof:

We will discuss Case a) $(t > 1)$ only. The similar discussion holds for Case a) $(t = 1)$ and Case b).

Assume that $(\widetilde{A}, \widetilde{B})$ is not *divertible-free*, that is, there exists verifier $\widetilde{B}$ succeeds the multi-verifier attack in the divertible scenario, where signer $\widetilde{A}$ convinces verifier $\widetilde{B}$ of the validity of $\widehat{A}$'s proof using the above protocol, and plural verifiers, $\widetilde{B}_i$ $(i = 1, \ldots, t)$, try to share the validity of $\widehat{A}$'s proof through $\widetilde{B}$, where $\widetilde{B}_i$ issues $y_i = \mathcal{Y}(x_i, b_i)$ to $\widetilde{B}$.

$\widetilde{B}$ could commit the value y^* calculating from $y_1, \ldots, y_t$ at step 1a, where we denote this function as $\mathcal{M}$, i.e., $y^* = \mathcal{M}(y_1, \ldots, y_t)$, and could open the values b^* and x^* calculating from $b_1, \ldots, b_t$, and $x_1, \ldots, x_t$, such that $y^* = \mathcal{Y}(b^*, x^*)$ at step 3a, where we denote these functions as $\mathcal{B}$ and $\mathcal{X}$, i.e., $b^* = \mathcal{B}(b_1, \ldots, b_t, y_1, \ldots, y_t)$ and $x^* = \mathcal{X}(x_1, \ldots, x_t, y_1, \ldots, y_t)$. [2]

Since $\widetilde{B}$ is a polynomial time probabilistic Turing machine, the triple $(\mathcal{M}, \mathcal{B}, \mathcal{X})$ is so. This is a contradiction of the definition of $\mathcal{Y}$.

(Q.E.D. Theorem)

3.3 Properties of Functions

We will clarify some properties of $\mathcal{X}, \mathcal{B}$ introduced in the above definition. The following properties come from the divertible scenario, where verifiers cannot trust each other. That is, verifier B_i is afraid of the conspiracy of false A and other verifiers B_j $(j \neq i)$. So, the values of b^* and x^* should depend on all b_i and x_i components equally.

Properties

1) $\mathcal{B}_i^{b_1, \ldots, b_{(i-1)}, b_{(i+1)}, \ldots, b_t} : b_i \longrightarrow \mathcal{B}(b_1, \ldots, b_{(i-1)}, b_i, b_{(i+1)}, \ldots, b_t)$: bijective for any i $(1 \leq i \leq t)$. We call this the bijective property of $\mathcal{B}$.

2) $\mathcal{X}_i^{x_1, \ldots, x_{(i-1)}, x_{(i+1)}, \ldots, x_t} : x_i \longrightarrow \mathcal{X}(x_1, \ldots, x_{(i-1)}, x_i, x_{(i+1)}, \ldots, x_t)$: bijective for any i $(1 \leq i \leq t)$. We call this the bijective property of $\mathcal{X}$.

Remark

These assumptions imply that $(t - 1)$ verifiers can neither guess the hidden bit, b_i, nor control the value b^* in a conspiracy.

4 Naor's Bit Commitment is Secure against Divertibility

We will discuss the security of the bit commitment function based on Naor's idea [N].

---

[2]Note that our framework, where $x^* = \mathcal{X}(x_1, \ldots, x_t, y_1, \ldots, y_t)$ and $b^* = \mathcal{B}(b_1, \ldots, b_t, y_1, \ldots, y_t)$, is slightly restricted than the case, where $x^* = \mathcal{X}((x_1, b_1), \ldots, (x_t, b_t))$ and $b^* = \mathcal{B}((x_1, b_1), \ldots, (x_t, b_t))$, since $y_i = \mathcal{Y}(x_i, b_i)$ $(i = 1, \ldots, t)$ hold. So the security in the latter case is an open problem.

4.1 Naor's Bit Commitment Function

Let G be a pseudo-random generator, and $G_i(x)$ be the i-th bit of the output of $G(x)$.

Commit stage:
step 1: Verifier B selects a random vector $r = (r_1, \ldots, r_{3n})$, where $r_i \in \{0, 1\}$ for $1 \leq i \leq 3n$, and sends it to committer A.
step 2: A selects a seed $x \in \{0, 1\}^n$ and sends to B the vector $y = (d_1, \ldots, d_{3n})$ where

$$d_i = \begin{cases} G_i(x) & \text{if } r_i = 0 \\ G_i(x) \oplus b & \text{if } r_i = 1 \end{cases}$$

and $b \in \{0, 1\}$ is the bit A is committed to, where $\oplus$ denotes the exclusive-or operation.
Reveal stage:
A sends x and B verifies that for all $1 \leq i \leq 3n$

$$\text{if } r_i = 0 \quad \text{then } d_i \overset{?}{=} G_i(x)$$
$$\text{if } r_i = 1 \quad \text{then } d_i \overset{?}{=} G_i(x) \ominus b$$

Notation: Hereafter, we denote $y = \mathcal{Y}^{<r>}(x, b) = G^{(3n)}(x) \oplus br$, where $G^{(3n)}$ means the first $3n$-bits output of $G(x)$.

Assumption 4.1 *There exists a pseudo-random generator* $G^{(3n)} : \{0, 1\}^n \longrightarrow \{0, 1\}^{3n}$ *satisfying the following property: for any polynomial time relation* $\mathcal{R} : \{0, 1\}^n \times \{0, 1\}^n \longrightarrow \{0, 1\}$, *there is no expected polynomial time probabilistic Turing machine* $\mathcal{D} : \{0, 1\}^{3n} \times \{0, 1\}^{3n} \longrightarrow \{0, 1\}$ *such that* $\mathcal{D}(g, g') = \mathcal{R}(x, x')$ *holds with non-negligible probability, where* $g = G^{(3n)}(x)$ *and* $g' = G^{(3n)}(x')$. *We call this assumption the independence of* G.

Remark
Generally PSRGs do not always satisfy this assumption. There are two directions to avoid this assumption, which are open problems:

(1) The construction of PSRG which satisfies the independence of PSRG: Note that it was proven that if there exists a one-way function, then there exists a pseudo-random generator (PSRG) [ILL, H]. So, there might be some construction technique based on a one-way function.

(2) Clarify the condition of $\mathcal{R}$ which is sufficient for the main theorem: The relation $\mathcal{R}$ used in the proof of main theorem is defined using a function $\mathcal{X}$. If we could prove the sufficient condition of $\mathcal{R}$ for the theorem based on the properties of $\mathcal{X}$, the above assumption is avoidable. [3]

---

[3] We conjecture that a sufficient condition of $\mathcal{R}$ as follows: for any x, $\#\{x' | \mathcal{R}(x, x') = 0\} / \#\{x'\}$ is negligible, where $\#S$ means the number of elements of a set S.

4.2 Main Theorem

Theorem 4.2 *If there exists a pseudo-random generator $\mathcal{G}$, then Naor's bit commitment function is secure against divertibility, assuming the independence of $\mathcal{G}$ and the bijective properties of $\mathcal{B}$ and $\mathcal{X}$, where $t = O(poly(|\mathcal{Y}(x,b)|))$. The random vector r is independently selected for each bit commitment in the case of $t = 1$.*

Sketch of Proof:

We will discuss the case where $t \geq 2$ at first, then discuss the case where $t = 1$.

Case of $t \geq 2$

The proof is by contradiction. Hereafter we prove the situation where the random vector r is fixed. Since a scheme with the independently selected vector r is more intractable than the one with a fixed r, this restriction to r component is not considered essential.

Assume that expected polynomial time probabilistic Turing machines, $\mathcal{M}(y_1, \ldots, y_t)$, $\mathcal{Y}^{<r>}(x_j, b_j)$ $(1 \leq j \leq t)$, $\mathcal{X}(x_1, \ldots, x_t, y_1, \ldots, y_t)$ and $\mathcal{B}(b_1, \ldots, b_t, y_1, \ldots, y_t)$ exist such that $\mathcal{M}(y_1, \ldots, y_t) = \mathcal{Y}^{<r>}(x^*, b^*)$ holds with non-negligible probability(ϵ), where $y_j = \mathcal{Y}^{<r>}(x_j, b_j)$, $x^* = \mathcal{X}(x_1, \ldots, x_t, y_1, \ldots, y_t)$, and $b^* = \mathcal{B}(b_1, \ldots, b_t, y_1, \ldots, y_t)$.

We will consider two cases with regard to the output of $\mathcal{M}$:

Case 1) The output y^* of $\mathcal{M}(y_1, \ldots, y_t)$ is coincident with some y_α or $y_\alpha \oplus r$, where $1 \leq \alpha \leq t$, (hereafter we denote the probability that Case 1) occurs under the existence of $(\mathcal{M}, \mathcal{Y}, \mathcal{X}, \mathcal{B})$ as δ.) and

Case 2) All other cases than Case 1).

We will construct an expected polynomial time probabilistic algorithm $\mathcal{A}$ which guesses b from the input y $(= \mathcal{G}^{(3n)}(x) \oplus br)$, where b is unknown to $\mathcal{A}$, for **Case 1)** using $(\mathcal{M}, \mathcal{Y}, \mathcal{X}, \mathcal{B})$. This is a contradiction for the difficulty of guessing a committed bit in Naor's protocol, when δ is non-negligible.

For **Case 2)**, we will construct an expected polynomial time probabilistic algorithm $\mathcal{D}$, which, given g and g', decides whether some relation, $\mathcal{R}(x, x')$, holds or not using $\mathcal{M}$, where $\mathcal{R}$ is defined with $\mathcal{X}$, $y = \mathcal{G}^{(3n)}(x)$ and $g' = \mathcal{G}^{(3n)}(x')$. This is a contradiction of the independence of $\mathcal{G}$, when δ is not non-negligible.

Case 1)

At first, we will construct $\mathcal{A}$, whose input is y $(= \mathcal{G}^{(3n)}(x) \oplus br)$ and output is a guessed bit of b, as follows:

[Algorithm $\mathcal{A}$]
For $i = 1$ to t do
    Repeat N times

Step 1: Put $y_i = y$.

Step 2: Select $x_j \in \{0,1\}^n$ and $b_j \in \{0,1\}$ randomly $(j \neq i)$.

Step 3: Calculate $y_{jb_j} = \mathcal{G}^{(3n)}(x_j) \oplus b_j r$ $(j \neq i)$ and

$$y^* = \mathcal{M}(\vec{y}) \quad \text{where} \quad \vec{y} = (y_{1b_1}, \ldots, y_{i-1b_{i-1}}, y_i, y_{i+1b_{i+1}}, \ldots, y_{tb_t}).$$

Step 4: If $\mathcal{M}$ succeeds, that is, there exist $\alpha \in \{1, \ldots, i-1, i+1, \ldots, t\}$ and $\beta \in \{0,1\}$ such that $y^* = y_{\alpha\beta}$, then output b' which satisfies $\mathcal{B}(b_1, \ldots, b_{i-1}, b', b_{i+1}, \ldots, b_t, \vec{y}) = \beta$ as a guessed bit of b such that $y = \mathcal{G}^{(3n)}(x) \oplus br$ and halt. Otherwise go to the next iteration.

End repeat

End do

[end of **Algorithm** $\mathcal{A}$]

Claim 1: Algorithm $\mathcal{A}$ is polynomial time computable, and it can guess the value of bit b such that $y = \mathcal{G}^{(3n)}(x) \oplus br$ with probability significantly better than $\frac{1}{2}$.

(Proof of Claim 1)

When there exist $(\mathcal{M}, \mathcal{Y}, \mathcal{X}, \mathcal{B})$, $y^* = \mathcal{G}^{(3n)}(x^*) \oplus \mathcal{B}(b_1, \ldots, b_{i-1}, b, b_{i+1}, \ldots, b_t, \vec{y})r$ always holds from their definitions, where $y = \mathcal{G}^{(3n)}(x_i) \oplus br$ and $x^* = \mathcal{X}(x_1, \ldots, x_{i-1}, x_i, x_{i+1}, \ldots, x_t, \vec{y})$.

Generally, the probability that x and x' $(x \neq x')$ exist and satisfy $\mathcal{G}^{(3n)}(x') - \mathcal{G}^{(3n)}(x) \oplus r$ is at most 2^{-n}, since $r \in \{0,1\}^{3n}$ is selected randomly, and both x and x' are elements of $\{0,1\}^n$. On the other hand, $\mathcal{B}(b_1, \ldots, b_{i-1}, 0, b_{i+1}, \ldots, b_t, \vec{y}) \neq \mathcal{B}(b_1, \ldots, b_{i-1}, 1, b_{i+1}, \ldots, b_t, \vec{y})$ holds (**Property 1**)). So the probability that $y^* = y_{\alpha 0} = y_{\alpha' 1}$ is at most 2^{-n}. Therefore, if $y^* = y_{\alpha\beta}$ holds, then $x^* = x_\alpha$ and b satisfies $\mathcal{B}(b_1, \ldots, b_{i-1}, b, b_{i+1}, \ldots, b_t, \vec{y}) = \beta$ with overwhelming probability $(\geq 1 - \frac{1}{2^n})$. Let us call the case, where $\mathcal{M}$ finds $y^* = y_{\alpha\beta}$ including the case where $\alpha = i$ at **Step 4**, **Case 1**), and denote the probability that **Case 1**) occurs under the existence of $(\mathcal{M}, \mathcal{Y}, \mathcal{X}, \mathcal{B})$ as δ. In the case where $\alpha = i$, since we don't know the value of β, **Step 4** does not decide the value of b. Since the probability that $\alpha = i$ holds at **Step 4** is $\frac{1}{t}$, so the probability $\mathcal{A}$ outputs b is $(1 - \frac{1}{t})\delta\epsilon$. Note that there is an error probability at most $\frac{1}{2^n}$ in the output of b at **Step 4**.

Since we assume that $(\mathcal{M}, \mathcal{Y}, \mathcal{X}, \mathcal{B})$ exist with non-negligible probability(ϵ), the algorithm $\mathcal{A}$ can guess the value of bit b such that $y = \mathcal{G}^{(3n)}(x) \oplus br$ with success probability significantly better than $\frac{1}{2}$ if δ is non-negligible. This is because for all sufficiently large n,

$$
\begin{aligned}
Prob[\mathcal{A}(y) &= b] \\
&= 1 - Prob[\mathcal{A} \text{ fails at all round}] \\
&= 1 - \{Prob[\mathcal{A} \text{ fails at round } i]\}^t \\
&> 1 - \{(1-\epsilon) + (1-\delta)\epsilon + \tfrac{1}{t}\delta\epsilon + \tfrac{1}{2^n}(1-\tfrac{1}{t})\delta\epsilon\}^{Nt} \\
&> 1 - \{1 - (1-\tfrac{2}{t})\delta\epsilon\}^{Nt} \\
&\to 1 - e^{-1} \quad (N = \tfrac{1}{(t-2)\delta\epsilon}),
\end{aligned}
$$

where N is selected as $O(\frac{1}{(t-2)\delta\epsilon})$ such that this probability is significantly better than $\frac{1}{2}$.

Note that $\mathcal{A}(y)$ is polynomial time computable, since $t = O(poly(|\mathcal{Y}(x,b)|))$ and $N = O(\frac{1}{(t-2)\delta\epsilon})$, where both ϵ and δ are non-negligible probabilities.

(q.e.d. Claim 1)

Remark

1) If the bijective property of $\mathcal{B}$ does not hold, then it might hold that $\mathcal{B}(b_1,\ldots, b_{i-1}, 0, b_{i+1}, \ldots, b_t, \vec{y}) = \mathcal{B}(b_1,\ldots, b_{i-1}, 1, b_{i+1}, \ldots, b_t, \vec{y}) = \beta$ or $\overline{\beta}$. Here the value of b is not decided uniquely or there is no value of b that satisfies this relation.

2) We can not guess the value of b given in both **Case 2)** and $\alpha = i$, where $y^* \neq y_{\alpha\beta}$, that is, $x^* \neq x_\alpha$ ($\alpha \in \{1, \ldots, i-1, i+1, \ldots, t\}$ in each round i). Thus, the following discussion is necessary for **Case 2)**, where δ is not non-negligible.

Case 2)

Let denote $\mathcal{R}_{x_1,\ldots,x_{i-1},x_{i+1},\ldots,x_t,\vec{y}}(x, x^*) = 0$ iff $\mathcal{X}(x_1, \ldots, x_{i-1}, x, x_{i+1}, \ldots, x_t, \vec{y}) = x^*$ holds, where $\vec{y} = (y_{1b_1}, \ldots, y_{i-1b_{i-1}}, y_{ib_i}, y_{i+1b_{i+1}}, \ldots, y_{tb_t})$.

Next we will construct $\mathcal{D}$ and the relation $\mathcal{R}_{x_1,\ldots,x_{i-1},x_{i+1},\ldots,x_t,\vec{y}}$, on input of g $(= \mathcal{G}^{(3n)}(x))$ and g^* $(= \mathcal{G}^{(3n)}(x^*))$, where x and x^* are not known to $\mathcal{D}$, as follows:

[Algorithm $\mathcal{D}$]
For $i = 1$ to t do
  Repeat N times

Step 1: Put $y_i = g$.
Step 2: Select $b_i \in \{0,1\}$, $x_j \in \{0,1\}^n$ and $b_j \in \{0,1\}$ randomly ($j \neq i$).
Step 3: Calculate $y_{ib_i} = y_i \oplus b_i$, $y_{jb_j} = \mathcal{G}^{(3n)}(x_j) \oplus b_j r$ ($j \neq i$) and

$$y^* = \mathcal{M}(\vec{y}) \quad \text{where} \quad \vec{y} = (y_{1b_1}, \ldots, y_{i-1b_{i-1}}, y_{ib_i}, y_{i+1b_{i+1}}, \ldots, y_{tb_t}).$$

Step 4: If the following equation holds, then output 0. Otherwise output 1.

$$y^* = g^* \oplus \mathcal{B}(b_1, \ldots, b_{i-1}, b_i, b_{i+1}, \ldots, b_t, \vec{y})r$$

  End repeat
End do

[end of Algorithm $\mathcal{D}$]

We will prove that $\mathcal{D}(g, g^*) = \mathcal{R}_{x_1,\ldots,x_{i-1},x_{i+1},\ldots,x_t,\vec{y}}(x, x^*)$ holds with non-negligible probability, where g $(= \mathcal{G}^{(3n)}(x))$ and y^* $(= \mathcal{G}^{(3n)}(x^*))$.

Claim 2: If there exist $(\mathcal{M}, \mathcal{Y}, \mathcal{X}, \mathcal{B})$, $\mathcal{R}_{x_1,\ldots,x_{i-1},x_{i+1},\ldots,x_t,\vec{y}}(x, x') = 0$ implies $\mathcal{D}(g, g') = 0$, where $g = \mathcal{G}^{(3n)}(x)$ and $g' = \mathcal{G}^{(3n)}(x')$.
(Proof of Claim 2)

By the definition of $\mathcal{R}$, $\mathcal{R}_{x_1,\ldots,x_{i-1},x_{i+1},\ldots,x_t,\vec{y}}(x, x') = 0$ means $x' = \mathcal{X}(x_1, \ldots, x_{i-1}, x, x_{i+1}, \ldots, x_t, \vec{y})$. The existence of $(\mathcal{M}, \mathcal{Y}, \mathcal{X}, \mathcal{B})$ implies that

$$y^* = \mathcal{G}^{(3n)}(x') \oplus \mathcal{B}(b_1, \ldots, b_t, \vec{y})r.$$

Since **Step 4** of algorithm $\mathcal{D}$ holds, $\mathcal{D}(g, g') = 0$ holds, where $g = \mathcal{G}^{(3n)}(x)$ and $g' = \mathcal{G}^{(3n)}(x')$.

(q.e.d. Claim 2)

Claim 3: If there exist $(\mathcal{M}, \mathcal{Y}, \mathcal{X}, \mathcal{B})$, $\mathcal{D}(g, g^*) = 0$ implies $\mathcal{R}_{x_1,\ldots,x_{i-1},x_{i+1},\ldots,x_t,\vec{y}}(x, x^*) = 0$, where $g = \mathcal{G}^{(3n)}(x)$ and $g^* = \mathcal{G}^{(3n)}(x^*)$ with overwhelming probability.
(Proof of Claim 3)

By the definition of $\mathcal{D}$, $\mathcal{D}(g, g^*) = 0$ means

$$y^* = g^* \oplus \mathcal{B}(b_1, \ldots, b_{i-1}, b_i, b_{i+1}, \ldots, b_t, \vec{y})r.$$

The existense of $(\mathcal{M}, \mathcal{Y}, \mathcal{X}, \mathcal{B})$ implies that

$$y^* = \mathcal{G}^{(3n)}(x') \oplus \mathcal{B}(b_1, \ldots, b_t, \vec{y})r,$$

where x is defined as $g = \mathcal{G}^{(3n)}(x)$ and $x' = \mathcal{X}(x_1, \ldots, x_{i-1}, x, x_{i+1}, \ldots, x_t, \vec{y})$.

These equations imply $g^* = \mathcal{G}^{(3n)}(x')$ with overwhelming probability. The reason is as follows: Generally, the probability that x and x' $(x \neq x')$ satisfy $\mathcal{G}^{(3n)}(x') = \mathcal{G}^{(3n)}(x)$ is at most 2^{-3n}, since pseudo-random generators pass the next bit test [Y].

Since x^* satisfies $g^* = \mathcal{G}^{(3n)}(x^*)$, the probability that $x^* \neq x'$ holds is negligible $(\leq \frac{1}{2^{3n}})$. Therefore, $x^* = x' = \mathcal{X}(x_1, \ldots, x_{i-1}, x, x_{i+1}, \ldots, x_t, \vec{y})$, that is, $\mathcal{R}_{x_1,\ldots,x_{i-1},x_{i+1},\ldots,x_t,\vec{y}}(x, x^*) = 0$ with overwhelming probability $(\geq 1 - \frac{1}{2^{3n}})$.

(q.e.d. Claim 3)

It is clear that if there exist $(\mathcal{M}, \mathcal{Y}, \mathcal{X}, \mathcal{B})$, then $\mathcal{D}(g, g^*) = \mathcal{R}_{x_1,\ldots,x_{i-1},x_{i+1},\ldots,x_t,\vec{y}}(x, x^*)$ holds by combining **Claim 2** and **Claim 3** with overwhelming probability (γ). Since we assume that $(\mathcal{M}, \mathcal{Y}, \mathcal{X}, \mathcal{B})$ exist with non-negligible probability (ϵ) and **Case 2)** occurs with probability $(1-\delta)$, $\mathcal{D}(g_1, g^*) = \mathcal{R}_{x_1,\ldots,x_{i-1},x_{i+1},\ldots,x_t,\vec{y}}(x, x^*)$ holds with non-negligible probability $(\gamma\epsilon(1 - \delta))$, where δ is not non-negligible. This is a contradiction of the independence of $\mathcal{G}$.

Case of $t = 1$

Finally we will prove the case where $t = 1$. The proof is also by contradiction. Differently from the case that $t > 1$, in this case we assume an additional condition that random vector r for Naor's bit commitment scheme is independently selected for each bit commitment.

Assume that there exists $x_0 \in \{0,1\}^*$ and expected polynomial time probabilistic Turing machines, $\mathcal{M}$, $\mathcal{X}$ and $\mathcal{B}$, such that $\mathcal{M}(y_1, x_0) = \mathcal{Y}^{<r^*>}(x^*, b^*)$ holds with non-negligible probability(ϵ), where $y_1 = \mathcal{Y}^{<r_1>}(x_1, b_1)$, $x^* = \mathcal{X}(x_1, x_0, y_1)$ and $b^* = \mathcal{B}(b_1, x_0, y_1)$.

We will also consider two cases with regard to the output of $\mathcal{M}$:

Case 1) The output y^* of $\mathcal{M}(y_1, x_0)$ is coincident with y_1 or $y_1 \oplus r^*$, and

Case 2) All other cases than Case 1).

We can construct an expected polynomial time probabilistic algorithm $\mathcal{A}'$ which guesses b from input y $(= \mathcal{G}^{(3n)}(x) \oplus br)$. The algorithm $\mathcal{A}'$ is as follows:

Step 1: Determine x_0 and calculate $y^* = \mathcal{M}(y, x_0)$.

Step 2: Determine b^* by checking whether y^* is coincident with y or $y \oplus r^*$. Then determine b by using $b^* = \mathcal{B}(b, x_0, y^*)$.

Similarly to Claim 1, we can show that algorithm $\mathcal{A}'$ can guess b correctly with non-negligible probability. This is a contradiction for the difficulty of guessing a committed bit in Naor's protocol.

For **Case 2)**, we can construct an expected polynomial time probabilistic algorithm $\mathcal{D}'$, which, given g and g', decides whether some relation, $\mathcal{R}(x, x')$, holds or not using $\mathcal{M}$, where $\mathcal{R}$ is defined with $\mathcal{X}$, $g = \mathcal{G}^{(3n)}(x)$ and $g' = \mathcal{G}^{(3n)}(x')$. Algorithm $\mathcal{D}'$ can be constructed in a manner similar to algorithm $\mathcal{D}$, and it can be shown similarly that $\mathcal{D}'$ works correctly with non-negligible probability. Hence, this is a contradiction of the independence of $\mathcal{G}$.

These results can be easily extended to any multiple bit case $(k > 1)$, because if a multiple bit relation $\mathcal{B}$ holds such that $\vec{b}^* = \mathcal{B}(\vec{b}_1, \vec{b}_2, \vec{y}_1, \vec{y}_2)$, then the first elements of $\vec{b}^*$, $\vec{b}_1$ and $\vec{b}_2$ satisfy a relation with the parameters of the remaining elements of $\vec{b}^*$, $\vec{b}_1$, $\vec{b}_2$, $\vec{y}_1$ and $\vec{y}_2$. Therefore, the result that there is no relation among single bit variables implies that there is no relation among multiple bit variables.

(Q.E.D. Theorem)

5 Conclusion and Remarks

This paper has proposed a new security concept, the *secure bit commitment function against divertibility*. We has shown that Naor's bit commitment function based on a pseudo-random generator satisfies this property under the independence of PSRG. Implementation of *divertible ZKIP* using the *secure bit commitment function against divertibility* ensures invulnerability against multi-verifier attacks in the *divertible scenario* where the *non-detectivity* and the *dishonest majority* of the verifiers are satisfied. Thus, any negative attributes of a divertible ZKIP, such as the mafia fraud attack and the multi-verifier attack, can be removed by using the secure bit commitment function against divertibility.

Note that this conversion from a *divertible ZKIP* to a *divertible-free-ZKIP* by using the secure bit commitment function against divertibility is also effective against the *meddler attack* described in [DY] since the intermediate node uses the homomorphic property of challenge generating function.

However, there are several open problems:

(1) Security in a more general situation, e.g. where $x^* = \mathcal{X}((x_1, b_1), \ldots, (x_t, b_t))$, and $b^* = \mathcal{B}((x_1, b_1), \ldots, (x_t, b_t))$.

(2) The construction of PSRG which satisfies the independence of PSRG.

(3) Clarify the condition of $\mathcal{R}$ which is sufficient for the main theorem.

Acknowledgments

This research was conducted while the first author was visiting the MIT Laboratory for Computer Science. He would like to acknowledge the generous support provided by MIT. The authors would like to thank anonymous referees for useful comments on our preliminary manuscript.

References

[BD]   M.Burmester and Y.Desmedt, "All Languages in NP Have Divertible Zero-Knowledge Proofs and Arguments under Cryptographic Assumptions," EUROCRYPT'90

[C]   D.Chaum,   "Zero-Knowledge   Undeniable   Signatures," EUROCRYPT'90

[D]   Y.Desmedt, "Subliminal-Free Authentication and Signature," EUROCRYPT'88

[DDN]   D.Dolev, C.Dwork and M.Naor, "Non-Malleable Cryptography," STOC'91

[DGB]   Y.Desmedt, C.Goutier and S.Bengio, "Special Uses and Abuses of the Fiat-Shamir Passport Protocol," CRYPTO'87

[DY]   Y.Desmedt and M.Yung, "Weaknesses of Undeniable Signature Schemes," EUROCRYPT'91

[FS]   A.Fiat and A.Shamir, "How to Prove Yourself," CRYPTO'86

[GMR]   S.Goldwasser, S.Micali and C.Rackoff, "The Knowledge Complexity of Interactive Proof Systems," STOC'85

[GMW]   O.Goldreich, S.Micali and A.Wigderson, "Proofs that Yield Nothing But their Validity and a Methodology of Cryptographic Protocol Design," FOCS'86

[H]   J.Håstad, "Pseudo-Random Generators under Uniform Assumptions," STOC'90

[ILL]   R.Impagliazzo, L.Levin and M.Ludy, "Pseudo-random generation from one-way functions," STOC'89

[ISS]   T.Itoh, K.Sakurai and H.Shizuya, "Any Language in IP has a Divertible ZKIP," ASIACRYPTO'91

[N]   M. Naor, "Bit Commitment Using Pseudo-Randomness," CRYPTO'89

[OO89] T.Okamoto and K.Ohta, "Divertible Zero-Knowledge Interactive Proofs and Commutative Random Self-Reducible," EUROCRYPT'89

[OO90] T.Okamoto and K.Ohta, "How to Utilize the Randomness of Zero-Knowledge Proofs," CRYPTO'90

[OOF1] K.Ohta, T.Okamoto and A.Fujioka, "Abuses of Undeniable Signature and Their Countermeasures," IEICE Transactions, Vol. E-74, No. 8, pp. 2109–2113, 1991

[OOF2] K.Ohta, T.Okamoto and A.Fujioka, "Multi-Verifier Digital Signature Scheme," (in Japanese) Japanese Patent, File No. Toku-gan-hei 3-24856 (Feb. 19, 1991)

[Y]   A.Yao, "Theory and Applications of Trapdoor Functions," FOCS'82

Non-Interactive Circuit Based Proofs
and
Non-Interactive Perfect Zero-knowledge with Preprocessing

Ivan Damgård

Aarhus University, Mathematical Institute
Ny Munkegade, DK 8000 Aarhus C, Denmark

Abstract. In the first part of this paper, we present a non-interactive zero-knowledge proof system for Circuit Satisfiability. With this protocol, we can prove an arbitrary NP-statement non-interactively without using Karp-reductions to 3-SAT or Graph Hamiltonicity. The proof system is based on the quadratic residuosity problem and allows processing of XOR and NOT gates at virtually no cost. It is significantly more efficient than previously known non-interactive proof systems. In the second part, we present protocols based on the existence of collision intractable hash functions, leading to a *statistical* zero-knowledge non-interactive argument with preprocessing for *any* NP-statement. Under the certified discrete log assumption, the protocol is perfect zero-knowledge. In the preprocessing, the parties need only exchange messages of length independent of the theorem to be proved later. This is the first protocol with such efficient preprocessing that does not need to assume oblivious transfer. Finally we present a perfect zero-knowledge non-interactive protocol based on discrete logarithms that may potentially remove the need for preprocessing.

1 Introduction

A non-interactive zero-knowledge proof system is a protocol that allows a prover to convince a verifier that some statement is true, simply by sending one message to the verifier. The prover should be unable to cheat, and the verifier should learn nothing more from receiving the message than the mere fact that the statement involved is true. Accomplishing this for non-trivial statements requires that prover and verifier share a random string, the randomness of which they both trust.

Such proof systems were introduced in [BFM], [BSMP], where a proof system for 3-SAT was presented, based on the Quadratic Residuosity Assumption (QRA). Later, a proof system for Graph Hamiltonicity (GH) was presented in [LS], which was based on the more general assumption that a one-way permutation exists, or if the prover has only polynomial computing power, that one-way trapdoor permutations exist.

R.A. Rueppel (Ed.): Advances in Cryptology - EUROCRYPT '92, LNCS 658, pp. 341-355, 1993.
© Springer-Verlag Berlin Heidelberg 1993

There has also been some work published on non-interactive zero-knowledge with preprocessing [SMP],[KMO], in which prover and verifier first execute an interactive phase, which allows the prover to later convince the verifier about some statement. This statement may be unknown when the preprocessing is done. The proofs in [SMP] were based on any one-way function, while [KMO] needed in addition oblivious transfer, but had a much smaller preprocessing step.

Since both 3-SAT and GH are NP-complete (and since sufficiently nice reductions to them are known), the results of [BSMP] and [LS] imply that any NP-statement can be proved in non-interactive zero-knowledge, if one-way trapdoor permutations exist, or in particular if quadratic residuosity decision is hard.

This, however, requires the use of Karp-reductions, such that from the original problem instance, one constructs for example a graph that is Hamiltonian, precisely if the original instance was a yes-instance. Thus the size of theorems one has to work with becomes larger than the original instance, implying a loss of efficiency.

A proof system for SAT, such as the one presented here, does not have this problem: for any NP-language L, by Cooks Theorem, there is a circuit C_L that when given a word w as fixed input on some of the wires, is satisfiable precisely if $w \in L$. Thus this circuit can be used directly in the proof system. Moreover, we are free to build ad hoc as small a circuit as possible for the given problem. Such a circuit will usually be far smaller than one conctructed from Cook's theorem. Although this circuit could also be handled by constructing from it a (linearly larger) 3-SAT instance and using the protocol of [BSMP], using our protocol directly will be significantly more efficient: we need the same amount of work and shared random bits per binary gate as the 3-SAT protocol needs per clause, and moreover all XOR and NOT gates can be handled in our protocol at virtually no cost (1 multiplication and no bits of the shared random string).

The known non-interactive zero-knowledge proofs for NP-complete problems, with or without preprocessing, are only computationally zero-knowledge. This is because the results were proved in the model where the prover may have unlimited computing power, while the verifier is polynomially bounded. In this model, perfect/statistical ZK seems to require special properties of the statement proved: if we were given a non-interactive perfect ZK proof for an NP-complete problem, we could use an unconditionally hiding bit commitment scheme (see USBCA, section 4) to "move" the proof system to the ordinary interactive model. Thus, the "impossibility" result of [Fo] would apply.

We therefore consider in stead the dual model (following [BCC]), where the verifier may be unlimited, while the prover must be polynomially bounded, i.e. we consider ZK arguments. We present a non-interactive statistical ZK argument with preprocessing for GH. This scheme will work if collision intractable hash functions exist, and has the property that the interaction required in the preprocessing phase does not depend on the size of theorem to be proved later. An alternative protocol for this problem could be derived from the techniques in [KMO] or [SP], but this would require an oblivious transfer subprotocol ([SP] gives a specific example based on the quadratic residuosity assumption), and oblivious transfer is not known to be (and probably is not) implementable based only on collision intractable hash functions.

Finally, we look at the problem of constructing a perfect (or statistical) zero-knowledge argument in the shared string model for an NP-complete problem. [SP] considered this problem and solved it in a modified scenario where, in addition to the shared random string, the prover has access to a message broadcast by the verifier, specifying a public key to use in the proof. To the best of the authors knowledge, the problem in the original shared string model is still open. We present a potential solution based on discrete logs that is provably perfect zero-knowledge, and is conjectured to be a proof system.

2 Notation and Definitions

The protocols we describe take place between a prover P and a verifier V. These are probabilistic Turing machines. In Section 3, the prover may have infinite computing power (but is not required to use it in the protocol), while the verifier is polynomially bounded. The roles are reversed in Section 4.

In Section 3, the *shared random string model* is assumed, i.e. both prover and verifier have read-only access to a one-way infinite tape containing independent random bits. This bit string is called α. P tries to convince V that some word w is in the NP-language L.

By $A(\cdot)$, we denote the random variable resulting from running probabilistic algorithm A on input $\cdot$. Thus $\sigma = P(w, \alpha)$ is the proof computed by P and sent to V. $V(\sigma, \alpha)$ is either *accept* or *reject*, and the proof system is said to accept or reject accordingly.

(P, V) is said to be a *proof system* for L, if it is *complete*: $w \in L$ implies that $Prob(V(w, \sigma, \alpha)) = reject$ is superpolynomially small in $|w|$; and *sound*: whenever $w \notin L$, for any (possibly infinitely powerful) P^*, $Prob(V(P^*(\alpha, w), \alpha)) = accept$ is superpolynomially small in $|w|$.

(P, V) is said to be *zero-knowledge* if there is a probabilistic polynomial time simulator M, which on input w only produces a random string and "proof" that looks just like the strings actually used in the protocol. More precisely, $M(w)$ is polynomially indistinguishable from α, σ (the simulator does not have to be relative to a verifier, because V does not take part in any interaction).

What we have described here is in fact what is called *bounded* non-interactive zero-knowledge in [BSMP].

In Section 4, we change the model, such that the verifier may now have infinite computing power, while the prover is polynomially bounded. Moreover, P and V are now interactive Turing machines, and there is no shared random string. By $(P, V)(w)$, we denote the verifiers decision after talking to P about common input w. In addition to w, P gets access to a witness x proving that $w \in L$.

The definition that (P, V) is a proof system for L is the similar to the above: the proof system must be *complete*: if $w \in L$, and x is a valid witness for w, then $Prob((P, V)(w) = accept)$ is superpolynomially large in $|w|$. It must be *sound*:

if $w \notin L$, then for any probabilistic polynomial time P^*, any auxiluary input x, and any set of coinflips for P^*, $Prob((P^*, V)(w) = accept)$ is superpolynomially small in $|w|$.

The definition of zero-knowledge in this model is the same as the standard one for interactive proof systems, except that we allow infinitely powerful verifiers, and hence are only interested in statistical or perfect zero-knowledge.

In addition, we would like the system to satisfy an additional requirement - (P, V) is said to be a *proof system with preprocessing*, if the protocol can be split in an interactive preprocessing phase, where neither party is given access to w, and a subsequent non-interactive proof-phase, where P gets w and x, while V gets only w. From this P computes in polynomial time a proof, which is sent to V. V checks the proof against w and the conversation from the preprocessing phase, and outputs *accept* or *reject*. Although the protocol should be secure against an infinitely powerful V, we will only look at protocols that a "real-life" verifier can execute, so we require that V can do both preprocessing and checking in polynomial time.

Note that w, x may be chosen as a function of the interaction that took place in the preprocessing phase.

3 Non-Interactive Proof System for SAT

In this section, we describe non-interactive proofs that are bounded in the sense of [BSMP]: the prover can only prove 1 SAT-instance of length $O(\sqrt[3]{l})$, where l is the length of the shared random string. In [BSMP], methods are given that allow transformation of their protocol to one where the prover can prove arbitrarily many instances of any size. These methods are also applicable to our protocol, with only trivial modifications.

The common input to prover and verifier is a one-output satisfiable Boolean circuit C. The prover may be thought of as a polynomial time machine, that knows an assignment of bits to the input wires that causes the output to be 1.

Without loss of generality, assume that C has gates $G_1, G_2, ..., G_k$, where all gates have two inputs and 1 output (which may be fanned out to several other gates). Thus each gate G_i may be described by T_i, a binary array with 4 rows and 3 columns, the truthtable of G_i. The gates receiving 1 or 2 input bits to C are called input gates. The circuit may specify that an input bit should enter several input gates. Such input bits are said to be shared between the gates.

The proof will use the quadratic residuosity assumption (QRA, introduced in [GM]) which says that for an n which is the product of two large primes, when given a random x with Jacobi symbol 1, it is hard to tell whether x is a quadratic residue, if the factorization of n is not known. We will say that x *hides the bit* b, if the quadratic character of x equals b. Clearly x_1 and x_2 hide the same bit precisely if $x_1 x_2$ is a square modulo n. If y is a non-square, then x hides b exactly if $y^b x$ is a square. When P includes a square root of $y^b x$ in his proof, this is referred to as *opening the number x*.

We can now describe the algorithm of P, which works by combining techniques from [BCC] and [BSMP]:

PROVER'S ALGORITHM

1. Choose random primes p, q, such that $n = pq$ is of length k bits, and a random non-square y.
2. From the first k^3 bits of α, produce a proof that y is a nonsquare modulo n, that n has two distinct prime factors, and that n is not a perfect square, exactly as in [BSMP].
3. Split the first $264k^3$ bits of the remainder of α in segments of length k bits, and consider the segments as $264k^2$ integers. Discard all numbers that are $\geq n$, or are $< n$, but have Jacobi symbol -1. Compute the bits hidden by all the remaining numbers.
4. The obtained sequence of numbers is split in groups of 3. A group x_1, x_2, x_3 is said to encode a row in the truth table T if T has a row b_1, b_2, b_3 such that x_i hides b_i for $i = 1, 2, 3$.
5. For $i = 1...k$ do the following:
   repeat for the first $11k$ unused groups in the sequence:
   If the group does not encode a row in T_i, prove this by opening all numbers in the group. Otherwise write information in the proof that this group encodes some row in T_i.
   Finally divide all the unopened groups assigned to T_i into 4 classes, such that all groups in a class encode the same row in T_i (only the subdivision is included in the proof, P does not reveal which row corresponds to which group). For each pair of groups belonging to the same class, prove this by displaying square roots of products of corresponding numbers in the groups.
6. The computation in C induced by using the satisfying assignment as input will select a row in each T_i. Now include in the proof a pointer to a group encoding this row. Let x_{1i}, x_{2i}, x_{3i} be this group, hiding the first and second input bit, resp. the output bit.
   For $i = 1..k$, do:
   If for some j, T_j receives its t'th input from T_i, prove consistency of the computation by displaying a square root of $x_{3i} x_{tj}$.
   If T_i is an input gate such that its t'th input bit is shared with the m'th input bit of gate T_j, where $j > i$, prove consistency by displaying a square root of $x_{ti} x_{mj}$.
   If T_i is the final output gate of C, prove that C outputs 1 by opening x_{3i}.

Note that this algorithm fails, if there are not $11k^2$ groups available for step 5, or if encodings of all 4 rows do not show up in step 5. In these cases P outputs a random string and stops. We proceed by describing how V should check the proof:

VERIFIER'S ALGORITHM

i. As in [BSMP], verify the proof that y is a non-square modulo n, that n has two distinct prime factors, and that n is not a square.

ii. Check that all numbers discarded by P in step 3 are $\geq n$ or have Jacobi symbol -1.

iii. For $i = 1..k$, check that P has correctly opened all numbers in groups discarded in step 5, and that indeed no opened group encodes a row of T_i.

iv. Check all the other sub-proofs produced by P in step 5. Check that the unopened groups assigned to each T_i have been divided into 4 classes.

v. Check all the square roots produced by P in step 6.

The first result about this protocol is:

Theorem 1 (P, V) is a non-interactive proof system for SAT.

Proof Completeness is obvious by inspection of the protocol: P only fails in the two cases mentioned after step 6, and their probability is exponentially small in k: for a fixed n, each k bit integer has a chance of at least $1/4$ of not being discarded, so we expect to have $66k^2$ integers left after step 3. It follows from Bernsteins law of large numbers that the probability that the actual number of remaining integers is less than half the expected value is exponentially small in k^2. Since there are at most 2^k possible n's, the probability that there exists an n leading to less than $33k^2$ remaining integers is exponentially small in k. The other failure case is handled below

For soundness, we have to argue that given C is not satisfiable, P^* manages to convince V with at most negligible probability. Assume that P^* has in fact produced a convincing proof.

First, by the proof in [BSMP], we may assume that the n, y given by P^* has the correct form. We may also assume that there were $11k^2$ groups available for step 5 of the prover.

Now observe that if C is not satisfiable, at least for one i, the group G used for T_i in step 6 does not encode a row of T_i. But this group comes from one of the classes of step 5. G cannot be proven equivalent to any of the proper encodings, so since there are only 4 classes and no proper encodings were opened, at least one row encoding of T_i never showed up during step 5.

Since α is random, for each given choice of n, y, the bits hidden by the numbers of Jacobi symbol 1 are independent and each bit is 1 with probability $1/2$. Therefore each group considered encodes each of the possible 8 rows with probability $1/8$. By elementary probability theory, we find that the probability that encodings of all possible 8 rows did not show up for one of the k gates is at most $8k(7/8)^{11k}$. Since there are at most 2^{2k} possibilities for n, y, the probability that there exists a choice of n, y allowing cheating by P^* (or failure for the honest prover) is at most

$$2^{2k} 8k(\frac{7}{8})^{11k} \leq 8k(0.93)^k$$

□

Theorem 2 Under QRA, (P, V) is a zero-knowledge non-interactive proof system.

Proof The simulator may be constructed as follows: first choose a modulus n of the correct form and a random *square* y, and then produce a simulated "proof" that y is a non-square modulo n, exactly as in [BSMP]. This will also produce a first segment of the simulated shared random string.

The rest of the shared random string is produced as follows: the simulator chooses a sufficient number of random bits, splits them in k bit integers, and marks as discarded those that are $\geq n$ or have Jacobi symbol -1. The rest of the integers are all replaced by random squares modulo n. This concludes the computation of the simulated shared random string. The simulator includes pointers to the discarded numbers in its proof, just as the prover would have done.

Note that since y is chosen as a square, the simulator can "open" each of the squares now produced, both as a 1 and as a 0.

The simulator now makes groups of the non-discarded numbers, and assigns $11k$ groups to each truth table, as in the prover's step 5. For each group, it decides with probability $1/2$ to "open" the group as a random group that does not properly encode a row in the truthtable in question. With probability $1/2$ it decides to leave the group unopened. Each unopened group is randomly put into 1 of 4 classes. If each truthtable does not own at least 1 group in all 4 classes, we fail and stop, as the prover would. Otherwise, for each truthtable, we "prove" equivalence of groups in the same class by displaying square roots of products of corresponding numbers in the groups. This is easy, since they are all squares.

Finally we simulate the prover's step 6 by choosing for each truthtable a random unopened group. We then display square roots of products of some of the numbers in the rows, exactly as required in step 6. Finally, we "open" the output bit as a 1.

It is easy to see that the only difference between the simulation and the real proof lies in the distribution of bits that are hidden in unopened numbers. Hence it is intuitively reasonable that a successful distinguisher would need the ability to distinguish squares from nonsquares.

This can be proved by contradiction: assume that for infinitely many k, there exists a satisfiable circuit C of size k for which the proofs constructed by our protocol are efficiently distinguishable from the simulation. We can then derive a contradiction with QRA by constructing an efficient algorithm A which on input n, y and a satisfying assignment for C will produce an output satisfying the following:

- If y is a square modulo n, the output is distributed as the simulator's output.
- If y is a non-square, the output is distributed exactly as the real random shared string and the prover's proof.

It is clear that this, together with the distinguisher, leads to a nonuniform algorithm violating QRA (non-uniform because we have to hardwire in the satisfying assignment for C).

To construct A, we proceed as follows: first run the simulation from [BSMP] of the proof that y is a nonsquare modulo n. It is proved in [BSMP] that this part has the property we require of A. We then construct the last part of the shared random string as in the simulator's algorithm above, EXCEPT that after constructing the non-discarded numbers as random squares, we decide at random for each number whether or not to multiply it by y. This means that A now can open each resulting number in only 1 way. But since A knows a satisfying assignment for C, A can still complete the proof: it simply follows the prover's algorithm. It should now be clear that if y is a square, nothing is changed compared to the simulation (all non-discarded numbers are still squares), while otherwise we get exactly the real prover's situation (the non-discarded numbers hide independent random bits). Note that, as mentioned in [BSMP], the verifier will obtain an indistinguishable view, no matter which satisfying assignment the prover uses. It therefore does not matter whether A uses the same assignment as the prover□

The protocol we have described is capable of handling arbitrary binary gates. However, from the homomorphic property of the mapping from numbers modulo n to the bits they hide, it is clear that XOR and NOT gates can be handled much more efficiently than general binary gates: given two numbers hiding input bits to an XOR gate, we simply multiply the numbers modulo n to get a bit hiding the output. Similarly, given a number hiding an input bit to a NOT gate, we multiply by y.

4 Perfect and Statistical Zero-Knowledge Arguments With Preprocessing

In this section we will be concerned with constructing a non-interactive perfect or statistical zero-knowledge protocol for an NP-complete problem (both with and without preprocessing). To get perfect zero-knowledge, however, it seems we have to change to the model where the prover is polynomially bounded, whereas the verifier may be unlimited, so we get zero-knowledge arguments. For ordinary interactive proofs, this is because of Forthnow's "impossibility result". The same result can be applied to non-interactive proofs, provided that unconditionally hiding bit commitments exist (assumption USBCA below).

One example of such commitments is the one presented in [CDG] and independently in [BKK]. Here, the prover receives a prime p, the factorization of $p-1$, a generator g of Z_p^*, and a random element $a \in Z_p^*$. The prover can check from these data that indeed g generates Z_p^*. A commitment to the bit b is computed as $a^b g^r$ where r is chosen uniformly in $[0..p-2]$. The prover is unable to change his mind unless he can compute the discrete log base g of a. On the other hand, commitments have distribution independent of the bits they hide. We will refer to this scheme as the discrete log scheme (DLS). The assumption that P cannot find the discrete log of a, even when given the factors of $p-1$ is known as the certified discrete log assumption (CDLA).

Rather than using a particular commitment scheme, it is natural to try to base the results on the general assumption that there exists a bit commitment scheme hiding bits unconditionally, since there is evidence to suggest that this is the minimal assumption that will support perfect zero-knowledge arguments in general [Da]. More precisely, we consider the following assumption:

Unconditionally Secure Bit Commitment Assumption (USBCA) There exists a infinite family of finite sets $\{I_k\}$, where an element in I_k is a function $BC : \{0,1\} \times \{0,1\}^k \to \{0,1\}^{s(k)}$, and $s(k)$ is polynomially related to k. The following should be satisfied:

- Given k, a random element (instance) of I_k can be selected in probabilistic polynomial time.
- Given $BC \in I_k$ selected according to the above condition, no probalistic polynomial time algorithm can find r, r' such that $BC(1,r) = BC(0,r')$.
- For any instance BC, the distribution of $BC(1,r)$ equals the distribution of $BC(0,r)$, when r is uniformly chosen.

In general, establishing and opening a commitment may be possible by some interaction between sender and receiver. Such schemes are not usable in this context, however.

In the third condition above, we may replace the requirement that the two distributions are equal with the requirement that they be statistically indistinguishable.

USBCA follows from many different intractability assumptions: hardness of discrete log, factoring or graph isomorphism; the existence of perfect zero-knowledge MA-proofs of knowledge [Da]; or hardness of some forms of the knapsack problem [NI].

To get more efficient protocols, we will need another assumption, namely that families of collision-intractable (or collision free) hash functions exist. Such a family has the property that it is easy to select at random a member function h with a k bit output, but although the input is longer than k bits, it is hard to find $x \neq y$ such that $h(x) = h(y)$. See [Da2] or [NY1] for a formal definition. The assumption on existence of collision intractable hash functions is at least as strong as USBCA, by a result of Naor and Yung [NY1]:

Proposition The existence of families of collision-intractable hash functions imply USBCA.

If the function $s(k)$ from USBCA is at most k, it is easy to see using the tecniques from [Da] that the inverse implication holds. Hence, for example, CDLA implies existence of collision intractable hash functions. The general inverse implication seems likely to be true also, but this is an open problem so far.

In the next two subsections, we look at protocols with preprocessing, and return to the shared string model in Section 4.3.

4.1 The protocol by Shamir and Lapidot

A simple approach to constructing an interactive argument with preprocessing for an NP-complete problem is to use the technique of Lapidot and Shamir [LS], which will give us a protocol for GH: they present a proof system with preprocessing which can be adapted to be based on USBCA.

For convenience, we briefly repeat the protocol here: in the preprocesing phase, P commits to k $k \times k$ incidence matrices $H_i, i = 1...k$, that each represent a graph consisting of one random Hamiltonian cycle. V responds with k random bits $b_1, ..., b_k$. In the proof phase, P sends a k-node graph G that he wants to prove is Hamiltonian, and also opens completely all those H_i for which $b_i = 0$, so that V can check that they were correctly constructed. Finally, for those i, where $b_i = 1$, P shows a permutation of the rows and columns of H_i and opens as a 0 all those entries in the permuted matrix that corresponds to 0's in the incidence matrix of G.

4.2 An Improvement Based on Hash Functions

One disadvantage with the protocol of Lapidot and Shamir is that it only works if G happens to have k nodes exactly, which is a problem if P does not know at preprocessing time which graph he will want to prove later. A simple solution is to let P supply sets of H's with j nodes for $j = 1..K$, where K is some maximum polynomially related to the security parameter k. Although this is still polynomial in k, it is hardly an attractive solution, since P and V have to exchange a number of bits corresponding to the value of K, even if G turns out to be much smaller.

Based on the assumption that collision intractable hash functions exist, we propose a different way to use the techniques of [LS], for which a much more efficient preprocessing and proof phase is possible. More precisely:

- The communication complexity of the preprocessing phase is independent of both K and the size of G.
- The communication complexity of the proof phase is $O(j^2 ks(k) + jk)$ bits, where j is the size of G and k is the security parameter. In particular, it is independent of K, and if G has k nodes, we get essentially the same complexity as the basic Lapidot/Shamir solution.

First, let us remark that USBCA implies that one-way functions exist: consider the function that maps r, b to $BC(r, b)$. This function must be hard to invert, since the prover could clearly use an inversion algorithm to cheat. Thus, by [Na], USBCA also implies the existence of a bit commitment scheme with the dual property: the committer unconditionally cannot cheat, but the receiver of a commitment may find the bits if he has enough computing power.

In [FS], a construction is presented which allows transforming any bit commitment scheme into one that is *chameleon*, assuming that a one-way function exists; i.e. using this result, we may assume that associated with an instance of the commitment scheme is some *trapdoor information*, which can be chosen by the verifier, and which allows changing the contents of commitments.

The final observation we need comes from [Da2], where it is shown how to construct a collision intractable hash function h that is defined for arbitrary length inputs, based on a collision intractable function f from m bits to k bits, where $m > k$. We repeat here a simplified version of the construction which will be sufficient for our purposes. Put $t = m - k$. To hash input M, split it in t-bit blocks $M_1, ..., M_n$, padding the last block with 0's if needed. Then we define:

$$h(M, Z) = f(M_n || f(M_{n-1} || f(\cdots f(M_2 || f(M_1 || Z)) \cdots))),$$

where Z denotes a string of k bits, and $||$ denotes concatenation. It is now quite easy to prove that if it is infeasible to find collisions for f, then it is infeasible to find M, M', Z, Z' such that $M \neq M'$, the length of M equals that of M', and $h(M, Z) = h(M', Z')$. Moreover, it always holds that

$$h(M, Z) = h(M_j || M_{j+1} \cdots || M_n, h(M_1 || \cdots || M_{j-1}, Z))$$

In other words, when the prover sends a hashvalue, this commits him to the preimage, but the construction of h allows him to convincingly reveal only part of it at some later time. For simplicity in the following, we will assume that $t = k$. It is easy to modify the construction to do without this condition.

Let $H^{(j)} = \{H_i | i = 1...k\}$ denote a set of matrices chosen by the prover as in the above description, where k is the security parameter of the protocol below, and let $BC(H^{(j)}, R)$ denote a set of commitments to the bits in $H^{(j)}$, computed with random input R. The preprocessing goes as follows:

PREPROCESSING PHASE

1. V chooses an instance of a chameleon bit commitment scheme BC according to USBCA and the above observations. Using commitments based on a one way function, he convinces P in zero-knowledge that he, V, knows the trapdoor for the commitment scheme (using for example the general protocol from [BCC]).
   He also chooses a random member h of a family of collision intractable hash functions (constructed as above), and sends h to P.

2. P chooses at random $H^{(j)}$ for $j = 1...K$. He then computes

$$A_j = h(BC(H^{(j)}, R_j), Z_j)$$

   for $j = 1...K$ and randomly chosen R_j, Z_j. Finally, he sends to V the hashed image of the A_j's $h(A_K || \cdots || A_1, Z)$, for randomly chosen Z.

3. V returns a string E consisting of k random bits.

PROVER'S ALGORITHM, PROOF PHASE

1. Given a Hamiltonian graph G with j nodes, where $1 \leq j \leq K$, the prover sends $A_1, ..., A_j$, $BC(H^{(j)}, R_j)$, Z_j, and $h(BC(H^{(j+1)}, R_{j+1})|| \cdots ||BC(H^{(K)}, R_K), Z)$.
2. From $BC(H^{(j)}, R_j)$ and E, the prover generates a proof that G is Hamiltonian, as described above.

VERIFIER'S ALGORITHM, PROOF PHASE

1. Given a Hamiltonian graph G with j nodes, where $1 \leq j \leq K$, and the data received from the prover, the verifier checks that $A_j = h(BC(H^{(j)}, R_j), Z_j)$, and that $h(A_j||A_{j-1} \cdots ||A_1, h(A_K|| \cdots ||A_{j+1}, Z))$ equals the hash value received in the preprocessing.
2. Use $BC(H^{(j)}, R_j)$ and E to verify the provers proof as described above.

We can now prove:

Theorem 3 Under the assumption that families of collision intractable hash functions exist, the above constitutes a statistical zero-knowledge non-interactive argument with preprocessing for GH.

Proof sketch Completeness is (as usual) quite trivial. For soundness, let $E1 \neq E2$ be any pair of E-values for which a cheating prover P^* has success, and for which P^* sends the same preimages under h in the two cases. $E1$ must be different from $E2$ in at least one bit position. This implies that there is an H_i in some $H^{(j)}$, which P^* can open and show Hamiltonian, but for which he can also show an appropriate relation to a non-Hamiltonian graph. It is easy to see that this cannot be the case, unless P^* opens at least one commitment to a bit in H_i in two different ways.

Let EE_j be the set of E-values for which P^* successfully produces a proof for a non-Hamiltonian graph with j nodes. If this set constitutes a polynomial fraction of the total number of possibilities, there exists a polynomial time algorithm that (by using rewinding of P^* to the start of step 3 above) generates several elements in EE_j together with the resulting proofs from P^*. By assumption on h, P^* will reveal the same preimages for h in nearly all cases from EE_j, since the preimages must have the same length in order for the proofs to be convincing. Then by the above, P^* will tell us how to open a commitment in two different ways, which contradicts USBCA. Since K is polynomial in k, this means that the union of all EE_j, $j = 1..K$ constitutes a negligible fraction of the total number of cases.

To prove the zero-knowledge property, observe that the simulator can use rewinding of a cheating verifier V^* to extract from the proof in step 1 the trapdoor of the bit commitment scheme. If the simulator manages to find this trapdoor, it can change the contents of any commitment, and the rest of the simulation becomes trivial. Therefore we can simulate all cases perfectly, except those exponentially few ones where V^* manages to cheat in step 1□

Corollary 1 If in the above construction DLS is used for bit commitments, then under CDLA, the resulting protocol is a perfect zero-knowledge argument with preprocessing for GH.

Proof As mentioned before, CDLA implies that collision intractable hash functions exist. Therefore, CDLA alone suffices to implement all the tools we need for the protocol, and soundness and completeness can be proved just as in Theorem 3.

For zero-knowledge, recall that from the data received by the prover for DLS, the prover can check that g generates all of Z_p^*. Therefore, the cases where the verifier cheats successfully in step 1 can still be simulated: the simulator simply finds the discrete log of a by exhaustive search, and then proceeds as above. This situation only occurs with exponentially low probability, and therefore the contribution to the expected running time is only polynomial□

Remark Note that since no message sent in the protocol depends on the upper bound on the theorem size, the fact that such a bound must be known by P is of little consequence: P can generate as many commitments off-line as he wishes, and does not even have to store all the random bits needed for the commitments, because he can generate them pseudorandomly (although the protocol is then only computationally zero-knowledge).

4.3 Perfect Zero-Knowledge Arguments in the Shared String Model

A natural question is of course whether one can construct perfect or statistical zero-knowledge non-interactive arguments in the shared string model, i.e. without preprocessing. For some problems the answer is yes, [BSMP] contains an example for quadratic non-residuosity. No such protocol is known for an NP-complete problem, however. We close this section with a protocol for SAT that partially solves the problem: it is perfect zero-knowledge, but we have not been able to reduce the question of soundness to a generally accepted intractability assumption.

First, recall that it is well-known that there exists a probabilistic polynomial time algorithm A that on input k selects a random k bit prime p and a generator g of Z_p^*, and only fails with negligible probability.

Given the random string α and a satisfiable circuit C of size k, the prover does the following: run A on input k, using the first bits of α as coinflips for A. Let the output be p, g. With the next unused bits, select a constant a uniformly in Z_p^* by repeatedly considering k bit segments of α until one is found that as an integer is less than p. Now encrypt k copies of C as in [BCC] using DLS.

Take the concatenation of all the encryptions thus produced through a one-way function f, and use the first k bits of the output in place of the challenges from the verifier of [BCC]. Include all of the encryptions and answers in the proof.

As for Fiat-Shamir, Schnorr and Guillou-Quisquater signatures (and [BCC] which also mentions the idea of using an f as described above), we can only conjecture the existence of a function f that would make this proof convincing. It is clear, however, that the protocol is perfect zero-knowledge: the simulator can generate p, g as the prover would do it, but choose a with known discrete log base g (and then let the corresponding portion of the shared random string be determined by a).

5 Conclusion and Open Problems

We have shown a non-interactive zero-knowledge proof for SAT, making non-interactive proofs for general NP-statements more direct and efficient.

Based on the assumption that collision intractable hash functions exist, we have shown a non-interactive statistical zero-knowledge argument with preprocessing for GH, in which the preprocessing phase can be made independent of the size of theorem to be proved later.

This protocol works, based on CDLA only, and is then perfect zero-knowledge.

Open problem: find a perfect or statistical zero-knowledge non-interactive argument for an NP-complete problem.

References

[BCC] Brassard, Chaum and Crépeau: "Minimum Disclosure Proofs of Knowledge", JCSS, vol. 37 (1988) pp. 156-189.

[BFM] Blum, Feldman and Micali: "Non-interactive Zero-Knowledge Proof Systems and Applications", Proceedings of STOC 88.

[BKK] Boyar, Krentel and Kurtz: "A Discrete Logarithm Implementation of Perfect Zero-Knowledge Blobs", J. Cryptology, vol 2, no.2.

[CDG] Chaum, Damgård and van de Graaf: "Multiparty Computations Ensuring Privacy of Each Party's Input and Correctness of the Result", Proc. of Crypto 87, Springer Verlag.

[Da] Damgård: "On the Existence of Bit Commitment Schemes and Zero-Knowledge Proofs", Proc. of Crypto 89.

[Da2] Damgård: "A Design Principle for Hash Functions", Proc. of Crypto 89.

[FFS] Fiat, Feige and Shamir: "Zero-Knowledge Proofs of Identity" J. Cryptology, vol 1, pp. 77-94.

[FS] Feige and Shamir: "Zero-Knowledge Proofs of Knowledge in two Rounds", Proc. of Crypto 90.

[FLS] Feige, Lapidot and Shamir: "Multiple Non-Interactive Zero-Knowledge Proof based on a Single Random String", Proc. of FOCS 90.

[Fo] Fortnow: "The Complexity of Perfect Zero-Knowledge", Proc. of STOC 87.

[GM] Goldwasser and Micali: "Probabilistic Encryption", JCSS, vol.28 (1984), pp.270-299.

[GMR] Goldwasser, Micali and Rackoff: "The Knowledge Complexity of Interactive Proof Systems", Proc. of STOC 85.

[KMO] Killian, Micali and Ostrovski: "Minimum Resource Zero-knowledge Proofs", Proc. of Crypto 89.

[LS] Lapidot and Shamir: "Publicly Verifiable Non-Interactive Zero-Knowledge Proofs", Proc. of Crypto 90.

[Na] Naor: "Bit Commitments using Pseudo-Randomness", Proc. of Crypto 89.

[NI] Naor and Impagliazzo: "Efficient Cryptographic Schemes Provavbly as Secure as Subset Sum", Proc. of STOC 89.

[NY1] Naor and Yung: "Universal One-Way Hash Functions and their Cryptographic Applications", Proc. of STOC 90.

[BSMP] Blum, De Santis, Micali and Persiano: "Non-Interactive Zero-Knowledge", SIAM J.Computing, Vol.20, no.6, 1991.

[SMP] De Santis, Micali and Persiano: "Non-Interactive Zero-knowledge with Preprocessing", Proc. of Crypto 88.

[SP] De Santis and Persiano: "Public-Randomness in Public-Key Cryptography", Proc. of EuroCrypt 90.

Tools for Proving Zero Knowledge

Ingrid Biehl, Johannes Buchmann,
Bernd Meyer, Christian Thiel, Christoph Thiel

Universität des Saarlandes, Fachbereich Informatik,
Im Stadtwald 15, 6600 Saarbrücken, Germany

Abstract. We develop general techniques that can be used to prove the
zero knowledge property of most of the known zero knowledge protocols. Those techniques consist in reducing the circuit indistinguishability
of the output distributions of two probabilistic Turing machines to the
indistinguishability of the output distributions of certain subroutines.

1 Introduction

It is an important result in the theory of zero knowledge proofs that assuming the
existence of a circuit secure encryption machine every language in NP has a zero
knowledge proof. This result can be obtained by constructing a zero knowledge
proof system for the NP-complete language 3C of three colourable graphs (see
[1, 2]). In this protocol the prover and the verifier repeat a certain subprotocol a
number of times which is polynomial in the length of the input. The encryption
machine is called in a subroutine used in that subprotocol. The protocol can
therefore be written in the form

$$S = (MNO)^{n_x} \tag{1}$$

where MNO is the subprotocol which is repeated n_x times, where x is the input
and N is the subroutine which calls the encryption machine, and where M and
O are the machines that carry out the computations before and after N is used.
In order to show that S has the zero knowledge property one must show that the
communication carried out in S can be simulated by a probabilistic polynomial
time Turing machine even if the verifier is replaced by a cheating verifier. After
replacing the verifier the protocol is still of the form (1). The simulator S' is
constructed by replacing N with a machine N' which has no knowledge of a three
colouring of the input graph. By virtue of the circuit security of the encryption
machine, the output distributions of N and N' are circuit indistinguishable.
It remains to be shown that the output distributions of S and S' are circuit
indistinguishable.

Protocols and simulators for other zero knowledge protocols are constructed
in the same way.

The goal of this paper is to unify the proofs for the zero knowledge property.
We show that replacing the subroutine N with N' in a probabilistic Turing
machine S of the form (1) yields (under certain conditions) a machine S' whose
output distribution is circuit indistinguishable from the output distribution of

R.A. Rueppel (Ed.): Advances in Cryptology - EUROCRYPT '92, LNCS 658, pp. 356-365, 1993.

S if the output distribution of N is circuit indistinguishable from the output distribution of N'. Another goal of this paper is to precisely define the notions used in this context.

2 Probabilistic Turing Machines

Throughout this paper we use the alphabets $\Sigma = \{0,\ 1,\ \#\}$ and $\Sigma_0 = \{0,\ 1\}$.

Definition 1. A *probabilistic Turing machine* is a pair $Z = (M, p)$ where

1. $M = (K, \Sigma, \Delta, s)$ is a k-tape nondeterministic Turing machine (see [3], pp. 204–211)
2. $p \colon \Delta \to [0, 1]$ is a function which determines the probability of each transition in Δ, i.e. for every $q \in K$ and $\mathbf{a} \in \Sigma^k$ we have

$$\sum_{d \in \Delta(q, \mathbf{a})} p(d) = 1 \ ,$$

where $\Delta(q, \mathbf{a}) = \Delta \cap (\{q\} \times \{\mathbf{a}\} \times (K \cup \{h\}) \times (\Sigma \cup \{L, R\})^k)$.

A probabilistic Turing machine with $p(\Delta) \subseteq \{0, \frac{1}{2}, 1\}$ is called *coin tossing machine*.

We adopt the input and output conventions of [3]. For $x, y \in \Sigma_0^*$ we denote by $\Pi_Z(x, y)$ the probability for Z to output y on input of x. It is easy to see that

$$\sum_{y \in \Sigma_0^*} \Pi_Z(x, y) \leq 1 \ .$$

For $x \in \Sigma_0^*$ we denote by $Z(x)$ the set of all elements in Σ_0^* that can with positive probability occur as an output of Z on input of x, and $Z(L) = \bigcup_{x \in L} Z(x)$. If the length of each computation of Z on input of x is bounded by $c \in \mathbb{N}$, we denote the maximal length of a computation of Z on input of x by time$(Z(x))$. We say that *the running time of Z is bounded by a function* $T : \mathbb{N} \to \mathbb{N}$ if for all $x \in \Sigma_0^*$ we have time$(Z(x)) \leq t(|x|)$.

If there is a function $\ell : \mathbb{N} \to \mathbb{N}$ such that, on input of strings of length u, the machine only outputs strings of length $\ell(u)$ with positive probability, then we call Z a *homogeneous* probabilistic Turing machine.

If there is a polynomial $f \in \mathbb{N}[X]$ such that, on input of strings of length u, the output length of the machine Z is bounded by $f(u)$, we call Z a probabilistic Turing machine with *polynomially bounded output length*.

Let Z_1 and Z_2 be probabilistic Turing machines. Then the *concatenation* $Z_1 Z_2$ of Z_1 and Z_2 is defined as the probabilistic Turing machine that first operates as Z_1. Whenever Z_1 terminates, Z_2 is called where the input of Z_2 is the output of Z_1. We also use the notation Z_1^n for $\underbrace{Z_1 Z_1 \cdots Z_1}_{n \text{ times}}$.

3 Probabilistic Circuits

A *probabilistic circuit* is a deterministic circuit (see [4], pp. 73) with a partition $\text{In} = \text{In}_D \cup \text{In}_P$, $\text{In}_D \cap \text{In}_P = \emptyset$, of the input nodes. The input nodes in In_D (the deterministic input nodes) receive the input of the computation. The nodes in In_P (the probabilistic input nodes) are assigned uniformly at random 0 or 1. The number of all nodes but the input nodes of a circuit C is called size(C).

For a probabilistic circuit C with m input nodes and n output nodes and for $y \in \{0, 1\}^m$, $z \in \{0, 1\}^n$ we denote by $\Pi_C(y, z)$ the probability for C to output z on input of y.

Let C_1 be a probabilistic circuit with n output nodes and C_2 be a probabilistic circuit with n input nodes. Then the *composition* of C_1 and C_2 is the circuit which results by connecting the output nodes of C_1 with the input nodes of C_2.

Let $L \subseteq \Sigma_0^*$. A family $\{C_x\}_{x \in L}$ of probabilistic circuits is called *polynomial* if size(C_x) is bounded by $|x|^k$ for some $k \in \mathbb{N}$.

In order to be able to prove our main theorems we need the following results.

Lemma 2. *There is a constant $c \in \mathbb{N}$ such that for all Turing-decidable languages $L \subseteq \{0, 1\}^*$ the following holds: if L is decided by a deterministic Turing machine $M = (K, \Sigma, \delta, s)$ in time $T: \mathbb{N} \to \mathbb{N}$, then there is a family $\{C_n\}_{n \in \mathbb{N}}$ of deterministic circuits which decides L and satisfies*

$$\text{size}(C_n) = (|K| \, |\Sigma|)^c T(n) \log T(n) \ .$$

Proof. See [4], pp. 84–91.

Lemma 3. *There are $c, d \in \mathbb{N}$ such that for all homogenous polynomial coin tossing machines $M = ((K, \Sigma, \delta, s), p)$ with output length $\ell : \mathbb{N} \to \mathbb{N}$ and running time bounded by $T \in \mathbb{N}[X]$ there is a polynomial family $\{C_n\}_{n \in \mathbb{N}}$ of probabilistic circuits such that $\{\Pi_{C_{|x|}}(x, \cdot)\}_{x \in L}$ and $\{\Pi_M(x, \cdot)\}_{x \in L}$ are equal and which satisfies*

$$\text{size}(C_n) = d\ell(n)(|K| \, |\Sigma|)^c T(n) \log T(n) \ .$$

Proof. Without loss of generality we assume that there is $q \in \mathbb{N}[X]$ such that on input of length n the machine M tosses the coin exactly $q(n)$ times. Moreover, we can construct a homogeneous polynomial time deterministic Turing machine $M' = (K', \Sigma, \delta', s')$ with the following property: suppose that on input of x the machine M carries out the sequence of coin tosses $\alpha = (\alpha_1, \dots, \alpha_{q(|x|)})$ and outputs y, then, one input of (x, α), the machine M' outputs y. There is a constant $r \in \mathbb{N}$ (independent of M) such that $|K'| \leq r|K|$. There is an other constant $s \in \mathbb{N}$ (independent of M') such that $T'(|(x, \alpha)|) \leq sT(|x|)$ for the running time T' of M'.

If we define for $x, y \in \Sigma_0^*$

$$\Pi_{M'}(x, y) = \frac{1}{2^{q(|x|)}} \left| \left\{ \alpha \in \{0, 1\}^{q(|x|)} | M'(x, \alpha) = y \right\} \right|$$

then we have $\Pi_{M'} = \Pi_M$.

In order to be able to apply Lemma 2 we consider the deterministic Turing machine M'_m $(m \in \mathbb{N})$ which on input of (x, α) outputs the mth bit of the output y which is defined to be 0 if $m > \ell(|x|)$.

For $x \in \{0, 1\}^*$, $y = (y_1, \ldots, y_{\ell(|x|)}) \in \{0, 1\}^*$ we have

$$\Pi_{M'}(x, y) = \frac{1}{2^{q(|x|)}} \left| \left\{ \alpha \in \{0, 1\}^{q(|x|)} | M'_i(x, \alpha) = y_i, \ \forall \ 1 \le i \le \ell(|x|) \right\} \right| .$$

The machine M'_m works exactly as M' and deletes at the end of its computation all but the mth bit of the output. Therefore there is a constant $t \in \mathbb{N}$ (independent of M') such that $T'_m(|(x, \alpha)|) \le tT'(|(x, \alpha)|)$ for the running time T'_m of M'_m. The number of states of M'_m is polynomial in the number of states of M.

We apply Lemma 2 to M'_m and thus obtain a polynomial family $\{C_n^{(m)}\}_{n \in \mathbb{N}}$ of deterministic circuits which simulates M'_m. The circuit $C_n^{(m)}$ has n deterministic and $q(n)$ probabilistic input vertices.

We construct the circuit C_n by connecting the $C_n^{(m)}$, $1 \le m \le \ell(|x|)$, in the natural order in parallel, that means all circuits have the same deterministic and probabilistic input.

Since C_n is constructed from $\ell(n)$ circuits whose size is polynomially bounded in n, the size of C_n itself is bounded by a polynomial in n, which means that $\{C_n\}_{n \in \mathbb{N}}$ is a polynomial family of circuits. Moreover, we have by construction that for every $x \in \{0, 1\}^*$

$$\Pi_{C_{|x|}}(x, \cdot) = \Pi_M(x, \cdot) .$$

$\square$

4 Indistinguishability

Let U and V be two probability distributions on Σ_0^*. The series

$$\delta_S(U, V) = \sum_{y \in \Sigma_0^*} |U(y) - V(y)|$$

is called the *statistical difference* between U and V. In general it is impossible to determine in polynomial time that two probability distributions have a non zero statistical difference. Therefore one uses tools like probabilistic circuits and probabilistic algorithms (i.e. probabilistic Turing machines) to distinguish between probability distributions.

For a probabilistic circuit with m input nodes and one output node we call

$$\delta_C(U, V) = \left| \sum_{y \in \Sigma_0^m} \Pi_C(y, 1)(U(y) - V(y)) \right|$$

the *circuit difference* between U and V with respect to C.

Finally, for a probabilistic Turing machine Z we call the series

$$\delta_Z(U, V) = \left| \sum_{y \in \Sigma_0^*} \Pi_Z(y, 1)(U(y) - V(y)) \right|$$

the *algorithmical difference* between U and V with respect to Z.

Definition 4. Let $L \subseteq \Sigma_0^*$, let $U = \{U_x\}_{x \in L}$ and $V = \{V_x\}_{x \in L}$ be two families of probability distributions on Σ_0^*.

1. The families U and V are called *perfectly indistinguishable* (*p*-indistinguishable) if $U = V$.
2. The families U and V are called *statistically indistinguishable* (*s*-indistinguishable) if for every $k \in \mathbb{N}$

$$\lim_{\substack{x \in L \\ |x| \to \infty}} |x|^k \delta_S(U_x, V_x) = 0 \ .$$

3. The families U and V are called *circuit indistinguishable* (*c*-indistinguishable) if for every polynomial family $\{C_x\}_{x \in L}$ of probabilistic circuits C_x and for every $k \in \mathbb{N}$ we have

$$\lim_{\substack{x \in L \\ |x| \to \infty}} |x|^k \delta_{C_x}(U_x, V_x) = 0 \ .$$

4. The families U and V are called *algorithmically indistinguishable* (*a*-indistinguishable) if for every polynomial time probabilistic Turing machine Z and for every $k \in \mathbb{N}$ we have

$$\lim_{\substack{x \in L \\ |x| \to \infty}} |x|^k \delta_Z(U_x, V_x) = 0 \ .$$

Lemma 5. *Let $L \subseteq \Sigma_0^*$. Let Z and Z' be homogeneous probabilistic Turing machines. Assume that Z has polynomial output length. If $\{\Pi_Z(x, \cdot)\}_{x \in L}$ and $\{\Pi_{Z'}(x, \cdot)\}_{x \in L}$ are circuit indistinguishable then for all $x \in L$ but a finite set the elements of $Z(x)$ and $Z'(x)$ are of the same length.*

Proof. The case $|L| < \infty$ is trivial. So assume $|L| = \infty$. Assume that there is an infinite subset $L' \subseteq L$ such that for every $x \in L'$ the elements of $Z(x)$ and $Z'(x)$ are of different length. For $x \in L$ let C_x be the circuit with m_x input nodes, m_x being the length of the elements in $Z(x)$, which always outputs 1. Then we have

$$\delta_{C_x}(\Pi_Z(x, \cdot), \Pi_{Z'}(x, \cdot)) = 1$$

for all $x \in L'$, hence

$$\lim_{\substack{|x| \to \infty \\ x \in L}} |x| \delta_{C_x}(U_x, V_x) \neq 0 \ .$$

□

5 The Main Theorems

Let $L \subseteq \Sigma_0^*$. A family $\{Z_x = ((K_x, \Sigma_x, \delta_x, s_x), p_x)\}_{x \in L}$ of probabilistic Turing machines is called *polynomial* if there are $p, q, r \in \mathbb{N}[X]$ such that for all $x \in L$ and $y \in \Sigma_0^*$ $\mathrm{time}(Z_x(y)) \leq p(|x|)q(|y|)$ and $|K_x| |\Sigma_x| \leq r(|x|)$.

Theorem 6. *Let $L \subseteq \Sigma_0^*$. Let $\{M_x\}_{x \in L}$ be a family of probabilistic Turing machines. Let N and N' be homogeneous probabilistic Turing machines, N' having polynomial output length. We define $\Gamma = \bigcup_{x \in L} M_x(x)$. Let $\{O_x\}_{x \in L}$ be a polynomial family of homogenous coin tossing machines. Assume that the following conditions hold:*

1. $|z| \geq |x|$ *for all $x \in L$ and $z \in M_x(x)$.*
2. *For $x \in L$ all elements of $M_x(x)$ are of the same length.*
3. $\{\Pi_N(u, \cdot)\}_{u \in \Gamma}$ *and* $\{\Pi_{N'}(u, \cdot)\}_{u \in \Gamma}$ *are c-indistinguishable.*

Then $\{\Pi_{M_x N O_x}(x, \cdot)\}_{x \in L}$ and $\{\Pi_{M_x N' O_x}(x, \cdot)\}_{x \in L}$ are c-indistinguishable.

Proof. For $x \in L$ we set $A_x = M_x N O_x$ and $B_x = M_x N' O_x$. Let $x \in L$ and let $\{O_{x,n}\}_{n \in \mathbb{N}}$ be a polynomial family of circuits simulating the probabilistic Turing machine O_x (see Lemma 3). Let $\overline{N}(u) = N(u) \cup N'(u)$.

Let $I \in \mathbb{N}$ such that for every $u \in \Gamma, |u| > I$ the elements of $N(u)$ and $N'(u)$ are of the same length. According to Lemma 5 such an I exists. Let $x \in L$, $|x| > I$, the elements of $A_x(x)$ and $B_x(x)$ are of the same length, say m_x. To measure a non zero circuit difference between $\Pi_{A_x}(x, \cdot)$ and $\Pi_{B_x}(x, \cdot)$ a circuit C must have m_x input nodes. Let $\{C_x\}_{x \in L}$ be a polynomial family of circuits. We assume that C_x has exactly m_x input nodes.

For $u \in M_x(x)$ all elements in $\overline{N}(u)$ have the same length $\ell(u)$, $\ell \in \mathbb{N}[X]$. Let $O'_{x,u} = O_{x,\ell(u)} C_x$ be the composition of $O_{x,\ell(u)}$ and C_x. Now we have:

$$\delta_{C_x} \left(\Pi_{A_x}(x, \cdot), \Pi_{B_x}(x, \cdot) \right)$$

$$= \left| \sum_{y \in \Sigma_0^{m_x}} \Pi_{C_x}(y, 1) \left(\Pi_{M_x N O_x}(x, y) - \Pi_{M_x N' O_x}(x, y) \right) \right|$$

$$= \left| \sum_{y \in \Sigma_0^{m_x}} \Pi_{C_x}(y, 1) \sum_{u \in M_x(x)} \sum_{v \in \overline{N}(u)} \Pi_{M_x}(x, u) \left(\Pi_N(u, v) - \Pi_{N'}(u, v) \right) \Pi_{O_x}(v, y) \right|$$

$$= \left| \sum_{u \in M_x(x)} \Pi_{M_x}(x, u) \sum_{v \in \overline{N}(u)} \sum_{y \in \Sigma_0^{m_x}} \Pi_{O_x}(v, y) \Pi_{C_x}(y, 1) \left(\Pi_N(u, v) - \Pi_{N'}(u, v) \right) \right|$$

$$= \left| \sum_{u \in M_x(x)} \Pi_{M_x}(x, u) \sum_{v \in \overline{N}(u)} \Pi_{O'_{x,u}}(v, 1) \left(\Pi_N(u, v) - \Pi_{N'}(u, v) \right) \right|$$

$$\leq \sum_{u \in M_x(x)} \Pi_{M_x}(x, u) \delta_{O'_{x,u}} \left(\Pi_N(u, \cdot), \Pi_{N'}(u, \cdot) \right) .$$

Let $u \in \Sigma_0^*$ and let $x' \in L$ such that $u \in M_{x'}(x')$ then $|x'| \leq |u|$. Hence there are only finitely many $x' \in L$ with $u \in M_{x'}(x')$. Among the finitely many circuits $O'_{x',u}$, $u \in M_{x'}(x')$, we denote by O'_u the circuit which maximizes $\delta_{O'_{x',u}}(\Pi_N(u, \cdot), \Pi_{N'}(u, \cdot))$. Then we have

$$\delta_{C_x}(\Pi_{A_x}(x, \cdot), \Pi_{B_x}(x, \cdot)) \leq \max_{u \in M_x(x)} \delta_{O'_u}(\Pi_N(u, \cdot), \Pi_{N'}(u, \cdot)) . \qquad (2)$$

Now we consider $\text{size}(O'_u)$:

$$\text{size}(O'_u) \leq \max_{\substack{x \in L \\ |x| < |u|}} \text{size}(O'_{x,u}) = \max_{\substack{x \in L \\ |x| \leq |u|}} \text{size}(O_{x,\ell(u)} C_x)$$

$$\leq \max_{\substack{x \in L \\ |x| \leq |u|}} (\text{size}(O_{x,\ell(u)}) + \text{size}(C_x)) .$$

Since $\{O_x\}_{x \in L}$ is a polynomial family of coin tossing machines and $|x| \leq |u|$ there is (according to Lemma 3) a $p \in \mathbb{N}[X]$ such that for all $u \in M_x(x)$ $\text{size}(O'_u) \leq p(|u|)$. This implies with (2) that

$$\lim_{\substack{x \in L \\ |x| \to \infty}} |x|^k \delta_{C_x}(\Pi_{A_x}(x, \cdot), \Pi_{B_x}(x, \cdot))$$

$$\leq \lim_{\substack{x \in L \\ |x| \to \infty}} \max_{u \in M_x(x)} |u|^k \delta_{O'_u}(\Pi_N(u, \cdot), \Pi_{N'}(u, \cdot))$$

$$= 0 .$$

$\square$

Let $L \subseteq \Sigma_0^*$ be a language and let $\{n_x\}_{x \in L}$ be a sequence in $\mathbb{N}$. That sequence is called *polynomially bounded* if there is $d > 0$ such that for every $x \in L$ we have $n_x \leq |x|^d$.

Theorem 7. *Let $L \subseteq \Sigma_0^*$. Let $\{n_x\}_{x \in L}$ be a polynomially bounded sequence in $\mathbb{N}$. Let S and T be homogeneous probabilistic Turing machines. Assume that the following conditions hold:*

1. *$|y| \geq |x|$ for all $x \in L$ and $y \in S(x)$.*
2. *$S(L) \subseteq L$.*
3. *T is a coin tossing machine such that for $q \in \mathbb{N}[X]$ and for all $i \in \mathbb{N}$ we have $\text{time}(T^i(z)) \leq iq(|z|)$.*
4. *$\{\Pi_S(x, \cdot)\}_{x \in L}$ and $\{\Pi_T(x, \cdot)\}_{x \in L}$ are c-indistinguishable.*

Then $\{\Pi_{S^{n_x}}(x, \cdot)\}_{x \in L}$ and $\{\Pi_{T^{n_x}}(x, \cdot)\}_{x \in L}$ are c-indistinguishable.

Proof. We have

$$\delta_{C_x}(\Pi_{S^{n_x}}(x, \cdot), \Pi_{T^{n_x}}(x, \cdot)) = \left| \sum_{y \in \Sigma_0^{m_x}} \Pi_{C_x}(y, 1)(\Pi_{S^{n_x}}(x, y) - \Pi_{T^{n_x}}(x, y)) \right| .$$

We can write

$$\Pi_{S^{n_x}}(x,y) - \Pi_{T^{n_x}}(x,y) = \sum_{i=0}^{n_x-1} \Pi_{S^{n_x-i}T^i}(x,y) - \Pi_{S^{n_x-i-1}T^{i+1}}(x,y)$$

and therefore

$$\delta_{C_x}\left(\Pi_{S^{n_x}}(x,\cdot), \Pi_{T^{n_x}}(x,\cdot)\right)$$

$$= \left| \sum_{y \in \Sigma_0^{m_x}} \Pi_{C_x}(y,1) \left(\sum_{i=0}^{n_x-1} \Pi_{S^{n_x-i}T^i}(x,y) - \Pi_{S^{n_x-i-1}T^{i+1}}(x,y) \right) \right|$$

$$= \left| \sum_{i=0}^{n_x-1} \sum_{y \in \Sigma_0^{m_x}} \Pi_{C_x}(y,1) \left(\Pi_{S^{n_x-i}T^i}(x,y) - \Pi_{S^{n_x-i-1}T^{i+1}}(x,y) \right) \right|$$

$$\leq \sum_{i=0}^{n_x-1} \left| \sum_{y \in \Sigma_0^{m_x}} \Pi_{C_x}(y,1) \left(\Pi_{S^{n_x-i}T^i}(x,y) - \Pi_{S^{n_x-i-1}T^{i+1}}(x,y) \right) \right|$$

$$\leq n_x \max_{0 \leq i \leq n_x-1} \delta_{C_x}\left(\Pi_{S^{n_x-i-1}ST^i}(x,\cdot), \Pi_{S^{n_x-i-1}TT^i}(x,\cdot)\right) ,$$

i.e. for $x \in L$ there is $0 \leq i_x \leq n_x - 1$ with

$$\delta_{C_x}\left(\Pi_{S^{n_x}}(x,\cdot), \Pi_{T^{n_x}}(x,\cdot)\right)$$
$$\leq n_x \delta_{C_x}\left(\Pi_{S^{n_x-i_x-1}ST^{i_x}}(x,\cdot), \Pi_{S^{n_x-i_x-1}TT^{i_x}}(x,\cdot)\right) . \qquad (3)$$

Let $M_x = S^{n_x-i_x-1}$, $N = S$, $N' = T$ and $O_x = T^{i_x}$. We know:

1. M_x being a concatenation of homogeneous Turing machines is also homogeneous. Since we have $|y| \geq |x|$ for $y \in S(x)$, we find $|y| \geq |x|$ for all $y \in M_x(x)$.
2. N' has polynomial output length.
3. The families $\{\Pi_N(x,\cdot)\}_{x \in L}$ and $\{\Pi_{N'}(x,\cdot)\}_{x \in L}$ are c-indistinguishable. The set $\Gamma = \bigcup_{x \in L} M_x(x)$ is a subset of L and therefore $\{\Pi_N(x,\cdot)\}_{x \in \Gamma}$ and $\{\Pi_{N'}(x,\cdot)\}_{x \in \Gamma}$ are c-indistinguishable, too.
4. O_x has the same alphabet as T. The set of states of O_x is at most n_x-times the size of the set of states of T. Being a concatenation of homogeneous Turing machines, O_x itself is a homogeneous Turing machine. Therefore $\{O_x\}_{x \in L}$ is a polynomial family of homogeneous coin tossing machines.

Therefore we can apply Theorem 6. Using (3) we have for every $k \in \mathbb{N}$

$$\lim_{\substack{|x| \to \infty \\ x \in L}} |x|^k \delta_{C_x}\left(\Pi_{S^{n_x}}(x,\cdot), \Pi_{T^{n_x}}(x,\cdot)\right)$$

$$\leq \lim_{\substack{|x| \to \infty \\ x \in L}} n_x |x|^k \delta_{C_x}\left(\Pi_{M_x N O_x}(x,\cdot), \Pi_{M_x N' O_x}(x,\cdot)\right)$$

($\{n_x\}_{x \in L}$ is polynomially bounded. Therefore there is $d \in \mathbb{N}$ with $n_x \leq |x|^d$)

$$\leq \lim_{\substack{|x| \to \infty \\ x \in L}} |x|^{d+k} \delta_{C_x}\left(\Pi_{M_x N O_x}(x,\cdot), \Pi_{M_x N' O_x}(x,\cdot)\right)$$

$$= 0 .$$

Thus $\{\Pi_{S^{n_x}}(x,\cdot)\}_{x\in L}$ and $\{\Pi_{T^{n_x}}(x,\cdot)\}_{x\in L}$ are c-indistinguishable. □

Theorem 8. *Let $L \subseteq \Sigma_0^*$. Let $\{n_x\}_{x\in L}$ be a sequence in $\mathbb{N}$. Let S and T be coin tossing machines. Assume that $S(L) \subseteq L$. If $\{\Pi_S(x,\cdot)\}_{x\in L}$ and $\{\Pi_T(x,\cdot)\}_{x\in L}$ are p-indistinguishable, then also $\{\Pi_{S^{n_x}}(x,\cdot)\}_{x\in L}$ and $\{\Pi_{T^{n_x}}(x,\cdot)\}_{x\in L}$.*

Proof.

$$\Pi_{S^{n_x}}(x,y) - \Pi_{T^{n_x}}(x,y)$$

$$= \sum_{i=0}^{n_x-1} \Pi_{S^{n_x-i}T^i}(x,y) - \Pi_{S^{n_x-i-1}T^{i+1}}(x,y)$$

$$= \sum_{i=0}^{n_x-1} \sum_{u\in\Sigma_0^*} \sum_{v\in\Sigma_0^*} \Pi_{S^{n_x-i-1}}(x,u) \underbrace{(\Pi_S(u,v) - \Pi_T(u,v))}_{=0} \Pi_{T^i}(v,y)$$

$$= 0 \; .$$

□

Theorem 9. *Let $L \subseteq \Sigma_0^*$. Let $\{n_x\}_{x\in L}$ be a polynomially bounded sequence in $\mathbb{N}$. Let S and T be coin tossing machines. Assume that the following conditions hold:*

1. $|y| \geq |x|$ *for all $x \in L$ and all $y \in S(x)$.*
2. $S(L) \subseteq L$.
3. $\{\Pi_S(x,\cdot)\}_{x\in L}$ *and* $\{\Pi_T(x,\cdot)\}_{x\in L}$ *are s-indistinguishable.*

Then $\{\Pi_{S^{n_x}}(x,\cdot)\}_{x\in L}$ and $\{\Pi_{T^{n_x}}(x,\cdot)\}_{x\in L}$ are s-indistinguishable.

Proof. As in the proof of Theorem 7 we obtain

$$\delta_S(\Pi_{S^{n_x}}(x,\cdot), \Pi_{T^{n_x}}(x,\cdot)) \leq n_x \delta_S(\Pi_{M_x N O_x}(x,\cdot), \Pi_{M_x N' O_x}(x,\cdot)) \; ,$$

where $M_x = S^{n_x-i_x-1}$, $N = S$, $N' = T$ and $O_x = T^{i_x}$. According to the conditions the families $\{\Pi_N(x,\cdot)\}_{x\in L}$ and $\{\Pi_{N'}(x,\cdot)\}_{x\in L}$ are s-indistinguishable.

$$\delta_S\left(\Pi_{M_x N O_x}(x,\cdot), \Pi_{M_x N' O_x}(x,\cdot)\right)$$

$$= \sum_{y\in\Sigma_0^*} |\Pi_{M_x N O_x}(x,y) - \Pi_{M_x N' O_x}(x,y)|$$

$$\leq \sum_{y\in\Sigma_0^*} \sum_{u\in\Sigma_0^*} \sum_{v\in\Sigma_0^*} |\Pi_{M_x}(x,u)(\Pi_N(u,v) - \Pi_{N'}(u,v))\Pi_{O_x}(v,y)|$$

$$\leq \sum_{u\in\Sigma_0^*} \Pi_{M_x}(x,u) \sum_{v\in\Sigma_0^*} |(\Pi_N(u,v) - \Pi_{N'}(u,v))| \underbrace{\sum_{y\in\Sigma_0^*} \Pi_{O_x}(v,y)}_{\leq 1}$$

$$\leq \underbrace{\sum_{u\in\Sigma_0^*} \Pi_{M_x}(x,u)}_{\leq 1} \sup_{u\in M_x(x)} \sum_{v\in\Sigma_0^*} |\Pi_N(u,v) - \Pi_{N'}(u,v)| \; .$$

Then we have for all $k \in \mathbb{N}$

$$\lim_{\substack{x \in L \\ |x| \to \infty}} |x|^k \delta_S \big(\Pi_{S^{n_x}}(x, \cdot), \Pi_{T^{n_x}}(x, \cdot) \big)$$

$$\leq \lim_{\substack{x \in L \\ |x| \to \infty}} n_x |x|^k \sup_{u \in M_x(x)} \delta_S \big(\Pi_N(u, \cdot), \Pi_{N'}(u, \cdot) \big)$$

$$\leq \lim_{\substack{x \in L \\ |x| \to \infty}} n |x|^{k+d} \sup_{u \in M_x(x)} \delta_S \big(\Pi_N(u, \cdot), \Pi_{N'}(u, \cdot) \big)$$

($\{n_x\}_{x \in L}$ is polynomially bounded. Therefore there is $d \in \mathbb{N}$ with $n_x \leq |x|^d$)

$$\leq \limsup_{\substack{u \in L \\ |u| \to \infty}} |u|^{k+d} \delta_S \big(\Pi_N(u, \cdot), \Pi_{N'}(u, \cdot) \big)$$

$$= 0 \; .$$

$\square$

References

1. M. Garey and D. S. Johnson. *Computers and Intractability, A Guide to the Theory of NP-Completeness.* Freeman, 1979.
2. O. Goldreich, S. Micali, and A. Widgerson. *Proofs that Yield Nothing But Their Validity or All Languages in NP Have Zero-Knowledge Proof Systems.* J. ACM Vol. 38, pp. 691–729, 1991.
3. Harry R. Lewis and Christos H. Papadimitriou. *Elements of the theory of Computation.* Prentice-Hall, 1981.
4. K. R. Reischuk. *Einführung in die Komplexitätstheorie.* Teubner, 1990.

How to Make Efficient Fail-stop Signatures

Eugène van Heyst

CWI Centre for Mathematics and Computer Science,
Kruislaan 413, 1098 SJ Amsterdam, The Netherlands.

Torben Pryds Pedersen[‡]

Aarhus University, Computer Science Department,
Ny Munkegade, DK-8000 Århus C, Denmark.

Abstract. Fail-stop signatures (introduced in [WP89]) have the very nice property that the signer is secure against unlimited powerful forgers. However, the known fail-stop signatures require very long keys, and they are quite inefficient, because messages are signed bit-wise. This paper presents a fail-stop signature scheme, in which signing a message block requires two modular multiplications and verification requires less than two modular exponentiations. Furthermore a construction is shown of an undeniable signature scheme, which is unconditionally secure for the signer, and which allows the signer to convert undeniable signatures into fail-stop signatures. This is the first published undeniable signature having this property.

1. Introduction

Digital signatures, as introduced in [DH76], allow a person who knows a secret key to make signatures that everybody can verify. These signatures are only computationally secure in the sense that a forger with unlimited computing power can always make false signatures of other persons. Hence, the security of these systems relies on the fact that a forger has not enough time and no efficient algorithm to carry out the computations needed to forge signatures.

With fail-stop signatures (see [WP89]), unforgeability also relies on a cryptographic assumption, but if nevertheless a signature is forged, then the presumed signer can prove that the signature is a forgery: he can prove that the underlying assumption of the system has been broken. This proof of forgery may fail with a very small probability, but the ability to prove a forgery does not rely on any cryptographic assumption and is independent of the computing power of the forger. Hence, the polynomially bounded signer is protected against all powerful forgers, because after the first forgery, all other participants in the system and the system operator know that the signature scheme has

---

[‡] Research done while visiting CWI.

R.A. Rueppel (Ed.): Advances in Cryptology - EUROCRYPT '92, LNCS 658, pp. 366-377, 1993.
© Springer-Verlag Berlin Heidelberg 1993

been broken, and the system will be stopped. That is why this system is called "fail-stop".

Fail-stop signatures are known to exist if claw-free permutation pairs exist (see for instance [BPW90] and [PW91]). These signatures use the idea of the one-time signatures of [La79]. In particular, this means that the signatures and keys are very long, and signing as well as verification require quite a lot of computation. Several tricks are therefore applied to make the signatures more efficient (see for example [PW91]).

However, these tricks are not needed in one of the applications of fail-stop signatures, called the 3-phase protocol, proposed in [PW91]. Here it is suggested that customers in an electronic payment system should use fail-stop signatures when signing a request to the bank (for example for the withdrawal of money). These signatures have the advantage over usual digital signatures that the customers need not worry about the bank (which in general has more computing power than the customer) being able to break the underlying assumptions of the signature scheme. In this 3-phase protocol for making such requests, the customer only has to sign one single bit with a fail-stop signature. Hence, this protocol is quite efficient, but it is interactive, and it relies on the assumption that the set of possible recipients is fixed.

It has been very unsatisfactory that the "general", known fail-stop signatures were much less efficient than digital signatures (see [PW91]).

The main objective of this paper is to present a fail-stop signature scheme whose complexity is comparable with that of RSA-signatures (see [RSA78]). More specifically, for each message the public key is approximately 1000 bits (assuming a modulus of size 500 bits), the secret key is approximately 2000 bits and the signature of a 500-bit message is 1000 bits (a longer message can be hashed to 500 bits before signing). The construction of a signature requires two multiplications and two additions modulo a prime, and the verification requires only a little more thane one exponentiation modulo a prime. Thus messages can be signed very efficiently, whereas the verification of signatures is a little slower than for RSA-signatures.

In the next section, we give an informal description of fail-stop signatures and discuss the properties of such signatures. After a brief description of the notation, Section 4 presents the new scheme and it is proven that it satisfies the requirements of fail-stop signatures. The scheme presented here can only be used to sign a single message, but in Section 5 it is shown how to generalize this scheme to sign a constant number of messages. Section 6 uses the ideas of Section 4 to construct an undeniable signature scheme which is unconditionally secure for the signer. This scheme has the property that the signer can convert these undeniable signatures into fail-stop signatures. Finally, Section 7 discusses some applications of the presented schemes.

2. Fail-stop Signatures

This section gives a brief and rather informal description of the properties of fail-stop signatures, which are required for the scheme in Section 4 (for a formal definition see [PW90]).

In a fail-stop signature scheme, (exponentially in a security parameter) many secret keys correspond to a given public key, and different secret keys will (with very high probability) give different signatures on the same message. However, the signer knows only one of these secret keys and can therefore construct only one of these signatures on a message. A signature is called *valid* if it passes the public test function.

Furthermore, given the public key and signatures on some messages, a forger must not be able to guess which signature the signer is able to construct on a new message, so that even if the forger (by using his unlimited power) succeeds in making a valid signature, this signature will with very high probability be different from the signer's signature. Given such a forged signature the signer must then be able to prove that it is different from his own signature, thereby proving that it is a forgery. After having discovered such a forgery and proved it, the signer should stop using the scheme.

More specifically, if SK is the secret key of the signer, and PK the public key, the signature S on a message, m, is denoted by $S=sign(SK, m)$. The recipient of such a signature can verify its correctness using the polynomial time computable predicate $test(PK, m, S)$ (here polynomial time means polynomial in a security parameter k). For fail-stop signatures this predicate must satisfy that for every secret key SK^* corresponding to PK

$$test(PK, m, sign(SK^*, m)) \text{ is } true.$$

A scheme is a fail-stop signature scheme if it satisfies the following three requirements:

(i) Let PK and the signature $S = sign(SK, m)$ on m be given. Then there are exponentially many (in k) possible secret keys SK^* corresponding to PK such that $S = sign(SK^*, m)$. Furthermore, if such a secret key SK^* is chosen at random, then the probability that $sign(SK, m^*) = sign(SK^*, m^*)$ is negligible, for every message $m^* \neq m$.

(Informally: it is not possible to compute the signer's signature on a new message, even with unlimited computing power.)

(ii) There is a polynomial-time computable function *proof*, which on input SK, PK, a message m, and a valid, forged signature $S' \neq sign(SK, m)$ on m, outputs a proof that S' is a forgery.

(Informally: the presumed signer is able to supply a proof of the forgery.)

(iii) No signer with polynomial-time computing power is able to construct a valid signature S on a given message m and also construct a proof that S is a forgery.

(Informally: the signer cannot make signatures which he can later prove to be forgeries.)

These requirements are for one-time keys. It is not hard to generalize this definition of fail-stop signatures to comprise schemes in which more than one message can be signed.

The first two requirements imply that fail-stop signatures are unconditionally secure for the signer, whereas the third requirement says that the scheme is secure for the recipients of the signatures. Unlike the security for the signer, the security for the recipient depends on a complexity theoretic assumption. The reader is referred to [PW90] for a thorough discussion of the properties of fail-stop signatures. (To avoid confusion, note that [PW90] considers different security parameters for the security of the signer and for the security of the recipient, while in our scheme these are equal.)

3. Notation

Throughout this paper p and q denote large primes such that q divides $p-1$, and G_q is the unique subgroup of $\mathbb{Z}_p^*$ of order q. As any element $b \neq 1$ of G_q generates the group, the discrete logarithm of $a \in G_q$ with respect to the base b is defined, and it is denoted by $\log_b(a)$.

4. Signature Scheme

In this section an efficient fail-stop signature based on the discrete logarithm assumption is described. Let g and h be elements of G_q such that no participant knows $\log_g(h)$. These elements can either be chosen by a trusted authority, when the system is initialized, or by some of the participants using a coin-flipping protocol.

Although (p, q, g, h) is part of the public key, it will not be mentioned in the following. To give a better idea of the scheme, we will in this section assume that a person issues only one signature. Let the secret key of person $\mathcal{A}$ be

$$SK = (x_1, x_2, y_1, y_2) \in \mathbb{Z}_q^4,$$

and $\mathcal{A}$ publishes the corresponding public key

$$PK = (p_1, p_2) = (g^{x_1} h^{x_2}, g^{y_1} h^{y_2}). \tag{1}$$

To sign a message $m \in \mathbb{Z}_q$, $\mathcal{A}$ computes the following numbers:

$$S = sign(SK, m) = (\sigma_1(SK, m), \sigma_2(SK, m)), \text{ where} \tag{2}$$
$$\sigma_1(SK, m) \equiv x_1 + m y_1 \pmod{q},$$
$$\sigma_2(SK, m) \equiv x_2 + m y_2 \pmod{q}.$$

The recipient of this signature verifies that

$$p_1 p_2^m = g^{\sigma_1} h^{\sigma_2}. \tag{3}$$

The proof of forgery is $\log_g(h)$.

The following three lemmas show that this signature scheme is a fail-stop signature scheme. First note that for every secret key, SK^*, corresponding to PK, the predicate $test(PK, m, sign(SK^*, m))$ is $true$ for all messages, m, i.e., for every tuple $(x_1, x_2, y_1, y_2) \in \mathbb{Z}_q^4$ that satisfies

$$PK = (g^{x_1} h^{x_2}, g^{y_1} h^{y_2}),$$
$$\sigma_1(SK, m) \equiv x_1 + my_1 \pmod{q},$$
$$\sigma_2(SK, m) \equiv x_2 + my_2 \pmod{q},$$

we have

$$p_1 p_2^m = g^{\sigma_1} h^{\sigma_2}.$$

Lemma 1. *The public key PK, together with a signature sign(SK, m) on m, contain no information about which of q possible secret keys are used for SK.*

Proof. This lemma is a special case of Thm 4.4 of [Ped91]. Another way to prove it is the following. Define $h = g^a, p_1 = g^{e_1}, p_2 = g^{e_2}$. This representation is possible because g is a generator of G_q. Then we can write Equations (1) and (2) as:

$$\begin{cases} e_1 \equiv x_1 + ax_2 \pmod{q}, \\ e_2 \equiv y_1 + ay_2 \pmod{q}, \\ \sigma_1 \equiv x_1 + my_1 \pmod{q}, \\ \sigma_2 \equiv x_2 + my_2 \pmod{q}. \end{cases}$$

The fact that equation (3) holds, follows immediately from equations (1) and (2) as noted above. The forger has to find a solution (x_1, x_2, y_1, y_2) to the Equations

$$\begin{pmatrix} 1 & a & 0 & 0 \\ 0 & 0 & 1 & a \\ 1 & 0 & m & 0 \\ 0 & 1 & 0 & m \end{pmatrix} \begin{pmatrix} x_1 \\ x_2 \\ y_1 \\ y_2 \end{pmatrix} \equiv \begin{pmatrix} e_1 \\ e_2 \\ \sigma_1 \\ \sigma_2 \end{pmatrix} \pmod{q}.$$

It is easy to see that this matrix has rank 3 (the rank is defined because q is prime), hence, since there is one solution, there are exactly q solutions to this equation. ❑

Lemma 2. *Let PK, the signature $S = sign(SK, m)$ on m and a valid signature $S' = (\tau_1, \tau_2)$ on $m' \neq m$ (so $p_1 p_2^{m'} \equiv g^{\tau_1} h^{\tau_2}$), be given. Then there exists a unique secret key, SK^*, corresponding to PK such that $S = sign(SK^*, m)$ and $S' = sign(SK^*, m')$*

Proof As in the proof of Lemma 1, a solution (x_1, x_2, y_1, y_2) to the matrix equation

$$\begin{pmatrix} 1 & a & 0 & 0 \\ 0 & 0 & 1 & a \\ 1 & 0 & m & 0 \\ 0 & 1 & 0 & m \\ 1 & 0 & m' & 0 \\ 0 & 1 & 0 & m' \end{pmatrix} \begin{pmatrix} x_1 \\ x_2 \\ y_1 \\ y_2 \end{pmatrix} = \begin{pmatrix} e_1 \\ e_1 \\ \sigma_1 \\ \sigma_2 \\ \tau_1 \\ \tau_2 \end{pmatrix}$$

has to be found. It is easy to see that this matrix has rank 4 (because $m' \neq m$), so there is exactly one solution. ❑

Lemma 1 says that there are q possible secret keys corresponding to a given public key and one signature. By Lemma 2, each of these will yield a different signature on a message $m' \neq m$. This shows that the first requirement for the security of the signer is satisfied.

Lemma 3. *If the presumed signer receives a valid, forged signature* $S' = (\tau_1, \tau_2)$ *on* m *(so* $p_1 p_2^m \equiv g^{\tau_1} h^{\tau_2}$ *), but* $S' \neq sign(SK, m)$*, then he can compute* $\log_g(h)$*.*

Proof By writing $S = sign(SK, m) = (\sigma_1, \sigma_2)$, we have that $p_1 p_2^m \equiv g^{\tau_1} h^{\tau_2} \equiv g^{\sigma_1} h^{\sigma_2}$ and thus that $g^{\sigma_1 - \tau_1} \equiv h^{\tau_2 - \sigma_2}$ (mod p). If $\sigma_2 = \tau_2$, then we also have that $\sigma_1 = \tau_1$ and thus $S = S'$. This is a contradiction, and therefore the presumed signer can compute $\log_g(h)$ as $(\sigma_1 - \tau_1)(\tau_2 - \sigma_2)^{-1} \bmod q$. ☐

Hence, under the assumption that the signer cannot compute $\log_g(h)$, he is not able to construct a valid signature and also a proof that it is a forgery. Thus we can define

$$proof(SK, S') := (\sigma_1 - \tau_1)(\tau_2 - \sigma_2)^{-1} \bmod q.$$

On the other hand the signer cannot compute a proof of forgery without being given a forged signature unless he is able to compute discrete logarithms. This shows that the scheme is computationally secure for the recipients if it is infeasible to compute $\log_g(h)$.

Note that this secret key is a one-time key: if two different messages are signed using the same secret key, then it is easy to compute the secret key from these signatures.

Remark. In the above scheme, the public key consists of two numbers modulo q. It is possible to reduce the size of this public key as follows. Let H be a collision-free hash function that maps the elements of G_q into numbers of a smaller size. Then the public key will be

$$PK^* = (H(p_1), p_2),$$

where p_1 and p_2 are defined as before. A signature (σ_1, σ_2) on the message m is constructed as before, and it is verified as

$$H(g^{\sigma_1} h^{\sigma_2} p_2^{-m}) = H(p_1).$$

By using this public key, the Lemmas 1, 2 and 3 have to be modified. For instance, Lemma 3 has to be modified as follows:

Lemma 3.a. *If the presumed signer receives a valid, forged signature* $S' = (\tau_1, \tau_2)$ *on* m *(so* $H(g^{\tau_1} h^{\tau_2} p_2^{-m}) = H(p_1)$*)), but* $S' \neq sign(SK, m)$*, then he can compute* $\log_g(h)$ *or he has found a collision for* H*.*

Long messages can first be hashed into smaller messages before signing. In this case a proof of forgery is either $\log_g(h)$ or a collision of the hash function H. Hence,

Lemma 3 has to be modified in a similar way as was done in Lemma 3.a. Even if no hash functions are used in the public key, it is more efficient to verify the signature by computing

$$g^{\sigma_1} h^{\sigma_2} p_2^{-m},$$

and comparing it with p_1. This requires less than $2|q|$ multiplications (where $|q|$ is the number of bits of q), if the products gh, gp_2, hp_2 and ghp_2 are precomputed.

5. More than one Signature per Public Key

As noted in the previous subsection, a person can use his public key (and secret key) only once. We now present three different ways to overcome this problem: the public key can be used to sign k messages, and the signer still has unconditional security. In the first two methods, the secret key consists of $2k$ numbers, while each signature consists of 2 numbers. These two schemes differ in the computations needed. In the third method, the secret key consists of at most k elements, while each signature consists of $\lceil \log k \rceil + 3$ numbers.

Method 1. Person A chooses as a secret key

$$SK = (x_1, y_1, x_2, y_2, \ldots, x_{k+1}, y_{k+1}),$$

and he publishes the corresponding public key

$$PK = (p_1, \ldots, p_{k+1}) = (g^{x_1} h^{y_1}, \ldots, g^{x_{k+1}} h^{y_{k+1}}).$$

To sign a message $m \in \mathbb{Z}_q$, A computes the following numbers:

$$sign(SK, m) = (\sigma_1, \sigma_2), \text{ where}$$
$$\sigma_1 \equiv x_1 + mx_2 + \ldots + m^k x_{k+1} \pmod{q},$$
$$\sigma_2 \equiv y_1 + my_2 + \ldots + m^k y_{k+1} \pmod{q}.$$

The recipient of this digital signature verifies that

$$p_1 p_2^m \ldots p_{k+1}^{m^k} \equiv g^{\sigma_1} h^{\sigma_2} \pmod{p}.$$

After issuing signatures on k different messages, the signer still has unconditional security. (This follows from Theorem 4.4 of [Ped91].) Given k signatures, there are q possible secret keys, and $k+1$ signatures determine the secret key uniquely. As before $\log_g(h)$ constitutes a proof of forgery, and it follows from similar arguments that the two last requirements to fail-stop signatures are satisfied.

Method 2. ([Pf91]) Person A chooses the same secret key and public key as in Method 1. In Method 2, the signature on the message depends on the number of messages that A has signed previously. If A has signed $i-1$ messages ($1 \leq i \leq k$), the signature on a message, m, will be

$$sign(SK, m, i) = (i, \sigma_1, \sigma_2), \text{ where}$$

$$\sigma_1 \equiv x_i + mx_{i+1},$$
$$\sigma_2 \equiv y_i + my_{i+1}.$$

The recipient of this signature verifies that

$$p_i p_{i+1}^m = g^{\sigma_1} h^{\sigma_2}.$$

Hence, at the cost of including a counter in the signatures, the computations of the signer as well as the recipient are easier here than in Method 1.

Again, the security of the recipient follows from the fact that $\log_g(h)$ is a proof of forgery. In order to prove that the signer has unconditional security we first note that any $k+1$ signatures determine the secret key. Hence it is sufficient to show that after issuing k different signatures there are many (q) possible secret keys. This will be done by showing that the rank of the following $(3k+1)\times(2k+2)$ matrix, A_k, is $2k+1$ (remember that $a = \log_g(h)$):

$$A_k = \begin{bmatrix} 1 & 0 & m_1 & & & & & \\ & 1 & 0 & m_1 & & & & \\ & & 1 & 0 & m_2 & & & \\ & & & 1 & 0 & m_2 & & \\ & & & & \ddots & & & \\ & & & & & 1 & 0 & m_k \\ & & & & & & 1 & 0 & m_k \\ 1 & a & & & & & & \\ & & 1 & a & & & & \\ & & & & \ddots & & & \\ & & & & & & 1 & a \end{bmatrix}.$$

This will be done by proving that matrix $\tilde{A}_k$ can be obtained from A_k by elementary row operations, where

$$\tilde{A}_k = \begin{bmatrix} 1 & 0 & m_1 & & & & \\ & 1 & 0 & m_1 & & & \\ & & 1 & 0 & m_2 & & \\ & & & 1 & 0 & m_2 & \\ & & & & \ddots & & \\ & & & & & 1 & 0 & m_k \\ & & & & & & 1 & 0 & m_k \\ & & & & & & & 1 & a \end{bmatrix}$$

The matrix $\tilde{A}_k$ clearly has rank $2k+1$. We change A_k to $\tilde{A}_k$ by using Lemma 1 for $i=1,\ldots,k-1$ as follows. Consider the following four rows of the matrix A_k.

$$\begin{array}{ccccccccccc} 0 & \ldots & 0 & 1 & 0 & m_i & 0 & 0 & \ldots & 0 \\ 0 & \ldots & 0 & 0 & 1 & 0 & m_i & 0 & \ldots & 0 \\ 0 & \ldots & 0 & 1 & a & 0 & 0 & 0 & \ldots & 0 \\ 0 & \ldots & 0 & 0 & 1 & a & 0 & 0 & \ldots & 0 \end{array}$$

By Lemma 1, this sub-matrix has rank 3 and the third row can be removed. By using this method for $i=1,\ldots,k-1$, we delete all rows $(0 \ldots 0\ 1\ a\ 0 \ldots 0)$, except the last one. The resulting matrix is $\tilde{A}_k$.

Hence, PK and k signatures $sign(SK, m_1, 1),...,sign(SK, m_k, k)$ contain no information about which of q possible secret keys are used for SK, and each of these possible secret keys will yield a different signature on a new message m^*.

Method 3. Use tree-authentication, which is described in [Merk80] and [PW91], for example.

6. Convertible Undeniable Signatures Unconditionally Secure for the Signer

Convertible undeniable signatures were introduced in [BCDP90]. Briefly, these signatures allow the signer of undeniable signatures to change his signatures to ordinary digital signatures. In [CHP91] an undeniable signature scheme was presented in which the signer is unconditionally secure. This section combines the ideas of [BCDP90], [CHP91] and Section 4 by constructing an undeniable signature scheme which is unconditionally secure for the signer and has the property that the signer can convert the undeniable signatures to fail-stop signatures.

Let p, q, g and h be as in Section 2. The secret key of a person $\mathcal{A}$ is

$$SK = (x_1, x_2, y_1, y_2) \in \mathbb{Z}_q^4,$$

and the corresponding public key is

$$PK = (p_1, p_2) = (g^{x_1} h^{x_2}, g^{y_1} h^{y_2}).$$

The undeniable signature of $\mathcal{A}$ on the message $m \in \mathbb{Z}_q$ is

$$\sigma_1(SK, m) \equiv x_1 + m y_1 \pmod{q},$$

This signature is undeniable, because given PK, the signature is just a random number in $\mathbb{Z}_q$. The signature scheme is unconditionally secure for the signer, because given σ_1, it is impossible for a forger with unlimited computing power to construct $\sigma_1' = sign(SK, m')$ for a new message $m' \neq m$. This follows from the same arguments as in Lemma 1 and 2. We mention that $\mathcal{A}$ can convert this signature to a fail-stop signature by publishing

$$\sigma_2 \equiv x_2 + m y_2 \pmod{q},$$

which changes the undeniable signature into the fail-stop signature of Section 4.

Verification

To verify the signature σ_1 on a message m, $\mathcal{A}$ and the recipient compute $u \equiv p_1 p_2^m g^{-\sigma_1} \pmod{p}$, and $\mathcal{A}$ convinces the recipient with a zero-knowledge protocol that he knows a number σ_2 such that $u \equiv h^{\sigma_2} \pmod{p}$. Perfect zero-knowledge protocols for this problem are well known (e.g. [CEvdG87]), hence

Theorem 4. *There is a perfect zero-knowledge protocol for convincing someone of having σ_2 satisfying $u \equiv h^{\sigma_2} \pmod{p}$.*

Disavowal of these signatures is slightly more complicated. A given number $\sigma \in \mathbb{Z}_q$ is not $\mathcal{A}$'s signature on m if

$$p_1 p_2^m \not\equiv g^\sigma h^{\sigma_2} \pmod{p},$$

where $\sigma_2 \equiv x_2 + my_2 \pmod{q}$. $\mathcal{A}$ can therefore convince someone that σ is not his signature by convincing him that he knows numbers s and t such that

$$p_1 p_2^m \equiv g^s h^t \pmod{p}, \quad \text{and} \quad \sigma \neq s.$$

(because if σ was his signature this would mean that $\mathcal{A}$ knew $\log_g(h)$). A perfect zero-knowledge proof for this was presented in [CHP91].

7. Applications

As mentioned in the introduction, fail-stop signatures have been suggested to be used in electronic payment systems, such that the customer does not need to rely on the assumption that the bank does not have sufficient computing power to forge his signatures. Using the previously known schemes, this application required a 3-phase protocol between the bank and the customer in which also usual digital signatures are used (see [PW91]). If the signature scheme presented here is used, this protocol can be avoided and the customer just needs to send a single message to the bank in order to sign the request.

8. Conclusion

This paper has described a fail-stop signature scheme which is very efficient from a computational point of view, and in which signatures are only twice as long as the messages (long messages can first be hashed into smaller messages before signing). This scheme makes it possible to avoid the 3-phase protocol of [PW91] when applied to payment systems.

Furthermore, it has been shown how to construct convertible, undeniable signatures, which are unconditionally secure for the signer.

The main disadvantage of the presented schemes is that they can only be used to construct a fixed number of signatures. This property cannot be avoided, because in a forthcoming paper (with Birgit Pfitzmann) it is shown that a signature scheme which is unconditionally secure for the signer, requires secret keys whose lengths are linear in the number of messages to be signed.

Acknowledgements

We wish to thank Birgit Pfitzmann for many discussions about fail-stop signatures and her comments to this paper.

References

[BCDP90] Joan Boyar, David Chaum, Ivan Damgård and Torben Pedersen, Convertible Undeniable Signatures, *Advances in Cryptology-CRYPTO'90*, LNCS 537, Springer Verlag, pp. 189-205.

[CEvdG87] David Chaum, Jan-Hendrik Evertse and Jeroen van de Graaf, An improved protocol for demonstrating possession of discrete logarithms and some generalizations, *Advances in Cryptology -EUROCRYPT '87*, LNCS 304, Springer-Verlag, pp. 127-141.

[BPW90] Gerrit Bleumer, Birgit Pfitzmann and Michael Waidner, A remark on a signature scheme where forgery can be proved, *Advances in Cryptology-EUROCRYPT '90*, LNCS 473, Springer Verlag, pp. 441-445.

[CHP91] David Chaum, Eugène van Heyst and Birgit Pfitzmann, Cryptographically strong undeniable signatures, unconditionally secure for the signer, *Advances in Cryptology -CRYPTO '91*.

[DH76] Whitfield Diffie and Martin Hellman, New directions in cryptography, *IEEE IT* 22 (1976), pp. 644-654.

[FFS88] Uriel Feige, Amos Fiat and Adi Shamir, Zero-Knowledge Proofs of Identity, Journal of Cryptology 1 (1988), pp. 77-94.

[GMR88] Shafi Goldwasser, Silvio Micali and Ronald Rivest, A digital signature scheme secure against adaptive chosen-message attacks, *SIAM J. Comp.* 17 (1988), pp. 281-308.

[La79] L. Lamport, Constructing Digital Signatures from a One-Way Function, SRI Intl. CSL-98 (October 1979).

[Merk80] Ralph C. Merkle, Protocols for public key cryptosystems; *Proceedings of the 1980 symposium on security and privacy*, April 1980, Oakland, California, pp. 122-134.

[Ped91] Torben Pedersen, Non-interactive and information-theoretic secure variable secret sharing, *Advances in Cryptology -CRYPTO '91*.

[Pf91] Birgit Pfitzmann, personal communication.

[PW90] Birgit Pfitzmann and Michael Waidner, Formal aspects of fail-stop signatures, Interner Bericht 22/90, Fakultät für Informatik, Universität Karlsruhe, December 1990.

[PW91] Birgit Pfitzmann and Michael Waidner, Fail-stop signatures and their applications, *SECURICOM '91*, Paris, 1991, pp. 145-160.

[RSA78] Ronald Rivest, Adi Shamir, and Leonard Adleman, A Method for Obtaining Digital Signatures and Public Key Cryptosystems, *Comm. of the ACM* 21 (1978), pp. 120-126.

[WP89]   Michael Waidner and Birgit Pfitzmann, The Dining Cryptographers in the Disco: Unconditional Sender and Recipient Untraceability with Computationally Secure Serviceability, *Advances in Cryptology-EUROCRYPT '89*, LNCS 434, Springer Verlag, p. 690.

Which new RSA Signatures can be Computed from RSA Signatures, Obtained in a Specific Interactive Protocol?

Jan-Hendrik Evertse [1]

Department of Mathematics and Computer Science, University of Leiden,
P.O. Box 9512, 2300 RA Leiden, The Netherlands.

Eugène van Heyst

CWI Centre for Mathematics and Computer Science,
Kruislaan 413, 1098 SJ Amsterdam, The Netherlands.

Abstract. We consider certain interactive protocols, based on RSA. In these protocols, a signature authority Z (which chooses the RSA-modulus N that is kept fixed) issues a fixed number of RSA-signatures to an individual A. These RSA-signatures consist of products of rational powers of residue classes modulo N; some of these residue classes are chosen by Z and the others can be chosen freely by A. Thus, A can influence the form of the signatures that he gets from Z. A wants to choose his residue classes in such a way that he can use the signatures he gets from Z to compute a signature of a type not issued by Z.

In previous literature, some special cases of our protocols were considered, namely that only A chooses the residue classes ([Dav82],[Denn84],[DO85]) and that only Z chooses the residue classes [EvH92]. The results in our paper are used under the following assumptions:

- A cannot compute RSA-roots on randomly chosen residue classes modulo N.
- In his computations, A uses only multiplications and divisions modulo N.

Our main result gives a necessary and sufficient condition under which A is able to influence the signatures he gets from Z in such a way that he can use these RSA-signatures to compute a signature of a type not issued by Z. It turns out that this condition is equivalent to the solvability of a particular quadratic equation in integral matrices. We also study a particular case of this problem in more detail.

1. Introduction

A challenging problem in cryptology is to study the security of certain classes of interactive protocols. To this end, one must investigate certain classes of attacks on these protocols.

For instance, consider the interactive *ping-pong protocols* (cf. [EGS85]). In such a

---

[1] This research has been made possible by a fellowship of the Royal Netherlands Academy of Arts and Sciences (K.N.A.W.)

protocol (which consists of several moves), one party generates a secret message, applies a sequence of operators to it, and sends it to the other party. This party also applies a sequence of operators to the message received, and sends back the result. In each move of the protocol, one of the parties applies a sequence of operators to the last message it received, and sends it back. The question is, whether an "active" third party can discover the initial message (by altering messages, impersonating other users, etc.).

In this chapter we consider interactive protocols based on the RSA-system ([RSA78]) as in Figure 1 below, in which only one party Z, called the *signature authority*, can create signatures and issues these to the other parties, called the *individuals*. Such protocols are used, for instance, in credential and payment systems, in which a signature represents a credential or money. In fact, in such credential systems or payment systems, the signature authority issues different types of signatures, corresponding to different credentials or different values of money. The security of these systems depends on whether an individual (or a group of conspiring individuals) is not able to compute a useful signature of a type not issued by the signature authority, by using the signatures which were issued before by the authority Z (for instance by using the multiplicative property of RSA).

Initially, Z chooses two large primes P, Q and computes their product N. Further, Z chooses two integers a,b coprime to $\varphi(N) = (P-1)(Q-1)$. Z makes N, a, b public, and keeps P and Q secret. Let c, d also be some integers coprime to $\varphi(N)$. In this protocol, Z chooses a residue class u, and $\mathcal{A}$ wants to choose h in such a way that after the execution of this protocol, he is able to compute from $\{u,h,u^{1/a}h^{1/b}\}$ a pair $\{t,k\}$ satisfying $t \equiv u^{1/c}k^{1/d} \pmod{N}$. The reason for considering such problems is that in the payment systems and credential systems mentioned above, the user gets so-called blind signatures from Z which contain residue classes chosen by Z and residue classes chosen by the user himself.

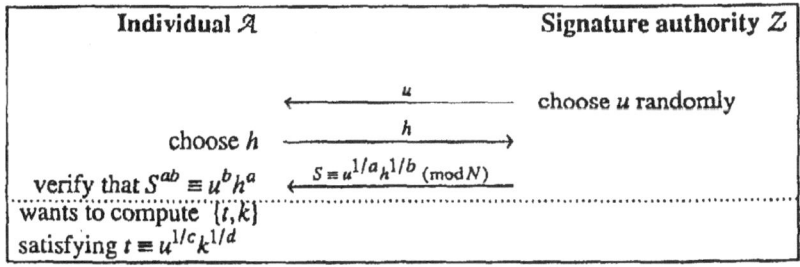

Fig. 1. An interactive signature-issuing protocol in which the signature authority Z issues a signature to individual $\mathcal{A}$.

In [EvH92] we studied the case in which $\mathcal{A}$ has no influence on the signature received, that is, $\mathcal{A}$ chooses no residue class (i.e., $b=1$ in Figure 1). A necessary and sufficient condition was given for the computation of this new signature to be feasible for $\mathcal{A}$.

In [Dav82], [Denn84] and [DO85] the case is studied in which Z chooses no residue class, that is, in which individuals were able to obtain signatures on messages of their choice (i.e., $a=1$ in Figure 1). [Dav82] states that $\mathcal{A}$ can decrypt ciphertext encrypted under Z's public key and can forge Z's signature on meaningful messages; [Denn84] can foil this attack by using hashing. [DO85] showed that if $\mathcal{A}$ can get $L(N)^{1/2}$ RSA-signatures modulo N on carefully chosen residue classes, where $L(N) = \exp\{(1 + o(1))(\log N \log\log N)^{1/2}\}$ as $N \to \infty$, then $\mathcal{A}$ can compute any RSA-signature modulo N of his choice in $L(N)^{1/2}$ bitoperations.

We consider more general interactive protocols in which Z issues a fixed number of RSA-signatures to $\mathcal{A}$. These RSA-signatures consist of products of rational powers of residue classes modulo the composite number N of the underlying RSA-scheme; some of these residue classes are chosen by Z and the others can be chosen freely by $\mathcal{A}$. We make the following two assumptions:

(i) $\mathcal{A}$ cannot compute RSA-roots of randomly chosen residue classes modulo N.
(ii) In his computations, the only operations that $\mathcal{A}$ uses are multiplications and divisions modulo N.

The problem whether assumption (ii) is necessary remains open. We formulate a necessary and sufficient condition under which $\mathcal{A}$ is able to influence the signatures he receives from Z in such a way that he can later use these signatures to compute a signature of a type not issued by Z. It turns out that this condition is equivalent to the solvability of a particular quadratic equation in integral matrices.

This paper is organized as follows. The notation used is introduced in the next section, while in Section 3 the interactive protocol considered and the problem we are facing are defined. In Section 4 we analyze this problem by assuming that the individual performs only multiplications and divisions modulo N (this is called an algebraic strategy). A special case of the considered protocol with these algebraic strategies is studied in Section 5, while some generalizations to the protocol of Section 3 are given in Section 6.

2. Notation

The following notation is used throughout this paper.

$\gcd(a_1,\ldots,a_t)$ the greatest common divisor of $a_1,\ldots,a_t$; defined for rational numbers by $\gcd(a_1,\ldots,a_t):=\frac{\gcd(a_1 d,\ldots,a_t d)}{d}$, where $d\in \mathbb{N}$ is such that $a_1 d,\ldots,a_t d\in \mathbb{Z}$; this definition is independent of the choice of d.

S^k the set of vectors $(a_1,\ldots,a_k)$ with $a_1,\ldots,a_k\in S$, for any set S; we use bold face characters to denote vectors.

$a\in {}_R S$ denotes the random selection of an element (that will be called a) from S according to the uniform probability distribution; for any set S.

$\mathbf{ab}$ $(a_1 b_1,\ldots,a_k b_k)$, if $\mathbf{a}=(a_1,\ldots,a_k)$ and $\mathbf{b}=(b_1,\ldots,b_k)$.

$\mathbf{a}\equiv \mathbf{b}\,(\bmod m)$ $m^{-1}(\mathbf{b}-\mathbf{a})\in \mathbb{Z}^k$; this is defined for $\mathbf{a},\mathbf{b}\in \mathbb{Q}^k$, $m,k\in \mathbb{N}$, $m>0$.

N a composite, odd number.

$\mathbb{Z}_N^*$ the set $\{a \mid a\in \mathbb{N}, 1\le a\le N, \gcd(a,N)=1\}$ of $\varphi(N)$ elements.

$\tilde{\mathbb{Q}}_N$ the ring $\{\frac{a}{d} \mid a,d\in \mathbb{Z}, d>0, \gcd(d,\varphi(N))=1\}$.

$x^{1/d}\,(\bmod N)$ the dth RSA-$root$ of x $(\bmod N)$: the unique solution $S\in \mathbb{Z}_N^*$ to $S^d\equiv x\,(\bmod N)$ for $x\in \mathbb{Z}_N^*$ and $d\in \mathbb{Z}$ with $\gcd(d,\varphi(N))=1$.

$\mathbf{x^a}\,(\bmod N)$ the number $S\in \mathbb{Z}_N^*$ with $S\equiv x_1^{a_1}x_2^{a_2}\ldots x_k^{a_k}\,(\bmod N)$, for $\mathbf{x}=(x_1,\ldots,x_k)$ $\in (\mathbb{Z}_N^*)^k$ and $\mathbf{a}=(a_1,\ldots,a_k)\in (\tilde{\mathbb{Q}}_N)^k$.

$[a_1\ldots a_l]$ the matrix with columns $\mathbf{a}_1,\ldots,\mathbf{a}_l$.

$\mathbf{x}^A\,(\bmod N)$ $(\mathbf{x^{a_1}},\ldots,\mathbf{x^{a_l}})\in (\mathbb{Z}_N^*)^l$, for $A=[\mathbf{a}_1\ldots \mathbf{a}_l]\in (\tilde{\mathbb{Q}}_N)^{k,l}$ and $\mathbf{x}\in (\mathbb{Z}_N^*)^k$; so $(\mathbf{x}^A)^B=\mathbf{x}^{AB}$.

$l(n)$ length of the binary representation of $n\in \mathbb{N}$; the length of a negative integer m, a rational number p/q $(q\neq 1)$, a vector $\mathbf{c}$, and a matrix $A=(a_{i,j})$ are defined by: $l(m)=l(-m)+1$, $l(p/q)=l(p)+l(q)+1$, $l(\mathbf{c})=\sum_i(l(c_i)+1)$, and $l(A)=\sum_{i,j}(l(a_{i,j})+1)$, respectively.

$length(A,B)$ $l(A)+l(B)$.

3. The Protocol and Problem under Consideration

In this paper we consider the following interactive Protocol 1 (see Figure 2), which is more general than the protocol of Figure 1. The signature authority Z has created an RSA-modulus N, and issues RSA-signatures that will be products of rational powers of residue classes modulo N. Let $M=\{A,B,C,D\}$ be a set of fixed rational matrices $A\in (\tilde{\mathbb{Q}}_N)^{k,l}$, $B\in (\tilde{\mathbb{Q}}_N)^{m,l}$, $C\in (\tilde{\mathbb{Q}}_N)^{k,n}$, $D\in (\tilde{\mathbb{Q}}_N)^{m,n}$. In Protocol 1, an individual $\mathcal{A}$ requests Z to create the RSA-signature (in fact it consists of l RSA-signatures)

$$\mathbf{s}_1\equiv \mathbf{u}^A \mathbf{h}_1^B \,(\bmod N),$$

where $\mathbf{u}\in (\mathbb{Z}_N^*)^k$ is chosen by Z, and $\mathbf{h}_1\in (\mathbb{Z}_N^*)^m$ is chosen by $\mathcal{A}$ ($\mathbf{h}_1$ may depend on

$N,M = \{A,B,C,D\}$ and **u**). But actually, $\mathcal{A}$ wants to have the RSA-signature

$$s_2 \equiv u^C h_2^D \pmod{N},$$

for some $h_2 \in (\mathbb{Z}_N^*)^m$ and $s_2 \in (\mathbb{Z}_N^*)^n$. Therefore he wants to choose h_1 in such a way that after the execution of Protocol 1, he can compute from $\{N,A,B,C,D,u,h_1,$ $s_1 \equiv u^A h_1^B\}$ a pair $\{s_2,h_2\}$ satisfying $s_2 \equiv u^C h_2^D$. This way of choosing h_1 (which may depend on $N,M = \{A,B,C,D\}$ and **u**) in order to be able to compute a pair $\{s_2,h_2\}$, is called an M-*strategy*. We assume that $\mathcal{A}$ uses a probabilistic Turing machine that can do random coin tosses. We neglect the computation time used by Z, so the running time of an M-strategy is the number of steps that $\mathcal{A}$'s machine needs to compute h_1,s_2 and h_2. We assume that the running time of an M-strategy depends only on N and M, i.e., is independent of the choice of u of Z and the random coinflips by $\mathcal{A}$'s machine. However, the output of an M-strategy is a stochastic variable on the probability space defined by the uniform choice of **u** from $(\mathbb{Z}_N^*)^k$ by Z, and the random coin tosses of $\mathcal{A}$'s machine. In general, the probability that the M-strategy outputs $\{s_2,h_2\}$ with $s_2 \equiv u^C h_2^D$ is smaller than 1.

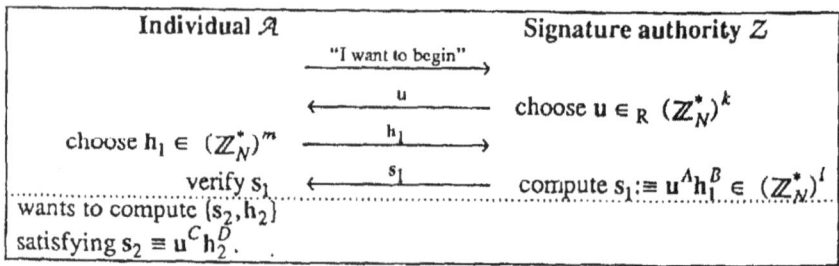

Fig. 2. The considered interactive Protocol 2

Problem 1. *For which systems of matrices* $M = \{A,B,C,D\}$ *are there feasible M-strategies, these are M-strategies with polynomial in length(N,A,B,C,D) running time that output with probability* $\geq \frac{1}{2}$*, say, a pair* $\{s_2,h_2\}$ *satisfying* $s_2 \equiv u^C h_2^D$?

This problem was solved in [EvH90] for the special case that B and D are matrices consisting of only ones, i.e., for the non interactive case.

If there are no restrictions on h_2 (e.g., h_2 must be an element from a special subset of $(\mathbb{Z}_N^*)^m$), then we can restrict ourselves in Protocol 1 (and thus also in Problem 1) to the case that D consists of only zeros, according to the next lemma.

Lemma 1. *Let* $C \in (\tilde{\mathbb{Q}}_N)^{k,n}$, $D \in (\tilde{\mathbb{Q}}_N)^{m,n}$ *and* $u \in (\mathbb{Z}_N^*)^m$. *Then there exists a matrix* $\tilde{C} \in (\tilde{\mathbb{Q}}_N)^{k,n}$ *(which is computable in polynomial time from C and D), such* *that computing a pair* $(s,h) \in (\mathbb{Z}_N^*)^n \times (\mathbb{Z}_N^*)^m$ *satisfying* $s \equiv u^C h^D$ *is polynomial-time equivalent to computing* $u^{\tilde{C}}$.

Proof. We can reformulate the identity $s \equiv u^C h^D$ as

$$u^C \equiv (s,h)^{\left[\begin{smallmatrix} I \\ -D \end{smallmatrix}\right]},$$

where the first n coordinates of vector (s,h) are those of s, the last m coordinates are those of h, the first n rows of $\left[\begin{smallmatrix} I \\ -D \end{smallmatrix}\right]$ are those of the identity matrix I, and the last m rows are those of $-D$. According to [KaBa79], we can find in polynomial time unimodular matrices P, Q and a matrix $\left[\begin{smallmatrix} G \\ 0 \end{smallmatrix}\right]$ in Smith normal form, such that $P\left[\begin{smallmatrix} G \\ 0 \end{smallmatrix}\right] = \left[\begin{smallmatrix} I \\ -D \end{smallmatrix}\right]Q$. Because $\left[\begin{smallmatrix} I \\ -D \end{smallmatrix}\right]$ has full column rank, G is invertible. Define $\tilde{C} = CQG^{-1}$. Then from (s,h) satisfying $s \equiv u^C h^D$ we can compute $u^{\tilde{C}}$ in polynomial time because

$$u^{\tilde{C}} = u^{CQG^{-1}} = (s,h)^{\left[\begin{smallmatrix} I \\ -D \end{smallmatrix}\right]QG^{-1}} \equiv (s,h)^{P\left[\begin{smallmatrix} G \\ 0 \end{smallmatrix}\right]G^{-1}} = (s,h)^{P\left[\begin{smallmatrix} I \\ 0 \end{smallmatrix}\right]}$$
$$= \text{first } n \text{ coordinates of } (s,h)^P.$$

If, on the other hand, $u^{\tilde{C}}$ is given, then we obtain the pair (s,h) satisfying $s \equiv u^C h^D$ by first computing the vector $\bar{s} = (u^{\tilde{C}}, 1, \ldots, 1)$ of length $n+m$ and then by defining s, h by $(s,h) :\equiv \bar{s}^{P^{-1}}$. □

4. Algebraic Strategies

As shown in the previous section, we may restrict ourselves to the case in which M = $\{A, B, C, [0]\}$. It seems a hard problem to determine the matrices A, B, C for which there exist *arbitrary* feasible M-strategies. Therefore, we consider only M-strategies of a special kind, so-called *algebraic M-strategies*. In an algebraic strategy, $\mathcal{A}$ applied to u only multiplications and divisions mod N in order to compute h_1, and the choice of these multiplications and divisions is independent of u. It is conceivable that algebraic strategies are the best, i.e., if there is no feasible algebraic M-strategy, then there is also no feasible M-strategy of another kind. But we have no insight in this matter

Let $A \in (\tilde{\mathbb{Q}}_N)^{k,l}, B \in (\tilde{\mathbb{Q}}_N)^{m,l}, C \in (\tilde{\mathbb{Q}}_N)^{k,n}$ be fixed rational matrices. $\mathcal{A}$ is assumed to follow an algebraic M-strategy, hence, in Protocol 1, h_1 must consist of products of integral powers of the entries of u, i.e. $h_1 \equiv u^X$, for some $X \in \mathbb{Z}^{k,m}$. So instead of analyzing the general Protocol 1, we will analyze Protocol 2 in this section (see Figure 3, in which we write h in stead of h_1).

We now assume also that it is computationally infeasible for $\mathcal{A}$ to compute RSA-roots modulo N, since otherwise he could forge all signatures. Under this assumption, the Corollary of [EvH92] implies the following for Protocol 2.

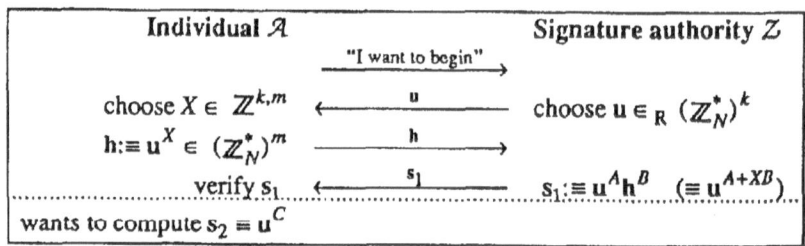

Fig. 3. Protocol 2, which is equivalent with Protocol 1, if $\mathcal{A}$ follows an algebraic M-strategy.

Proposition. *Let A,B,C be matrices with rational entries. Then the following two statements are equivalent*

(i) *There is a probabilistic polynomial in length(A,B,C) algorithm to compute integral matrices X,Y,Z such that $C = (A+XB)Y + Z$.*

(ii) *There is a feasible algebraic M-strategy for the individual $\mathcal{A}$ to compute $\mathbf{h}$ from $\{N,A,B,C,\mathbf{u}\}$ and $\mathbf{u}^C$ from $\{N,A,B,C,\mathbf{u},\mathbf{u}^A\mathbf{h}^B\}$.*

Because of this result, we are interested in the following problem.

Problem 2. *Let $A \in (\tilde{\mathbb{Q}}_N)^{k,l}$, $B \in (\tilde{\mathbb{Q}}_N)^{m,l}$, $C \in (\tilde{\mathbb{Q}}_N)^{k,n}$ be rational matrices. Find a polynomial (in length(A,B,C))-time algorithm that decides whether the equation*

$$C = (A+XB)Y + Z$$

is solvable in integral matrices $X \in \mathbb{Z}^{k,m}$, $Y \in \mathbb{Z}^{l,n}$, $Z \in \mathbb{Z}^{k,n}$, and if so, find a solution X,Y,Z.

We have not been able to solve Problem 2 in full generality. We have proven that there exists such a polynomial-time algorithm for Problem 2 in the case that $n=1$, but this is not included in this paper. In the next section, we solve Problem 2 in the special case that $k=l=m=n=1$.

5. A Special Case of Protocol 2

Let $a,b,c \in \tilde{\mathbb{Q}}_N$ be fixed and assume that the denominators of a,b, and c are coprime to $\varphi(N)$. We analyze the following Protocol 3 between Z and $\mathcal{A}$ (see Figure 4). In this protocol, $\mathcal{A}$ receives from Z the RSA-signature u^{a+xb} (mod N), which $\mathcal{A}$ can verify. Note that $\mathcal{A}$ cannot compute this signature on a randomly chosen residue class u himself, because in general $a+xb \in \mathbb{Q} \setminus \mathbb{Z}$.

The next lemma states when it is feasible for $\mathcal{A}$ to compute u^c after the execution of this protocol.

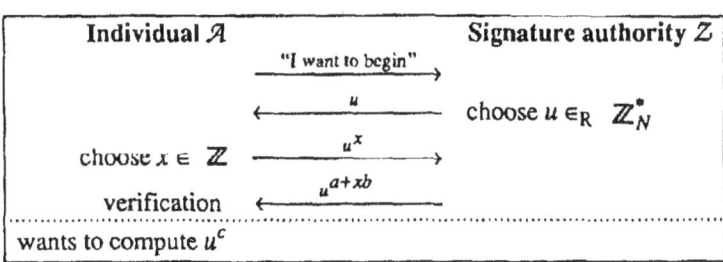

Fig. 4. Protocol 3.

Lemma 2. *$\mathcal{A}$ can choose x in Protocol 3 in such a way (and in polynomial time) that it is feasible for him to compute u^c after the execution of the protocol if and only if* $\gcd(1,a,b)|c$.

Note that $\gcd(1,a,b)$ is in general not 1, because a and b are rational numbers. This lemma can be proved by using Corollary 1 and the following two lemmas.

Lemma 3. *Let $a,b,c \in \mathbb{Q}$, $c \neq 0$. Then there exists an integer λ such that* $\gcd(a+\lambda b,c) = \gcd(a,b,c)$, *and this λ can be computed in polynomial (in* $length(a,b,c))$ *time.*

Proof. Define $\bar{a} = a/\gcd(a,b,c), \bar{b} = b/\gcd(a,b,c), \bar{c} = c/\gcd(a,b,c)$. Thus $\bar{a},\bar{b},\bar{c}$ are integers with $\gcd(\bar{a},\bar{b},\bar{c}) = 1$. It suffices to show that we can compute in polynomial time a $\lambda \in \mathbb{N}$ that satisfies $\gcd(\bar{a} + \lambda\bar{b}, \bar{c}) = 1$.

For each prime number p and each $a \in \mathbb{Z}, a \neq 0$, let $\text{ord}_p(a)$ be the integer such that $a \cdot p^{-\text{ord}_p(a)}$ is an integer not divisible by p. Take

$$\lambda = \prod_{p|\bar{c},\, p\nmid \bar{a}} p^{\text{ord}_p(\bar{c})}.$$

Let p be a prime dividing $\bar{c}$. If $p|\bar{a}$, then $p\nmid \bar{b}$ (by $\gcd(\bar{a},\bar{b},\bar{c}) = 1$) and $p\nmid \lambda$ (by definition of λ), hence $p\nmid (\bar{a} + \lambda\bar{b})$. If $p\nmid \bar{a}$, then $p|\lambda$ (by definition of λ and $\text{ord}_p(\bar{c}) \geq 1$) and hence also $p\nmid (\bar{a} + \lambda\bar{b})$. We conclude that no prime divides both $\bar{c}$ and $(\bar{a} + \lambda\bar{b})$; therefore $\gcd(\bar{a} + \lambda\bar{b}, \bar{c}) = 1$.

Define the sequence $c_0 := |\bar{c}|$ and $c_{i+1} := c_i/\gcd(\bar{a},c_i)$ for $i = 0,1,2,\dots$. Let i_0 be the smallest integer such that $\gcd(\bar{a},c_{i_0})=1$. It is easy to see that $c_{i_0}=\lambda$ and that $i_0\le l(\bar{c})$; thus λ can be computed in polynomial time. $\square$

Lemma 4. *Let $a,b,c\in\mathbb{Q}$. Then there are $x,y,z\in\mathbb{Z}$ such that $c=(a+xb)y+z$ if and only if $\gcd(1,a,b)|c$. Further, if such $x,y,z\in\mathbb{Z}$ exist, then they can be computed in polynomial (in $length(a,b,c)$) time.*

Proof. Note that $a,b,1$ are integral multiples of $\gcd(1,a,b)$. Hence, if there exist $x,y,z\in\mathbb{Z}$ such that $c=(a+xb)y+z$, then c is also an integral multiple of $\gcd(1,a,b)$. Hence $\gcd(1,a,b)|c$.

On the other hand, assume that $\gcd(1,a,b)|c$. By Lemma 3 we can compute in polynomial time an $x\in\mathbb{Z}$ such that $\gcd(a+xb,1)=\gcd(1,a,b)$. Further we can compute in polynomial time $y,z\in\mathbb{Z}$ with $c=(a+xb)y+z$ (e.g., let $d\in\mathbb{N}$ such that $da,db,dc\in\mathbb{N}$, and use $\gcd(a+xb,1)|c$ and Euclid's algorithm to compute $y,z\in\mathbb{Z}$ with $dc=(da+xdb)y+dz$). This proves Lemma 4. $\square$

Proof of Lemma 2.

(i) Suppose that $\gcd(1,a,b)|c$. According to Lemma 4, $\mathcal{A}$ can compute in polynomial time numbers $x,y,z\in\mathbb{Z}$ such that $c=(a+xb)y+z$. $\mathcal{A}$ will use the obtained number x during the execution of Protocol 2. Afterwards, $\mathcal{A}$ can compute u^c from $\{N,a,b,c,u,u^{a+xb}\}$ as follows:

$$(u^{a+xb})^y\cdot u^z\equiv u^c\ (\mathrm{mod}\,N).$$

(ii) Suppose that $\mathcal{A}$ can choose x in Protocol 2 in such a way that it is feasible for him to compute u^c after the execution of the protocol. Corollary 1 states that computing u^c from $\{N,a,b,c,u,u^{a+xb}\}$ for uniformly chosen $u\in\mathbb{Z}_N^*$ is feasible for $\mathcal{A}$ if and only if $c\in\mathbb{Z}\{1,a+xb\}$. That is, if and only if there are $y,z\in\mathbb{Z}$ such that $c=(a+xb)y+z$. $\square$

Individual $\mathcal{A}$		Signature authority $\mathcal{Z}$
choose $v\in_R\mathbb{Z}_N^*$	$\xrightarrow{\text{"I want to begin"}}$	
	$\xleftarrow{\quad u\quad}$	choose $u\in_R\mathbb{Z}_N^*$
choose $x,y\in\mathbb{Z}$	$\xrightarrow{h\equiv u^x v^y}$	
verify S	$\xleftarrow{\quad S\quad}$	$S:\equiv u^a h^b\ (\equiv u^{a+xb}v^{by})$
wants to compute u^c		

Fig. 5. Protocol 4.

We generalize Protocol 3 to Protocol 4 (see Figure 5), in which $\mathcal{A}$ initially chooses some residue class, but we will prove that doing so does not influence the feasibility of computing the signature u^c after the execution of the protocol.

We are interested in how $\mathcal{A}$ should choose v, x, y so that it is feasible for him to compute u^c from $\{u, v, u^{a+xb}v^{by}, a, b, c, x, y\}$. According to the Corollary of [EvH92] this computation is feasible if and only if there is an integral solution z_1, z_2, z_3 to

$$\begin{cases} z_1 + z_3(a + xb) = c, \\ z_2 + z_3 by = 0. \end{cases}$$

According to Lemma 2, a necessary condition for the first equation is that $\gcd(1, a, b) | c$. But the number z_3 obtained there does not need to be a solution of the second equation. If $y=0$, then the z_3 obtained is also a solution of the second equation. Hence a *necessary* condition for the simultaneous solvability of the two equations is that $\gcd(1, a, b) | c$; and if $y=0$, then this condition is also *sufficient*. Therefore, the best strategy for $\mathcal{A}$ is to choose $y=0$ and to take x according to Lemma 3, and thus this algebraic strategy "works" if and only if $\gcd(1, a, b) | c$.

6. Generalizations

In Protocol 1 (see Figure 2) the system of matrices used is $M = \{A, B, C, D\}$, so the individual will receive one type of signature (s_1), and wants to compute a second type (s_2).

We now assume that there are $(t+1)$ types of signatures, so $\mathcal{Z}$ creates a public system of matrices $M = \{A_1, B_1, \ldots, A_{t+1}, B_{t+1}\}$, where the matrices (A_i, B_i) are used for the ith type. We assume that $\mathcal{Z}$ issues only signatures of type $1, \ldots, t$ to $\mathcal{A}$, and that $\mathcal{A}$ tries to compute a signature of type $t+1$ from these. Therefore, we want to consider the serial Protocol 5 (see Figure 6), in which we assume that $\mathcal{Z}$ uses the same u in every signature, that the individual chooses $h_1, \ldots, h_t$ (where h_i may depend on M, u and $s_1, \ldots, s_{i-1}$), and receives the signatures

$$s_i \equiv u^{A_i} h_i^{B_i} \quad (i=1, \ldots, t).$$

$\mathcal{A}$ will not receive signatures of type $(t+1)$, so he tries to choose $\{h_1, \ldots, h_t\}$ in such a way that after receiving $\{s_1, \ldots, s_t\}$, he is able to compute a pair (s_{t+1}, h_{t+1}) such that

$$s_{t+1} \equiv u^{A_{t+1}} h_{t+1}^{B_{t+1}}.$$

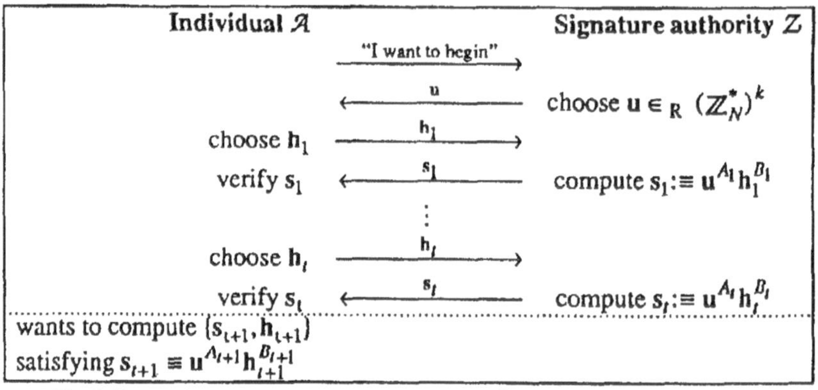

Fig. 6. The serial signature issuing Protocol 5, in which $\mathcal{A}$ receives the signatures $s_1,...,s_t$.

If we assume that $\mathcal{A}$ will use an algebraic M-strategy, then we can prove it suffices to consider algebraic strategies on protocols with $t=1$, i.e., we can reduce Protocol 5 in polynomial time to Protocol 1 as follows:

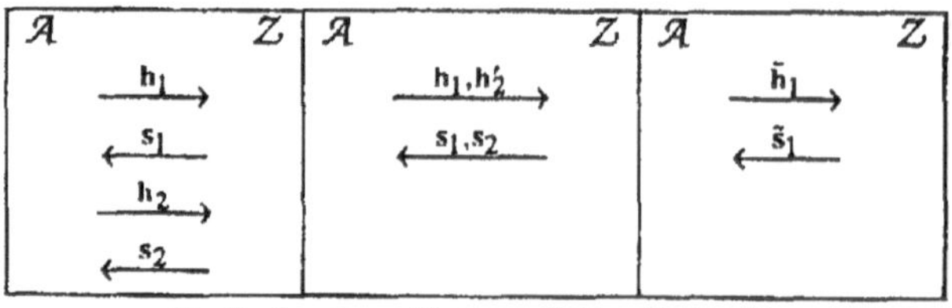

Fig. 7. How to modify Protocol 5 into Protocol 1.

Moves 3 up to 6 of Protocol 5 are shown in the left-hand side of Figure 7. Let d be the smallest positive integer such that dA_1 and dB_1 are integral matrices (so d can be the lcm of all the denominators of A_1 and B_1). Hence s_1^d can be computed by $\mathcal{A}$ *without* knowing s_1, because $s_1^d \equiv u^{dA_1} h_1^{dB_1}$, and the used exponents are integral. In order to create h_2, $\mathcal{A}$ might use s_1. But $\mathcal{A}$ only applies multiplications and divisions on s_1, so $\mathcal{A}$ is able to compute $h_2':=h_2^d$ *without* knowing s_1 ($\mathcal{A}$ will only use s_1^d, which he could compute without knowing s_1). By defining the new matrix $B_2':=\frac{1}{d}B_2$, we have that $h_2'^{B_2'} \equiv h_2^{B_2}$, so $\mathcal{A}$ does not need to know s_1 in order to compute $h_2^{B_2}$ (by using matrix B_2'). The possibility that $\mathcal{A}$ can compute s_{t+1} at the end of the protocol remains the same if we carry out the first four moves in parallel instead of serially (see Figure 7, middle). By defining $\tilde{s}_1:=(s_1,s_2)$, $\tilde{h}_1:=(h_1,h_2')$, $\tilde{A}_1:=[A_1\ A_2]$, $\tilde{B}_1:=[B_1\ B_2']^T$, we have that $\tilde{s}_1 \equiv u^{\tilde{A}_1} h_1^{\tilde{B}_1}$; thus we can combine the first four moves into two (see Figure 7, right-hand side). In this way we obtain a protocol with 2 moves less. By repeating this argument, we only have to analyze a protocol with 2+2 moves, i.e., Protocol 1.

References

[Dav82]   George Davida, Chosen signature cryptanalysis of the RSA (MIT) public key cryptosystem, Tech. rept. TR-CS-82-2, Dept of Electrical Engineering and Computer Science, Univ. of Wisconsin, October 1982.

[Denn84]   Dorothy Denning, "Digital signatures with RSA and other public-key cryptosystems", *Comm. of the ACM*, 27 (1984) pp. 388-392.

[DO85]   Yvo Desmedt and Andrew Odlyzko, "A chosen text attack on the RSA cryptosystem and some discrete logarithm schemes", *Advances in Cryptology-CRYPTO 85*, H.C. Williams ed., LNCS 218, Springer-Verlag, pp. 516-522.

[EGS85]   Shimon Even, Oded Goldreich and Adi Shamir, "On the security of ping-pong protocols when implemented using the RSA", *Advances in Cryptology-CRYPTO 85*, H.C. Williams ed., LNCS 218, Springer-Verlag, pp. 58-72.

[EvH92]   Jan-Hendrik Evertse, Eugène van Heyst, "Which new RSA signatures can be computed from certain given RSA signatures?", *Journal of Cryptology*, 5 (1992), pp. 41-52.

[KaBa79]   R. Kannan and A. Bachem, "Polynomial algorithms for computing the Smith and Hermite normal forms of an integer matrix", *SIAM Journal on Computing*, 8 (1979) pp. 499-507.

[RSA78]   R.L. Rivest, A. Shamir, and L. Adleman, A Method for Obtaining Digital Signatures and Public Key Cryptosystems, *Comm. of the ACM* 21 (1978) pp. 120-126.

Transferred Cash Grows in Size

David Chaum
CWI
The Netherlands

Torben Pryds Pedersen*
Aarhus University
Denmark

Abstract

All known methods for transferring electronic money have the disadvantages that the number of bits needed to represent the money after each payment increases, and that a payer can recognize his money if he sees it later in the chain of payments (forward traceability). This paper shows that it is impossible to construct an electronic money system providing transferability without the property that the money grows when transferred. Furthermore it is argued that an unlimited powerful user can always recognize his money later. Finally, the lower bounds on the size of transferred electronic money are discussed in terms of secret sharing schemes.

1 Introduction

Transferability of electronic cash means that the payee in one payment transaction can spend the received money in a later payment to a third person without contacting the bank or another central authority between the two transactions. As on-line electronic payment systems require communication with a central authority during the payment transaction, transferability is only an issue for off-line systems. Although the ability to transfer "normal" money (coins, notes) is very important in our daily life, this property has only received very little attention in relation to electronic money. To the knowledge of the authors, transferability of electronic money has only been described in [vA90], [OO90] and [OO92].

This paper first sketches the (generic) method for transferring electronic cash proposed in [vA90]. At a first glance this method is not ideal, because extra bits are appended to the transferred coins, and a person, whose coin has been transferred a number of times, can always recognize this coin if he sees it later (this property will be referred to as forward traceability).

Intuitively, it is not surprising that the size of transferred money increases, because it must be possible for the bank to identify people, who spends a coin twice. Hence, a transferred coin must contain some information about every person, who has spent it.

This paper formalizes this argument, as it gives lower bounds on the number of bits needed to represent transferred money. These lower bounds depend on whether the systems provide unconditional untraceability or computational untraceability. In

---

*Research done while visiting CWI

R.A. Rueppel (Ed.): Advances in Cryptology - EUROCRYPT '92, LNCS 658, pp. 390-407, 1993.

particular it is shown that in case of unconditional untraceability the size of a transferred coin must increase by the number of bits needed to identify the payer, whereas a computational untraceable coin grows with approximately half this number of bits.

It is furthermore argued that a payer with unlimited computing power can always recognize his own money, if he sees it later in the chain of payments.

All statements in this paper are with respect to coin-systems, but it is not hard to see that the results are valid for electronic checks as well. The first section describes a general model of off-line electronic coins and presents the notation which will be used in this paper. In Section 3 it is shown how to add transferability to all known payments systems, and Section 4 and 5 give lower bounds on the size of transferred electronic coins. Section 4 considers payment systems providing unconditional payer untraceability, and Section 5 gives a lower bound for computationally untraceable money. In Section 6 it is argued that these lower bounds are optimal, and Section 7 concludes the paper.

2   The Model

This section presents a basic model for off-line electronic cash which will be used in the following.

The results in this paper are independent of whether the payments system provides a protocol for refunding unspent parts of the money. For simplicity, we will therefore assume that the payer always spends his electronic money for its total value (coins). Hence, we consider an off-line electronic payment system involving a bank (B) and K individuals $(p_1, \ldots, p_K)$ providing protocols for:

1. Withdrawal of money from the bank;

2. Payment transactions from one individual to another; and

3. Deposit (at B) of received money.

The system is said to provide transferability, if the payee in on payment transaction can use the received money as a payer in a later payment transaction without talking with the bank (or anybody else) between these two transactions.

The "life-cycle" of an electronic coin in such a system looks like:

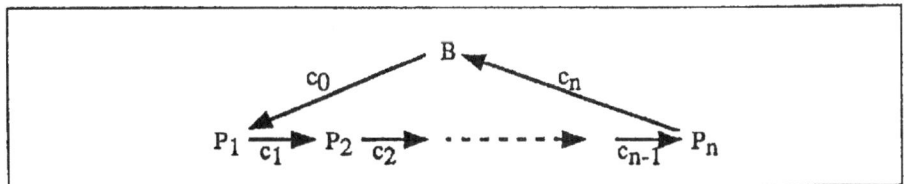

Figure 1: Life-cycle of a coin

Figure 1 illustrates that a person, p_1, first withdraws c_0 from B and then spends the coin in a payment transaction. During this transaction c_0 is changed to c_1. In general, for $i = 2, 3, \ldots, n-1$, p_i receives c_{i-1}, and when he later spends it, it is transformed into c_i. Finally, p_n deposits the coin at the bank, who receives c_n.

Unconditional payer untraceability means that even an unlimited powerful bank cannot identify any of $p_1, p_2, \ldots, p_{n-1}$ from c_n. Computational untraceability means that given c_n, the bank cannot identify the payers unless it can make a computation, which is thought to be infeasible.

To prevent a payer from using a coin twice, it most be possible for the bank (with very high probability) to discover the identity of such a double-spender. This must be possible even if the coins have been transferred a number of times after the double-spending occurred.

3 How to Transfer Electronic Money

This section gives a brief description of the method for transferring electronic money presented in [vA90]. As this method is generic in the sense that it works for a large class of electronic payment systems, it will only be described in general terms here. First a general coin-system is sketched, and then it is extended with transferability (for more details see [CFN90] and [vA90]). It is not difficult to apply this method to electronic checks as well (see [vA90]).

Let the bank have two secret keys S_0 and S_1 with corresponding public keys P_0 and P_1. A signature with secret key S_1 is worth a fixed amount (say \$1), whereas a signature with S_0 is worth nothing. Both signature schemes must have the property that it is possible to make blind signatures: A user can get a signature $S_i(m)$ on the message m, but the signer gets no information about m (for $i = 0, 1$). See [Cha83] and [Cha84] for examples of such signature schemes.

An electronic coin system can now be constructed as follows.

Withdrawal

1. User P constructs a message m_P of a special form (see later), and proves that m_P is constructed correctly (without giving the bank any information about m_P).

2. The bank makes a blind signature on m_P and withdraws \$1 from P's account.

3. P recovers the signature on m_P.

P can later pay another person, R, one dollar using the following protocol

Payment

1. P sends m_P and $S_1(m_P)$ to R.

2. R verifies that $S_1(m_P)$ is a signature on m_P, chooses a random challenge, c_P, and sends it to P.

3. P sends back an answer r_P.

4. R verifies that r_P is correct (using c_P and m_P).

In order to prevent double-spending, m_P must be constructed such that if P can send correct answers corresponding to two different challenges, then the bank can find P's identity from these two answers and m_P. However, a single correct answer must not give the bank any (Shannon) information about P's identity (see [CFN90] and [vA90] for details about how P can construct m_P and prove to the bank that it was constructed correctly).

The receiver, R, can at any time after the payment deposit the electronic coin at the bank (and get \$1):

Deposit

1. R sends m_P, $S_1(m_P)$, c_P and r_P to the bank.

2. The bank verifies the signature and that r_P is a correct response to the challenge c_P. Then the bank increases B's account with the amount \$1.

3. Finally, the bank searches through its database to see if m_P has been deposited previously, and in that case it finds the identity of P (provided the challenges in the two payments are different).

In [CFN90] it is discussed how it can be ensured that P will always get different challenges, if she tries to spend a coin more than once.

In order to add transferability to this scheme the signature scheme given by (S_0, P_0) is needed. Furthermore, a one-way function, f, is required.

Before R acts as a payee in a payment, he goes to the bank and performs a protocol corresponding to the withdrawal except that the bank gives R a signature $S_0(m_R)$, where m_R has the same properties as m_P (in practice, R would get signatures on many different messages, m_R, in an initial transaction).

When R receives the coin given by $S_1(m_P)$ from P, he does not choose the challenge at random, but as

$$c_P = f(m_R, \rho_R)$$

where ρ_R is formed in a special way (to ensure that R can later deposit the money if he wants to, and to ensure that P gets different challenges, if he tries to spend the same coin twice — even if P and R cooperate).

Later R can pay the received coin to a third person, S, without contacting the bank between the two payments:

Payment of a transferred coin

1. R sends m_P, $S_1(m_P)$, r_P, m_R, $S_0(m_R)$ and ρ_R to S.

2. S verifies the two signatures.
   S verifies, that r_P is a correct answer to the challenge $f(m_R, \rho_R)$.

3. S computes a challenge, c_R, and sends it to R.

4. R sends an answer r_R.

5. S verifies that r_R is correct (using c_R and m_R).

S can in particular compute the challenge c_S such that she can later transfer the coin. This method has the advantage, that it is easy to implement in known electronic cash systems, but it has two drawbacks:

1. The money grows in size when transferred (because transcripts of the previous payments are appended and must be verified in each payment).

2. If a payer sees his coin later in the chain of payments, he can recognize it.

It is intuitively clear that the bank can identify double-spenders even if the coins has been transferred a number of times, but it is outside the scope of this paper to state and prove this property formally.

4 Unconditionally Untraceable Money

In this section it is shown that the size of electronic money has to grow each time it is transferred, if unconditional payer untraceability is provided.

Consider the tree of payments constructed as follows. A payer, p_1, withdraws a coin, c_0', from the bank and uses it to pay the user p_2. During this transaction c_0' is changed to c_1'. Later p_2 pays this coin to p_3, and after this transaction the coin has been changed to c_2'. In general, we consider n such payments ($n \in I\!N$), in which p_i receives the coin c_{i-1}', and when he spends it again later, it is transformed into c_i'. In the following it will be assumed that the prescribed payment protocol is executed *correctly* in all these transactions.

Assume furthermore that each p_i spends c_{i-1}' again in another *correct* execution of the payment protocol completely *independent* of p_i's first payment. The resulting coin (the payee's output from this transaction) is denoted c_i. In the following we are going to look at these c_i's and forget about the c_i''s except c_n' which we will also denote by c_{n+1}. Figure 2 shows the relation between the payers and the coins.

The c_i's and p_i's depend on the random choices in the transactions and the choices of payees. Let therefore C_i be a random variable whose value is c_i for $i = 1, 2, \ldots, n, n+1$, and let P_i be a random variable whose value is p_i for $i = 1, 2, \ldots, n$.

In this section, a lower bound will be given for the entropies of the random variables representing coins. The results are based on elementary information theory as presented in [Wel88] for example. In the following $H(X)$ denotes the entropy of the finite random variable, X:

$$H(X) = -\sum_x Prob(X = x) \log(Prob(X = x))$$

(all logarithms are with the base 2).

Let U, V and W be three vectors of finite, random variables. The following rules will be used repeatedly:

$$
\begin{align}
H(U) &\geq 0 & (1) \\
H(U, V) &= H(U \mid V) + H(V) & (2) \\
H(U, V \mid W) &= H(U \mid V, W) + H(V \mid W) & (3) \\
H(U \mid V, W) &\leq H(U \mid V) & (4)
\end{align}
$$

The following lemma will also be used several times.

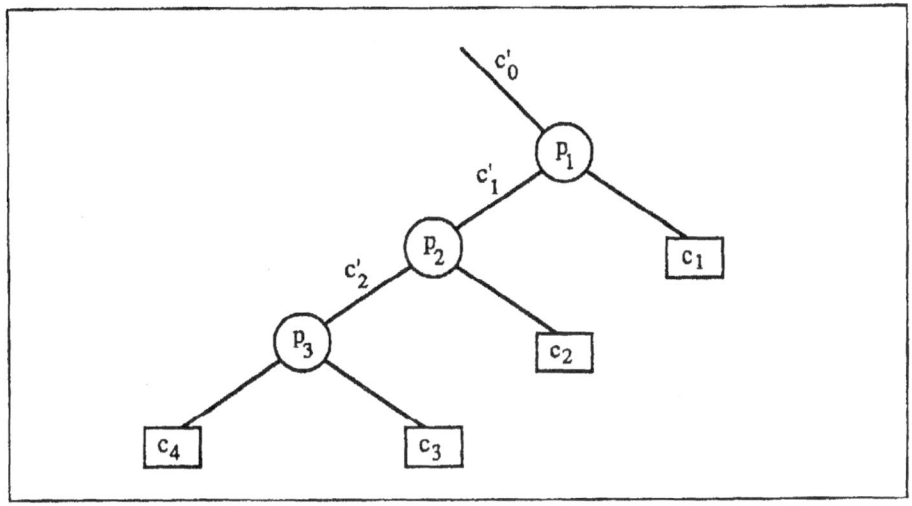

Figure 2: Payment tree for $n = 3$ — unconditional untraceability

Lemma 4.1

Let $\epsilon > 0$, and let U, V, W and Z be four vectors of finite, random variables. If

$$H(V \mid W, Z) \leq \epsilon$$

then

$$H(U \mid W, Z) \leq H(U \mid V, Z) + \epsilon.$$

Proof

$$
\begin{aligned}
H(U \mid V, Z) &\geq H(U \mid V, W, Z) &&\text{by (4)} \\
&= H(U, V \mid W, Z) - H(V \mid W, Z) &&\text{by (3)} \\
&\geq H(U, V \mid W, Z) - \epsilon \\
&= H(V \mid U, W, Z) + H(U \mid W, Z) - \epsilon \\
&\geq H(U \mid W, Z) - \epsilon &&\text{by (1)}
\end{aligned}
$$

$\blacksquare$

If the payment system provides unconditional payer untraceability then the conditional entropy of P_i given C_i equals $H(P_i)$:

$$H(P_i \mid C_i) = H(P_i) \qquad \text{for } i = 1, 2, \ldots, n.$$

This property can be further strengthened to

$$H(P_i \mid C_1, C_2, \ldots, C_i) = H(P_i) \qquad \text{for } i = 1, 2, \ldots, n,$$

because of the independence of the payment transactions.

The fact that the bank can identify double-spenders with probability at least $1-p$, where p is the probability that a double-spender is not detected, implies that

$$H(P_i \mid C_i, C_j) \leq p \log K \qquad \text{for } 1 \leq i < j \leq n+1,$$

where K is the number of possible payers. Let

$$\epsilon = p \log K.$$

In practice p must be very small (negligible as a function of a security parameter) and then ϵ is very small as well.

Theorem 4.2
$$H(C_{n+1}) \geq H(P_1) + H(P_2) + \ldots + H(P_n) - 2n\epsilon.$$

Proof
Claim: For $1 \leq i \leq n$:

$$H(C_{n+1} \mid C_1, C_2, \ldots, C_i) \geq H(C_{n+1} \mid C_1, C_2, \ldots, C_i, C_{i+1}) + H(P_i) - 2\epsilon.$$

From this claim it follows by simple induction that

$$
\begin{aligned}
H(C_{n+1}) &\geq H(C_{n+1} \mid C_1) \\
&\geq H(C_{n+1} \mid C_1, C_2) + H(P_1) - 2\epsilon \\
&\cdots \\
&\geq H(C_{n+1} \mid C_1, C_2, \ldots, C_i) + \sum_{j=1}^{i-1} H(P_j) - 2\epsilon(i-1) \\
&\cdots \\
&\geq H(C_{n+1} \mid C_1, C_2, \ldots, C_{n+1}) + \sum_{j=1}^{n} H(P_j) - 2\epsilon n \\
&= \sum_{j=1}^{n} H(P_j) - 2\epsilon n
\end{aligned}
$$

In order to prove the claim, let $i \in \{1, 2, \ldots, n\}$ be given, and let A_i denote the vector $(C_1, C_2, \ldots, C_i)$. Then

$$
\begin{aligned}
H(C_{n+1} \mid A_i) &= H(P_i, C_{n+1} \mid A_i) - H(P_i \mid C_{n+1}, A_i) &&\text{by (3)} \\
&\geq H(P_i, C_{n+1} \mid A_i) - \epsilon \\
&= H(C_{n+1} \mid P_i, A_i) + H(P_i \mid A_i) - \epsilon &&\text{by (3)} \\
&= H(C_{n+1} \mid P_i, A_i) + H(P_i) - \epsilon \\
&\geq H(C_{n+1} \mid C_{i+1}, A_i) + H(P_i) - 2\epsilon
\end{aligned}
$$

where last inequality follows from Lemma 4.1 and the fact that $H(P_i \mid C_{i+1}, A_i) \leq \epsilon$.
∎

As the entropy is a measure of the number of bits in optimal encodings, Theorem 4.2 implies that the number of bits needed to represent an electronic coin grows each time the coin is transferred. Furthermore, the increase is the number of bits needed to identify the new payer. In particular, if $H(P_i) = k$ for every i (the bank needs k bits of information to identify each payer) we see that

$$H(C_{n+1}) \geq kn - 2n\epsilon,$$

and by the symmetry of C_n and C_{n+1} we get

$$H(C_n) \geq n(k - 2\epsilon).$$

As $\epsilon << k$ the coin grows with the number of bits needed to identify the payer each time the coin is transferred.

We conclude this section with a short remark on forward traceability. If all secret keys of the bank are uniquely determined by the bank's public key, then a payer with unlimited computing power can alway determine, given a transferred coin, if he has previously had this coin in his possession:

1. Simulate another payment of his original coin.

2. Compute the secret key of the bank (using the unlimited power).

3. Determine the identity of the double-spender (as the bank would have done).

5 Computationally Untraceable Money

In the previous section we saw that unconditionally untraceable, electronic money must grow in size when transferred. In this section a similar result is proven for computationally untraceable money.

The tree considered in the previous section is not sufficient to give an interesting lower bound on the size of computational untraceable money. This is due to the fact that in such a system each c_i could, in principle, contain all information needed to identify p_i, but no additional information about the previous payers.

The proof in this section is therefore based on a tree of payments constructed as follows. First, p_0^1 receives a coin, c, from the bank, and then he chooses two payees, p_0^2 and p_1^2, at random among all individuals in the system and pays both of them in *correct and independent* executions of the payment protocol. As the money system provides transferability p_0^2 and p_1^2 can later, independently of each other, spend the received coin twice in a similar way. In general, for $j \geq 1$ and $0 \leq i < 2^{j-1}$, p_i^j transfers a received coin to p_{2i}^{j+1} and p_{2i+1}^{j+1}. After some time, the original coin has been changed to $c_0, c_1, \ldots, c_{2^n-1}$, where p_i^n is the payer of c_{2i} and c_{2i+1} for $i = 0, 1, \ldots, 2^{n-1} - 1$.

Let C_i be a random variable with value c_i, and let P_i^j be a random variable whose value is the identity of p_i^j. Figure 3 shows how C_i is related to each P_i^j for $n = 3$.

For any $i = 0, 1, \ldots, 2^n - 1$, let $P_{i,1}, P_{i,2}, \ldots, P_{i,n}$ be the path from the root to C_i in the tree of payments ($P_{i,1} = P_0^1$ for all i). Hence, P_i^j ($0 \leq i < 2^{j-1}$) denotes the i'th payer (from left) at depth $j - 1$, whereas $P_{i,j}$ ($0 \leq i < 2^n$) denotes then j'th payer of the coin which is finally transferred to C_i. In both cases $1 \leq j \leq n$.

Furthermore, for any pair (i, j) where $0 \leq i < 2^n$ and $1 \leq j \leq n$, let $T_{i,j}$ be the sub-tree of height $n - (j - 1)$ having $P_{i,j}$ as root. The leaves in this subtree can be numbered from 0 to $2^{n+1-j} - 1$ from left to right. Let C_i have number k in this

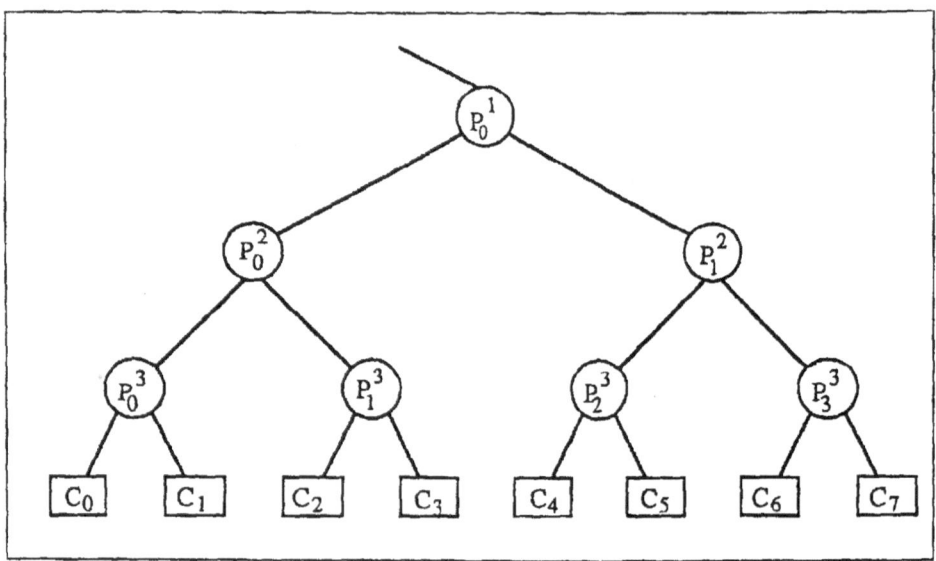

Figure 3: Payment tree for random variables ($n = 3$) — computational untraceability

enumeration. Then $f(i,j)$ is defined as the index of the leaf with number $2^{n+1-j} - 1 - k$ in this enumeration ($C_{f(i,j)}$ is "symmetric" to C_i in the sub-tree $T_{i,j}$). For example for $n = 3$

$$\begin{aligned} f(0,1) &= 7 \\ f(2,1) &= 5 \\ f(4,2) &= 7 \\ f(0,3) &= 1 \end{aligned}$$

The function $i \mapsto f(i,j)$ is a bijection for every $j \in \{1, 2, \ldots, n\}$ because

$$f(f(i,j),j) = i.$$

Let p be the maximal probability with which a user can spend a coin twice without being identified, and let

$$\epsilon = p \log K,$$

where K is the number of possible participants. The property that the bank can identify double-spenders can be expressed in terms of entropies as follows. Given two leaves C_i and C_k ($i \neq k$), let j be maximal such that C_k is a leaf in $T_{i,j}$. Then C_i and C_k are both (transferred) results of double-spending by $p_{i,j}$. Hence

$$H(P_{i,j} \mid C_i, C_k) \leq \epsilon \qquad (*).$$

In particular this implies that

$$H(P_{i,j} \mid C_i, C_{f(i,j)}) \leq \epsilon.$$

Now consider the subtree, $\overline{T}_{i,j}$, defined as the entire tree, but with the tree $T_{i,j}$ removed ($\overline{T}_{2,3}$ for $n = 3$ is shown in figure 4).

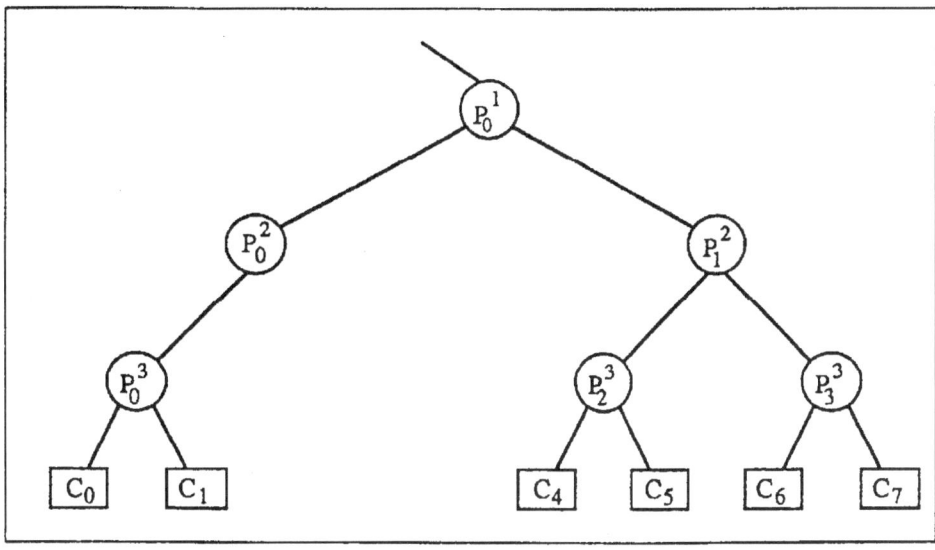

Figure 4: $\overline{T}_{2,3}$ for $n = 3$

In the construction of the tree it was required that each payer spends the received coin twice in two independent payments to two independently chosen payees. This means that for any vector, V, of random variables in $\overline{T}_{i,j}$ (nodes and leaves):

$$H(P_{i,j} \mid V) = H(P_{i,j})$$

(the independence property). As this section considers payment systems, which do not (necessarily) offer unconditional untraceability, the proof of the following theorem is based on $(*)$ and the independence property.

Theorem 5.1
In a tree of depth n:

$$\sum_{i=0}^{2^n-1} H(C_i) \geq \sum_{j=1}^{n} \left[\sum_{i=0}^{2^{j-1}-1} 2^{n-j} H(P_i^j) \right] - 2^{n-1} 3n\epsilon$$

As entropies are always positive this shows that electronic coins must grow in size when transferred. To be more concrete, consider the case where the uncertainty about each payer is k (k bits of information are needed to uniquely identify a payer). Then the theorem implies that

$$\sum_{i=0}^{2^n-1} H(C_i) \geq \sum_{j=1}^{n} \left[\sum_{i=0}^{2^{j-1}-1} 2^{n-j} k \right] - 2^n n \frac{3}{2} \epsilon$$
$$= n 2^n \left(\frac{k - 3\epsilon}{2} \right)$$

In particular this means that the entropy of some C_i is at least $n\frac{k-3\epsilon}{2}$, and if the entropies of all C_i's are equal then

$$H(C_i) \geq n \left(\frac{k - 3\epsilon}{2} \right).$$

This shows that the entropy of the coins grows linearly in the number of transfers, and furthermore, that the coins grow by approximately $\frac{k}{2}$ bits when transferred (since $k \gg 3\epsilon$).

This is less than for unconditionally secure money. The difference is due to the fact that a coin, c_i, in principle may contain all information needed to identify a payer as long as the bank cannot compute this identity from the coin. Hence, the uncertainty about another coin spent by the same person can be very small. Furthermore, Theorem 5.1 is tight in a sense to be discussed in Section 6.

The theorem is proven by combining two lemmas. The first lemma gives a lower bound on $H(C_i)$, and the second gives an upper bound on $H(P_i^j)$. The proofs use the same four rules as in the previous section. For convenience these are repeated here:

$$H(U) \geq 0 \tag{1}$$
$$H(U,V) = H(U \mid V) + H(V) \tag{2}$$
$$H(U,V \mid W) = H(U \mid V,W) + H(V \mid W) \tag{3}$$
$$H(U \mid V,W) \leq H(U \mid V) \tag{4}$$

Lemma 5.2
For every i ($0 \leq i < 2^n$)

$$H(C_i) \geq \sum_{j=1}^{n} H(P_{i,j} \mid C_{f(i,1)}, \ldots, C_{f(i,j)}) - 2n\epsilon$$

Proof
The proof is very similar to that of Theorem 4.2.
Claim: For $1 \leq j < n$:

$$H(C_i \mid C_{f(i,1)}, \ldots, C_{f(i,j)}) \geq H(C_i \mid C_{f(i,1)}, \ldots, C_{f(i,j+1)}) + H(P_{i,j} \mid C_{f(i,1)}, \ldots, C_{f(i,j)}) - 2\epsilon$$

From this claim it follows by simple induction that

$$
\begin{aligned}
H(C_i) \geq\ & H(C_i \mid C_{f(i,n)}, C_{f(i,n-1)}, \ldots, C_{f(i,1)}) + \\
& \sum_{j=1}^{n-1} H(P_{i,j} \mid C_{f(i,j)}, \ldots, C_{f(i,1)}) - 2(n-1)\epsilon \\
\geq\ & H(P_{i,n} \mid C_{f(i,n)}, C_{f(i,n-1)}, \ldots, C_{f(i,1)}) + \\
& \sum_{j=1}^{n-1} H(P_{i,j} \mid C_{f(i,j)}, \ldots, C_{f(i,1)}) - (2n-1)\epsilon \\
\geq\ & H(P_{i,n} \mid C_{f(i,n)}, C_{f(i,n-1)}, \ldots, C_{f(i,1)}) + \\
& \sum_{j=1}^{n-1} H(P_{i,j} \mid C_{f(i,j)}, \ldots, C_{f(i,1)}) - 2n\epsilon
\end{aligned}
$$

because

$$H(P_{i,n} \mid C_i, C_{f(i,n)}) \leq \epsilon$$

implies that

$$H(C_i \mid C_{f(i,n)}, \ldots, C_{f(i,1)}) \geq H(P_{i,n} \mid C_{f(i,n)}, \ldots, C_{f(i,1)}) - \epsilon.$$

The proof of this is very similar to that of Lemma 4.1. Now we just have to prove the claim. Let i and j be given, and let

$$A_j := (C_{f(i,1)}, \ldots, C_{f(i,j)}).$$

Then

$$
\begin{aligned}
H(C_i \mid C_{f(i,1)}, \ldots, C_{f(i,j)}) &= H(C_i \mid A_j) \\
&= H(P_{i,j}, C_i \mid A_j) - H(P_{i,1} \mid C_i, A_j) \quad \text{by (3)} \\
&\geq H(P_{i,j}, C_i \mid A_j) - \epsilon \\
&= H(C_i \mid P_{i,j}, A_j) + H(P_{i,j} \mid A_j) - \epsilon \quad \text{by (4)}
\end{aligned}
$$

By Lemma 4.1

$$H(P_{i,j} \mid C_{f(i,j+1)}, A_j) \leq \epsilon$$

implies that

$$H(C_i \mid P_{i,j}, A_j) \geq H(C_i \mid C_{f(i,j+1)}, A_j) - \epsilon.$$

Thus

$$
\begin{aligned}
H(C_i) &\geq H(C_i \mid C_{f(i,j+1)}, A_j) + H(P_{i,j}) \mid A_j - 2\epsilon \\
&= H(C_i \mid C_{f(i,1)}, \ldots, C_{f(i,j+1)}) + H(P_{i,j} \mid C_{f(i,1)}, \ldots, C_{f(i,j)}) - 2\epsilon
\end{aligned}
$$

This completes the proof. ∎

In order to give an upper bound on $H(P_i^j)$ it is necessary to introduce some more notation. Consider the sub-tree $T_{k,j}$ of height $n - j + 1$. This tree has two sub-trees $T_{k,j+1}$ and $T_{f(k,j),j+1}$ of height $n - j$. We define $B_{k,j}$ to be the set of leaves in the subtree, which *does not* contain C_k. Hence for $0 \leq k < 2^n$ and $1 \leq j < n$ is $B_{k,j}$ defined as the set of leaves in the subtree $T_{f(k,j),j+1}$. For example, for $n = 3$:

$$
\begin{aligned}
B_{2,1} &= \{C_4, C_5, C_6, C_7\} \\
B_{3,2} &= \{C_0, C_1\} \\
B_{6,2} &= \{C_4, C_5\} \\
B_{4,3} &= \{C_5\}
\end{aligned}
$$

$B_{k,j}$ has the property that each element in $B_{k,j}$ is a leaf in $T_{k,j+1}$, and the set of all leaves in $T_{k,j+1}$ for $1 \leq j \leq n - 1$ is

$$A_{k,j} := B_{k,1} \cup B_{k,2} \cup \ldots \cup B_{k,j}.$$

Furthermore, $T_{k,j}$ is the smallest subtree containing C_k and C for every element $C \in B_{k,j}$. Thus

$$H(P_{k,j} \mid C, C_k) \leq \epsilon.$$

Lemma 5.3
For $0 \leq k < 2^n$:

$$\sum_{j=2}^{n} H(P_{k,j}) \leq H(C_k \mid P_0^1) + \sum_{j=2}^{n} H(P_{k,j} \mid C_k, B_{k,1}, B_{k,2}, \ldots, B_{k,j-1}) + n\epsilon$$

Proof

Let k be given. Since $A_{k,j-1}$ is the set of leaves in $T_{k,j}$ for $2 \le j \le n$, the independence property implies that

$$H(P_{k,j}) = H(P_{k,j} \mid P_{k,j-1}, A_{k,j-1})$$

(this is the only time the independence property is used). Now

$$
\begin{aligned}
\sum_{j=2}^{n} H(P_{k,j}) &= \sum_{j=2}^{n} H(P_{k,j} \mid P_{k,j-1}, A_{k,j-1}) \\
&= \sum_{j=2}^{n} [H(C_k, P_{k,j} \mid P_{k,j-1}, A_{k,j-1}) - H(C_k \mid P_{k,j}, P_{k,j-1}, A_{k,j-1})] \quad \text{by (3)} \\
&= \sum_{j=2}^{n} [H(P_{k,j} \mid C_k, P_{k,j-1}, A_{k,j-1}) + H(C_k \mid P_{k,j-1}, A_{k,j-1})] - \\
&\qquad \sum_{j=2}^{n} H(C_k \mid P_{k,j}, P_{k,j-1}, A_{k,j-1}) \qquad \text{by (3)} \\
&= \sum_{j=2}^{n} H(P_{k,j} \mid C_k, P_{k,j-1}, A_{k,j-1}) + H(C_k \mid P_{k,1}, A_{k,1}) + \\
&\qquad \sum_{j=2}^{n-1} [H(C_k \mid P_{k,j}, A_{k,j}) - H(C_k \mid P_{k,j}, P_{k,j-1}, A_{k,j-1})] - \\
&\qquad H(C_k \mid P_{k,n}, P_{k,n-1}, A_{k,n-1}) \\
&\le \sum_{j=2}^{n} H(P_{k,j} \mid C_k, A_{k,j-1}) + H(C_k \mid P_{k,1}, A_{k,1}) + \\
&\qquad \sum_{j=2}^{n-1} [H(C_k \mid P_{k,j}, A_{k,j}) - H(C_k \mid P_{k,j}, P_{k,j-1}, A_{k,j-1})] - \\
&\qquad H(C_k \mid P_{k,n}, P_{k,n-1}, A_{k,n-1}) \qquad \text{by (4)} \\
&\le \sum_{j=2}^{n} H(P_{k,j} \mid C_k, A_{k,j-1}) + H(C_k \mid P_{k,1}, B_{k,1}) + \\
&\qquad \sum_{j=2}^{n-1} [H(C_k \mid P_{k,j}, A_{k,j}) - H(C_k \mid P_{k,j}, P_{k,j-1}, A_{k,j-1})] \qquad \text{by (1)} \\
&\le \sum_{j=2}^{n} H(P_{k,j} \mid C_k, A_{k,j-1}) + H(C_k \mid P_0^1) + \\
&\qquad \sum_{j=2}^{n-1} [H(C_k \mid P_{k,j}, A_{k,j}) - H(C_k \mid P_{k,j}, P_{k,j-1}, A_{k,j-1})] \qquad \text{by (4)}
\end{aligned}
$$

Due to the facts that the elements in $B_{k,j}$ are leaves in $T_{k,j}$, and $B_{k,j-1}$ is the set of leaves in $T_{f(k,j-1),j}$, and $P_{k,j-1} = P_{f(k,j-1),j-1}$, (∗) implies that for every $C \in B_{k,j}$ and $D \in B_{k,j-1}$

$$H(P_{k,j-1} \mid C, D) \le \epsilon.$$

Hence,

$$H(P_{k,j-1} \mid B_{k,j}, B_{k,j-1}) \le \epsilon$$

for $2 \le j \le n - 1$. Using (4), we get

$$H(P_{k,j-1} \mid B_{k,j}, A_{k,j-1}, P_{k,j-1}) \le \epsilon,$$

and Lemma 4.1 implies

$$\begin{aligned}
H(C_k \mid P_{k,j}, A_{k,j}) &= H(C_k \mid B_{k,j}, A_{k,j-1}, P_{k,j}) \\
&\leq H(C_k \mid P_{k,j}, P_{k,j-1}, A_{k,j-1}) + \epsilon.
\end{aligned}$$

Hence

$$\sum_{j=2}^{n} H(P_{k,j}) \leq H(C_k \mid P_0^1) + \sum_{j=2}^{n} H(P_{k,j} \mid C_k, A_{k,j-1})) + n\epsilon.$$

∎

We are now ready to present a proof of Theorem 5.1:

Proof
By using Lemma 5.2 for all even indices we get

$$\begin{aligned}
\sum_{i=0}^{2^n-1} H(C_i) &= \sum_{i=0}^{2^{n-1}-1} H(C_{2i}) + \sum_{i=0}^{2^{n-1}-1} H(C_{2i+1}) \\
&\geq \sum_{i=0}^{2^{n-1}-1} \sum_{j=1}^{n} H(P_{2i,j} \mid C_{f(2i,1)}, \ldots, C_{f(2i,j)}) - 2^{n-1}2n\epsilon + \\
&\qquad \sum_{i=0}^{2^{n-1}-1} H(C_{2i+1}) \\
&= \sum_{i=0}^{2^{n-1}-1} \sum_{j=2}^{n} H(P_{2i,j} \mid C_{f(2i,1)}, \ldots, C_{f(2i,j)}) - 2^{n-1}2n\epsilon + \\
&\qquad \sum_{i=0}^{2^{n-1}-1} H(P_{2i,1} \mid C_{f(2i,1)}) + \sum_{i=0}^{2^{n-1}-1} H(C_{2i+1})
\end{aligned}$$

Since $f(2i, 1)$ is always odd, and since $f(\cdot, 1)$ is a permutation

$$\sum_{i=0}^{2^{n-1}-1} H(C_{2i+1}) = \sum_{i=0}^{2^{n-1}-1} H(C_{f(2i,1)})$$

Using (2) twice this implies

$$\begin{aligned}
\sum_{i=0}^{2^n-1} H(C_i) &\geq \sum_{i=0}^{2^{n-1}-1} \sum_{j=2}^{n} H(P_{2i,j} \mid C_{f(2i,1)}, \ldots, C_{f(2i,j)}) - 2^{n-1}2n\epsilon + \\
&\qquad \sum_{i=0}^{2^{n-1}-1} H(P_{2i,1}, C_{f(2i,1)}) \\
&= \sum_{i=0}^{2^{n-1}-1} \sum_{j=2}^{n} H(P_{2i,j} \mid C_{f(2i,1)}, \ldots, C_{f(2i,j)}) - 2^{n-1}2n\epsilon + \\
&\qquad \sum_{i=0}^{2^{n-1}-1} H(C_{f(2i,1)} \mid P_{2i,1}) + \sum_{i=0}^{2^{n-1}-1} H(P_{2i,1}) \qquad (**)
\end{aligned}$$

Now consider the sum

$$\sum_{i=0}^{2^{n-1}-1} H(P_{2i,j} \mid C_{f(2i,1)}, \ldots, C_{f(2i,j)})$$

for a fixed j $(2 \leq j \leq n)$. Let $k = f(2i, j)$. Then

$$P_{2i,j} = P_{k,j}$$

and hence

$$C_{f(2i,s)} \in B_{k,s},$$

for $1 \leq s \leq j-1$, but C_k is not in $B_{k,j}$. By (4) this implies

$$
\begin{aligned}
H(P_{2i,j} \mid C_{f(2i,1)}, \ldots, C_{f(2i,j)}) &= H(P_{k,j} \mid C_{f(2i,1)}, \ldots, C_{f(2i,j)}) \\
&\geq H(P_{k,j} \mid B_{k,1}, \ldots, B_{k,j-1}, C_{f(2i,j)}) \\
&= H(P_{k,j} \mid B_{k,1}, \ldots, B_{k,j-1}, C_k)
\end{aligned}
$$

Since $f(2i, j)$ is odd and $f(\cdot, j)$ is a permutation we obtain that (writing $k = f(2i, j)$ as $2l + 1$)

$$\sum_{i=0}^{2^{n-1}-1} \sum_{j=2}^{n} H(P_{2i,j} | C_{f(2i,1)}, \ldots, C_{f(2i,j)}) \geq \sum_{l=0}^{2^{n-1}-1} \sum_{j=2}^{n} H(P_{2l+1,j} | B_{2l+1,1}, \ldots, B_{2l+1,j-1}, C_{2l+1})$$

and by Lemma 5.3

$$
\begin{aligned}
& \sum_{i=0}^{2^{n-1}-1} \sum_{j=2}^{n} H(P_{2i,j} \mid C_{f(2i,1)}, \ldots, C_{f(2i,j)}) \\
\geq \;& \sum_{l=0}^{2^{n-1}-1} [\sum_{j=2}^{n} H(P_{2l+1,j}) - H(C_{2l+1} \mid P_0^1) - n\epsilon] \\
= \;& \sum_{l=0}^{2^{n-1}-1} [\sum_{j=2}^{n} H(P_{2l+1,j}) - H(C_{2l+1} \mid P_0^1)] - 2^{n-1} n\epsilon \qquad (***)
\end{aligned}
$$

Combining $(**)$ and $(***)$ results in

$$
\begin{aligned}
\sum_{i=0}^{2^n-1} H(C_i) \;\geq\;& \sum_{i=0}^{2^{n-1}-1} \sum_{j=2}^{n} H(P_{2i,j} \mid C_{f(2i,1)}, \ldots, C_{f(2i,j)}) - 2^{n-1} 2n\epsilon + \\
& \sum_{i=0}^{2^{n-1}-1} H(C_{f(2i,1)} \mid P_{2i,1}) + \sum_{i=0}^{2^{n-1}-1} H(P_{2i,1}) \qquad \text{by } (**) \\
\geq\;& \sum_{l=0}^{2^{n-1}-1} [\sum_{j=2}^{n} H(P_{2l+1,j}) - H(C_{2l+1} \mid P_0^1)] - 2^{n-1} n\epsilon - 2^{n-1} 2n\epsilon + \\
& \sum_{i=0}^{2^{n-1}-1} H(C_{f(2i,1)} \mid P_{2i,1}) + \sum_{i=0}^{2^{n-1}-1} H(P_{2i,1}) \qquad \text{by } (***) \\
=\;& \sum_{l=0}^{2^{n-1}-1} [\sum_{j=2}^{n} H(P_{2l+1,j}) - H(C_{2l+1} \mid P_0^1)] - 2^{n-1} 3n\epsilon + \\
& \sum_{l=0}^{2^{n-1}-1} H(C_{2l+1} \mid P_0^1) + \sum_{l=0}^{2^{n-1}-1} H(P_0^1) \\
=\;& \sum_{l=0}^{2^{n-1}-1} \sum_{j=2}^{n} H(P_{2l+1,j}) + \sum_{l=0}^{2^{n-1}-1} H(P_0^1) - 2^{n-1} 3n\epsilon
\end{aligned}
$$

$$=^{*)} \sum_{j=2}^{n} \sum_{i=0}^{2^{j-1}-1} 2^{n-j} H(P_i^j) + 2^{n-1} H(P_0^1) - 2^{n-1} 3n\epsilon$$

$$= \sum_{j=1}^{n} \sum_{i=0}^{2^{j-1}-1} 2^{n-j} H(P_i^j) - 2^{n-1} 3n\epsilon$$

At $*$) we use the fact that for every $j = 1, 2, \ldots, n$

$$\sum_{l=0}^{2^{n-1}-1} H(P_{2l+1,j}) = \sum_{i=0}^{2^{j-1}-1} 2^{n-j} H(P_i^j).$$

This completes the proof. ∎

6 Applications to Secret Sharing

Theorem 4.2 and 5.1 give lower bounds on the size of transferred electronic money. In this section these theorems will be discussed from a different point of view.

Consider the tree in figure 2. If $p_1, p_2, \ldots, p_n$ are considered as k-bits secrets and $c_1, c_2 \ldots, c_{n+1}$ as shares of these secrets, then this tree depicts a situation, where a person (dealer) has n secrets and wants to distribute them among $n + 1$ persons in an information theoretic secure way, such that for every $i = 1, 2, \ldots, n$, it is possible to find p_i from c_i and c_j, where $i < j \leq n+1$. Theorem 4.2 says that the share c_{n+1} must be at least nk bits. It is not hard to generalize the theorem to show that each c_i must be at least ik bits long for $1 \leq i \leq n$. These lower bounds on the sizes of shares are also optimal as they can be achieved by choosing $r_1, \ldots, r_n \in \{0,1\}^k$ at random and letting

$$\begin{aligned} c_1 &= r_1 \oplus p_1 \\ c_i &= (r_1, \ldots, r_{i-1}, r_i \oplus p_i) \qquad \text{for } i = 2, \ldots, n \\ c_{n+1} &= (r_1, r_2, \ldots, r_n) \end{aligned}$$

The result in Section 5 can also be described in terms of secret sharing schemes — although this time the schemes are somewhat unusual. Again we consider the identities of the payers to be secrets and the coins to be shares of the secrets. Hence, there are $2^n - 1$ secrets and 2^n persons, who get shares $c_0, \ldots, c_{2^n-1}$. The access structure is defined by the tree described in Section 5.

It is required that the share, c_i may only contain Shannon information about the secrets on the path from the root to itself. Hence, c_2 may for instance contain all Shannon information about p_0^1, p_0^2 and p_1^3, but it may not contain information about other secrets. If each p_i^j is a uniformly chosen k-bits secret Theorem 5.1 says that

$$\sum_{i=0}^{2^n-1} H(C_i) \geq n2^{n-1}k - 2^{n-1}n3\epsilon.$$

Again, this result is optimal, as it is possible to give values to each c_i such that $\epsilon = 0$ and

$$\sum_{i=0}^{2^n-1} H(C_i) = n2^{n-1}k.$$

Assume that k is even. Let $\lfloor x \rfloor$ denote the first $\frac{k}{2}$ bits of $x \in \{0,1\}^k$, and let $\lceil x \rceil$ denote the last $\frac{k}{2}$ bits of x. For the case $n = 8$, the lower bound can now be achieved as follows:

$$
\begin{aligned}
c_0 &:= (\lfloor p_0^1 \rfloor, \lfloor p_0^2 \rfloor, \lfloor p_0^3 \rfloor) \\
c_1 &:= (\lfloor p_0^1 \rfloor, \lfloor p_0^2 \rfloor, \lceil p_0^3 \rceil) \\
c_2 &:= (\lfloor p_0^1 \rfloor, \lceil p_0^2 \rceil, \lfloor p_1^3 \rfloor) \\
&\cdots \\
c_7 &:= (\lceil p_0^1 \rceil, \lceil p_1^2 \rceil, \lceil p_3^3 \rceil)
\end{aligned}
$$

Here each share consists of $\frac{3k}{2}$ bits as required.

The secrets p_i^j can also be shared in an information theoretic secure way. This can be done by letting each share c_i consist of $3k$ bits. This shows that the lower bound from Section 4 can not be improved using the tree from Section 5.

7 Conclusion

This paper has demonstrated that it is not possible to construct off-line electronic payment systems without allowing extra bits for transferred money. On one hand this limits the practical use of electronic money, and on the other it shows that the general method described in Section 3 is close to optimal, because the transferred money in this scheme only increases by the number of bits needed to identify double-spenders with high probability. However, the known payment systems require more bits for this purpose than the actual number of bits needed to uniquely describe a payer. It would therefore be interesting to construct an off-line payment for which fewer bits are needed to compute the identity of double-spenders.

It was mentioned that the method suggested in [vA90] has the problem of forward traceability, and it was further argued in Section 4 that a person with unlimited computing power can probably always trace his money forwards. This leaves open the following problems:

- Prove that an unlimited powerful payer can always trace his transferred money;

- Construct a payment system in which forward traceability, although possible, is not feasible (under some assumption).

Acknowledgements

We wish to thank Eugène van Heyst for many useful comments during the construction of the lower bounds. We also wish to thank David B. Wilson for some comments on an earlier version of this paper.

References

[CFN90] D. Chaum, A. Fiat, and M. Naor. Untraceable electronic cash. In *Advances in Cryptology - proceedings of CRYPTO 88*, Lecture Notes in Computer Science, pages 319 – 327. Springer-Verlag, 1990.

[Cha83] D. Chaum. Blind signatures for untraceable payments. In *Advances in Cryptology - proceedings of CRYPTO 82*, pages 199 – 203, 1983.

[Cha84] D. Chaum. Blind signature systems. In *Advances in Cryptology - proceedings of CRYPTO 83*, 1984.

[OO90] T. Okamoto and K. Ohta. Disposable zero-knowledge authentication and their applications to untraceable electronic cash. In *Advances in Cryptology - proceedings of CRYPTO 89*, pages 481 – 496, 1990.

[OO92] T. Okamoto and K. Ohta. Universal electronic cash. In J. Feigenbaum, editor, *Advances in Cryptology - proceedings of CRYPTO 91*, Lecture Notes in Computer Science, pages 324 – 337. Springer-Verlag, 1992.

[vA90] Hans van Antwerpen. Electronic cash. Master's thesis, CWI, 1990.

[Wel88] Dominic Welsh. *Codes and Cryptography*. Oxford University Press, 1988.

Local Randomness in Candidate
One–Way Functions

H. Niederreiter[1] and C.P. Schnorr[2]

[1] Österreichische Akademie der Wissenschaften, Institut für
Informationsverarbeitung, Sonnenfelsgasse 19, A–1010 Wien, Austria
e–mail: nied@qiinfo.oeaw.ac.at
[2] Fachbereich Mathematik / Informatik, Universität Frankfurt, Postfach 111932,
D–6000 Frankfurt/M., Germany
e–mail: schnorr@informatik.uni-frankfurt.de

Abstract. We call a distribution on n–bit strings (ε, e)–locally random,
if for every choice of $e \leq n$ positions the induced distribution on e–bit
strings is in the L_1–norm at most ε away from the uniform distribution
on e–bit strings. We establish local randomness in polynomial random
number generators (RNG) that are candidate one–way functions. Let N
be a squarefree integer and let $f_1, \ldots, f_\ell$ be polynomials with coefficients in $\mathbb{Z}_N = \mathbb{Z}/N\mathbb{Z}$. We study the RNG that stretches a random
$x \in \mathbb{Z}_N$ into the sequence of least significant bits of $f_1(x), \ldots, f_\ell(x)$.
We show that this RNG provides local randomness if for every prime di-
visor p of N the polynomials $f_1, \ldots, f_\ell$ are linearly independent modulo
the subspace of polynomials of degree ≤ 1 in $\mathbb{Z}_p[x]$. We also establish
local randomness in polynomial random function generators. This yields
candidates for cryptographic hash functions. The concept of local randomness in families of functions extends the concept of universal families
of hash functions by Carter and Wegman (1979). The proofs of our
results rely on upper bounds for exponential sums.

1 Introduction and Summary

A major open problem in cryptography is to establish one–way functions. While
we cannot prove one–wayness it makes sense to analyse candidate one–way functions and to prove properties of these functions that are useful in cryptographic
applications. We call a distribution on n–bit strings (ε, e)–locally random if for
every choice of $e \leq n$ positions the induced distribution on e–bit strings is in
the L_1–norm at most ε away from the uniform distribution on e–bit strings. We
prove (ε, e)–local randomness for large classes of candidate one–way functions
and candidate cryptographic hash functions.

We show that ℓ–tuples of polynomials $(f_1, \ldots, f_\ell) \in \mathbb{Z}_N[x]^\ell$ with fixed coefficients in $\mathbb{Z}_N$ and for arbitrary odd squarefree N provide local randomness if for
every prime divisor p of N the polynomials $f_1, \ldots, f_\ell$ are linearly independent
modulo the subspace of polynomials of degree ≤ 1 in $\mathbb{Z}_p[x]$. To give an example
let N be prime $N > 2^n$, let $f_1, \ldots, f_\ell \in \mathbb{Z}_N[x]$ be any polynomials that

R.A. Rueppel (Ed.): Advances in Cryptology - EUROCRYPT '92, LNCS 658, pp. 408-419, 1993.
© Springer-Verlag Berlin Heidelberg 1993

are linearly independent modulo the subspace of polynomials of degree ≤ 1 in $\mathbb{Z}_N[x]$. We prove in Corollary 2 that for random $x \in \mathbb{Z}_N$ the bit string

$$(f_1(x)[1, \ldots, f_\ell(x)[1)$$

consisting of the parity bits $f_i(x)[1$ of the residues $f_i(x) \mod N$ in $[0, N-1]$ is (ε, e)–locally random provided that ε, n, ℓ and e satisfy the inequality

(1) $2^{-n/2}(2n \log 2)^{e+1} 2\ell \leq \varepsilon$,

where $\log$ denotes the natural logarithm. E.g. we can choose $n \geq 64$, $\ell = \lfloor 2^{n/7} \rfloor$, $\varepsilon = 2^{-n/7}$, $e = \lfloor n/(7 \log n) \rfloor$. Our main result comprises the case that N is an arbitrary odd squarefree integer, that the output contains several bits from each of the residues $f_i(x) \mod N$, $i = 1, \ldots, \ell$, and that x is chosen to be random in a subinterval $[0, M-1]$ of $[0, N-1]$.

Note that the above function

(2) $[0, N-1] \ni x \longmapsto (f_1(x)[1, \ldots, f_\ell(x)[1)$

is a candidate one–way function. No inversion algorithm is known that is polynomial time in $\min(\ell, \log_2 N)$. So far the one–wayness of the function (2) has only be proved for random RSA–moduli N and RSA–polynomials $f_i = x^{e^i}$ (see below) provided that the RSA–scheme is secure. It is however possible that this one–way function is more secure than the RSA–scheme. We are not aware of any inversion algorithm which for RSA–moduli N runs in time $\min(2^\ell, N)^{o(1)}$. On the other hand the RSA–scheme can be broken by factoring N using only $\exp(\sqrt{\log N \log \log N})$ many steps. Is there any inversion algorithm that uses knowledge of the factorization of RSA–numbers N ? Is there any inversion algorithm that uses the structure of particular odd moduli N and of particular non–constant polynomials f_i ? Of course the function (2) can easily be inverted for $N = 2$ since $f_i(x)[1$ only depends on $x[1 = x \mod 2$. Also the problem of inverting is trivial for constant functions as $f_i(x) = x^{N-1} (\mod N)$ with N prime. Are there more exceptions? Almost nothing is known about the problem to invert (2). However if we cannot even find inverting algorithms for particular cases given the factorization of the modulus then this may be a sign that the function (2) is a truly one–way function.

It is important that the source of randomness in $(f_1(x)[1, \ldots, f_\ell(x)[1)$ is the random argument x while the coefficients of $f_1, \ldots, f_\ell$ are all fixed. Such functions are cryptographically interesting. A well known example is the random number generator (RNG) related to the RSA–scheme by ALEXI, CHOR, GOLDREICH and SCHNORR (1988) and MICALI, SCHNORR (1991). E.g. let N be the product of two large random primes and let the integer $e \geq 3$ be relatively prime to $\varphi(N)$. Then the mapping

$$[0, N-1] \ni x \longmapsto (x^e[1, x^{e^2}[1, \ldots, x^\ell[1)$$

where x^{e^i} is taken modulo N, is a perfect (in the sense of YAO (1982) and BLUM, MICALI (1982)) RNG provided that the RSA–scheme is secure.

The functions $x \mapsto (f_1(x)[1, \ldots, f_\ell(x)[1)$ extend the class of polynomial random number generators (RNG) proposed by MICALI and SCHNORR (1991) which stretch a random seed $x \in [1, N2^{-k}]$ into a polynomial residue $P(x)(\mathrm{mod}\,N)$. Micali and Schnorr prove that the m least significant bits of $P(x)(\mathrm{mod}\,N)$ are in the L_1–norm at most $O(N^{-1/2}2^{k+m}(\log N)^2 \deg_N(P))$ away from the uniform distribution provided that N is prime and $\deg_N(P) \geq 2$ where $\deg_N(P)$ is the degree of P when P is considered modulo N.

So far local randomness has mainly been studied in functions that are easy to invert, see ALON, BABAI, ITAI (1986), LUBY (1986), SCHNORR (1988), MAURER, MASSEY (1989), NAOR, NAOR (1990), NISAN (1990) and ALON, GOLDREICH, HASTAD, PERALTA (1990). Most of these constructions are methodically simple and are not directed towards cryptographic applications. They aim at minimizing the number of random bits that are used in randomised algorithms. Merely the quadratic character construction by ALON et alii (1990) is similar to our generator, it relies on Weil's theorem. Our proof of local randomness relies on upper bounds for exponential sums and an inequality on quantitative Fourier inversion. We use upper bounds for the discrepancy of polynomial residues from NIEDERREITER (1977) and we extend these bounds from prime moduli to arbitrary squarefree moduli.

We also establish random function generators, associated with fixed polynomials, that provide local randomness. These generators are candidates for cryptographic hash functions. We associate with a polynomial $P \in \mathbb{Z}_N[x]$ of degree d a polynomial function family $P_z(y) = P(z+y)$ where z is the function name and y is the input. For fixed $k, m \leq \log_2 N$ we associate with a random $z \in \mathbb{Z}_N$ a random function

$$P_z^m : [0, 2^k - 1] \longrightarrow \{0,1\}^m , \quad y \longmapsto P(z+y)\,[m$$

where $P(z+y)[m$ denotes the bit string consisting of the m least significant bits of the residue $P(z+y) \bmod N$ in $[0, N-1]$.

We call a function family $\{P_z\}$ (ε, e)–*locally random* if for random z and for any e distinct points $y_1, \ldots, y_e$ the distribution of the em–bit string $P_z(y_1)\cdots P_z(y_e)$ is in the L_1–norm at most ε away from the uniform distribution on em–bit strings.

We prove in Theorem 6 that the above family of functions $\{P_z^m\}$ is (ε, e)–locally random, if N is prime, $d = \deg P$ satisfies $e + 1 \leq d < N$ and if

$$N^{-1/2}(\log N)^{e+1}2^{em+2}d \leq \varepsilon.$$

A family of functions is an e–universal family of hash functions as introduced by CARTER and WEGMAN (1979) if and only if it is $(0, e)$–locally random. Our hash functions require fewer random bits than those of Carter and Wegman since we only randomize the input of the polynomial whereas Carter and Wegman randomize all its coefficients. The main point however is that our hash–functions are – if $\deg P$ is sufficiently large – candidates for cryptographically secure hashing whereas the Carter–Wegman hash functions are easy to invert. Thus for the first time we establish local randomness in families of cryptographic hash functions.

2 Random Number Generators that Provide Statistical Local Randomness

We present in Theorem 1 our main result and we derive from it RNG's that are locally random. In order to prove Theorem 1 we establish in Theorem 3 an upper bound on the discrepancy for multidimensional polynomial number sequences. This upper bound relies on an upper bound for exponential sums given in Lemma 4 and on an inequality of Niederreiter (1977) on quantitative Fourier inversion.

Notation. Let $p_1, \ldots, p_r$ be r distinct primes, $N = p_1 \cdots p_r$ (i.e. N is squarefree) and $\mathbb{Z}_N = \mathbb{Z}/N\mathbb{Z}$. Let $\mathbf{F} = (f_1, \ldots, f_\ell)$ be an ℓ-tuple of polynomials $f_j \in \mathbb{Z}[x]$, $j = 1, \ldots, \ell$. We denote by $d_i(f_j)$ the degree of f_j when f_j is considered $\bmod\, p_i$ and we put $d_i(\mathbf{F}) = \max_{1 \le j \le \ell} d_i(f_j)$. We define $c_i(\mathbf{F}) = \min(d_i(\mathbf{F}) - 1, \sqrt{p_i})$ for $i = 1, \ldots, r$ and $c(\mathbf{F}) = \prod_{i=1}^{r}(c_i(\mathbf{F}) + 1)$. We call $\mathbf{F}$ N-*admissible* if for every prime divisor p_i of N the polynomials $f_1, \ldots, f_\ell$ are linearly independent modulo the subspace of polynomials of degree ≤ 1 in $\mathbb{Z}_{p_i}[x]$. In this case we also call the set of polynomials $f_1, \ldots, f_\ell$ N-admissible. Thus $f_1, \ldots, f_\ell$ are N-admissible if for $i = 1, \ldots, r$ and for all $a_1, \ldots, a_\ell \in \mathbb{Z}$ either the polynomial $\sum_{j=1}^{\ell} a_j f_j (\bmod\, p_i)$ is non–linear or $a_1 = \cdots = a_\ell = 0 (\bmod\, p_i)$.

We let $\log N$ denote the natural logarithm of N. We identify $\mathbb{Z}_N$ with the integer interval $[0, N-1]$. We abbreviate the set $\{0,1\}^n$ as I_n and we identify the integer interval $[0, 2^n - 1]$ with I_n . If $y \in [0, N-1] = \mathbb{Z}_N$ and $n \le \log_2 N$ we let $y[n \in I_n$ denote the bit string consisting of the n least significant bits of y. Let $\mathbb{N}$ denote the set of positive integers.

A collection of m least significant output bits. We associate with $\mathbf{F} = (f_1, \ldots, f_\ell) \in (\mathbb{Z}[x])^\ell$, $N \in \mathbb{N}$ and $\mathbf{m} = (m_1, \ldots, m_\ell) \in \mathbb{N}^\ell$ the mapping

$$\mathbf{F}^{\mathbf{m}} : [0, N-1] \to I_m , \quad x \mapsto \prod_{j=1}^{\ell}(f_j(x)[m_j]) \quad \text{with} \quad f_j(x) \in \mathbb{Z}_N,$$

where $m = \sum_{j=1}^{\ell} m_j$ and $\prod$ is the concatenation of strings. The mapping $\mathbf{F}^{\mathbf{m}}$ outputs a collection of m least significant bits of $\mathbf{F}(x)$, where $\mathbf{F}(x)$ is taken modulo N.

Our main theorem provides explicit estimates for the max–norm difference between the distribution induced by $\mathbf{F^m}(x)$ for random $x \in [0, M-1] \subset [0, N-1]$, N–admissible $\mathbf{F}$ and the uniform distribution on $\{0,1\}^m$.

Theorem 1. *Let N be odd and squarefree, let $\mathbf{F}, \mathbf{m}, m, \mathbf{F^m}$ be as above and let $\mathbf{F}$ be N–admissible. Then for $N > 148$, $1 \le M \le N$ and random $x \in [0, M-1]$ we have that*

$$\max_{z \in \{0,1\}^m} \left|\mathrm{prob}[\mathbf{F^m}(x) = z] - 2^{-m}\right| \le \frac{4}{M}\sqrt{N}(\log N)^{l+1} c(\mathbf{F}) \ .$$

The condition that $\mathbf{F}$ is N–admissible cannot be completely removed from Theorem 1. Theorem 1 does not hold for linear polynomials $f_1, \ldots, f_\ell$ with $\ell \ge 2$. This is because the least significant bits in two linear polynomials are highly correlated. On the other hand our proof shows that Theorem 1 holds for a single polynomial of degree 1 in the case that $N = M$.

For example let $N > 2^{512}$ be prime and let $d = 2^{32}$. Then the polynomials $x^2, \ldots, x^d$ are N–admissible. Consider for random $x \in [0, N-1]$ the bit string $(x^2[1], \ldots, x^d[1]) \in I_{d-1}$. For any choice of 24 bit positions $2 \le i_1 < i_2 \cdots < i_{24} \le 2^{32}$ and every $z \in \{0,1\}^{24}$ we have that

$$\left|\mathrm{prob}(x^{i_1}[1] \cdots x^{i_{24}}[1] = z) - 2^{-24}\right| < 2^{-44} \ .$$

This follows from Theorem 1 with $\ell = 24$, $f_j = x^{i_j}$ for $j = 1, \ldots, 24$, $N = M$ and $c(\mathbf{F}) \le 2^{32}$.

Definition. A random variable y ranging over a finite set S is called *statistically random* within ε (in S) if $\sum_{s \in S} |\mathrm{prob}(y = s) - 1/\#S| \le \varepsilon$, i.e. the L_1–norm statistical difference of y from the uniform distribution on S is at most ε.

Definition. A probability distribution D on I_n is called (ε, e)–*locally random* if for any sequence of positions $1 \le j_1 < j_2 < \cdots < j_e \le n$ the substring $(y_{j_1}, \ldots, y_{j_e}) \in I_e$ of a D–random string $y = (y_1, \ldots, y_n)$ is statistically random within ε.

Using Theorem 1 we can stretch a short random seed into a long bit string that is "locally random".

Corollary 2. *Let $N = p_1 \cdots p_r$ be a product of r distinct odd primes, $1 \le M \le N$ and $N > 148$. Let $f_1, \ldots, f_\ell \in \mathbb{Z}[x]$ be polynomials of degree at most d that are N–admissible. Then for random $x \in [0, M-1]$ the bit string $(f_1(x)[1], \ldots, f_\ell(x)[1]) \in I_\ell$ with $f_j(x) \in \mathbb{Z}_N$ is (ε, e)–locally random with $\varepsilon = 2\frac{\sqrt{N}}{M}(2 \log N)^{e+1} d^r$ for $e = 1, \ldots, \ell$.*

Proof. Let $1 \leq j_1 < j_2 < \cdots < j_e \leq \ell$ be any sequence of e output bit positions. We apply Theorem 1 with $\mathbf{F} = (f_{j_1}, \ldots, f_{j_e})$, $\mathbf{m} = (1, \ldots, 1) \in \mathbb{N}^e$ and $m = \ell = e$. The L_1–norm difference between the distribution induced by $\mathbf{F^m}(x) \in \{0, 1\}^e$ and the uniform distribution is at most 2^e–times the max–norm difference. We have $c(\mathbf{F}) \leq d^r$, and thus by Theorem 1 $\mathbf{F}(x)$ is statistically random within $2\frac{\sqrt{N}}{M}(2 \log N)^{e+1} d^r$. $\qquad\square$

The *discrepancy* $D_M^{(\ell)} = D_M^{(\ell)}(\mathbf{y}_1, \ldots, \mathbf{y}_M)$ of M points $\mathbf{y}_1, \ldots, \mathbf{y}_M \in [0, 1)^\ell$ is defined to be

$$D_M^{(\ell)}(\mathbf{y}_1, \ldots, \mathbf{y}_M) = \sup_{\mathcal{I}} |F_M(\mathcal{I}) - V(\mathcal{I})|$$

where $\mathcal{I}$ ranges over all half open subintervals $\mathcal{I}$ of $[0, 1)^\ell$, i.e.

$$\mathcal{I} = \{(z_1, \ldots, z_\ell) \in [0, 1)^\ell \,\big|\, a_i \leq z_i < b_i \text{ for } i = 1, \ldots, \ell\}$$

with $0 < a_i < b_i \leq 1$ for $i = 1, \ldots, \ell$. $V(\mathcal{I})$ is the volume of $\mathcal{I}$ and $F_M(\mathcal{I}) = M^{-1}\#\{k \,|\, \mathbf{y}_k \in \mathcal{I}\}$.

The proof of Theorem 1 relies on the following upper bound for the discrepancy of multidimensional polynomial sequences. For a real number a we let $\{a\}$ denote the residue of $a \bmod \mathbb{Z}$ in the real interval $[0, 1)$.

Theorem 3. *Let N be squarefree and let $D_M^{(\ell)}$ be the discrepancy of the M points $\left(\left\{\frac{f_1(k)}{N}\right\}, \ldots, \left\{\frac{f_\ell(k)}{N}\right\}\right) \in [0, 1)^\ell$ for $k = 1, \ldots, M$. If $\mathbf{F} = (f_1, \ldots, f_\ell)$ is N–admissible then $D_M^{(\ell)} \leq \frac{4B}{M}\sqrt{N}(\log N)^{\ell+1}c(\mathbf{F})$ for $1 \leq M \leq N$ and $N > 148$. Here $B = \sqrt{2}$ if N is even and $B = 1$ if N is odd.*

The proof is based on a bound for exponential sums. For $f \in \mathbb{Z}[x]$ and $n \in \mathbb{N}$ define

$$S(f, n) = \sum_{x=1}^{n} e\left(\frac{f(x)}{n}\right) \text{ with } e(u) = e^{2\pi\sqrt{-1}u} \text{ for } u \in \mathbb{R}.$$

The proofs for Lemmata 4 and 5 are omitted due to lack of space. They are contained in the complete paper that is to appear in Siam J. Computing.

Lemma 4. *If $N = p_1 \ldots p_r$ is squarefree, B is as in Theorem 3, and $f \in \mathbb{Z}[x]$ is arbitrary, then $|S(f, N)| \leq B\sqrt{N}\prod_{i=1}^{r} c_i'(f)$, where $c_i'(f) = \sqrt{p_i}$ if $d_i(f) < 1$ and $c_i'(f) = \min(d_i(f) - 1, \sqrt{p_i})$ if $d_i(f) \geq 1$.*

Lemma 5. *Let $D_M^{(\ell)}$ be the discrepancy of the M points $\mathbf{y}_k \in [0, 1)^\ell$ for $k = 1, \ldots, M$ and let $D_N^{(\ell+1)}$ be the discrepancy of the N points $(\mathbf{y}_k, \frac{k-1}{N})$ for $k = 1, \ldots, N$. Then $D_M^{(\ell)} \leq \frac{N}{M}D_N^{(\ell+1)}$ for $1 \leq M \leq N$.*

Proof of Theorem 3. Let N have r distinct prime factors. Put $C_\ell(N) = (-N/2, N/2]^\ell \cap \mathbb{Z}^\ell$, $C_\ell^*(N) = C_\ell(N)\backslash\{0\}$ (here we use, as in NIEDERREITER (1977), the interval $(-N/2, N/2]$ rather than $[0, N))$. For $\mathbf{h} = (h_1, \ldots, h_\ell) \in C_\ell(N)$ we put

$$r(\mathbf{h}, N) = \prod_{j=1}^{\ell} r(h_j, N) \text{ with } r(h_j, N) = \begin{cases} 1 & \text{if } h_j = 0 \\ N \sin \frac{\pi|h_j|}{N} & \text{if } h_j \neq 0. \end{cases}$$

By Lemma 2.2 of Niederreiter (1977), we get

$$(3) \quad D_N^{(\ell+1)} \leq \frac{\ell+1}{N} + \frac{1}{N} \sum_{\mathbf{h} \in C_{\ell+1}^*(N)} \frac{1}{r(\mathbf{h}, N)} \left| S(h_1 f_1 + \ldots + h_\ell f_\ell + h_{\ell+1}x, N) \right|,$$

where $\mathbf{h} = (h_1, \ldots, h_{\ell+1})$ and $f_1, \ldots, f_\ell \in \mathbb{Z}[x]$. By Lemma 4

$$(4) \quad |S(h_1 f_1 + \ldots + h_\ell f_\ell + h_{\ell+1}x, N)| \leq BN^{1/2} \prod_{i=1}^{r} c_i'(h_1 f_1 + \ldots + h_\ell f_\ell + h_{\ell+1}x).$$

If $\mathbf{h} \in C_{\ell+1}^*(N)$ with $(h_1, \ldots, h_\ell) = 0$, then $c_i'(h_1 f_1 + \ldots + h_\ell f_\ell + h_{\ell+1}x) = c_i'(h_{\ell+1}x) = 0$ for some i, namely when $h_{\ell+1} \neq 0 \bmod p_i$, and so

$$\prod_{i=1}^{r} c_i'(h_1 f_1 + \ldots + h_\ell f_\ell + h_{\ell+1}x) = 0.$$

Thus we only have to consider those $\mathbf{h} \in C_{\ell+1}^*(N)$ with $(h_1, \ldots, h_\ell) \neq 0$. We split up the set of $(h_1, \ldots, h_\ell) \in C_\ell^*(N)$ according to the set of i's for which $d_i(h_1 f_1 + \ldots + h_\ell f_\ell) \leq 1$. For $I \subseteq A_r := \{1, 2, \ldots, r\}$ we put $H(I) = \{(h_1, \ldots, h_\ell) \in C_\ell^*(N) : d_i(h_1 f_1 + \ldots + h_\ell f_\ell) \leq 1 \text{ if and only if } i \in I\}$. If $(h_1, \ldots, h_\ell) \in H(I)$ and $i \in A_r \backslash I$, then for any $h_{\ell+1} \in C_1(N)$ we have

$$d_i(h_1 f_1 + \ldots + h_\ell f_\ell + h_{\ell+1}x) = d_i(h_1 f_1 + \ldots + h_\ell f_\ell) \geq 2.$$

Since $d_i(h_1 f_1 + \ldots + h_\ell f_\ell) \leq d_i(\mathbf{F})$, it follows that

$$c_i'(h_1 f_1 + \ldots + h_\ell f_\ell + h_{\ell+1}x) \leq c_i(\mathbf{F}).$$

Using the trivial bound $c_i'(f) \leq p_i^{1/2}$, we obtain

$$\prod_{i=1}^{r} c_i'(h_1 f_1 + \ldots + h_\ell f_\ell + h_{\ell+1}x) \leq \prod_{i \in I} p_i^{1/2} \cdot \prod_{i \in A_r \backslash I} c_i(\mathbf{F})$$

for any $(h_1, \ldots, h_\ell) \in H(I)$ and $h_{\ell+1} \in C_1(N)$. Together with (3) and (4) this yields

$$D_N^{(\ell+1)} \leq \frac{\ell+1}{N}$$

$$+ BN^{-1/2} \sum_{I \subseteq A_r} \prod_{i \in I} p_i^{1/2} \cdot \prod_{i \in A_r \backslash I} c_i(\mathbf{F}) \sum_{\mathbf{h} \in H(I)} \frac{1}{r(\mathbf{h}, N)} \sum_{h_{\ell+1} \in C_1(N)} \frac{1}{r(h_{\ell+1}, N)}.$$

Using the inequality

$$(5) \quad \sum_{h \in C_1^*(m)} \frac{1}{r(h,m)} < \frac{2}{\pi} \log m + \frac{2}{5} \quad \text{for} \quad m \geq 2$$

from Niederreiter [14, (2.7)] this yields

$$D_N^{(\ell+1)} < \frac{\ell+1}{N}$$

$$+ BN^{-1/2} \left(\frac{2}{\pi} \log N + \frac{7}{5} \right) \sum_{I \subseteq A_r, i \in I} \prod_{i \in I} p_i^{1/2} \cdot \prod_{i \in A_r \setminus I} c_i(\mathbf{F}) \sum_{h \in H(I)} \frac{1}{r(h,N)}.$$

By the assumption of the theorem $(f_1, \ldots, f_\ell)$ is N-admissible. Therefore if $(h_1, \ldots, h_\ell) \in H(I)$ we get $h_k = 0 \bmod p_i$ for $i \in I$ and $1 \leq k \leq \ell$, thus $h_k = 0 \bmod \prod_{i \in I} p_i$ for $1 \leq k \leq \ell$. Therefore with $L = \prod_{i \in I} p_i$ we obtain

$$\sum_{h \in H(I)} \frac{1}{r(\mathbf{h},N)} \leq \sum_{\substack{h \in C_\ell^*(N) \\ h=0 \bmod L}} \frac{1}{r(\mathbf{h},N)} = \sum_{h \in C_\ell(N/L)} \frac{1}{r(L\mathbf{h},N)} - 1$$

$$= \left(\sum_{h \in C_1(N/L)} \frac{1}{r(L h,N)} \right)^\ell - 1$$

$$= \left(1 + \sum_{h \in C_1^*(N/L)} \frac{1}{r(Lh,N)} \right)^\ell - 1$$

$$= \left(1 + \frac{1}{L} \sum_{h \in C_1^*(N/L)} \frac{1}{r(h,N/L)} \right)^\ell - 1$$

$$< \left(1 + \frac{1}{L} \left(\frac{2}{\pi} \log \frac{N}{L} + \frac{2}{5} \right) \right)^\ell - 1 \quad \text{(by the inequality (5))}$$

$$< \frac{\ell}{L} \left(\frac{2}{\pi} \log N + \frac{7}{5} \right)^\ell,$$

where we applied the mean-value theorem in the last step. It follows that

$$D_N^{(\ell+1)} < \frac{\ell+1}{N} + B\ell N^{-1/2} \left(\frac{2}{\pi} \log N + \frac{7}{5} \right)^{\ell+1} \sum_{I \subseteq A_r, i \in I} \prod_{i \in I} p_i^{-1/2} \cdot \prod_{i \in A_r \setminus I} c_i(\mathbf{F})$$

$$= \frac{\ell+1}{N} + B\ell N^{-1/2} \left(\frac{2}{\pi} \log N + \frac{7}{5} \right)^{\ell+1} \prod_{i=1}^r (c_i(\mathbf{F}) + p_i^{-1/2})$$

$$< BN^{-1/2} (\log N)^{\ell+1} c(\mathbf{F}) \left(\frac{\ell+1}{N^{1/2} (\log N)^{\ell+1}} + \ell \left(\frac{2}{\pi} + \frac{7}{5 \log N} \right)^{\ell+1} \right)$$

$$< BN^{-1/2} (\log N)^{\ell+1} c(\mathbf{F}) \left(\frac{1}{60} (\ell+1) 5^{-\ell} + \ell (\frac{2}{\pi} + \frac{7}{25})^{\ell+1} \right)$$

$$< 4BN^{-1/2}(\log N)^{\ell+1}c(\mathbf{F})$$

provided that $\log N \geq 5$, i.e. that $N > 148$. Together with Lemma 5 we get the result of Theorem 3. □

Proof of Theorem 1 Let N be an odd squarefree integer and $\overline{f}_j \in \mathbb{Z}[x]$ be polynomials such that $\overline{f}_j(x) = 2^{-m_j} f_j(x)(\bmod N)$ for $j = 1, \ldots, \ell$. Application of Theorem 3 to $\overline{\mathbf{F}} = (\overline{f}_1, \ldots, \overline{f}_\ell)$ shows that the discrepancy $\overline{D}_M^{(\ell)}$ of $(\{\overline{f}_1(k)/N\}, \ldots, \{\overline{f}_\ell(k)/N\})$ for $k = 1, \ldots, M$ satisfies

$$\overline{D}_M^{(\ell)} \leq \frac{4}{M} \sqrt{N}(\log N)^{\ell+1}c(\mathbf{F}),$$

where we use that $c(\mathbf{F}) = c(\overline{\mathbf{F}})$. We apply to this inequality the equivalence

$$\{\overline{f}_j(x)/N\} \in [k_j 2^{-m_j}, (k_j + 1)2^{-m_j}) \quad \Longleftrightarrow$$

$$[f_j(x)]_N = -k_j N(\bmod 2^{m_j}) \quad \text{for} \quad j = 1, \ldots, \ell,$$

where $[f_j(x)]_N$ is the residue of $f_j(x) \bmod N$ in $[0, N-1]$, and $0 \leq k_j < 2^{m_j}$. To see the equivalence we note that $\{\overline{f}_j(x)/N\} \in [k_j 2^{-m_j}, (k_j+1)2^{-m_j})$ implies that there is an integer y satisfying

$$k_j N \leq y < (k_j + 1)N, \; y = f_j(x) \bmod N, \; y = 0 \bmod 2^{m_j},$$

and thus $[f_j(x)]_N = -k_j N(\bmod 2^{m_j})$. This proves one direction of the equivalence and the converse direction is an immediate consequence.

We see from the above inequality and the equivalence that for every $y \in \{0, 1\}^m$

$$\left| \frac{1}{M} \#\{x \subset [1, M] : \mathbf{F^m}(x) - y\} - \frac{1}{2^m} \right| \leq \frac{4}{M} \sqrt{N}(\log N)^{\ell+1}c(\mathbf{F}).$$

□

The above proof of Theorem 1 extends to the following larger class of functions $\mathbf{F^u}$. Let the polynomials $f_1, \ldots, f_\ell \in \mathbb{Z}_N[x]$ be N-admissible and let $u_1, \ldots, u_\ell$ be integers that are relatively prime to N, $\mathbf{F} = (f_1, \ldots, f_\ell)$ and $\mathbf{u} = (u_1, \ldots, u_\ell)$. Define $\mathbf{F^u}$ as

$$\mathbf{F^u} : [0, N-1] \ni x \mapsto (\, (f_i(x) \bmod N) \bmod u_i \mid \text{for } i = 1, \ldots, \ell).$$

Corollary 6. *For $N > 148$, $1 \leq M \leq N$ and random $x \in [0, M-1]$ the max-norm difference between the distribution induced by $\mathbf{F^u}(x)$ and the uniform distribution on $[0, u_1 - 1] \times \cdots \times [0, u_\ell - 1]$ is at most $\frac{4}{M} \sqrt{N}(\log N)^{\ell+1}c(\mathbf{F})$.*

Theorem 1 deals with the particular case that the integers u_i are powers of 2. It is necessary that $u_1, \ldots, u_\ell$ are relatively prime to N. The proof of the Corollary uses the polynomials $\tilde{f}_j = u_j^{-1} f_j(\bmod N)$ and thus requires a division by u_j modulo N.

3 Random Function Generators that Provide Statistical Local Randomness

Let $H_{k,\ell} = I_\ell^{I_k} =$ "the set of functions $f : I_k \to I_\ell$". A *random function generator* F is an efficient algorithm that generates from names $x \in I_n$ a function $f_x = F(x, *) \in H_{k,\ell}$.

We call a probability distribution D on $H_{k,\ell}$ (ε, e)–*locally random* if for random f, $f \in_D H_{k,\ell}$, for any set of e distinct inputs $y_1, \ldots, y_e \in I_k$ the concatenated output $f(y_1)f(y_2) \cdots f(y_e) \in I_{e\ell}$ is statistically random within ε.

The concept of (ε, e)–locally random distribution D on $H_{k,\ell}$ extends the concept of *universal hash functions* of Carter and Wegman (1979). If D is $(0, e)$–locally random then for any distinct inputs $y_1, \ldots, y_e \in I_k$ the bit string $f(y_1)f(y_2) \cdots f(y_e) \in I_e$ is truly random, i.e. D is the probability distribution of an e–universal family of hash functions in the sense of Carter and Wegman.

Carter and Wegman show how to generate an e–universal family of hash functions in $H_{k,k}$ from ke random bits. Let $K = GF(2^k)$ be the field with 2^k elements. If $(a_0, \ldots, a_{e-1}) \in K^e$ is random then the polynomial $P = \sum_{i=0}^{e-1} a_i x^i \in K[x]$ yields an e–universal family of hash functions in $H_{k,k}$.

Let N be a prime and $P \in \mathbb{Z}_N[x]$ be a polynomial with coefficients in the field $\mathbb{Z}_N$. We associate with P and $k, \ell \in \mathbb{N}$, $k, \ell \leq \log_2 N$, the function

$$P^\ell : \mathbb{Z}_N \times [0, 2^k - 1] \to I_\ell, \ (z, y) \mapsto P(y + z)\lceil \ell.$$

Here we let $P(y + z)\lceil \ell$, for $\ell \leq \log_2 N$, denote the bit string consisting of the ℓ least significant bits of the residue of $P(y + z) \bmod N$ that is in $\mathbb{Z}_N$. We let $P_z^\ell : I_k \to I_\ell$ denote the function $P^\ell(z, *)$.

Theorem 7. *Let* N *be prime,* $N > 148$, $P \in \mathbb{Z}_N[x]$, $k, \ell \leq \log_2 N$, *let* $P_z^\ell : I_k \to I_\ell$ *be as above and let* $e + 1 \leq \deg P < N$. *Then for random* $z \in \mathbb{Z}_N$ *the family of functions* $\{P_z^\ell\}$ *is* (ε, e)–*locally random with* $\varepsilon = N^{-1/2}(\log N)^{e+1}2^{e\ell+2} \deg P$.

Proof. Let $d = \deg P$, let $y_1, \ldots, y_e \in \mathbb{Z}_N$ be pairwise distinct and let $f_i \in \mathbb{Z}_N[x]$ be the polynomial $f_i(x) = P(y_i + x)$ for $i = 1, \ldots, e$. We next show that the polynomials $f_1, \ldots, f_e$ are linearly independent modulo the subspace of polynomials of degree ≤ 1 in $\mathbb{Z}_N[x]$. For suppose that there are $b_1, \ldots, b_e \in \mathbb{Z}_N$ such that

$$\deg \left(\sum_{i=1}^{e} b_i P(y_i + x) \right) \leq 1.$$

Then for $j = d - e + 1, \ldots, d$ the j-th derivative of this linear combination vanishes at $x = 0$, hence

$$\sum_{i=1}^{e} b_i P^{(j)}(y_i) = 0 \quad \text{for} \quad j = d - e + 1, \ldots, d.$$

It is sufficient to prove that the coefficient matrix $[P^{(j)}(y_i)]_{\substack{1 \leq i \leq e \\ d-e+1 \leq j \leq d}}$ is non-singular since this implies that $b_1 = \cdots = b_e = 0$. Suppose that there exist $h_{d-e+1}, \ldots, h_d \in \mathbb{Z}_N$ such that

$$\sum_{j=d-e+1}^{d} h_j P^{(j)}(y_i) = 0 \quad \text{for} \quad 1 \leq i \leq e.$$

Put $g(x) = \sum_{j=d-e+1}^{d} h_j P^{(j)}(x)$, then $g(y_i) = 0$ for $1 \leq i \leq e$. Since $y_1, \ldots, y_e$ are distinct and $\deg(g) \leq d - (d - c + 1) = e - 1$ we have $g = 0$, so

$$\sum_{j=d-e+1}^{d} h_j P^{(j)}(x) = 0.$$

Comparing coefficients of x^{e-1} we get $h_{d-e+1} = 0$ (the coefficient of x^{e-1} in $P^{(d-e+1)}$ is nonzero since $d < N$). Continuing in this manner, we obtain $h_{d-e+1} = \cdots = h_d = 0$.

Since $f_1, \ldots, f_e$ are linearly independent modulo $\mathbb{Z}_N + x\mathbb{Z}_N$ we can apply Theorem 1 to $\mathbf{F} = (f_1, \ldots, f_e)$. Since $\prod_{j=1}^{e} f_j(z)[\ell \in I_{e\ell}$ the m in Theorem 1 is $e\ell$. The ℓ in Theorem 1 is e. Hence $\prod_{j=1}^{e} P(y_j + z)[\ell \in I_{e\ell}$ is statistically random within $\varepsilon = N^{-1/2}(\log N)^{e+1} d\, 2^{e\ell+2}$. Therefore $\{P_z^\ell\}$ is (ε, e)-locally random. $\qquad \square$

References

[1] ALEXI, W., CHOR, B., GOLDREICH, O. and SCHNORR, C.P.: RSA and Rabin Functions: certain parts are as hard as the whole. SIAM J. Comput., 17, 2 (1988), pp. 194 – 208.

[2] ALON, N., BABAI, L. and ITAI, A.: A fast and simple randomised parallel algorithm for the maximal independent set problem. J. of Alg. 7 (1986), pp. 567 – 583.

[3] ALON, N., GOLDREICH, O., HASTAD, J. and PERALTA, R.: Simple constructions of almost k-wise independent random variables. Proceedings of the 31st IEEE Symposium on Foundations of Computer Science (1990) pp. 544 – 552.

[4] BLUM, L, BLUM, M., and SHUB, M.: A simple unpredictable pseudo-random number generator. SIAM J. Comput. 15 (1986), pp. 364 – 383.

[5] BLUM, M. and MICALI, S.: How to generate cryptographically strong sequences of pseudo-random bits. Proceedings of the 23rd IEEE Symposium on Foundations of Computer Science, IEEE, New York (1982); also SIAM J. Comput. 13 (1984), pp. 850–864.

[6] CARLITZ, L. and UCHIYAMA, S.: Bounds for exponential sums. Duke Math. J. 24, (1957), pp. 37 – 41.

[7] CARTER, L. and WEGMAN, M.: Universal hash functions. J. Comp. and Syst. Sci. 18, (1979) pp. 143 – 154.

[8] GOLDREICH, O., GOLDWASSER, S. and MICALI, S.: How to Construct Random Functions. Proceedings of the 25th IEEE Symposium on Foundations of Computer Science, New York, (1984); also Journal ACM 33, 4 (1986), pp. 792–807.

[9] LIDL, R., and NIEDERREITER, H.: Finite Fields. Reading: Addison–Wesley 1983.

[10] LUBY, M.: A simple parallel algorithm for the maximal independent set problem. SIAM J. Comput., 15 (1986), pp. 1036 – 1053.

[11] MAURER, U. M., and MASSEY, J.L.: Perfect local randomness in pseudo–random sequences. Proceedings Crypto '89, Lecture Notes in Computer Science, Vol. 435, Springer–Verlag 1990, pp. 100–112.

[12] MICALI, S. and SCHNORR, C.P.: Efficient, perfect polynomial random number generators. J. of Cryptology 3, (1991), pp. 157 – 172.

[13] NAOR, J. and NAOR, M.: Small–bias Probability Spaces: Efficient Constructions and Applications. Proceedings of the 22nd ACM Symposium on Theory of Computing (1990), pp. 213–223.

[14] NIEDERREITER, H.: Pseudo–random numbers and optimal coefficients. Advances in Math. 26, (1977) pp. 99 – 181.

[15] NISAN, N.: Pseudorandom generators for space-bounded computation. Proceedings of the 22nd ACM Symposium on Theory of Computing (1990), pp. 204–208.

[16] SCHNORR, C.P.: On the construction of random number generators and random function generators. Proc. EUROCRYPT '88, Lecture Notes in Computer Science, Vol. 330, Springer–Verlag 1988, pp. 225–232.

[17] WEIL, A.: On some exponential sums. Proc. Nat. Acad. Sci. USA 34, (1948), pp. 204 – 207.

[18] YAO, A.C.: Theory and applications of trapdoor functions. Proceedings of the 23rd IEEE Symposium on Foundations of Computer Science, IEEE, New York (1982), pp. 80–91.

How Intractable Is the Discrete Logarithm for a General Finite Group?

Tatsuaki Okamoto*   Kouichi Sakurai[†]   Hiroki Shizuya[‡]

* NTT Laboratories
Nippon Telegraph and Telephone Corporation
1-2356, Take, Yokosuka-shi, Kanagawa-ken, 238-03 Japan
Email: okamoto@sucaba.ntt.jp

[†]Computer & Information Systems Laboratories
Mitsubishi Electric Corporation
5-1-1, Ofuna, Kamakura, 247 Japan
Email: sakurai@isl.melco.co.jp

[‡]Education Center for Information Processing
Tohoku University
Kawauchi, Aoba-ku, Sendai, 980 Japan
Email: shizuya@ecip.tohoku.ac.jp

Abstract

GDL is the discrete logarithm problem for a general finite group G. This paper gives a characterization for the intractability of GDL from the viewpoint of computational complexity theory. It is shown that GDL $\in$ NP $\cap$ co-AM, assuming that G is in NP $\cap$ co-NP, and that the group law operation of G can be executed in a polynomial time of the element size. Furthermore, as a natural probabilistic extension, the complexity of GDL is investigated under the assumption that the group law operation is executed in an expected polynomial time of the element size. In this case, it is shown that GDL $\in$ MA $\cap$ co-AM if $G \in$ NP $\cap$ co-NP. Finally, we show that GDL is less intractable than NP-complete problems unless the polynomial time hierarchy collapses to the second level.

R.A. Rueppel (Ed.): Advances in Cryptology - EUROCRYPT '92, LNCS 658, pp. 420-428, 1993.

1 Introduction

The discrete logarithm problem has played an important role in the construction of some cryptographic protocols. The problem is usually defined over the multiplicative group of a finite field, but it has some varieties with respect to its underlying finite group such as the multiplicative group over a finite ring modulo a composite, the elliptic curve group over a finite field [Mi, Ko1], the jacobian of the hyperelliptic curve over a finite field [Ko2], and so on. Since the computational complexity of each version of the discrete logarithm problem has some cryptographic or structural complexity-theoretic implications, it is important to characterize their complexity as well as to find efficient algorithms for solving them.

Our interest in this paper is to characterize the intractability of the general discrete logarithm problem that does not depend on a specific underlying finite group, from a viewpoint of structural complexity theory. The discrete logarithm problem for a general finite group G can be stated as follows: Given $a \in G$ and $b \in G$, find the smallest integer x such that $b = a^x$, provided that such an integer exists, where a^x denotes $a \circ a \circ \cdots \circ a$ (x times), and $\circ$ denotes the group law of G. The integer x is called the discrete logarithm of b to the base a.

Since the general discrete logarithm problem is a computing problem, we introduce a language GDL such that the complexity of membership problem in GDL is equivalent to that of the general discrete logarithm problem. Our goal is to show the class of the language GDL.

As an instance of GDL, when G is a multiplicative group over a finite field, we call GDL "MDL". When G is a hyperelliptic discrete logarithm [SIS], we call GDL "HEDL". Brassard has pointed out that MDL $\in$ NP $\cap$ co-NP [Br]. Shizuya, Itoh and Sakurai showed that HEDL $\in$ NP $\cap$ co-AM [SIS], after which Okamoto and Sakurai showed that HEDL $\in$ NP $\cap$ co-NP, provided that the jacobian of the hyperelliptic curve is non-half-degenerate [OS], where, for example, the jacobians with the most general (complicated) structure satisfy this condition (or non-half-degeneracy), and any jacobian with genus one (elliptic curve) also satisfies this condition. However, it has not been shown which class GDL belongs to, if G is a general finite group.

In this paper, we show that the result for HEDL by [SIS] can be generalized to GDL for a general finite group. That is, we show that GDL $\in$ NP $\cap$ co-AM, assuming that G is in NP $\cap$ co-NP, and that the group law operation of G can be executed in a polynomial time of the element size. If the group law operation is executed in an expected polynomial time and $G \in$ NP $\cap$ co-NP, we show that

GDL $\in$ MA $\cap$ co-AM.

2 Preliminaries

Throughout this paper, all strings will be over the finite alphabet $\Sigma = \{0, 1\}$. We use $|x|$ to represent the length of string x. We let Σ^* designate the set of all possible strings including zero-length string λ. A language is a set of strings. A class is a set of languages. For a language L, we use co-L to denote $\Sigma^* \setminus L$. For a class $\mathcal{C}$, we use co-$\mathcal{C}$ to denote its complement, i.e. the set of any L such that co-L is in $\mathcal{C}$.

We define the following language GDL to investigate the complexity of the general discrete logarithm problem for a general finite group G. It is clear that the complexity of the general discrete logarithm problem is equivalent to that of the membership problem in GDL.

Definition 2.1 *Let $\mathcal{N}$ be a countably infinite set, and let $N \in \mathcal{N}$ specify a finite group G_N. Define $G = \{G_N \mid N \in \mathcal{N}\}$. Given N and $a \in G_N$, a^x is calculated by the group law of G_N.*

$$\text{GDL} = \{(a, b, N, k) \mid a, b \in G_N \wedge k \in Z \wedge \exists x[b = a^x \wedge 0 \leq x \leq k]\}.$$

Here, we assume that, given N, for any $a, b \in G_N$, $|a| = |b|$. [1]

Furthermore, we take into account the complexity of the decision problem that, given N and x, asks whether x is in G_N or not. Such a decision is often required to check the validity of input strings, or the correctness of output strings in some computation process. For this purpose, we alternatively regard the general group G as a language over Σ^*, and will sometimes say that, for example, G is in P.

We will assume in subsequent sections that G is in P or in NP $\cap$ co-NP because it is reasonable to consider only finite groups such that there is a witness for $x \in G$ or $x \notin G$. (Another research topics would be to extend the discussion to cover groups out of this assumption, for example, G in a class beyond NP or in some probabilistic class.) We will also assume that there exists a program which calculates the group law, runs in a (deterministic or expected) polynomial time, and outputs only a correct answer.

Example 1: Let Z_p be a finite prime field of characteristic p, and let Z_p^* be its multiplicative group. Suppose $G = \{Z_p^*\}$ (or $G_p = Z_p^*$), which is true in the

---

[1] From this assumption, when (a, b, N, k) is an input of GDL, $|\text{ord}(a)|$ can be bounded by $|a|$ or the input size, $|(a, b, N, k)|$.

case of MDL. Then, it is known that MDL is in NP ∩ co-NP [Br]. Here, the multiplication over Z_p can be executed in deterministic polynomial time in $|p|$. G is clearly in P because given x, we can immediately check that x is a positive integer less than p.

Example 2: Let $E(C, F_q)$ be an elliptic curve group over a finite field of characteristic p, where C is the equation that gives the curve, and $q = p^n$. Suppose $G = \{E(C, F_q)\}$ (or $G_{(C,q)} = E(C, F_q)$), which is the special case of HEDL called EDL. Then, it is known that EDL is in NP ∩ co-NP [SIS, OS]. Here, the addition of two (possibly distinct) points on the elliptic curve can be executed in deterministic polynomial time. G is in P because given a point Q, we can check in deterministic polynomial time in $|p|$ that Q satisfies C over F_q.

Example 3: Let Z_n be a finite residue class ring modulo n, and let QR_n be the group of quadratic residues over Z_n. Suppose $G = \{QR_n\}$ (or $G_n = QR_n$). The group law is then simply the multiplication over Z_n, and it can be executed in deterministic polynomial time in $|n|$. However, unlike other examples, G is in NP ∩ co-NP, known as quadratic residuosity modulo n. In this case, we must take into account the complexity of G in order to characterize the complexity of the discrete logarithm problem over QR_n because checking the validity of input strings contains an NP-statement rather than an easily decidable P-statement. This is a typical example that indicates why we consider the complexity of G in our characterization for GDL.

3 Main Result

Theorem 3.1 *GDL* ∈ *NP* ∩ *co-AM, assuming that G is in P, and that the group law operation of G can be executed in a deterministic polynomial time of the element size.*

Proof: It is trivial that GDL ∈ NP, since the witness of GDL is x such that $b = a^x \wedge x \leq k$. To show that GDL ∈ co-AM, it is sufficient to show that co-GDL ∈ AM. There are two cases when $(a, b, N, k) \notin$ GDL. One is the case that there exists x such that $b = a^x$ and that $k < x < \text{ord}(a)$. The other is the case that there does not exist x such that $b = a^x$. In the former case, co-GDL ∈ NP ⊆ AM, since x, $\text{ord}(a)$ and all prime factors of $\text{ord}(a)$ are the witness of $(a, b, N, k) \notin$ GDL.

Hence, in the remaining part of this proof, we will show that co-GDL ∈ AM in the latter case. To show that there does not exist x such that $b = a^x$, it is sufficient to show that $b \notin <a>$, where $<a>$ denotes the subgroup generated by a.

The following protocol is the constant round interactive proof system that shows $b \not< a >$. Combining this and the result by [GS], we can conclude that co-GDL $\in$ AM in the latter case. Thus, GDL $\in$ NP $\cap$ co-AM.

Protocol:

Step 1 Prover P computes t (the order of a), $(s_1, \ldots, s_k)$ (all prime factors of t), and $(u_1, \ldots, u_k)$ (the witnesses of the primility of $s_1, \ldots, s_k$ [Pr]). P sends them to verifier V.

Step 2 V checks the correctness of t, by checking $a^t = e$ and $a^{t/s_i} \neq e$ for any $i = 1, \ldots, k$. V also checks the correctness of the primality of $(s_1, \ldots, s_k)$ using $(u_1, \ldots, u_k)$ through Pratt's algorithm [Pr]. If it is not correct, V halts. Otherwise, V selects a random bit $c \in \{0, 1\}$ and a random integer $r \in Z_t$. Then, V computes $d = a^r \circ b^c$. V sends d to P.

Step 3 P sets $c' = 0$ if $d \in < a >$, and sets $c' = 1$ if $d \not< a >$. P sends c' to V.

Step 4 V checks whether $c = c'$. If $c \neq c'$, V rejects and halts. Otherwise, V continues the protocol.

Step 5 After the protocol above is repeated in a constant round, V accepts the proof if $c = c'$ for all rounds.[2].

Finally, we show that the above protocol is an interactive proof system for $b \not< a >$.

(Completeness:) When $b \not< a >$, assume that $a^r \circ b \in < a >$. Then, there exists l such that $a^r \circ b = a^l$. Hence, $b = a^{l-r}$. This is a contradiction. Therefore, if $b \not< a >$, then $a^r \circ b \not< a >$. Thus, when $b \not< a >$, $d \in < a >$ if $c = 0$, and $d \not< a >$ if $c = 1$. That is, $c = c'$ for all rounds. Thus, when $b \not< a >$, V accepts the proof with probability 1.

(Soundness:) If $b \in < a >$ or $b = a^u$, then $a^r \circ b$ distributes uniformly over $< a >$, since $a^r \circ b = a^{r+u}$ and $r + u \bmod t$ distributes uniformly over Z_t. Clearly, a^r distributes uniformly over $< a >$. Therefore, when $b \in < a >$, no P' can guess the value of c with a probability of more than $1/2$. Thus, the probability that P' convinces V with a constant $(K > 1)$ round repetition is at most $1/2^K < 1/3$. $\square$

---

[2] This procedure can be parallelized

Since we assume in the above that G is in P, we do not explicitly consider the validity check of strings. Whereas, if G is not known to be in P, it is necessary to take into account the complexity of G as shown in Example 3. However, the following theorem shows that the complexity of GDL is *not* affected if G is in NP ∩ co-NP.

Corollary 3.2 *GDL* ∈ *NP* ∩ *co-AM, assuming that G is in NP* ∩ *co-NP, and that the group law operation of G can be executed in a deterministic polynomial time of the element size.*

Proof: The strategy of our proof is the same as for the previous theorem except that we use G as an oracle set to check the validity of input strings or the correctness of output strings in some computation process.

It is clear that GDL is recognized by a polynomial time bounded nondeterministic oracle Turing machine with G oracle. Since G is in NP ∩ co-NP, GDL is at most in $NP^{NP∩co-NP}$. However, by the result on low and high hierarchies within NP [Sch], $NP^{NP∩co-NP} = NP$. Thus, GDL ∈ NP.

To prove GDL ∈ co-AM, we can show almost the same constant round interactive protocol for $b \not\in < a >$. The only one difference is that the verifier V is allowed to make queries to the oracle set G. Thus, co-GDL is at most in $AM^{NP∩co-NP}$. However, by the result on lowness in probabilistic complexity classes [Kl], $AM^{AM∩co-AM} = AM$. Thus, co-GDL ∈ AM, and we conclude that GDL ∈ NP ∩ co-AM. □

The following corollaries are the consequences of natural probabilistic extension for the group law operation.

Corollary 3.3 *GDL* ∈ *MA* ∩ *co-AM, assuming that G is in P, and that the group law operation of G can be executed in an expected polynomial time of the element size.*

Proof: ¿From the assumption, the operation of a^x can be executed in expected polynomial time. Therefore, GDL ∈ MA, since the prover (Merlin, M) sends the verifier (Arthur, A) the witness of GDL, x ($b = a^x \land x \leq k$), and a polynomial-time machine, A, can check the correctness of x with a probablity of more than 2/3.

Similarly, to show that GDL ∈ co-AM, it is sufficient to show that co-GDL ∈ MA when there exists x such that $b = a^x$ and that $x > k$, and to show that co-GDL ∈ AM when there does not exist x such that $b = a^x$. In the former case, clearly co-GDL ∈ MA ⊆ AM. In the latter case, co-GDL ∈ AM, since a modification of the protocol shown in the proof of Theorem 2.1 can become the constant round IP to show that $b \notin < a >$, where in Step 5 V accepts the proof if $c = c'$ for 2/3 of all rounds. □

Corollary 3.4 *GDL $\in$ MA $\cap$ co-AM, assuming that G is in NP $\cap$ co-NP, and that the group law operation of G can be executed in an expected polynomial time of the element size.*

Proof: It suffices to show that GDL is in MA under these assumptions. In the MA protocol in the previous corollary, we allow Arthur (A) to make queries to the oracle set G. Since G is in NP $\cap$ co-NP, GDL is at most in $\text{MA}^{\text{NP} \cap \text{co-NP}}$. Note in the proof that it is not known whether $\text{MA}^{\text{NP} \cap \text{co-NP}} = \text{MA}$. However, since Arthur's query is made only once in order to check the validity of input strings from the assumption, GDL still remains in MA. Because, by the robustness of NP, the response to the query can be merged with Merlin's strings and sent to Arthur. □

The following result is obtained directly from the above results and the result by [BHZ]. This result implies that, for any complicated finite group G (under the reasonable assumption), GDL cannot be so intractable as NP-complete problems unless the polynomial time hierarchy collapses to the second level.

Corollary 3.5 *Assume that G is in NP $\cap$ co-NP, and that the group law operation of G can be executed in an expected polynomial time of the element size. Then, if GDL $\in$ NP-complete, the polynomial time hierarchy collapses to the second level.*

4 Related Open Problems

An open question is to find an instance of the finite group G such that (i) G is not known to be in P but is in NP $\cap$ co-NP, and (ii) a deterministic polynomial time algorithm to compute its group law operation is not known to exist, but an expected polynomial time algorithm is available. We let GDL* designate GDL for such a group. The complexity of GDL* is characterized as MA $\cap$ co-AM, which contrasts to the fact that MDL, GDL for a multiplicative group over a finite field, and HEDL, GDL for the hyperelliptic discrete logarithm problem, are both in NP $\cap$ co-NP [Br, OS].

A related question arises as to whether GDL* is in SZK, the class of languages that have statistical zero-knowledge interactive proof systems. It is known that both MDL and HEDL have perfect zero-knowledge interactive proof systems, respectively [TW, SIS]. By the results of [AH, Fo], SZK $\subseteq$ AM $\cap$ co-AM, but it is not known whether the converse holds.

5 Conclusion

In this paper, we have shown that GDL $\in$ NP $\cap$ co-AM, assuming that G is in NP $\cap$ co-NP, and that the group law operation of G can be executed in a polynomial time of the element size. We extended the discussion to the case where the group law operation is executed in an expected polynomial time of the element size, and have shown that GDL is in MA $\cap$ co-AM if G is in NP $\cap$ co-NP. Finally we have shown that GDL cannot be NP-complete unless the polynomial time hierarchy collapses to the second level.

Acknowledgments

Authors wish to thank Claude Crépeau for his initial suggestion which led to this paper. They would also like to thank Birgit Pfitzmann and Osamu Watanabe for their invaluable comments and discussions on the preliminary version. They also thank Lance Fortnow for informing the correctness of Klapper's result on $\text{AM}^{\text{AM}\cap\text{co-AM}} = \text{AM}$. The third author would like to thank Gilles Brassard for his encouragement on this work. They thank anonymous referees for their useful comments.

References

[AH]   W. Aiello, J. Håstad, "Perfect zero-knowledge languages can be recognized in two rounds," Proc. 28th FOCS, pp.439-448 (1987).

[BHZ]   R. Boppana, J. Håstad, and S. Zachos, "Does co-NP have short interactive proofs?" Inform. Proc. Lett., vol.25, pp.127-132 (1987).

[Br]   G. Brassard, "A note on the complexity of cryptography," IEEE Trans. Inf. Theory, vol.IT-25, no.2, pp.232-233 (1979).

[Fo]   L. J. Fortnow, "The complexity of perfect zero-knowledge," Proc. 19th STOC, pp.204-209 (1987).

[GS]   S. Goldwasser and M. Sipser, "Private coins versus public coins in interactive proof systems," Proc. 18th STOC, pp.59-68 (1986).

[Kl]   A. Klapper, "Generalized lowness and highness and probabilistic complexity classes," Math. Syst. Theory, vol.22, pp.37-45 (1989).

[Ko1] N. Koblitz, "Elliptic curve cryptosystems," Math. Comp., vol.48, no.177, pp.203-209 (1987).

[Ko2] N. Koblitz, "Hyperelliptic cryptosystems," J. Cryptology, vol.1, no.3, pp.139-150 (1989).

[Mi]   V. S. Miller, "Use of elliptic curves in cryptography," *Advances in Cryptology: Proceedings of Crypto'85*, LNCS 218, pp.417-426 (1985).

[OS]   T. Okamoto, K. Sakurai, "Efficient algorithms for the construction of hyperelliptic cryptosystems," *Advances in Cryptology: Proceedings of Crypto'91*, LNCS 576, Springer-Verlag, pp.265-278 (1992).

[Pr]   V. R. Pratt, "Every prime has a succinct certificate," SIAM J. Comput., vol.4, pp.214-220 (1975).

[Sch] U. Schöning, "A low and high hierarchy within NP," J. Comp. Syst. Sci., vol.27, pp.14-28 (1983).

[SIS] H. Shizuya, T. Itoh, K. Sakurai, "On the complexity of hyperelliptic discrete logarithm problem", *Advances in Cryptology: Proceedings of Eurocrypt'91*, LNCS 547, Springer-Verlag, pp.337-351 (1991).

[TW]   M. Tompa and H. Woll, "Random self-reducibility and zero knowledge interactive proofs of possession of information," Proc. 28th FOCS, pp.472-482 (1987).

Factoring with an Oracle

Ueli M. Maurer

Institute for Theoretical Computer Science
ETH Zürich
CH-8092 Zürich, Switzerland
Email address: maurer@inf.ethz.ch

Abstract. The problem of factoring integers in polynomial time with the help of an (infinitely powerful) oracle who answers arbitrary questions with yes or no is considered. The goal is to minimize the number of oracle questions. Let N be a given composite n-bit integer to be factored. The trivial method of asking for the bits of the smallest prime factor of N requires $n/2$ questions in the worst case. A non-trivial algorithm of Rivest and Shamir requires only $n/3$ questions for the special case where N is the product of two $n/2$-bit primes. In this paper, a polynomial-time oracle factoring algorithm for general integers is presented which, for any $\epsilon > 0$, asks at most ϵn oracle questions for sufficiently large N. Based on a conjecture related to Lenstra's conjecture on the running time of the elliptic curve factoring algorithm it is shown that the algorithm fails with probability at most $N^{-\epsilon/2}$ for all sufficiently large N.

1. Introduction

An interesting direction of research in complexity theory is to determine the complexity of a problem under the assumption that an oracle is available who answers questions about the particular instance of the problem to be solved. Clearly, to introduce such an oracle can make sense only if some restrictions are placed on the questions that may be asked; otherwise every problem could trivially be solved by asking the oracle for the solution. For a given problem

R.A. Rueppel (Ed.): Advances in Cryptology - EUROCRYPT '92, LNCS 658, pp. 429-436, 1993.
© Springer-Verlag Berlin Heidelberg 1993

that is believed to be difficult there exists a trade-off between the restriction on the questions and the running time of an algorithm solving the problem with such restricted use of the oracle.

A natural restriction is to allow arbitrary questions with a binary answer (yes/no) but to restrict the number of questions. Note that every question having a d-ary answer can easily be simulated by $\lceil \log_2 d \rceil$ questions with binary answer. Clearly, in order to be of interest, the number of questions must be smaller than the size in bits of the solution.

It is common practice in theoretical computer science to distinguish as a coarse classification for the feasibility of an algorithm between polynomial and superpolynomial running time. The goal of the research described in this paper is to find a polynomial-time algorithm for a given problem (here integer factorization) asking as few questions as possible.

One motivation for considering this problem is to determine whether or not the difficulty of a certain problem can be concentrated in a few difficult bits, leading to a new complexity-theoretic classification of problems. Another motivation is the fact that if the number of questions could be reduced to $O(\log n)$ where n is the input size, then all possible oracle answers could be checked in polynomial time; this would result in a polynomial-time algorithm (without access to the oracle) for the original problem.

Motivated by a paper by Rivest and Shamir [5], this paper is concerned with the problem of factoring integers, which is widely believed to have no polynomial-time algorithm for its solution. In fact, several cryptographic systems (e.g., [6]) rely on the difficulty of factoring. A non-trivial factor of every n-bit integer N can easily be determined by asking $n/2$ questions, namely, "What is the i-th bits of the smallest prime factor of N?", for $i = 1, \ldots, n/2$. For the special case of integers that are the product of two primes of roughly equal size, Rivest and Shamir [5] described a polynomial-time algorithm based on integer programming which asks at most $n/3$ question. In this paper, a polynomial-time algorithm is presented that, for any given $\epsilon > 0$, asks at most ϵn questions. The claim that the algorithm fails only with exponentially small probability is based on a plausible number-theoretic conjecture about the distribution of smooth numbers in certain intervals and is closely related to a conjecture by Lenstra that he used in the (heuristic) running time analysis of his elliptic curve factoring algorithm [3].

The major motivation of Rivest and Shamir for investigating this problem was that an adversary often has some side-information about the secret parameters of a cryptographic system. Our analysis corresponds to a worst-case scenario in which the adversary can choose precisely what side-information he would like to obtain. This paper demonstrates that cryptosystems whose security is based on the difficulty of factoring a modulus (e.g., the RSA system [6]) could be broken if an adversary were allowed to obtain only ϵn bits of information of his choice about the modulus. However, because oracles do not exist in reality, the results of this paper have no direct implication on the security of existing cryptographic systems.

2. Preliminaries

The following lemma shows that in a sequence of k pairwise independent events, each having probability p of occurring, the probability that none of these events occurs is at most $(1-p)/(kp)$. The events are pairwise independent if for any two events A and B, $P[A \cap B] = P[A] \cdot P[B]$.

Lemma. *Let $X_1, \ldots, X_k$ be pairwise independent binary random variables where $P[X_i = 1] = p$ for $1 \leq i \leq k$. Then*

$$P[X_1 = X_2 = \cdots = X_k = 0] \leq \frac{1-p}{kp}.$$

Proof. Note that the expected value and the variance of X_i are given by $E[X_i] = p$ and $\mathrm{Var}[X_i] = p(1-p)$, respectively. Let S be the integer sum of $X_1, \ldots, X_k$, i.e., $S = X_1 + \cdots + X_k$. Hence $S = 0$ if and only if $X_1 = \cdots = X_k = 0$, and $E[S] = kp$. It is not difficult to prove (cf. [2]) that the variance of the sum of several pairwise independent random variables is equal to the sum of the individual variances. Thus $\mathrm{Var}[S] = kp(1-p)$. For every real-valued random variable Y we have

$$\mathrm{Var}[Y] \geq P[Y = 0] \cdot E[Y]^2$$

since the right-hand side is only one of several positive terms summing to the variance. We conclude that $P[S = 0] \leq \mathrm{Var}[S]/E[S]^2 = (1-p)/(kp)$. $\square$

Remarks. It is well-known that the expected value of the sum of several random variables is equal to the sum of their expected values, and that the variance of the sum is equal to the sum of the variances if the random variables are statistically independent. It is less well-known that a sufficient condition for the rth moment of the sum to be equal to the sum of the rth moments is that the random variables be only r-wise statistically independent. For fixed p and $k \to \infty$ the proved bound on the probability that $X_1 = \cdots = X_k = 0$ is $O(1/k)$, which is optimal.

It is well-known that a polynomial of degree at most d over a field can be interpolated from any set of $d+1$ distinct arguments and their corresponding polynomial values. For the case of a finite field $GF(q)$ with q elements (where q is a prime power) this observation leads to a construction of a sequence of q $(d+1)$-wise independent random variables: When the $d+1$ coefficients of the polynomial are selected randomly and independently from $GF(q)$ with uniform distribution, then the polynomial's values for any set of $d+1$ arguments are also statistically independent and uniformly distributed. We will make use only of the special case $d = 1$ (pairwise independence).

For a prime $p > 3$ the elliptic curve over $GF(p)$ with parameters a and b satisfying $4a^3 + 27b^2 \neq 0$ is defined as the set of points (x, y) with $x, y \in GF(p)$ satisfying the congruence equation

$$y^2 \equiv x^3 + ax + b \pmod{p}, \tag{1}$$

together with a special element denoted $\mathcal{O}$ and called the point at infinity. This curve is denoted as $E_{a,b}(p)$. It is well-known that a group operation, which is called addition, can be defined on the set of points of the elliptic curve $E_{a,b}(p)$. Let P and Q be two points on $E_{a,b}(p)$. The point $P + Q$ is defined according to the following rules. $P + \mathcal{O} = \mathcal{O} + P = P$ for all P on E (i.e., $\mathcal{O}$ is the identity element of $E_{a,b}(p)$). Let $P = (x_1, y_1)$ and $Q = (x_2, y_2)$. If $x_1 = x_2$ and $y_1 = -y_2$, then $P + Q = \mathcal{O}$ (i.e., the negative of the point (x, y) is the point $(x, -y)$). In all other cases the coordinates of $P + Q = (x_3, y_3)$ are computed as follows. Let λ be defined by

$$\lambda = \begin{cases} \dfrac{y_2 - y_1}{x_2 - x_1} & \text{if } x_1 \neq x_2 \\[2mm] \dfrac{3x_1^2 + a}{2y_1} & \text{if } x_1 = x_2, \end{cases}$$

where all operations are to be computed modulo p. (When $P + Q \neq \mathcal{O}$ then the denominator is not zero and thus the quotient is defined.) The resulting point $P + Q = (x_3, y_3)$ is defined by

$$\begin{aligned} x_3 &= \lambda^2 - x_1 - x_2 \\ y_3 &= \lambda(x_1 - x_3) - y_1. \end{aligned}$$

The prime p can be replaced by a composite N in the above definition and equations. However, $E_{a,b}(N)$ defined in this manner is not a group, but it can be extended to form a group by adjoining a small fraction of additional elements. (In the case where $N = p_1 \cdots p_r$ is the product of distinct primes > 3, $E_{a,b}(N)$ is the direct product of the corresponding elliptic curves over $GF(p_1), \ldots, GF(p_r)$.) Nevertheless, the addition operation, which is in this case called pseudo-addition, can be performed as long as it is defined, i.e., when the denominator is relatively prime to N, and it corresponds in fact to the addition operation on the extended curve. We refer to [3] for further information on elliptic curves. Note that in [3] points (x, y) are represented in projective coordinates as triples $(x : y : 1)$, and $\mathcal{O}$ is represented as $(0 : 1 : 0)$.

Unless stated otherwise, logarithms in this paper are to the natural base e. The cardinality of a set S is denoted by $\#S$.

3. The Oracle Factoring Algorithm

Let N be a given composite n-bit integer and let $\epsilon < 0.5$ be an arbitrary given positive constant. If N is not known to be composite, a simple probabilistic compositeness test such as the Miller-Rabin test [4] can be used to prove the compositeness of N. In the sequel a polynomial-time (in n) algorithm is described for finding a non-trivial divisor d of N ($1 < d < N$) which, for all sufficiently large N, succeeds with probability at least $1 - N^{-\epsilon/2}$ and asks at most ϵn oracle questions.

The algorithm consists of four steps.

(i) (Special cases.) If 2 or 3 divides N or if N is a prime power $N = q^t$, output 2, 3 or q, respectively, and stop.

(ii) (Setup.) Choose δ with $0 < \delta < \epsilon$ as an arbitrary positive constant and let

$$c = \frac{1}{\epsilon - \delta}$$

and

$$w = (\log N)^c.$$

Let further

$$h = \prod_{r \le w,\, r \text{ prime}} r^{e(r)}, \tag{2}$$

where $e(r)$ is the largest integer m with $r^m \le N^{1/2} + 2N^{1/4} + 1$. Choose s and t randomly from $GF(2^{3n})$. Fix a natural enumeration of the elements of $GF(2^{3n})$: $\alpha_1, \alpha_2, \ldots, \alpha_{2^{3n}}$. For a given natural representation of the elements of $GF(2^{3n})$ as triples of n-bit integers, let $(a_k, x_k, y_k) \in Z_{2^n} \times Z_{2^n} \times Z_{2^n}$ be the triple corresponding to $s\alpha_k + t$ where α_k is the k-th element of $GF(2^{3n})$, and let $b_k \in Z_N$ be defined by

$$b_k \equiv y_k^2 - x_k^3 - a_k x_k \pmod{N}.$$

Remarks. $N^{1/2} + 2N^{1/4} + 1$ is an upper bound on the order of an elliptic curve over $GF(p)$, where p is the smallest prime factor of N. As mentioned in Section 2 the above construction guarantees that the triples (a_k, x_k, y_k) are pairwise statistically independent. Instead of the field $GF(2^{3n})$ any other finite field with cardinality greater than N^3 could be used to create an appropriate list of pairwise independent triples (a_k, x_k, y_k). Only triples for which all three components are smaller than N will actually be of possible use.

(iii) (Oracle questions.) Now ask the oracle the following question. If there exists a positive integer $k < 2^{\lfloor \epsilon n \rfloor}$ such that

(1) for the smallest prime factor p of N,

$$4a_k^3 + 27b_k^2 \not\equiv 0 \pmod{p},$$

and each prime factor r dividing $\#E_{a_k, b_k}(p)$ satisfies $r \leq w$,

(2) and for some prime factor $q \neq p$ of N,

$$4a_k^3 + 27b_k^2 \not\equiv 0 \pmod{q}$$

and $\#E_{a_k, b_k}(q)$ is not divisible by the largest prime number dividing the order of the point $(x_k : y_k : 1)$ on the elliptic curve $E_{a_k, b_k}(p)$,

where a_k, x_k, y_k and b_k are defined in step (ii), then output (the binary representation of) the smallest such k, else output 0.

Remark. Of course, this question can easily be transformed into $\lfloor \epsilon n \rfloor$ questions with a yes/no answer.

(iv) (Factorization.) If the oracle's answer is 0, stop. In this case, the algorithm fails. If the oracle's answer is some $k > 0$, proceed as follows. Compute (a_k, x_k, y_k) and b_k as described in step (ii). Let $P = (x_k : y_k : 1)$ be a point on $E_{a_k, b_k}(N)$ (which is not a group). Try to compute $h \cdot P$ using the pseudo-addition method described in (2.4) of [3], pretending that N is prime. At some point during this computation the addition of two points $(x' : y' : 1)$ and $(x'' : y'' : 1)$ will fail because $\gcd(x' - x'', N) > 1$. Output this divisor of N.

4. Analysis of the Algorithm

We need to prove first that the algorithm works and runs in polynomial time and second that the failure probability is upper bounded by $N^{-\epsilon/2}$.

Theorem 1. *If the oracle's answer is $k > 0$ then the algorithm runs in polynomial time and always finds a non-trivial divisor of N.*

Proof. That the algorithm runs in polynomial time follows from the facts that the pseudo-addition can be performed in time $O(n^2)$ and that the number of pseudo-additions required for computing $h \cdot P$ is at most $2\lceil \log_2 h \rceil - 1$ which is polynomial in n since according to (2),

$$\log_2 h = \sum_{r \leq w,\, r \text{ prime}} e(r) \log_2 r \leq w \log_2 w$$

and $w = O(n^c)$. N is guaranteed to have a prime factor smaller than $\sqrt{N}$ and hence Proposition (2.6) in [3] for $\nu = \sqrt{N}$ implies that the algorithm always succeeds. □

It follows from the Corollary to Theorem 3.1 of Canfield, Erdös and Pomerance [1] that the probability that a random positive integer $s \le x$ has all its prime factors $\le L(x)^\alpha$, where $L(x) = e^{\sqrt{\log x \log \log x}}$, is $L(x)^{-1/(2\alpha)+o(1)}$, for $x \to \infty$. In the analysis of the elliptic curve factoring algorithm [3], Lenstra stated the conjecture that the same result is valid if s is a random integer in the interval $(x + 1 - \sqrt{x}, x + 1 + \sqrt{x})$. We will need a similar conjecture with a smaller smoothness bound. One can prove that the mentioned result of Canfield et al. implies that the probability that a random positive integer $s \le x$ has all its prime factors $\le (\log x)^c$ for $c > 1$ is greater than $x^{-1/c-\beta}$ for all $\beta > 0$ and for all sufficiently large x. The conjecture we will need is that the same result is valid if $1/c + \beta < 1/2$ and s is a random integer in the interval $(x + 1 - \sqrt{x}, x + 1 + \sqrt{x})$. We believe that our conjecture appears to be equally plausible as Lenstra's conjecture. Note that $c > 2$ in our algorithm but that for $c < 2$ the conjecture cannot be true since the expected number of smooth integers in the given interval is less than 1.

Theorem 2. *If the described conjecture is true, then the oracle outputs 0 (and hence the oracle factoring algorithm fails) with probability at most $N^{-\epsilon/2}$.*

Proof. Let p be the smallest prime divisor of N, and let U be the number of integers in the interval $(p + 1 - \sqrt{p}, p + 1 + \sqrt{p})$ for which no prime factor is greater than $w = (\log N)^c$. According to our conjecture with $\beta = \delta/2$, U is lower bounded by

$$U > (2\lfloor \sqrt{p} \rfloor + 1)p^{-1/c-\delta/2},$$

for all sufficiently large p. Note that $-1/c - \delta/2 = -\epsilon + \delta/2$. It follows from proposition (2.7) of [3] that the number T of triples $(a, x, y) \in Z_N \times Z_N \times Z_N$ that are successful in step (iii) of our algorithm is, for sufficiently large p, lower bounded by

$$T > N^3 \frac{C_1}{\log p} \cdot \frac{U - 2}{2\lfloor \sqrt{p} \rfloor + 1}$$

$$> N^3 \frac{C_2}{\log p} \cdot p^{-\epsilon+\delta/2},$$

where C_1 and C_2 are positive constants. Hence the probability that a triple selected randomly from $Z_{2^n} \times Z_{2^n} \times Z_{2^n}$ is successful is equal to $T/2^{3n}$. Because the triples (a_k, x_k, y_k), $1 \le k \le 2^{3n}$, are pairwise independent, it follows from the lemma in Section 2 that the probability Q that none of the triples (a_k, x_k, y_k), for $1 \le k < 2^{\lfloor \epsilon n \rfloor} - 1$, is successful (and therefore the oracle answers 0) is upper bounded by

$$Q < \frac{1}{(T/2^{3n}) \cdot (2^{\lfloor \epsilon n \rfloor} - 1)}$$

$$< \frac{1}{\frac{1}{8} \frac{C_2}{\log p} \cdot p^{-\epsilon + \delta/2} \cdot (\frac{1}{2} N^\epsilon - 1)},$$

where we have made use of $N^3/2^{3n} > 1/8$ and $2^{\lfloor \epsilon n \rfloor} > 2^{\epsilon n - 1} > \frac{1}{2} N^\epsilon$. Since $p \leq N^{1/2}$ the last expression is smaller than $N^{-\epsilon/2}$ for all sufficiently large N.
□

Acknowledgments

It is a pleasure to thank Uri Feige and Andrew Odlyzko for helpful discussions and an anonymous program committee member for a comment on our conjecture. Claus Schnorr has pointed out that a result similar to ours could be obtained by using the class group factoring algorithm of [7] instead of the elliptic curve factoring algorithm.

References

[1] E.R. Canfield, P. Erdös and C. Pomerance, On a problem of Oppenheim concerning "Factorisatio Numerorum", *J. Number Theory*, Vol. 17, pp. 1-28, 1983.

[2] B. Chor and O. Goldreich, On the power of two-point based sampling, *Journal of Complexity*, Vol. 5, No. 1, pp. 96-106, 1989.

[3] H.W. Lenstra, Jr., Factoring integers with elliptic curves, *Annals of Mathematics*, Vol. 126, pp. 649-673, 1987.

[4] M.O. Rabin, Probabilistic algorithm for testing primality, *Journal on Number Theory*, Vol. 12, pp. 128-138, 1980.

[5] R.L. Rivest and A. Shamir, Efficient factoring based on partial information, *Advances in Cryptology - EUROCRYPT '85*, Lecture Notes in Computer Science, Vol. 219, Berlin: Springer-Verlag, pp. 31-34, 1986.

[6] R.L. Rivest, A. Shamir, and L. Adleman, A method for obtaining digital signatures and public-key cryptosystems, *Communications of the ACM*, Vol. 21, No. 2, pp. 120-126, 1978.

[7] C.P. Schnorr and H.W. Lenstra, A Monte Carlo factoring algorithm with linear storage, *Mathematics of Computation*, Vol. 43, No. 167, pp. 289-311, July 1984.

Secure Audio Teleconferencing:
A Practical Solution

Rafi Heiman

Bellcore

Abstract

Secure audio teleconferencing is a multi-point communication service which uses encryption to prevent eavesdroppers from listening to the speech signals. Its greatest vulnerability is the audio bridge — that component which combines the conferees' speech signals and returns the result to them.

A new secure teleconferencing system is proposed here. It fits the public telephone network by eliminating the need for the conferees to share their secrets with the bridge. It combines a simplified ('instantaneous') bridging technique with secure bridging ideas previously suggested in the literature, overcoming their main practical disadvantages. In particular, it is not restricting the audio signals to be coded by linear PCM, a technique which is wasteful in terms of bit-rate. Rather, it enables the use of conventional μ-law and A-law PCM, as well as vector quantized PCM, thus can be used with a conventional 64kb/s digital channel.

1   Introduction

1.1   The Problem

Consider three or more parties that talk together over the telephone. This is called an audio teleconference and it is established by using special conferencing equipment called bridge. Conferees transmit their speech signals to the bridge. The bridge in turn, detects which signals are active (e.g., contain speech), and returns to each conferee the sum of some of the active signals. This creates the illusion that each conferee hears all others simultaneously, just as if they were all talking in the same room.

A problem arises when the conferees wish to use encryption to guarantee the privacy of their conference. In all commercial systems for secure teleconferencing, either conferees 'trust' the bridge or only a 'simplex' mode is allowed. Trusting the bridge means that conferees let it know the secure keys so that it can decrypt the signals, combine the clear signals, and encrypt the combined signals before returning

R.A. Rueppel (Ed.): Advances in Cryptology - EUROCRYPT '92, LNCS 658, pp. 437-448, 1993.
© Springer-Verlag Berlin Heidelberg 1993

them to the conferees. This results in the need to own and guard the bridge. In a simplex link all the bridge does is switching: only single and uncombined signal is returned to each conferee. This causes the inability to hear more than one conferee even when two or more conferees simultaneously speak.

To avoid these limitations, a non standard encryption scheme is needed. It is desirable that the bridge would not be able to listen to the conference contents but still perform its combining function. Therefore a special method is needed — a method that reveals enough information to enable bridging, but does not reveal information about what is being said, even if eavesdroppers have access to the bridge.

In Section 1.3 we mention previously suggested solutions and explain their practical disadvantages. In order to better understand these disadvantages, and see how our new solution overcomes them, a word on digital audio processing is due.

1.2   Digital Audio and Bridging

In order to better understand the problem, we briefly review some standard ways of encoding to digital and bridging audio signals. Further details may be found in the textbooks [RS79] and [JN84] and in bridging articles, [PC83], [PC85] and [MDS91].

Digital encoding of analog signals (such as voice) involves two parts. First, the analog (continuous-time and continuous-amplitude) signal is sampled, yielding a discrete-time and continuous-amplitude signal. Second, quantization converts the continuous-amplitude into a discrete-amplitude.

In Pulse Code Modulation (PCM) a so called 2^R-level quantizer is used and each sample is encoded independently to an R-bit number. In Linear PCM the range of possible amplitude levels is divided into segments of equal size, thus the mapping between amplitude levels and their codes is linear. Linear PCM introduces high distortion power relative to the signal power in low input signal levels. Logarithmic PCM techniques (μ-law and A-law) provide a more constant signal-to-noise ratio due to the code being logarithmically dependent on the sample amplitude. It is generally assumed that to achieve the adequate (so called 'toll') quality of logarithmic PCM with 7- or 8-bits per sample, linear PCM would require 11-12 bits per sample (see [JN84] section 5.3.2 and [RS79] section 5.3.2). Consequently, the logarithmic PCM techniques fit into 56kbit/sec or 64kbit/sec rates, became standards (μ-255 in North America, A-87.56 in Europe [CCI72]), and are widely used in practice.

A less known PCM technique, but one which further compresses the audio, is vector quantized PCM [MRG85]. In vector quantization a few consecutive samples, constituting a vector, are quantized jointly. The quantization process results in a sequence of numbers (or codes) over a previously selected set of possible quantization levels. When quantizing, each sample vector is replaced by that vector from a pre-established code-book of vectors which best matches it. The index of the code-book vector is then transmitted. At the receiver side each such index (or number) is decoded to its corresponding code-book vector of amplitude levels.

Bridging of audio signals typically consists of three stages: energy detection, selection, and addition. In the first stage some 'long term' statistics of each incoming signal are computed in order to determine whether this signal is active or silent. In the second stage a limited number of the active signals is selected. If there are only very few active signals (say, one or two) they are all selected. If there are too many active signals the bridge selects only few of them. The selection algorithm vary from bridge to bridge and it may depend, for example, on predetermined priorities between the conferees and on the exact time at which each conferee became active. In the third stage, the selected active signals are added, and the bridge returns the sum of the signals to each conferee. If a receiving conferee is active and selected, its signal is excluded from the sum that this conferee receives, avoiding echo effects. Among other computations, these stages involve converting the received streams of codes to linear PCM, summing, and converting back to the specific source coding used.

1.3 Previously Suggested Solutions

A few solutions were previously suggested for the problem of secure audio teleconferencing using an untrusted bridge. The first, by Brickell, Lee and Yacobi [BLY87], suggests a few ways to compute addition in the encrypted domain. That is, given the encryptions of a few numbers, to compute the encryption of the numbers' sum (without being able to compute the clear numbers or their clear sum). This provides a solution to our problem with the following assumptions: (1) Conferees provide the bridge with their activity levels. (2) Incoming signals are synchronized. (3) The operation the bridge has to perform on the speech samples is indeed addition. We view assumption 1 and 2 as not too demanding. Computing activity levels can be added to the encryption/decryption 'black box' conferees have anyway, and the activity information does not carry enough intelligibility to understand what is being said. A way to synchronize the incoming signals is described in Section 4. The third assumption, however, means that speech must be encoded by linear PCM, and this is probably the main barrier for the Brickel et al solution to become practical, as can be understood from the discussion above. A solution that uses e.g. logarithmic PCM, and fit into a 64kb/s bit-rate is thus desired.

A simple solution is to let one of the (trusted) conferees own the bridge. All conferees send their encrypted signals to that conferee, who then combines the decrypted signals and returns the results encrypted. This trivially solves the problem of trusting the bridge. However, owning a private bridge is not always economically attractive. A more efficient use of the bridge would be if it is offered by the public telephone network, thus be a shared resource. Even in large private networks, where there are many users to share the private bridge, the need to guard it carries a significant cost and would rather be avoided.

Two other solutions, in which all or part of the bridging process is done by the conferees were suggested by Steer, Strawczynski, Diffie and Weiner [SSDW88]. In the first, conferees are connected in a chain (as opposed to a star whose center is the bridge). Each conferee adds its speech (if active) to the prefix-sums received from its two neighbors and forwards these extended prefixes. The main barrier for this solution to become practical is probably the doubling effect: In this solution each conferee has two incoming and two outgoing voice signals thus bit-rate is doubled. This solution also introduces delay and noise problems.

The second solution proposed in [SSDW88] partially maintains the bridge as part of the communication network. Conferees send their encrypted signals as well as clear activity indications to the bridge. The bridge returns to each of them two selected signals without adding them. Conferees then decrypt them and add them by themselves. This solution too suffers from the doubling effect: one single link (bridge-to-conferee) is carrying two signals, thus, a special channel is needed. The new ISDN standard does allow two audio channels in a single link, but still, it is desirable to have the secure teleconference using the same bit-rate (and single channel) that the non-secured teleconference is using.

A simplification of the last solution that avoids the doubling effect is the so called 'simplex mode' in which the bridge returns only one signal to each conferee. All conferees get the loudest signal, presumed to be the active speaker's signal, except for the loudest speaker who gets the second loudest conferee's signal. In a typical conference most of the time only one conferee is speaking. This may be even more typical of conferences with a high demand for privacy: In business conferences, as opposed to informal private talks, conferees do not tend to interrupt or talk simultaneously with each other that much, and in many cases a chairman controls the conference. In any case interruption usually consists of a single word in order to get attention, and only when the active speaker becomes silent does the interrupting conferee begin his 'real' talk. However, a solution that does allow third parties to hear and be aware of interruptions is preferred. Human factor experts say that third parties prefer to hear interruptions, even at the cost of slightly decreasing the speech quality during these interruptions, a compromise we will adopt here.

1.4 The New Solution

Our new solution enables conferees to hear more than one speaking conferee at a time, it avoids the doubling effect mentioned above, it keeps the increase of bit-rate very low — one bit per sample, and it minimizes the extent to which conferees are involved in bridging. It can use all source coding techniques which are monotone in a sense defined below, among them are the conventional μ-law and A-law PCM and vector quantized PCM [JN84], [MRG85]. For a 64 kbit/sec channel, 7-bit μ-law or A-law can be used, maintaining adequate ('toll') quality using simple inexpensive terminals.

The new solution is very simple. In essence, it combines the main ideas of [BLY87] with a certain bridging technique, called max-bridging or instantaneous bridging [PR71]. These tools are described in Section 2. Their proposed combination and the system's security are discussed in Section 3. A possible concrete design of the system is described in Section 4.

2 Tools

2.1 Max-Bridging

A typical bridge, called sum-bridge was described above. A max-bridge differs from a sum-bridge only in the third stage, the addition. Once the max-bridge selects the (limited number of) active signals to be combined, it combines them by computing the maximum rather than addition. Namely, if m active signals were selected to be combined, and their amplitude levels at some time instance are $\{a_i\}_{i=1}^m$, then their combination is defined to be that a_i with maximal energy, or equivalently, with maximal absolute value, $|a_i|$. This is in contrast to the ordinary sum-bridge whose output is $b = \sum_{i=1}^m a_i$. We call it sample-by-sample max-bridging.

If only one conferee is speaking, that is $m = 1$, there is clearly no difference between the outputs of the two bridges. Only at those times when two or more conferees speak simultaneously do the max-bridge and the sum-bridge have different outputs. The interesting (and perhaps surprising) fact is that this difference is hardly noticeable when listening to these two outputs.

Pushing this phenomenon one step further, the bridge may operate on vectors of samples at a time rather on a sample-by-sample basis. It views the input signals as sequences of vectors, where each vector consists of a few consecutive samples. Each vector it outputs is the maximal one of the corresponding m input vectors. The definition of maximal vector can be relative to the energy of the whole vectors, yielding a bridge we call vector-max-bridge, or the energy of the central samples of the vectors, yielding a bridge we call center-max-bridge.

Preliminary simulations we made for evaluating these three types of max-bridging showed that the difference between signals obtained by max-bridging and signals obtained by sum-bridging is small (relative to the signals themselves). A few colleagues who listened to the different bridges' output found it hard to distinguish between them. The reason is perhaps that the effect of hearing two persons talking simultaneously dominates the distortion caused by max-bridging.

In order to quantize this we introduce a measure signal-to-difference ratio of a signal s relative to another signal s', defined as the power of s divided by the power of the difference $s - s'$ of the two signals. The signal-to-difference ratios of the output of sum-bridge relative to the outputs of various max-bridges are given in Figure 1. In the worst case studied, center-max-bridging of 5-length vectors, the signal-to-difference ratio was 9.2 db. All numbers were calculated for a few seconds of bridging two active signals (containing simultaneous speech of two persons).

The advantage of the max-bridge is clear. Given various kinds of digital PCM signals all it has to do is compute the maximum. Consider for example log-PCM. While

type of max-bridge	vector length	signal-to-difference
sample-by-sample		11.6 db
vector-max-bridge	3	11.4 db
	5	10.9 db
center-max-bridge	3	10.3 db
	5	9.2 db

Figure 1: Signal-to-difference ratio for max-bridges

a sum-bridge has to translate the logarithmic codes to linear, sum them, and then translate back to the logarithmic scale, a max-bridge can directly compare the codes (excluding their sign bits). This is so because the coding of an amplitude level in log-PCM consists of a sign bit and of a few more bits whose value (interpreted as a binary number) encodes the absolute value of the given amplitude in a monotone manner. Thus, any source coding technique that associates codes to sampled amplitude levels in a monotone manner requires only max computation in max-bridging.

The max-bridge is also suitable for vector quantized PCM. One only has to make sure that the code-book used for the vector quantization is sorted by increasing energy. This way vectors with higher energy would get higher indices in the code-book, i.e., get higher codes.

2.2   Secure-Sum and Secure-Max Calculations

The basic idea of Brickell, Lee and Yacobi [BLY87] for summing numbers given their encryptions is to use an encryption scheme of an additive nature: To encrypt two n-bit numbers a_1 and a_2, a randomly and uniformly chosen $(n + 1)$-bit number r is added to them modulo N where $N = 2^{n+1}$. This yields the ciphers

$$\bar{a}_i = a_i + r \pmod{N}, \quad i = 1, 2.$$

Now, $\bar{a} = \bar{a}_1 + \bar{a}_2 \pmod{N}$ can be computed by an untrusted authority (the bridge), from which $a_1 + a_2 = \bar{a} - 2r \pmod{N}$ can be deciphered (by conferees).

In the context of a sequence of sums that have to be carried, like that of summing sequences of audio samples, pseudo random number generator is used to produce the r-s. All conferees have identical generators (and identical seeds). The generators must be synchronized in the transmitting terminals so that the bridge can sum samples encrypted by the same value of r, and the conferees in turn, can subtract the appropriate value.

The secure sum calculation can be easily modified to calculate the maximum of two numbers securely. Given the two encryptions, $\bar{a}_1$ and $\bar{a}_2$ of a_1 and a_2, their difference,

$$\bar{d} = \bar{a}_1 - \bar{a}_2 = a_1 - a_2 \pmod{N}$$

can be computed by the untrusted authority (the bridge). This allows the maximum to be computed: Since $a_i < 2^n$ for $i = 1, 2$, clearly $a_1 \geq a_2$ if $\bar{d} < 2^n$, and $a_1 < a_2$ if $2^n \leq \bar{d}$. Therefore, the bridge can determine which of the a_i-s is maximal and return the corresponding $\bar{a}_i$. To decrypt,

$$a_i = \bar{a}_i - r \pmod{2^n}$$

is computed. Note that the most significant bit of the returned $\bar{a}_i$ is not needed. Note also that the difference between the two numbers is revealed, not just the knowledge of which is larger. This security weakness, which is inherited by our solution, is considered below.

3 The Proposed System and Its Security

Having these two tools in mind, the solution is rather simple. Conferees encode their speech in any monotone PCM technique. They encrypt their code-streams for secure-max computation using synchronized pseudo random number generators. A way to synchronize the generators is described in the next section. A max-bridge is used to combine the signals by computing the max on the given encrypted sequences.

The long term energy detection, the first stage of bridging, is done by the conferees. They provide the bridge with a clear and low-bit-rate signal that contains this information.

Self-detection of activity is done also for the sake of improving security. Clearly, a non active signal, say a signal with a constant PCM code that encodes zero amplitude, would reveal the pseudo random sequence and thus the active speaker's content. Therefore, a conferee that is not active transmits an idle signal to the bridge in clear (without adding the pseudo random numbers). This way, the pseudo random generator will be used in a one-time-pad manner except when at least two conferees are simultaneously speaking. The leakage of information will be limited to short and infrequent periods of time. A signal that the bridge is able to compute when, say, two conferees are simultaneously speaking, is the difference between the two encrypted signals. Each of its sample-codes (in case of sample-by-sample max-bridge) is the difference of the absolute values of the two incoming signals' sample-codes. Listening to this difference signal in an experiment we made, gave us no idea about what is being said. Speech signals contain much redundancy, though. A more sophisticated attack might exist. However, such attack, if exists, is applicable during simultaneous speech only.

Nevertheless, to reduce the amount of information that is revealed, vector quantized PCM can be used. This is more secure since vector quantized PCM contains less redundancy and the difference between the code-book indices of two vectors seems to be harder to interpret than the difference between absolute values of samples. When standard log-PCM is used center-max-bridge would be more secure. It would allow encrypting in the max-calculation manner only one sample in each vector. Thus the bit-rate of the 'parasitic' signal available to the bridge would be reduced by a factor equal to the vector length. Another way to reduce this bit-rate is by letting the bridge compare only some of the most significant bits of the numbers to be compared. This way fewer bits are encrypted in the max-calculation manner. Increasing the vector length, likewise decreasing the number of bits per sample-code that are involved in the max-calculation improve the security of the system but degrade the audio quality of the signal the bridge returns during simultaneous speech.

All samples in a vector other than the central one are encrypted by other and more secure means due to the fact that the bridge only has to forward them. Keeping in mind that conferees may be switched from vector to vector, conventional encryption methods, e.g. DES [DES77] or RSA [RSA78] can not be used here in a straight forward way without significantly expanding the bit-rates. The reason is that their block-lengths are too big. They can however be used for generating a pseudo random bit (or number) stream to be exclusive-ored (or to scramble in some other way). By the considerations above, different conferees will use different pseudo random bits to encrypt their non-central samples. This however implies that in addition to each vector, the bridge communicates some information to identify the conferee that has encrypted this vector.

On top of this, a system implementing this technique should have a user option for simplex bridging. In this mode each conferee can hear only one other conferee at a time, but no information other than activity indication leaks, provided a 'good' pseudorandom generator is used. Another possible option is to add a second layer of encryption for which the conferees do share the key with the bridge. This will cause the minor leakage of information during interruptions to be revealed to the bridge only but not to others who may eavesdrop to the conferees-to-bridge links.

4 A Concrete Design

In this section we suggest a detailed design for implementing the secure audio tele-conference system. This suggestion is made for the sake of concreteness, and some decisions made here might be changed elsewhere, depending on the specific implementation. We assume that audio signals consist of 7-bit PCM at sampling rate of 8000 samples per second and describe a system that fits a 64kbit/sec channel.

An octet is eight consecutive bits, consist of a 7-bit PCM word and an additional bit, used for the side information required by our scheme. A vector consists of five consecutive octets. The term vector also refers here to the five PCM words only, excluding the five overhead bits. The five overhead bits are used for framing (f-bit), pseudo random bit generator indexing (i-bit), extra bit for the secure-max scheme (BLY-bit), conferee activity reporting (a-bit), and conferee identification (id-bit). A frame consists of 16 vectors, which are 80 octets, and corresponds to 10 milliseconds.

The 16 f-bits per frame are used for frame synchronization and other channel signaling. The method suggested in [CCI90] for its FAS and BAS signals might be used here.

The bridge switches vector by vector based on comparing the absolute values of the third PCM-word received from each active conferee (center-max-bridge). The third PCM word in each vector is encrypted as in Section 2.2, using eight pseudo random bits. The extra eighth bit introduced is the above called BLY-bit. Prior to encrypting the third PCM word, this word is left-cyclic-shifted so that its sign bit becomes the least significant bit. This is done because the bridge has to return that conferee's sample of maximum absolute value. The other four PCM words are encrypted each by using seven pseudo random bits in a bit-by-bit exclusive-or (Xor) manner.[1]

The eight bits used to encrypt the third (central) PCM word are the same for all conferees. The seven bits used to Xor other PCM words vary from conferee to conferee. DES [DES77] can be used to generate all these pseudo random bits. A single computation of DES generates 64 pseudo-random bits. These are used either to encrypt the third PCM word of eight consecutive vectors or to encrypt other four PCM words in two consecutive vectors (using only 56 pseudo-random bits). Thus, to encrypt a frame, each conferee performs ten DES computations, two of them for the secure-max scheme and eight for the Xor-s. Before the conference takes place, conferees agree upon a secret DES key, k, to be used during the conference.[2]

Since non central samples are encrypted in a way that is conferee dependent, the signal returned from the bridge to conferees must identify the source of each vector. For this reason the number of conferees is bounded. In fact the number of listeners to the conference can be arbitrary, but the number of conferees that are allowed to speak is bounded. We first describe the system assuming only 7 conferees are allowed to speak. The pseudo random bits to encrypt frame number i of conferee j, $1 \leq j \leq 7$ are the output of DES when using the key k to encrypt the numbers $64i$ and $64i + 32$ (for the secure-max scheme) and $64i + j$, $64i + 8 + j$, ... , $64i + 56 + j$ (for the Xor's).

The 16 i-bits of a frame carry the frame number i (modulo 2^{16}). This repeats itself every $2^{16} \cdot 10$ milliseconds, which is over 10 minutes. The bridge needs this information to synchronize incoming signals and conferees need this for decryption. When a new conferee joins the conference, it performs a hand-shake protocol with the bridge, so that they can measure their transmission delays and the conferee can synchronize its pseudo random generator with those of the previously joined conferees. This synchronization should only be very rough, say with uncertainty of fifty milliseconds. To further synchronize the signals incoming to the bridge, and in order to compensate over transmission delays which may vary from conferee to conferee, the bridge buffers the incoming signals. It then can compare central samples encrypted with same

---

[1]In fact it would be more secure to encrypt this way only the seven bit absolute value of the third PCM word, and to encrypt the sign bit in a one time pad manner, similar to the other four PCM words. This is because the secure sum/max scheme reveals during interruptions the parity of the least significant bits of the two clear numbers, a_1 and a_2, as pointed out by J. Massey [Mas92]. Our concrete design does not follow this line for the sake of the system's simplicity.

[2]This key exchange is done using some other private channel or a public key system.

pseudo random bits. The i-bits in the returned signal allow conferees to synchronize the pseudo random bits for the decryption.

Every four of the 16 activity bits in a frame contain a 4-bit activity level of a conferee. Activity level is computed every four vectors (2.5 milliseconds) by averaging the signal's energy during that time. If the activity level represents a too low energy average, this conferee does not transmit its PCM samples but rather, an idle signal along with valid overhead bits. This allows the bridge to maintain synchronization with the conferee's signal, but prevents gaining information about the pseudo random bits used to encrypt the central samples, as discussed in Section 3.

When the bridge returns a vector, say generated by conferee number j, $1 \leq j \leq 7$, it also communicates the value j to identify the specific DES bits that should be used to decrypt the non-central samples of the vector. If no conferee is active the bridge returns idle vector (or white noise) and the value $j=0$. There are three bits per vector that are used to carry the value j. These are the id-bit, the a-bit, recalling that the a-bit is used only in the conferee-to-bridge link, and the BLY-bit, recalling that the most significant bit is not needed for the secure-max decryption. For this reason we assumed that at most seven conferees are allowed to speak in the conference.

A possible way to increase this number is by letting additional conferees use only 6 bits per sample, 'robbing' their seventh bit of each PCM sample for the additional conferee id-numbers, and by reassigning the pseudo random numbers generated by the DES machine. A better possibility is to dynamically assign the seven conferees that are allowed to speak and to maintain a list of the identities of these seven conferees. A conferee which is not in the list but starts talking and being bridged replaces one of the seven conferees that occupy the list at that moment. The bridge does it by sending the corresponding update of the list to all conferees. 'Robbing' a few bits, e.g., 16 consecutive i-bits for such update is enough.

5 Open Problems

- A system for secure bridging is proposed. Its major advantages are that the bridge need not be trusted, that it allows conventional PCM source coding and standard channels, and still it is a duplex system — more than one conferee can be heard at a time. During simultaneous speech a negligible degradation of the combined signal quality occurs as well as a minor leakage of information. Is there any attack on this system that we are not aware of? Known attacks on speech ciphers seem to be irrelevant here (cf. [CM86], [CR87] and [GDS91]).

- The [BLY87] solution which we adopt here enables the bridge to compare two numbers given their encryptions, while revealing their difference and requiring an extra bit. Is there a method for comparing encrypted numbers without one or both of these disadvantages?

- A technical need we have in our system is that for encryption scheme of small block size, say 8–32 bits, which is secure under known message attack of a small number, say 7, of messages. Commonly used encryption schemes that encrypt multiple messages by same key have block size of 64 bits or more. One-time-pad is secure for any block size, even of one bit. Can there be anything in between?

- A problem along the line of this paper that comes to mind is to develop a method to securely bridge signals encoded by more sophisticated source coding techniques. Examples are ADPCM, a method for which standards for 3 kHz audio over 32kbit/sec channel and for 7kHz audio over 64kbit/sec channel exist, LPC, and vector-quantized LPC.

Acknowledgements

I thank Milton Anderson, Alex Gelman, Sharad Singhal and Yacov Yacobi for many discussions on this subject, all of which were crucial for me.

References

[BLY87]   E.F. Brickell, P.J. Lee, and Y. Yacobi. *Secure audio teleconference*, volume 293 of *Lecture notes in computer science: Advances in cryptology - CRYPTO '87*, pages 418–426. Springer-Verlag, 1987.

[CCI72]   International Telegraph and Telephone Consultative Committee. *Pulse Code Modulation (PCM) of voice frequencies*. CCITT recommendation G.711, Geneva, 1972.

[CCI90]   International Telegraph and Telephone Consultative Committee. *Frame structure for a 64 to 1920 kbit/s channel in audiovisual systems*, 1990. CCITT recommendation H.221.

[CM86]   J.M. Carrol and S. Martin. The automated cryptanalysis of substitution ciphers. *Cryptologia*, 10:193–209, 1986.

[CR87]   J.M. Carrol and L.E. Robbins. The automated cryptanalysis of polyalphabetic ciphers. *Cryptologia*, 11:193–205, 1987.

[DES77]   The National Bureau of Standards. *Data Encryption Standard*, January 1977. U.S. department of Commerce, FIPS pub. 46.

[GDS91]   B. Goldburg, E. Dawson, and S. Sridharan. The automated cryptanalysis of analog speech scramblers. In *Advances in Cryptology, Eurocrypt*, pages 422–430, 1991.

[JN84]   N.S. Jayant and P. Noll. *Digital coding of waveforms: principles and applications to speech and video*. Prentice-Hall Inc., Englewood Cliffs, N.J., 1984.

[Mas92]   J.L. Massey. *private communication.*

[MDS91] Electrospace Systems Inc. *MDS-1 video conferencing bridge,* manuscript. Mail stop 4000, p.o.box 831359, Richardson, Texas 75083-1359, 1991.

[MRG85] J. Makhoul, S. Roucos, and H. Gish. Vector quantization in speech coding. *Proc. of the IEEE,* 73(11), November 1985.

[PC83]   M. V. Pitke and T. Chandrasekaran. An improved conference circuit. *Proc. of IEEE,* 71(12):1460–1461, 1983.

[PC85]   M. V. Pitke and T. Chandrasekaran. A digital conferencing technique. *Proc. of IEEE,* 73(11):1687–1688, 1985.

[PR71]   S. G. Pitroda and B. J. Rekiere. A digital conference circuit for an instant speaker algorithm. *IEEE transactions on communication technology,* com-19(6):1069–1076, December 1971.

[RS79]   L.R. Rabiner and R.W. Schafer. *Digital processing of speech signals.* Prentice-Hall, 1979.

[RSA78]   R. Rivest, A. Shamir, and L. Adelman. A method for obtaining digital signatures and public-key cryptosystems. *Communications of the ACM,* 21:120–126, 1978.

[SM87]   A. Shimizu and S. Miyaguchi. Fast data encryption algorithm (FEAL). In *Advances in Cryptology, Eurocrypt,* pages 267–267, 1987.

[SSDW88] D.G. Steer, L. Strawczynski, W. Diffie, and M. Wiener. *A secure audio teleconference system,* volume 403 of *Lecture notes in computer science: Advances in cryptology - CRYPTO '88,* pages 520–528. Springer-Verlag, 1988.

Secure Conference Key Distribution Schemes for Conspiracy Attack

Kenji Koyama

NTT Communication Science Laboratories
Seika-cho, Soraku-gun, Kyoto 619-02 Japan

Abstract

At the Eurocrypt'88 meeting, we proposed three identity-based conference key distribution schemes. At the Asiacrypt'91 meeting, Shimbo and Kawamura presented a conspiracy attacking method which worked against our schemes to disclose a user's secret information. This paper proposes an improved identity-based conference key distribution scheme to counter this attack.

1. Introduction

Since Diffie and Hellman proposed the public key distribution system (DH scheme), several advanced schemes and problems related to the DH scheme have been presented [ITT82, S85, O86, KO87, Y87, KO88, M88, LLH89, Y90, FM90, CI90]. One direction which the advanced schemes have taken is to authenticate exchanged messages with each user's identification information. This is called an identity-based system. Another direction being taken is to generate a common key among two or more users called a conference key. Several conference key distribution schemes have previously been presented [ITT82, KO88, LLH89, CI90]. These schemes can be regarded as examples of general multiparty protocols [B91, MR91], in which each of m members in a network has a private input x_i. Together, the members would like to compute, *correctly*, *privately* and *fairly*, any computable function $F(x_1, \cdots, x_m)$. In particular, multiparty protocols must be robust (secure) to guard against cheating members.

At the Eurocrypt'88 meeting, we proposed three identity-based conference key distribution schemes, constructed for star, complete graph and ring networks [KO88]. At the Asiacrypt'91 meeting, Shimbo and Kawamura presented a method for attacking our schemes for star and complete graph networks [SK91]. They pointed out that a pivot user's secret information could be revealed by a conspiracy between two legal users' conspiracy by using the Euclidean algorithm. In order to counter their attack, we propose improving the identity-based conference key distribution schemes by introducing new random variables.

2. Improved Conference Key Distribution Schemes

All identity-based conference key distribution schemes consist of a center procedure and a user procedure as follows.

[Center Procedure]
A trusted center generates the following information:
- Three large primes (p, q, r) and the partial product $N = pq$.
- Integers (e, d) satisfying the congruence:

$$ed \equiv 1 \bmod L, \quad \text{where } L = \operatorname{lcm}(p-1, q-1, r-1).$$

R.A. Rueppel (Ed.): Advances in Cryptology - EUROCRYPT '92, LNCS 658, pp. 449-453, 1993.
© Springer-Verlag Berlin Heidelberg 1993

- An integer g which is a primitive element over $\mathrm{GF}(p)$, $\mathrm{GF}(q)$ and $\mathrm{GF}(r)$.
- An integer S_i which is derived from user i's identification information I_i as follows:

$$S_i = I_i^d \bmod Nr, \quad I_i = h(ID_i),$$

where ID_i is user i's original identifier, and h is a public one-way hash function.

The above information is classified into three categories: a secret system key (p,q,d), a public system key (N,r,g,c), and a secret user key S_i for user i.

[User Procedure]
Let m be the number of users in a group sharing a common conference key. For simplicity, a user procedure for a star network is described here. One user becomes a "pivot user", who communicates with the other $(m-1)$ users belonging to the group. The procedures for interactions between the pivot user, user 1, and one of the other users, user j $(2 \leq j \leq m)$ are summarized as follows.

Step 1: User j's procedure.

Step 1.1 Choose a random number P_j and compute its reciprocal $\overline{P}_j$:

$$P_j\overline{P}_j \equiv 1 \;(\bmod\;(r-1))$$

Step 1.2 Compute the following (X_j, Y_j):

$$X_j = g^{eP_j} \bmod Nr,$$

$$Y_j = S_j g^{h(X_j \| Time)P_j} \bmod Nr.$$

Step 1.3 Send $(I_j, X_j, Y_j, Time)$ to user 1.

Step 2: User 1's procedure.

Step 2.1 Check the time and whether the following congruence holds:

$$Y_j^e / X_j^{h(X_j \| Time)} \equiv I_j \;(\bmod\;Nr).$$

If the congruence holds, user 1 is able to verify that the message is from user j.

Step 2.2 Choose random numbers R_1 $(0 < R_1 < r)$ and Q_{1j} $(0 < Q_{1j} < N, 2 \leq j \leq m)$.

Step 2.3 Compute the following (A_{1j}, B_{1j}):

$$A_{1j} = (X_j + Q_{1j}r)^{eR_1} \bmod Nr,$$

$$B_{1j} = S_1(X_j + Q_{1j}r)^{h(A_{1j} \| Time)R_1} \bmod Nr.$$

Step 2.4 Send $(I_1, A_{1j}, B_{1j}, Time)$ to user j.

Step 2.5 Compute the common key K with

$$K = g^{e^2 R_1} \bmod r.$$

Step 3 User j's procedure.

Step 3.1 Check the time and whether the following congruence holds:

$$B_{1j}^e / A_{1j}^{h(A_{1j}\|Time)} \equiv I_1 \pmod{Nr}.$$

If the congruence holds, user j is able to verify that the message is from user 1.

Step 3.2 Compute the common key K with

$$K = A_{1j}^{\overline{P}_j} \bmod r$$
$$= g^{e^?R_1} \bmod r.$$

Remarks:

1. The term $(X_j + Q_{1j}r)$ in variables A_{1j} and B_{1j} in this new version was represented by the term X_j in the previous version [KO88]. This improvement renders Shimbo and Kawamura's attack ineffective. The details will be discussed in Section 4.

2. The exponents $h(X_j\|Time)$ and $h(A_{1j}\|Time)$ in this new version were previously represented by the exponents X_j and A_{1j}, respectively, where $\|$ denotes concatenation. The usage of a time stamp with a public one-way hash function h is effective in preventing a replay attack.

3. Shimbo and Kawamura's Attacking Method

Here, a brief description is given of Shimbo and Kawamura's attacking method [SK91] for the previous version where the term X_j was used instead of the terms $(X_j + Q_{1j}r)$ in the variables A_{1j} and B_{1j}. Their attack requires a conspiracy between two users, other than user 1, belonging to the group. The attackers' aim is to disclose the pivot user's secret information S_1. Note that if this attack succeeds, the attackers can pretend to be user 1 in the subsequent key generation procedure. A concrete attacking procedure is as follows. Assume that user 2 and user 3 conspire and user 2 becomes a "pivot conspirator". First, user 3 sends (P_3, A_{13}, B_{13}) to user 2. Next, user 2 computes T:

$$T = B_{12}^{P_3 A_{13}} / B_{13}^{P_3 A_{12}} \bmod Nr$$
$$= S_1^{P_3 A_{13} - P_2 A_{12}} \bmod Nr.$$

Since user 2 knows the value of the exponent $(P_3 A_{13} - P_2 A_{12})$, denoted by c, and the relation:

$$S_1^e \equiv I_1 \bmod Nr,$$

the "Euclidean Attack" [SS3] can be applied as follows: if e and c are coprime, the integer solution (x, y) satisfying $ex + cy = 1$ can be easily obtained by the Euclidean algorithm. Then, S_1 is derived from (I_1, T, x, y) with

$$I_1^x T^y \bmod Nr = S_1^{ex + cy} \bmod Nr$$
$$= S_1 \bmod Nr.$$

Finally, user 2 sends S_1 to user 3.

The probability of a successful attack being carried out can be estimated by the probability that e and c are coprime. If P_1 and P_2 are chosen as a coprime pair and $e = 3$, then the probability of a successful attack is about 0.67.

It should be noted that this conspiracy attacking method is vulnerable because the non-pivot conspirator (user 3) discloses his secret information P_3. Once user 2 obtains the value of P_3, he can easily compute the value of S_3 with

$$S_3 = Y_3/g^{X_3 P_3} \bmod Nr.$$

Thus, the conspiracy attack is based on maintaining "trust" between the conspirators.

4. Security of New Schemes

The security of the improved schemes is based on the difficulty of deriving secret information from public keys, transmitted messages and the other user's secret keys. The secrecy of (p, q, d) is based on the difficulty of factoring a large number N, while the secrecy of $(P_i, \overline{P}_i, R_i, K, K')$ is based on the difficulty of computing a discrete logarithm over $GF(r)$. In the new version, the secrecy of S_i is based on the difficulty of computing P_i or extracting the e-th roots mod N when the factors of N are unknown.

As pointed out in [SK91], the previous version was attacked because only the fixed common random number R_1 was used to compute A_{1j} and B_{1j} $(2 \leq j \leq m)$ for each user. As a result, $S_1^Z \bmod Nr$ with a known integer Z $(\neq e)$ could easily be computed by canceling the random number R_1 through the conspiracy of two users.

An effective means of countering this attack is to introduce distinct random numbers Q_{1j} $(2 \leq j \leq m)$ into old variables (A_{1j}, B_{1j}). It is clear that new variables (A_{1j}, B_{1j}) satisfy the completeness properties which are needed to authenticate user 1's identity and to generate a common conference key. Even if the conspirators compute the variables $T = B_{12}^{P_3 A_{13}}/B_{13}^{P_2 A_{12}} \bmod Nr$ for the new version, T cannot be expressed by $S_1^Z \bmod Nr$ with a known integer Z $(\neq e)$.

Our proposed new schemes can be regarded as variants of the parallel version of the extended Fiat-Shamir scheme [FS86,FFS87,OO88,GQ88]. Although the value of $(X_j + Q_{1j}r)$ mod r is known by user j, the value of $(X_{1j} + Q_{1j}r)$ mod N is random and unknown to user j $(j \neq 1)$. Thus, the transmitted messages B_{1j} are independent of the secret S_1 and there are no additional information leaks about S_i in our schemes. Formally speaking, the parallel version of the extended Fiat-Shamir is a non-transferable (weak zero-knowledge) interactive proof system [OO88,GQ88]. Thus, we have the following lemma.

Lemma (Non Transferability) *In the new version of the identity-based conference key distribution schemes, no transferable information about a secret S_i is revealed.*

5. Conclusion

Improved interactive conference key distribution schemes were proposed to counter Shimbo and Kawamura's conspiracy attack. The introduction of new random variables was shown to be effective in preventing the disclosure of a user's secret key in the interactive protocols. The new schemes require additional time for the generation of $(m-1)$ random variables and $(m-1)$ additions modulo Nr. The transmission efficiency of the new schemes is the same as that of the previous schemes.

Acknowledgements

We wish to thank Kazuo Ohta (presently at MIT) and Michael Wang (University College, U.N.S.W.) for their valuable comments.

References

[B91] D. Beaver: "Foundations of Secure Interactive Computing", Proc. of CRYPTO'91, pp.9-1-9-7 (1991).

[CI90] T. Chikazawa and T. Inoue: "A new key sharing system for global telecommunications", Proc. of GLOBCOM'90, pp.1069-1072 (1990).

[FM90] W. Fumy and M. Munzert: "A modular approach to key distribution", Proc. of CRYPTO'90, pp.274-283 (1990).

[FS86] A. Fiat and A. Shamir: "How to prove yourself: Practical solutions to identification and signature problems", Proc. of CRYPTO'86, pp.186-194 (1986).

[FFS87] U. Feige, A. Fiat and A. Shamir: "Zero knowledge proofs of identity", Proc. of STOC, pp.210-217 (1987).

[GQ88] L.C. Guillou and J. J. Quisquarter: "A practical zero-knowledge protocol fitted to security microprocessor minimizing both transmission and memory", Proc. of Eurocrypt'88, pp.123-128 (1990).

[ITT82] I. Ingemarson, D.T. Tang and C.K. Wong: "A conference key distribution system", IEEE Trans. on Information Theory, Vol. IT-28, pp.714-720, (1982).

[KO88] K. Koyama and K. Ohta: "Security of Improved Identity-based Conference Key Distribution Systems", Proc. of Eurocrypt'88, pp.11-19 (1989).

[LLH89] C.S. Laih, J.Y. Lee and L. Harn: "A new threshold scheme and its application in designing the conference key distribution cryptosystem", Information Processing Letters, Vol.32, No.3, pp.95-99 (1989).

[M88] K.S. McCurley: "A key distribution system equivalent to factoring", J. of Cryptology, Vol.1 , No. 2, pp.95-106, (1988).

[MR91] S. Micali and P. Rogaway: "Secure computation", Proc. of CRYPTO'91, p9-8 (1991).

[O86] E. Okamoto: "Proposal for identity-based key distribution systems", Electronics Letters Vol.22 pp.1283-1284 (1986).

[OO88] K. Ohta and T. Okamoto: "A modification of the Fiat-Shamir scheme", Proc. of CRYPTO'88, pp.232-243 (1988).

[S83] G. J. Simmons: "A 'weak' privacy protocol using the RSA crypto algorithm", Cryptologia 7, 2, pp.180-182 (1983).

[S85] Z. Shmuely: "Composite Diffie-Hellman public-key generating systems are hard to break", TR. NO. 356, Computer Science Dept. Technion, IIT, Feb. (1985).

[SK91] A. Simbo and S. Kawamura "Cryptanalysis of several conference key distribution schemes", Proc. of Asiacrypt'91, pp.155-160 (1991).

[Y90] Y. Yacobi: "A key distribution "paradox"", Proc. of CRYPTO'90, pp.268-273 (1990).

A Note on
Discrete Logarithms with Special Structure

Rafi Heiman

Bellcore, 445 South Street, Morristown NJ 07962, USA

Abstract

Many cryptographic systems assume the computational difficulty of the discrete logarithm (DL) problem. In order to accelerate such practical systems, it was proposed to use logarithms of special structure, such as small Hamming weight. How difficult is the underlying restricted DL problem? By rephrasing Shanks' method we provide a close to square-root algorithm for such problems.

1 Introduction

Many cryptographic systems assume the computational difficulty of the discrete logarithm (DL) problem which is defined for a prime modulo as follows. Let p be a prime number, let Z_p be the additive group modulo p of all integers between 0 and $p-1$, let Z_p^* be the multiplicative group of order $p-1$ of the integers between 1 and $p-1$ modulo p, and let g be a generator of Z_p^*. The discrete logarithm (DL) problem (modulo p) is to compute $x \in Z_{p-1}$ for a given $y \in Z_p^*$, such that $g^x = y \pmod{p}$.

In practical cryptographic systems that are based on the difficulty of DL, logarithms of special structure are sometimes used. The idea is to choose a subset $X \subseteq Z_{p-1}$ of some special structure, which makes the use of the system (namely, the exponentiation) more efficient. Examples include the suggestion of Agnew et. al. [AMOV91] in which the special structure is small Hamming weight, and the suggestion of Yacobi [Yac90], in which the special structure is Lempel-Ziv compressibility. This however defines a restricted DL problem: it is guaranteed that the solution x of $g^x = y \pmod{p}$ satisfies $x \in X$. While for the general DL problem sub-exponential algorithms are known (see [LL90]), it is not clear whether one can compute restricted discrete logarithms faster than exhaustive search in the special structured set, X.

In this note we show a close to square-root algorithm for several such restricted DL problems, including that of [AMOV91]. This algorithm was independently noticed by Odlyzko [Odl]. We do not know of a similar result regarding the special structure suggested in [Yac90].

R. A. Rueppel (Ed.): Advances in Cryptology - EUROCRYPT '92, LNCS 658, pp. 454–457, 1993.

2 Rephrasing Shanks' Method

Shanks' method for computing DL (see [Knu73], pp. 9, 575-576) can be rephrased in the following slightly more general form. Assume, as above, that a subset $X \subseteq Z_{p-1}$ is known such that the required solution of $g^x = y \pmod{p}$ satisfies $x \in X$. Choose 'small' sets $A, B \subset Z_{p-1}$ such that $X \subseteq A + B$ where the sum of the sets is defined by $A + B = \{a + b \pmod{p-1} : a \in A, b \in B\}$. Rewrite $g^x = y \pmod{p}$ as $g^{a+b} = y \pmod{p}$ or as $g^a = yg^{-b} \pmod{p}$. Create the lists $\{g^a \pmod{p}\}_{a \in A}$ and $\{yg^{-b} \pmod{p}\}_{b \in B}$, sort (or hash) them, and find a common member, $g^a = yg^{-b} \pmod{p}$. The corresponding a and b define the required solution, $x = a + b \pmod{p-1}$.

The method has time complexity $O(s \log(s))$ (or $O(s)$ if hashing is used) and space complexity $O(s)$ where $s = \max\{|A|, |B|\}$. Clearly, $s \geq \sqrt{|X|}$ for any choice of A and B that satisfies $X \subseteq A + B$.

3 Applications

Many sets X of special structured logarithms can be decomposed as above into sets A and B of sizes not much greater than $\sqrt{|X|}$. Some examples follow. In these examples $n = \lceil \log p \rceil$ denotes the number of bits in p, t is some number between 0 and n, $[n]$ denotes the set $\{0, 1, ..., n-1\}$, and $\|x\|$ denotes the Hamming weight of a number x, that is, the number of 1's in the binary representation of X. Also, $X_{=t} = \{x \in Z_{p-1} : \|x\| = t\}$ denotes the set of logarithms with Hamming weight exactly t, and $X_{\leq t} = \{x \in Z_{p-1} : \|x\| \leq t\}$ denotes the set of logarithms with Hamming weight at most t.

Examples:

1. For $X = X_{=t}$ with $t < \frac{n}{2}$ choose $A = B = X_{=\frac{t}{2}}$.

2. For $X_{\leq t}$ with $t < \frac{n}{2}$ guess the Hamming weight of x and solve as above. Alternatively, choose $A = B = X_{\leq \frac{t}{2}}$.

3. For $X = X_{=t}$ with $t > \frac{n}{2}$ a cosmetic change in the method is convenient. choose $A = X_{=\frac{n+t}{2}}$ and $B = X_{=\frac{n-t}{2}}$, note that $X \subseteq A \setminus B$, and solve $g^{a-b} = y$, that is $g^a = yg^b \pmod{p}$.

4. An interesting structure is when some subset $I \subseteq [n]$ is known to project every $x \in X$ in the same way. Namely, there are some fixed values $c_i \in \{0, 1\}$ for $i \in I$, such that every $x = \sum_{i=0}^{n-1} x_i 2^i \in X$ satisfies $x_i = c_i$ for all $i \in I$. For this case, pick $I_A \subseteq [n]$ and $I_B \subseteq [n]$ of (roughly) the same size which are disjoint and satisfy $I_A \cup I_B = [n] \setminus I$. Then choose $A = X \cap \{x : \forall i \in I_A, x_i = 0\}$ and $B = \{x : \forall i \in I \cup I_B, x_i = 0\}$.

We would like to mention that this example is a generalization of the set-up for Pollard's λ-method for catching kangaroos [Pol78], in which X is some segment within Z_{p-1}, or an arithmetic sequence in Z_{p-1}. A segment that starts at some multiple of 2^{i_0} and is of length 2^{i_0} corresponds to this example with I consisting of all but the i_0 least significant bits, $I = \{i : i_0 \leq i < n\}$. An arithmetic sequence with jumps of size 2^{i_0}, that starts at some multiple of 2^{i_1}, and that is of length $2^{i_1-i_0}$, corresponds to this example with I consisting of both the i_0 least significant bits and the $n - i_1$ most significant bits, $I = \{i : 0 \leq i < i_0\} \cup \{i : i_1 \leq i < n\}$.

5. If X has restricted Hamming weight and in addition is restricted by some subset $I \subset [n]$ with fixed values c_i for $i \in I$ as above, the decomposition is easily obtained by 'merging' the two corresponding decompositions above.

Complexity:
Note that the complexity of example 4 is exactly the square-root of the size of the structured set, X. The complexity of small Hamming weight DL is worse than this, say $|X|^\beta$ for some $\beta > 1/2$. We compute the value β for example 1. In this example β satisfies $\binom{n}{t}^\beta = \binom{n}{t/2}$, where as above, n is the number of bits in the logarithms and t is their Hamming weight. As an example, for the concrete values $n = 512$ and $t = 50$ we have $\beta = .60$. Asymptotically, we look at a fixed ratio $\alpha = t/n$. By Stirling formula, $n! = \Theta(\frac{1}{\sqrt{n}} \cdot (n/e)^n)$ we have

$$\binom{n}{\alpha n} = \Theta\left(\frac{\sqrt{n}}{\alpha^{\alpha n}(1 - \alpha)^{(1-\alpha)n}}\right).$$

Thus,

$$\beta = \frac{\log(\binom{n}{(\alpha/2)n})}{\log(\binom{n}{\alpha n})} = \frac{H(\alpha/2)}{H(\alpha)} + o(1),$$

where H is the entropy function, $H(x) = -x \log(x) - (1 - x) \log(1 - x)$, and $o()$ is the 'little-oh' notation. For $\alpha = 1/4$, $1/10$ and $1/50$, β approaches $.67$, $.61$ and $.57$, respectively. Same asymptotical values of β are valid for examples 2, 3 and 5, where for the latter example, n should be replaced by $n - |I|$ and t should be replaced by the number of 1's that are allowed among the indices not belonging to I.

4   Concluding Remarks

This method can be clearly used over any finite group, e.g., Z_n^* for composite n, $GF(2^n)$ with the multiplication of polynomials as the group operator, or elliptic curve groups. The Agnew et. al. system [AMOV91] uses small Hamming weight of the secret exponent in $GF(2^n)$. This note suggests that not only $\binom{n}{t}$ should be large enough to prevent exhaustive search, but already $\binom{n}{t/2}$ should be 'large'. We would like to emphasize that the actual parameters chosen for the implementation of that system [Ros89], do seem to be so.

An interesting structured logarithm-set for accelerating exponentiations for which we do not know a better than exhaustive search algorithm is that of Lempel-Ziv compressibility, suggested by Yacobi [Yac90].

It would also be very interesting to obtain a faster than square-root attack for any of the structured DL problems mentioned above.

Acknowledgement

I thank Yacov Yacobi for asking the question, and I thank him as well as Stuart Haber and Arjen Lenstra for their comments.

References

[AMOV91] G.B. Agnew, R.C. Mullin, I.M. Onyszchuk, and S.A. Vanstone. An implementation for a fast public-key cryptosystem. *Journal of Cryptology*, 3(2):63–79, 1991.

[Knu73] D.E. Knuth. *The art of computer programming, vol. 3: sorting and searching.* Addison-Wesley, Reading, Mass., 1973.

[LL90]   A.K. Lenstra and H.W. Lenstra. *Algorithms in Number Theory*, volume A, chapter 12, pages 673–716. MIT press, 1990.

[Odl]   A. Odlyzko. Private communication.

[Pol78] J.M. Pollard. Monte Carlo methods for index computation (mod p). *Mathematics of computation*, 32(143):918–924, July 1978.

[Ros89] T. Rosati. A high speed data encryption processor for public key cryptography. In *IEEE Custom Integrated Circuits Conference*, pages 12.3.1–12.3.5, May 1989.

[Yac90] Y. Yacobi. Discrete-log with compressible exponents. In *Advances in Cryptology, Crypto90*, pages 639–643, 1990.

A Remark on a Non-interactive Public-Key Distribution System

Ueli M. Maurer

Inst. for Theoretical Computer Science
ETH Zurich
CH-8092 Zurich, Switzerland

Yacov Yacobi

Bellcore
445 South St.
Morristown, NJ 07962

An identity-based non-interactive public key distribution system was presented by these authors at Eurocrypt '91 [2]. It is based on the observation that for an appropriate choice of m, computing discrete logarithms in $\mathbf{Z}_m^*$ is feasible if and only if the factorization of m is known. This observation allows one to set up an exponential key distribution system in which a user's (say Alice's) public key is equal to her identity I_A. A trusted authority who knows the factorization of m is required for computing (during registration) Alice's secret key S_A as the discrete logarithm of I_A to some base g, where g is an element of $\mathbf{Z}_m^*$ with maximal order.

Because in some applications the users' identities (e.g. their email address) can be assumed known to other users, the public keys of the above scheme need not be transmitted. Moreover, the public keys are self-authenticated and require no further authentication by certificates. One can therefore build a non-interactive public-key cryptosystem having, for example, applications for electronic mail where only a one-way, non-interactive communication takes place when messages are sent. When a user Bob (knowing I_A) wants to send a message to Alice, he computes the cipher key

$$K_{AB} = I_A{}^{S_B}$$

and encrypts the message using this key and a conventional cryptosystem. When receiving the cryptogram, Alice computes the same cipher key according to $K_{AB} = I_B{}^{S_A}$ and deciphers the message using this key.

R.A. Rueppel (Ed.): Advances in Cryptology - EUROCRYPT '92, LNCS 658, pp. 458-460, 1993.

One problem with the described approach is that the group $\mathbf{Z}_m^*$ is not cyclic and hence there exists no base g to which every element has a discrete logarithm. One solution to this problem is to transmit a small offset which, when added to the identity, results in an element of $\mathbf{Z}_m^*$ having a discrete logarithm to a public base. However, this solution requires interaction: the sender of a message must receive (either from the receiver or from a trusted server) the offset before encrypting a message.

A different solution was proposed in [2] where it was observed that every square modulo m has a discrete logarithm to a base g with maximal order in $\mathbf{Z}_m^*$. It was therefore suggested that the square modulo m of a user's identity, $I_A{}^2$, rather than the identity I_A itself be used as her public key, allowing to retain the non-interactive feature. Unfortunately, this solution is insecure [1] because a square root modulo m of the squared identity $I_A{}^2$ can be obtained when given the secret key $S_A = \log_g(I_A{}^2)$ by computing $g^{S_A/2}$ (note that S_A is even). If for at least one of the prime factors p of m,

$$\log_g I_A \ (\mathrm{mod}\ p) < (p-1)/2$$

while for at least some other prime factor q of m,

$$\log_g I_A \ (\mathrm{mod}\ q) \geq (q-1)/2,$$

then the obtained square root of $I_A{}^2$ is different from I_A and $-I_A$ and thus allows one to find a non-trivial factor of m. This condition is satisfied by a fraction $1 - 2^{-r+1} \geq 1/2$ of all identities, where r is the number of distinct (odd) prime factors of m.

The purpose of this note is to point out the described weakness and to suggest a simple remedy. The trusted authority chooses once and for all a secret multiplier t randomly from $\mathbf{Z}_{\varphi(m)}^*$. Instead of issuing the discrete logarithms of squared identities as users' secret keys, the trusted authority conceals these logarithms by multiplying them with t before issuing them to users. Hence

$$S_A \equiv t \cdot \log_g(ID_A{}^2) \ (\mathrm{mod}\ \varphi(m)).$$

This modification of the secret key generation requires no change of the communication phase because the mutual cipher key computed by both users is according to the above described formulas

$$K_{AB} = I_A{}^{2S_B} = g^{vS_AS_B} = I_B{}^{2S_A}$$

where $v \equiv t^{-1} \ (\mathrm{mod}\ \varphi(m))$.

An alternative solution presented in [2] to the problem of achieving the non-interactive feature is based on the observation that, if m is the product of two odd primes, $m = pq$, then the (efficiently computable) Jacobi symbol (x/m) of an integer x is equal to 1 if and only if x has a discrete logarithm modulo m (to a base that is primitive both in $GF(p)$ and $GF(q)$). If $p \equiv 3 \pmod 8$ and $q \equiv 7 \pmod 8$ then $(2/m) = -1$ and Alice's modified identity can therefore be defined as

$$I'_A = \begin{cases} I_A & \text{if } (I_A/m) = 1 \\ 2I_A & \text{if } (I_A/m) = -1 \end{cases}$$

I'_A is obtained easily from I_A even without knowledge of the trapdoor and is guaranteed to have a discrete logarithm. However, this solution only works for the special case where m has only two prime factors. Note that for this choice computing discrete logarithms modulo p and q is feasible only when $p - 1$ and $q - 1$ are chosen to have only moderate prime factors [2].

When $m = p_1 \cdots p_r$ is the product of $r > 2$ prime factors (in which case general primes of appropriate size can be used), the following solution, which appears to be less attractive than the use of the secret multiplier t described above and is only briefly sketched, could alternatively be used. The trusted authority also publishes numbers $C_2, \ldots, C_r$, where C_i is a quadratic nonresidue modulo p_i but a quadratic residue modulo all the other prime factors of m. When Bob sends a message to Alice, he uses the secret key $K_{AB} = I_A{}^{s_B}$ and also sends $C_i^{S_B}$, $i = 2, \ldots, r$, to Alice. Alice knows for which primes her identity is a quadratic residue (non-residue), i.e., which of the C_i she must multiply her identity with to obtain an element with a discrete logarithm, which she obtained as her secret key. She can compute the cipher key K_{AB} by multiplying $I_B^{S_A}$ by an appropriate subset of the received $C_i^{S_B}$.

Acknowledgments

Some results presented in this paper were found independently by Pil Lee.

References

[1] A.J. Lenstra, private communication with Y. Yacobi.

[2] U.M. Maurer and Y. Yacobi, Non-interactive public-key cryptography, *Advances in Cryptology – EUROCRYPT '91*, D. Davies, Ed., Lecture Notes in Computer Science, vol. 547, pp. 498-507, Springer-Verlag, 1991.

Security Bounds for Parallel Versions of Identification Protocols

(Extended Abstract)

Lidong Chen, Ivan Damgård

Department of Mathematics

Aarhus University

Denmark

Abstract   The security bounds we will define and discuss in this paper is an universal security measure for parallel versions of identification protocols. From this bound we can judge which of the security measures defined in [FFS],[FeS],[OO] are satisfied. The bounds are controllable in the sense that they are connected with a security parameter. When the bound is a "sharp-threshold" security bound, it is tight enough to describe the security of the protocol precisely. Using this bound, we discuss the generalized Fiat-Shamir identification scheme ID(L,k,t,n) which is defined in [CDL]. Under the assumption that there is no polynomial time algorithm of factoring, the parallel version of the scheme is secure in the sense that even cheating verifier B can get some information from the interacting with the prover, the information he get is absolutely useless for cheating.

1. Introduction

The zero-knowledge property is a perfect measure for the security of identification scheme. But in most cases, especially for parallel versions, we can not establish this property. Therefore many researchers have been trying to define other security measures to explore its security([FFS],[FeS],[OO]). The security bound we will give is a more universal one. From this bound we can judge that which of the security measures defined before is satisfied.

Because most discussions are developed for the generalized Fiat-Shamir identification scheme with four parameters ID(L,k,t,n) defined in [CDL], we first introduce it here.

The scheme assumes the existence of a trusted center as in Fiat-Shamir scheme. The center chooses and makes public a modulus n, where $n=pq$, p and q are odd primes, and a pseudorandom function f. The center produces $v_j = f(I, j)$, $j=1,2,...,k$ for user U_I, such that for every v_j there exists s_j satisfying $s_j^L \cdot v_j \equiv 1 \bmod n$. $J=(v_1, v_2, ..., v_k)$ is public key. $S=(s_1, s_2, ..., s_k)$ is secret key. In the identification scheme, A want to prove to B that he is valid A, because he knows secret key S. The basic protocol is as follows:

1. A picks a random $r \in Z_n$, and sends

R.A. Rueppel (Ed.): Advances in Cryptology - EUROCRYPT '92, LNCS 658, pp. 461-466, 1993.

$$x \equiv r^L \bmod n$$

to B;

2. B sends a random vector $d=(e_1, e_2, \ldots, e_k)$, $0 \leq e_j \leq L\text{-}1$, $j=1,2,\ldots,k$, to A;

3. A sends to B

$$y \equiv r \cdot \prod_{j=1}^{k} s_j^{e_j} \bmod n;$$

4. B checks that if

$$x \equiv y^L \cdot \prod_{j=1}^{k} v_j^{e_j} \bmod n.$$

Repeat the basic protocol t times. This 4 parameter identification scheme is called the ID(L,k,t,n) identification scheme.

Among all the security measures for identification schemes, what we are most interested in is the concept of security level defined in [OO]. Here after, we use $\bar{A}$, $\bar{B}$ to represent honest prover and verifier, and $\tilde{A}$, $\tilde{B}$, cheating prover and verifier, separately. A protocol (A,B) is said to release no transferable information with security level ρ, if after a polynomial number of executions of $(\bar{A}, \tilde{B})$, the probability of $(\tilde{A}, \bar{B})$ success is not larger than ρ.

First, ρ is a bound for security of the parallel version of identification protocol. Generally in a protocol, we take $|n|$ as security parameter. The larger $|n|$ is, the smaller the probability of an impersonation event is.

Second, ρ is a measure of the amount of information leaking during executing($\bar{A}$, $\tilde{B}$) in parallel, which is meaningful only when we compare it with the probability of one cheating prover's success. Generally, what a coalition can do is better than one cheating prover $\tilde{A}$ can do, because for a coalition, $\tilde{A}$ may use the information which was extracted by $\tilde{B}$ from executing $(\bar{A}, \tilde{B})$ polynomial number of times. But for one cheating prover, what he can do is just guessing the verifier's query in advance. Of course, we can consider $\tilde{A}$ and $\tilde{B}$ to be one person but play different roles at different protocols. The question is how much it is better. From the security point of view, it should not be "much" better, if we require that the protocol release no transferable information. The extreme situation is another concept given by Ohta and Okamoto " the protocol (A,B) releases no transferable information with a strict sharp-threshold security level ρ " which means that what a coalition can do is "not" better than one cheating prover can. This concept may have some significance. But it is hard to find a general identification protocol ID(L,k,t,n) satisfying this condition. So we would like to give a more general concept and discuss it.

Now we give the definition of security bound.

Definition 1.1 The protocol $(\bar{A}, \bar{B})$ releases no transferable information with a security bound $\rho(|n|)$ if

1. It succeeds with probability 1;

2. For any coalition of polynomial time machines $\tilde{A}$, $\tilde{B}$,after a polynomial number of executions of ($\bar{A}$, $\bar{B}$), the probability of ($\tilde{A}$, $\bar{B}$) success is not larger than or equal to $\rho(|n|)$.

The protocol $(\bar{A}, \bar{B})$ releases no transferable information with a sharp-threshold security bound $\rho(|n|)$ if it satisfies condition 1 and 2 above as well as the following condition:

3. If the probability of $\tilde{A}$ cheating $\bar{B}$ is $\rho_0(|n|)$, then

$$\lim_{|n| \to \infty} \frac{\rho(|n|)}{\rho_0(|n|)} - 1 = 0 .$$

2. Protocols with Controlable Security Bound

For protocol ID(L,k,t,n), we can prove that even the parallel versions of it can not be proved to be zero-knowledge protocol, if some information is leaked during the execution of it, the information the cheating verifier can get is useless for successfully cheating. More precisely, the information leaked during the execution can not make the probability of success increase nonnegligibly.

Theorem 2.1 Let $p' = (L,p-1) \geq q' = (L,q-1)$, $p'q' > 1$, $k = O(\log |n|)$. Then if one of the conditions C1, C2, (C3) is satisfied, the parallel version of ID(L,k,1,n) releases no transferable information with security bound

$$\frac{1}{(p')^k} + \frac{1}{|n|^c} \qquad (\frac{1}{(q')^k} + \frac{1}{|n|^c})$$

for any positive constant c, if there is no probabilistic polynomial time algorithm of factoring n.

C1. $p' = \prod_{i=1}^{N} p_i$, where p_i is a prime, $p_i \ne p_j$ ($i \ne j$), and $N \ge 1$.

C2. $(p' , q') = 1$.

C3. $q' = \prod_{i=1}^{M} q_i$, where q_i is a prime, $q_i \ne q_j$ ($i \ne j$), and $M \ge 1$.

Definition 1.1 can also be used to describe other protocols than those of Fiat-Shamir type. For example, we can use this definition to discuss the modified Schnorr scheme[BM] which can be proved releasing no transferable information with security bound $2^t + |n|^{-c}$ for any constant c if there is no polynomial time algorithm of computing discrete logarithm. Here, t is the number of bits in the verifier's challenge.

As a security bound, $\rho(|n|)$ could be very loose so that no protocol can really be close to it. **Theorem 2.2** will point out which kind of the bounds are tight enough to describe the security of the protocols.

Theorem 2.2 The protocol ID(L,k,1,n) releases no transferable information with sharp-threshold security bound

$$\frac{1}{(p')^k} + \frac{1}{|n|^c} \qquad (\frac{1}{(q')^k} + \frac{1}{|n|^c})$$

for any constant c satisfying

$$\lim_{|n| \to \infty} \frac{(p')^k}{|n|^c} = 0 \qquad (\lim_{|n| \to \infty} \frac{(q')^k}{|n|^c} = 0) \qquad (1) ,$$

if there is no polynomial time algorithm of factoring n.

3. Connections to Other Definitions of Security

In this section , we would like to explore the relationship between security bounds and other security measures defined in [FFS], [FeS] and [OO]. From the following three theorems, it is reasonable to claim that the security bounds we have given are more universal compared with other security measures.

Theorem 3.1 If for any positive constant c, (A,B) releases no transferable information with security bound

$$\rho(|n|) = \frac{1}{|n|^k},$$

then (A,B) is secure according to the definition given by Feige, Fiat and Shamir[FFS].

Theorem 3.2 If protocol (A,B) is a proof of knowledge, and satisfy the condition given in **Theorem 3.1**, then (A,B) is witness hiding according to the definition given by Feige and Shamir[FeS].

Theorem 3.3 If ρ is a constant, such that for any positive constant d, protocol (A,B) releases no transferable information with security bound

$$\rho(|n|) = \rho + \rho \cdot \frac{1}{|n|^d},$$

then it releases no transferable information with strict security level ρ according to the definition given by Ohta and Okamoto[OO].

4. Discussion

For convenience, we give a representation of the probability we will discuss here. Let

$$\rho_0 = \text{Prob} \{ \ \bar{A} \ \text{cheats} \ \bar{B} \ \}.$$

And let

$$\rho_{\bar{A}\bar{B}} = \text{Prob} \{ \ (\ \tilde{A} \ , \ \bar{B} \ ) \text{ succeeds} \mid \text{after executing} (\ \bar{A} \ , \ \tilde{B} \ )$$
$$\text{polynomial number of times} \ \}$$

Obviously, if (A,B) is a zero knowledge protocol, $\rho_0 = \rho_{\bar{A}\bar{B}}$ for any $\bar{A}$ and $\tilde{B}$.

But in most cases $\rho_0 < \rho_{\bar{A}\bar{B}}$. The security bound $\rho(|n|)$ is a upper bound for $\rho_{\bar{A}\bar{B}}$, i.e. which describes the amount of usable information leaking during excuting (A,B).

For protocol ID(L,k,1,n), under the assumption that there is no polynomial time algorithm of factoring n, for any $\bar{A}, \bar{B}$, $\rho_{\bar{A}\bar{B}} - \rho_0 < \frac{1}{|n|^c}$ for any constant c, i.e. the increasing of probability gained by borrowing the information from executing ($\bar{A}, \tilde{B}$) is negligible.

Acknowledgements We would like to thank Peter Landrock for reading the manuscript and valuable comments.

References

[BM]   E.F.Brickell, K.S.McCurley "An Interactive Identification Scheme Based on Discrete Logarithms and Factoring" The oringinal one is from Proc. of EUROCRYPT'90 pp 63-71. Here we refer to a new version of it.

[CDL]   L.Chen, I.Damgård, P.Landrock "Extension and Analysis of Fiat-Shamir Identification Scheme" to appear.

[FFS]   U.Feige, A.Fiat and A.Shamir "Zero-Knowledge Proofs of Identity" Journal Cryptology 1(2), 1988

[FeS]   U.feige, A. Shamir " Witness Indistinguishable and Witness Hiding Protocols" Proc. of the 22nd ACM Symposium on Theory of Computing pp416-424

[OO]   K.Ohta and T.Okamoto "A Modification of the Fiat Shamir Scheme" Proc. of CRYPTO'88 pp 232-243.

INFORMATION-THEORETIC BOUNDS
FOR AUTHENTICATION FRAUDS

Andrea Sgarro

Dipartimento di Scienze Matematiche
Università di Trieste, 34100 TRIESTE (Italy)

Short version

Abstract. Several properties of authentication codes depend on a mathematical structure, called below a fraud scheme, which is much simpler than the one originally given. Relying on this fact, we present a powerful lower bound, which is a sort of mould to painlessly derive a whole range of information-theoretic bounds to fraud probabilities in authentication coding.

1. **Introduction**. A Shannon-theoretic frame for authentication theory has been put forward by G. Simmons ([1]). The main attacks to an authentication code are impersonation and substitution, but several variants of these can be considered. In the literature, information-theoretic bounds to fraud probabilities are provided which require a lot of boring computations to be repeated each time. We will show that one can dispose of all this drudgery.

An *authentication code* is a finite random triple XYZ (X: *source state* or *source message*, Y: *authenticated message* or *codeword*, Z: *encoding rule* or *key*). Under each key (encoding rule), decoding is assumed to be deterministic; instead probabilistic encoding (*splitting*) is allowed. Key and source state are independent random variables. In the *authentication matrix* χ of the code one has $\chi(z,y)=1$ iff key z authenticates codeword y, that is iff there exists a source state x which is encoded to y under key z. (Capital letters denote random variables, the corresponding small letters denote their values.)

R.A. Rueppel (Ed.): Advances in Cryptology - EUROCRYPT '92, LNCS 658, pp. 467-471, 1993.

A deterministic code is completely described by giving the encoding matrix, the key probability distribution and the source state probability distribution. In the case of splitting, there are entries in the encoding matrix which contain several codewords, and so one must further specify the random "splitting strategy" for each such entry. Since a zero-error decoding scheme is prescribed, each codeword can appear at most once in each row of the encoding matrix. So, the number of ones in each row of χ is at least $|X|$ (exactly $|X|$ for codes without splitting, $s|X|$ for s-balanced codes, i.e. codes such that each entry of χ contains s codewords).

In the case of an *impersonation attack* the mischievous opposer chooses a codeword y hoping it to be authenticated by the current key Z. The probability of fraud for the opposer's optimal strategy is:

(1)
$$P_I = \max_y \ \mathrm{Prob}\{\chi(Z,y)=1\}$$

In the case of *substitution* the opposer grabs the legal codeword c and replaces it by a fake codeword y hoping that y be decoded to a source state different from $g(Z,c)$ (g denotes deterministic decoding and so $g(z,c)$ represents the unique source state which is encoded to codeword c under key z). The relevant fraud probabilities are:

$$P_S(c) = \max_y \ \mathrm{Prob}\{\chi(Z,y)=1, \ g(Z,y)\neq g(Z,c) \ |Y=c\}$$

$$P_S = \Sigma_c \ \mathrm{Prob}\{Y=c\} \ P_S(c)$$

(The vertical bar denotes probabilistic conditioning.) In the case of codes without splitting, $P_S(c)$ can be written more simply as:

$$P_S(c) = \max_{y\neq c} \ \mathrm{Prob}\{\chi(Z,y)=1|Y=c\}$$

The most popular lower bounds to P_I are *Simmons bound* which involves the mutual information $I(Z;Y)$ (in terms of Shannon entropies $I(Z;Y)$, which is a measure of stochastic dependence, is defined as $H(Y)+H(Z)-H(YZ)$):

$$P_I \geq 2^{-I(Y;Z)}$$

and the combinatorial bound for deterministic codes:

$$P_I \geq \frac{|X|}{|Y|}$$

2. **Fraud schemes**. We find it convenient to define a more abstract notion than authentication codes, namely *fraud schemes* (T,ρ). Let two finite (non-empty) sets be given, the set of *tokens* and the set of *fakes* (in the applications, tokens will stand

for keys, or conditional keys, fakes for fraudulently inserted codewords). Let T be a random token and let ρ be a binary matrix row-indexed in the set of tokens and column-indexed in the set of fakes; when $\rho(t,f)=1$ we say that token t is *deceived* by fake f. The only requirement we make on (T,ρ) is that for each token fraud is possible (in each row of ρ there is at least a one). We find it convenient to allow for "impossible" (zero-probability) tokens and "unusable" fakes (whose column is all-zero). A *row-balanced* scheme is one where the number of ones is the same for each positive-probability row of ρ, i.e. each "possible" token is deceived by the same number of fakes, k, say.

Further we define the *fraud probability* as:

(2) $P = \max_f \ \mathrm{Prob}\{\rho(T,f)=1\}$

Notice that each authentication code XYZ yields a fraud scheme by simply setting T=Z, $\rho=\chi$. If XYZ is s-balanced, (T,ρ) is row-balanced with k=s|X|. Of course different codes can yield the very same scheme, while, if one is given the scheme (T,ρ) to start with, it may well happen that this scheme *cannot* be embedded into any authentication code XYZ with Z=T, $\rho=\chi$, |X|≥2. So, in spite of the formal coincidence of (1) and (2), the notion of a fraude scheme is *strictly more general* than that of an authentication code. What one can always do, however, is to complete (T,ρ) to a random couple TF such that $\rho(t,f)=1$ iff $\mathrm{Prob}\{F=f|T=t\}\neq0$. To do this, convert ρ into a stochastic matrix by replacing the ones in each row of ρ by positive numbers which sum to one. For any such completion TF one has

(3) $P \geq 2^{-I(T;F)}$

To prove this inequality one can literally repeat any of the standard proofs of Simmons bound to impersonation, e.g. those in [2,3]. As the "Simmons bound" (3) holds for any admissible completion TF, one soon obtains a result formally identical with the strengthened Simmons bound for impersonation given in [2]:

(4) $P \geq 2^{-\inf I(T;F)}$

the infimum being taken w.r. to all random couples TF yielding the given fraud scheme (T,ρ). The information theorist might appreciate the fact that inf I(T;F) as in the exponent of (4) is the *rate-distortion function* for source (*sic*) Z, distortion matrix equal to the binary complement of ρ, and distortion level equal to zero, $R(Z,C(\rho),0)$. Conditions for equality in this bound will be discussed in the full version of this paper. We shall be contented here with the following fact:

Equality criterium. For row-balanced authentication schemes, bound (4) holds with equality iff Prob$\{\rho(Z,f)=1\}$ does not depend on f, where f spans the set of usable fakes. In this case $P = \frac{k}{|F|}$ (|F| is the number of usable fakes).

(The fact that Prob$\{\rho(Z,f)=1\}$ does not depend on f means that any usable fake f, and consequently any *fraud strategy*, is equally good, or equally bad, from the point of view of the opposer.)

Inequality (3) gives actually a whole class of bounds, out of which (4) is the best. Even weak bounds can be, however, meaningful. For example, in the case of row-balanced schemes, one can take (3) with the stochastic matrix obtained from ρ by filling the free entries with the constant $\frac{1}{k}$. A simple computation shows that in this case: $I(T;F) = H(F) - \log k \leq \log |F| - \log k$. So, *for row-balanced fraud schemes* one gets the following counterpart of the combinatorial bound for deterministic codes:

(5)
$$P \geq \frac{k}{|F|}$$

Comparing with the equality criterium, we see that, rather surprisingly, this naive combinatorial bound is a *tight* bound for *good* row-balanced schemes (for schemes which attain the severe bound (4) with equality).

Up to now, the reader can have the unpleasant impression that we are just changing names: instead of *key, codeword, impersonation*, we say *token, fake, fraud*. In a way, this is it: old material used to a higher degree of abstraction. It is precisely this higher degree of abstraction, however, which will allow us to appreciate the real bearing of the result: in a way, *impersonation is a much more general notion than usually realized*. The next section 3 is meant to convince the reader of this fact.

3. Applications.

The most obvious application is impersonation; for a *meaningful* application, however, we go directly to substitution. Take T with the distribution of Z|Y=c. Construct $\rho=\chi_c$ from χ by specifying which codewords do cause a fraud when used by the opposer. To this end set to zero the entries in χ for which $g(z,y)=g(z,c)$, including the column corresponding to c. Then

$$P_S(c) = \max_y \text{Prob}\{\chi(Z,y)=1, g(Z,y) \neq g(Z,c) \mid Y=c\} =$$
$$= \max_y \text{Prob}\{\chi_c(Z,y)=1 \mid Y=c\} = \max_y \text{Prob}\{\rho(T,y)=1\}$$

and so one obtains the powerful bound which was first given in [3]:

$$P_S(c) \geq 2^{-R(Z,\mathcal{L}(\chi_c),0 \mid Y=c)}$$

For s-balanced codes XYZ, χ_c is clearly row-balanced, and so (5) becomes:

$$P_S(c) \geq \frac{s(|X|-1)}{|Y|-1}$$

Assume XYZ is deterministic and set $\text{Prob}\{Y^*=y \mid Z=z, Y=c\} = \text{Prob}\{Y=y \mid Z=z, Y \neq c\}$; as the authentication matrix derived from the conditional distribution of ZY* given Y=c is precisely χ_c, (3) and Jensen's inequality give:

$$P_S(c) \geq 2^{-I(Y^*;Z \mid Y=c)}, \quad P_S \geq 2^{-I(Y^*;Z \mid Y)}$$

the latter being "Simmons bound for substitution".

We could multiply our examples, by taking into account the various attacks introduced in literature. By now, however, our point should be clear. In all cases *what one has to do is to resort to the fraud scheme* (T,ρ), *where T gives the distribution of the key conditional to the information possessed by the opposer, and* ρ *specifies, for each value* t *of T, which are the codewords that can be used successfully as "fakes", that is which are the codewords that do cause a fraud against the system.* Once this is done, the abstract bounds of Section 2 are painlessly converted into bounds for the attack under consideration.

References.

[1] G. Simmons, "A survey of information authentication", Proceedings of the IEEE, may 1988, 603-620

[2] R. Johannesson, A. Sgarro, "Strengthening Simmons' bound on impersonation", IEEE Transactions on Information Theory, vol.IT-37, n.4 (1991) 1182-1185

[3] A. Sgarro, "Lower bounds for authentication codes with splitting", in "Advances in Cryptology - Eurocrypt '90", ed. by I. B. Damgård, Springer Verlag, Lecture Notes in Computer Science 473 (1991) 283-293

A GENERALIZED CORRELATION ATTACK WITH A PROBABILISTIC CONSTRAINED EDIT DISTANCE

Jovan Dj. Golić and Slobodan V. Petrović

Institute of Applied Mathematics and Electronics, Belgrade
School of Electrical Engineering, University of Belgrade
Bulevar Revolucije 73, 11001 Beograd, Yugoslavia

Abstract: For a noisy clock-controlled shift register statistically optimal probabilistic constrained edit distance a recursive algorithm for its efficient computation are derived. corresponding generalized correlation attack is proposed.

1. Introduction

Consider the initial state reconstruction of a binary r clock-controlled shift register depicted in Fig. 1, see [1].

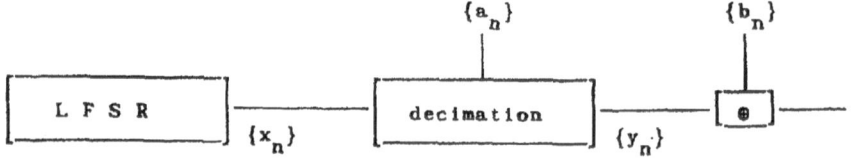

$\{a_n\}$ $\{b_n\}$

| L F S R | | decimation | | ⊕ |

$\{x_n\}$ $\{y_n\}$

Fig. 1. A noisy clock-controlled shift register.

Linear feedback shift register (LFSR) produces a binary seq $\{x_n\}$ as usual. In a statistical model, the decimation sequence is regarded as a realization of a sequence of indepe identically distributed (i.i.d.) integer variables $\{A_n\}$ with a probability distribution $Pr(A_n=k)=P_k$, $0 \leq k \leq E$. A binary noise seq

This research was supported by the Science Fund of Se grant #0403, through Institute of Mathematics, Serbian Acade Arts and Sciences.

$\{b_n\}$ is a realization of a sequence of i.i.d. binary variables $\{B_n\}$ with $Pr(B_n=1)=p<0.5$. The output sequence $\{z_n\}$ is defined as the modulo 2 sum

$$z_n = x_{t_n} \oplus b_n, \qquad t_n = n + \sum_{j=1}^{n} a_j, \qquad n \geq 1. \tag{1}$$

The objective is to reconstruct the LFSR initial state given the output segment $\{z_n\}_{n=1}^{N}$ along with the feedback polynomial and the noise probability. A solution to this problem is proposed in [1]. Essentially, it is a statistical procedure based on the minimum constrained Levenshtein distance (CLD) decision rule. However, as noted in [1], the problem remains to show how close the minimum CLD decision rule is to the maximum posterior probability one, which is optimal given the statistical model. For a slightly modified statistical model in which the LFSR sequence is assumed to be a realization of a sequence of balanced i.i.d. binary variables, we here derive the maximum posterior probability decision rule. Namely, we introduce the statistically optimal probabilistic constrained edit distance (PCED) and develop an efficient recursive algorithm for its computation.

2. Probabilistic Constrained Edit Distance

Let $U=\{u_i\}_{i=1}^{M}$ and $V=\{v_i\}_{i=1}^{N}$ denote two finite length sequences over a finite alphabet A. Let an edit transformation of U into V be defined as a series of edit operations of deletions and substitutions. It can be uniquely represented by a two-dimensional edit sequence $(U,V')=\{(u_i,v_i')\}_{i=1}^{M}$, where $V'=\{v_i'\}_{i=1}^{M}$ is a sequence over $\tilde{A} = A \cup \phi$, $\phi \notin A$, such that by deleting all the ϕ symbols from V' one obtains V. Namely, if $v_i'=\phi$, then u_i is deleted and if $v_i'\neq\phi$, then v_i' is substituted for u_i, $1 \leq i \leq M$. Let G_{UV} denote the set of all possible edit sequences that transform U into V subject to the constraint that there are no more than E consecutive ϕ symbols in V' and that the last symbol of V' is different from ϕ. It follows that $N \leq M \leq (E+1)N$.

In accordance with the noisy clock-controlled shift register statistical model in which the LFSR sequence is assumed to be a realization of a sequence of balanced i.i.d. binary variables, we now associate a probability distribution with the set of permitted edit sequences, that is, the union of G_{UV} over all U and V. To this end, for an arbitrary permitted edit sequence (U,V') define the decimation sequence $D(U,V')=\{d_i\}_{i=1}^{N}$ so that d_i is the length of the series of deletions between the $(i-1)$-th and the i-th substitution, $0 \leq d_i \leq E$, $1 \leq i \leq N$. An edit sequence (U,V') can then be regarded as a sequence of N blocks each composed of a substitution preceded by a series of deletions, allowing a series of zero deletions. Assume that the blocks are produced independently and that within each block the substitution and the series of deletions are also independent. Further, let $P_k A^{-k}$, A being the cardinality of A, denote the probability of a series of deletions of length k, $0 \leq k \leq E$, and let $p(u,v)$ denote the probability of substituting v for u, $u,v \in A$. Thus, P_k is the probability that a series of deletions has length k, $0 \leq k \leq E$. The probability of an edit sequence (U,V') is then given by

$$Pr(U,V') = \prod_{i=1}^{N} A^{-d_i} P_{d_i} p(u_{i+\sum_{j=1}^{i} d_j}, v_i) = A^{N-M} \prod_{i=1}^{N} P_{d_i} p(u_{i+\sum_{j=1}^{i} d_j}, v_i). \quad (2)$$

The probability that a sequence U can be transformed into a sequence V is then given by

$$Pr(U,V) = \sum_{(U,V') \in G_{UV}} Pr(U,V'). \quad (3)$$

The problem is how to compute $Pr(U,V)$ efficiently. To this end we define the partial probability $Pr(U_{e+s}, V_s)$ that a prefix $U_{e+s}=\{u_i\}_{i=1}^{e+s}$ of U can be transformed into a prefix $V_s=\{v_i\}_{i=1}^{s}$ of V, under the same constraints. Using an abbreviated notation $G_{es}=G_{U_{e+s} V_s}$ and $P(e,s)=Pr(U_{e+s}, V_s)$ we thus have

$$P(e,s) = \sum_{(U_{e+s}, V_s') \in G_{es}} \prod_{i=1}^{s} A^{-d_i} P_{d_i} p(u_{i+\sum_{j=1}^{i} d_j}, v_i) \quad (4)$$

and

$$Pr(U,V) = P(M-N,N). \tag{5}$$

The set of all the permitted values for (e,s) is clearly given by $1 \leq s \leq N$ and $0 \leq e \leq \min(M-N,sE)$.

Interestingly enough, using a similar technique as in [1] and bearing in mind the recursion [2] for unconstrained edit probability, it is not difficult to prove the following result which yields a recursion for $P(e,s)$.

Theorem 1: The partial probability $P(e,s)$ satisfies the recursion

$$P(e,s) = \sum_{e_1 \in \Psi_{es}} P(e-e_1,s-1)A^{-e_1} P_{e_1} p(u_{e+s},v_s) \tag{6}$$

for $1 \leq s \leq N$ and $0 \leq e \leq \min(M-N,sE)$, where Ψ_{es} is the set of all e_1 such that

$$\max(0, e-\min(M-N,(s-1)E)) \leq e_1 \leq \min(e, E), \tag{7}$$

with the initial value $P(0,0)=1$ for $s=0$ and $e=0$.

Now, Theorem 1 together with (5) enables the efficient computation of the probability $Pr(U,V)$, given U and V, which has the same space and time complexities as the recursive procedure [1] for the efficient computation of the CLD. Since the value of $Pr(U,V)$ is very close to zero, for practical reasons we define the probabilistic constrained edit distance (PCED) as

$$D(U,V) = -\log Pr(U,V). \tag{8}$$

Note that in the optimal statistical decision procedure about U given V, $D(U,V)$ has to be minimized rather than maximized.

We now discuss the relation between the CLD and the PCED. Instead of the overall probability that a sequence U can be transformed into a sequence V, consider the probability of the most likely edit transformation of U into V. Then the negative logarithm of this probability reduces to a CLD with elementary edit distances defined as the negative logarithms of the corresponding probabilities. Unlike the definition in [1], this CLD is modified in a sense that the elementary edit distance is associated with a series of deletions rather than a single deletion.

3. Generalized Correlation Attack

Given the noisy clock-controlled shift register statistical model, Fig. 1, with the additional assumption that the LFSR sequence is a realization of a sequence of balanced i.i.d. binary variables, the probability that a LFSR sequence $U=\{x_n\}_{n=1}^{M}$ can be transformed into an output sequence $V=\{z_n\}_{n=1}^{N}$ is given by (3) and (2) with $A=2$, $P_k=Pr(A_n=k)$, $0 \leq k \leq E$, and $p(u,v)=p$ if $u \neq v$ and $p(u,v)=1-p$ if $u=v$. Accordingly, applying Theorem 1 one can compute the probabilistic constrained edit distance, PCED, which is statistically optimal for a given model. Then, using the PCED instead of the constrained Levenshtein distance, CLD, the generalized correlation attack goes along the same lines as in [1]. Unlike the CLD procedure, the PCED procedure is statistically optimal and enables to take into account the probability distribution of the decimated sequence.

Comparative experimental analysis is underway. Preliminary results show a slight but clear advantage of the PCED procedure over the CLD one.

REFERENCES

[1] Jovan Dj. Golić, Miodrag J. Mihaljević, "A generalized correlation attack on a class of stream ciphers based on the Levenshtein distance," *J. Cryptology* vol. 3, pp. 201-212, 1991.

[2] P.A.V. Hall, G.R. Dowling, "Approximate string matching," *Computing surveys*. vol. 12, pp. 381-402, Dec. 1980.

Systolic-Arrays for Modular Exponentiation Using Montgomery Method

— Extended Abstract —

Keiichi Iwamura[†],   Tsutomu Matsumoto[††],   and   Hideki Imai[††]

Abstract — This paper proposes two ideas for modular exponentiation using Montgomery method. (1) A novel algorithm for modular exponentiation without operation of subtracting N for every Montgomery's modular multiplication (MMM). (2) Two types of systolic-array for MMM which can realize more efficient and flexible chip implementation than the array in [1].

1 Introduction

We have proposed a systolic-array for modular multiplication [1], which we will refer to as Array-A in the following. Array-A is practical for modular exponentiation and is very suitable for chip implementation. However Array-A does not achieve the ultimate efficiency in the wide range of processing speed, when the efficiency is defined as (processing speed)/(circuit scale). In this paper, we propose a novel algorithm for modular exponentiation using simple repetition of Montgomery's modular multiplication (MMM) [2] in Section 2, and in Section 3 we propose the structures and actions of more efficient systolic-arrays (Array-B, Array-C) suitable for MMM in the wide range of processing speed than Array-A. Combining these proposals, we have efficient and fast hardware algorithms for modular exponentiation. In Section 4, we show that compared with the use of Array-A the use of Array-B and that of Array-C respectively achieves more efficient high-speed processing and more flexible implementation for modular exponentiation.

2 A Novel Algorithm for Modular Exponentiation

Definition 1 For integers N, R, A, and B, let $\mathrm{Mont}_{N,R}(A, B)$ denote the rational number

$$\mathrm{Mont}_{N,R}(A, B) = \frac{A \cdot B + ((A \cdot B \cdot (-N^{-1} \bmod R)) \bmod R) \cdot N}{R}.$$

We have the following fact due to Montgomery [2].

Proposition 2 For relatively prime integers N and R, and for integers A and B, $\mathrm{Mont}_{N,R}(A, B)$ is an integer such that

$$\mathrm{Mont}_{N,R}(A, B) \equiv ABR^{-1} \bmod N \quad (\bmod N).$$

We call $\mathrm{Mont}_{N,R}(A, B)$ the Montgomery's Modular Multiplication (MMM).

The condition that the maximum number of bits in A, in B and in $\mathrm{Mont}_{N,R}(A, B)$ are the same is described as follows:

† Canon Research Center, 21 Laboratory,

5-1 Wakamiya, Morinosato, Atsugi-shi, Kanagawa 243-01, Japan,

Fax +81-462-48-0306, Tel +81-462-47-2111

†† Yokohama National University, Division of Electrical & Computer Engineering,

156 Tokiwadai, Hodogaya, Yokohama, 240 Japan,

Fax +81-45-338-1157, Tel +81-45-335-1451, Internet tsutomu@mlab.dnj.ynu.ac.jp

R.A. Rueppel (Ed.): Advances in Cryptology - EUROCRYPT '92, LNCS 658, pp. 477-481, 1993.
© Springer-Verlag Berlin Heidelberg 1993

Theorem 3 Let N, R, A, B, n, r, and l be integers and assume

$$0 < N < 2^n, \quad 0 < R \le 2^r, \quad \text{and} \quad \gcd(N, R) = 1.$$

The necessary and sufficient condition for that

$$0 \le A < 2^l, \quad 0 \le B < 2^l, \quad \text{and} \quad 0 \le \text{Mont}_{N,R}(A, B) < 2^l$$

is

$$n + 1 \le l \le r - 1. \quad (1)$$

(Proof) Let $0 \le A < 2^l$ and $0 \le B < 2^l$. Since it is straightforward that

$$0 \le \text{Mont}_{N,R}(A, B) < \max\{2^{2l-r+1}, 2^{n+1}\}$$

the necessary and sufficient condition for

$$0 \le \text{Mont}_{N,R}(A, B) < 2^l$$

is that

$$\max\{2^{2l-r+1}, 2^{n+1}\} \le 2^l$$

which is equivalent to

$$2^{2l-r+1} \le 2^l \quad \text{and} \quad 2^{n+1} \le 2^l$$

which is equivalent to (1).

Thus we let $n + 1 \le l \le r - 1$ so that the output of MMM can be directly used as the next input to MMM. The smallest possible value of l is $l = n + 1$ and of r is $r = n + 2$. Using this condition and introducing $R_R = R^2 \bmod N$, we propose the following algorithm which evaluates modular exponentiation $C = M^e \bmod N$ with repeated MMMs and a final modular reduction $\bmod\, N$.

Algorithm 4

```
(Input :     M, e = (e_k, ···, e_1)_2, N, R, R_R = R² mod N)
(Output :    C = M^e mod N)
             M_R = Mont_{N,R}(M, R_R)
             C_R = Mont_{N,R}(1, R_R)
             FOR i = k TO 1
                 IF e_i = 1 THEN C_R = Mont_{N,R}(C_R, M_R)
                 IF i > 1  THEN C_R = Mont_{N,R}(C_R, C_R)
             NEXT
             C_R = Mont_{N,R}(1, C_R)
             C = C_R mod N
```

Algorithm 4 has the following useful features:

(1) Algorithm 4 employs ordinary modular multiplication only at the precomputation of R_R which can be done independently with M. All the rest of modular multiplications are MMM. This feature helps to obtain simple structure.

(2) Except for the last step, the maximum length of an output of each MMM used in Algorithm 4 is not greater than that of each of the inputs to the MMM. Thus, the output can be directly fed into the next MMM without compensation like Fig.1, which is to obtain $ABR^{-1} \bmod N$ from $\text{Mont}_{N,R}(A, B)$ and often used in conventional algorithms. This feature greatly simplifies the control structure.

(2') For implementation by systolic-array, feature (2) can be effectively used to avoid idle processing elements. If such compensations are required, any bit of the input to a systolic-array for MMM cannot been fed before getting all the bits of the previous value of the output from the MMM. However such a loss does not emerge in Algorithm 4 since the previous output can be directly input to the next MMM without delay.

In contrast, any of the previously known methods for modular exponentiation using MMM (see [3][4][5]) does not simultaneously satisfy the above features.

3 Systolic-Arrays for Montgomery's Modular Multiplication

In the following we propose systolic arrays for computing $\text{Mont}_{N,R}(A, B)$ under the condition that N is odd, $n = \lfloor \log_2 N \rfloor + 1$, $n + 1 \leq l \leq r - 1$, $0 \leq A < 2^l$, $0 \leq B < 2^l$, and $R = 2^r$.

We express A in radix $Y = 2^v$ and B, N and $T_R = \text{Mont}_{N,R}(A, B)$ in radix $X = 2^d$ as follows,

$$
\begin{aligned}
A &= A_{k-1} \cdot Y^{k-1} &+ A_{k-2} \cdot Y^{k-2} &+ \cdots &+ A_1 \cdot Y &+ A_0 \\
B &= B_{m-1} \cdot X^{m-1} &+ B_{m-2} \cdot X^{m-2} &+ \cdots &+ B_1 \cdot X &+ B_0 \\
N &= N_{m-1} \cdot X^{m-1} &+ N_{m-2} \cdot X^{m-2} &+ \cdots &+ N_1 \cdot X &+ N_0 \\
T_R &= T_{m-1} \cdot X^{m-1} &+ T_{m-2} \cdot X^{m-2} &+ \cdots &+ T_1 \cdot X &+ T_0
\end{aligned}
$$

where $A_i \in \{0,1\}^v$ ($i = 0, \cdots, k-1$), and B_j, N_j and $T_j \in \{0,1\}^d$ ($j = 0, \cdots, m-1$), and $v \leq d$.

$T_R = \text{Mont}_{N,R}(A, B)$ can be calculated by the consecutive execution of the following operation from $i = 0$ to $i = k$.

$$
\begin{aligned}
T(i) &= (T(i-1) + A_i \cdot B \cdot Y + M_{i-1} \cdot N)/Y \quad (2) \\
\text{where} \quad M_{i-1} &= (T(i-1) \bmod Y) \cdot N_0' \bmod Y, \\
T(-1) &= 0, \ N_0' = N' \bmod Y
\end{aligned}
$$

3.1 Array-B

Expression (2) can be realized by the processing element (PE) described in Fig.2, when B and N are synchronously fed into the port $[B_{in}, N_{in}]$ of the PE as $[B_0, N_0]$, $[B_1, N_1]$, $[B_2, N_2]$, ..., $[B_{m-1}, N_{m-1}]$. The multiplication of $Y = 2^v$ is realized by the v-bit shift. For example, $A_i \cdot B_j \cdot Y$ is obtained by shifting the value $A_i \cdot B_j$ v bits into the direction of more significant bits.

If $v = 1$ the PE in Fig.2 can be easily realized as follows. Each of the multipliers M1 and M2 is constructed with only d AND gates. Each of the registers R1 and R2 is a 1-bit register respectively holding A_i and M_{i-1}. Since $N_0' = 1$, register R3 and multiplier M3 can be omitted, so that M_{i-1} can be the least significant bit of the output from adder A1. R4 and R5 are d-bit registers which respectively transfers B_j and N_j to the next PE with a delay of one clock cycle. A1 is a 5-input adder whose output is received by register R6 of $d + 3$ bits. The most and the second significant bits of R6 are fed back to adder A1 as carry bits. The least significant bit of R6 is fed into terminal $L2_{in}$ of the next PE but one to this PE. The rest of the bits of R6 are transferred to the terminal T_{in} of the next PE.

To obtain T_R we construct Array-B shown in Fig.4 which consists of $k + 1$ tandem PEs described in Fig.2 for the repetition of Expression (2) and of the last and the last second PEs described in Fig.3. The $k + 1$ PEs described as Fig.2 are connected in tandem by respectively tying terminals B_{out}, $L2_{out}$, T_{out}, $L1_{out}$, M_{out}, N_{out} to the corresponding terminals B_{in}, $L2_{in}$, T_{in}, $L1_{in}$, M_{in}, N_{in} of the next PE. And for $i = 0, 1, \ldots, k - 1$, register R1 of the i-th PE in Fig.2 is preset by value A_i. Values of $L1_{in}$, T_{in}, $L2_{in}$ and M_{in} of the first PE are set to 0.

3.2 Array-C

T_R can be also evaluated by systolic-array Array-C described in Fig.6. For the sake of simplicity we assume in this section that $k = m$, i.e., $v = d$.

Array-C consists of m copies of the same PE shown in Fig.5. Each PE is connected in tandem by respectively tying terminals A_{out}, B_{out}, T_{out}, M_{out}, N_{out} to the corresponding terminals A_{in}, B_{in}, T_{in}, M_{in}, N_{in} of the next PE. Values for T_{in} and M_{in} in the first PE are set to 0.

The multiplication by Y is realized by timing shift. In Array-C A_i is input one-clock-cycle prior to B_j for $i, j = 0, 1, \ldots, m-1$. Therefore we can set register R1 in PE# 0 the value of A_0 before computing $A_0 \cdot B_j$. At each PE A_i delays one clock cycle while B_j delays two clock cycle. Thus if A_i can be set in PE# i before computing $A_i \cdot B_j$ ($j = 0, 1, \ldots, m-1$) then A_{i+1} can be set in PE# $i+1$ before computing $A_{i+1} \cdot B_{j+1}$ ($j = 0, 1, \ldots, m-1$). Therefore values of A_i can be input serially like as B_j and N_j. That is, we do not have to preset the values of A_i.

If $v = d = 1$, the PE in Fig.5 can be readily realized as follows. Each of multipliers M1 and M2 is constructed with only one AND gate, and multiplier M3 can be omitted. Each of registers R1 $\sim$ R8 is a 1-bit register and register R3 can be also omitted. A1 is a 4-input adder and R9 is a 3-bit register, and the least significant bit of R9 is fed into the terminal T_{in} of the next PE and the rest of the bits of R9 is fed back to adder A1 as carry bits.

4 Conclusion

Since Array-A, Array-B, and Array-C is respectively constructed with 3 types of PE, 2 types of PE, and 1 type of PE, Array-C is simpler than Array-B, which is simpler than Array-A.

For modular reduction, Array-A uses ROM tables while Array-B and Array-C use multipliers, which can be constructed with only AND gates and can be implemented faster than ROMs. Thus the circuit scale and the processing time of Array-B and Array-C are less than those of Array-A. While the size of operands acceptable by Array-A is bounded by the available ROM size, that of Array-B and that of Array-C have no such severe restriction. Therefore Array-B and Array-C can flexibly cope with changing sizes of operands.

And since Array-B contains less number of registers than Array-C, the former is more efficient than the latter which is more efficient than Array-A.

In conclusion Array-B and Array-C are more useful than Array-A when we realize efficient and flexible implementations for modular exponentiation.

References

[1] K.Iwamura, T.Matsumoto and H.Imai, "High-speed implementation methods for RSA scheme," to appear in Advances in Cryptology — EUROCRYPT'92, Springer-Verlag (a primary version has appeared in EUROCRYPT'92 EXTENDED ABSTRACTS, pp.215-224).

[2] P.L.Montgomery, "Modular multiplication without trial division," Math.of Computation, Vol.44, pp.519-521, 1985.

[3] S.R.Dussé and B.S.Kaliski Jr., "A cryptographic library for Motorola DSP56000," Advances in Cryptology — EUROCRYPT'90, pp.230-244, Springer-Verlag.

[4] S.Even, "Systolic modular multiplication, " Advances in Cryptology — CRYPTO'90, pp.619-624, Springer-Verlag.

[5] B.Dixon and A.K.Lenstra, "Massively parallel elliptic curve factoring," EUROCRYPT'92 EXTENDED ABSTRACTS, pp.169-179.

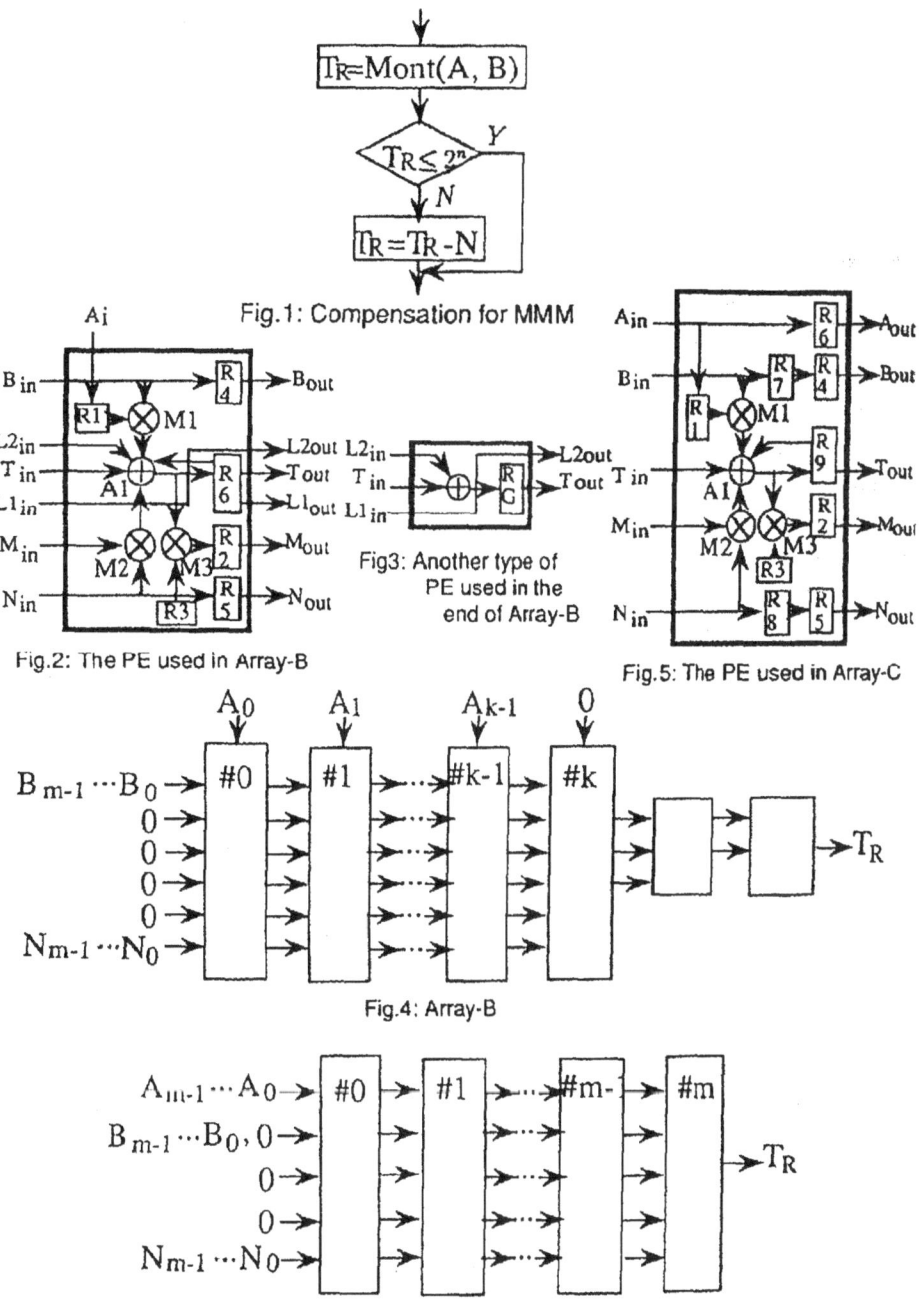

Fig.1: Compensation for MMM

Fig.2: The PE used in Array-B

Fig3: Another type of PE used in the end of Array-B

Fig.5: The PE used in Array-C

Fig.4: Array-B

Fig.6: Array-C

On the Development of a Fast
Elliptic Curve Cryptosystem[1]

G.B. Agnew, R.C. Mullin, S.A. Vanstone
University of Waterloo

Abstract

In this paper, we look at the development of a high speed elliptic curve cryptosystem based on a 40 Mhz. Motorola M68030 processor and a high speed optimal normal basis coprocessor for the ground field $GF(2^{155})$. The advantage of this system is the relatively small block size required for high security applications such as key management and digital signatures. In addition, the design is very compact and efficient and can be easily fit onto a standard Smart Card wafer (the coprocessor core requires less than 1 sq.mil. or < 4% of the area available on a Smart Card).

Introduction

Since their introduction by Diffie and Hellman in 1976 [1], researchers have been searching for practical implementations of public key cryptographic systems. The two most popular systems are based on the complexity of factoring the product of two large primes (RSA) or the difficulty of taking logarithms over a large field. To attain acceptable levels of security, both of these systems must use very large block sizes (the current trend is to use blocks of over 1000 bits for long term security). The complexity of performing calculations over these large block sizes generally makes the performance of software implementations unacceptable. Thus, hardware implementation in VLSI have been fabricated. At present, both public key systems have been realized with commercially available VLSI devices [2], [3], [4].

Even with the availability of high speed implementations of these systems, two issues still remain. The requirements that large block sizes be used leads to large storage requirements especially in such applications as financial transactions where digital signatures may be used for long term verification. In addition, the high levels of complexity of such devices makes transfer to such applications as smart cards difficult.

In 1985, Koblitz [5] and independently Miller [6], proposed the use of elliptic curves as the basis of a public key cryptographic system. It was believed at that time that no subexponential algorithm existed for solving the elliptic logarithm problem (the only known attack was Shank's Giant-Step-Baby-Step). More recently, Menezes, Vanstone and Okamoto [7], have discovered a new method of computing logarithms on the small, very special class of super-singular curves. Use of non-super singular curves does not have any such drawbacks.

---

[1] For a more complete treatment of this topic the reader is refered to reference [11].

R.A. Rueppel (Ed.): Advances in Cryptology - EUROCRYPT '92, LNCS 658, pp. 482-487, 1993.
© Springer-Verlag Berlin Heidelberg 1993

Non-Super Singular Curves Using Affine Co-ordinates

The use of elliptic curves in public key cryptographic systems is analogous to the use of discrete exponentiation substituting addition of points on the curve for multiplication in the field. The elliptic logarithm problem can thus be expressed as:

given a known starting point on the curve $P = (x_1, y_1)$ and a second point

$$Q = kP$$

find the integer k.

This is thought to be a hard problem even for relatively small curves.

To implement a system in hardware, we will only consider curves of characteristic 2. The curves we are interested in are of the form:

$$y^2 + xy = x^3 + ax + b$$

Operations on the curve are defined in the following way. For points $P = (x_1, y_1)$ and $Q = (x_2, y_2)$,

$$-P = (x_1, y_1 + x_1)$$

and, for $Q \neq -P$

$$x_3 = \begin{cases} \left(\dfrac{y_1+y_2}{x_1+x_2}\right)^2 + \left(\dfrac{x_1+y_2}{x_1+x_2}\right) + x_1 + a & P \neq Q \\[3ex] \dfrac{b}{x_1^2} + x_1^2 & P = Q \end{cases}$$

$$y_3 = \begin{cases} \left(\dfrac{y_1+y_2}{x_1+x_2}\right)(x_1+x_3) + x_3 + y_1 & P \neq Q \\[3ex] x_1^2 + \left(x_1 + \dfrac{y_1}{x_1}\right)x_3 + x_3 & P = Q \end{cases}$$

As mentioned above, using the elliptic curve as a cryptosystem involves choosing a secret integer k and forming $Q = kP$. In practice, this process would be done by using the binary expansion of k and forming the successive "squares" of the known point P, that is,

$$\underline{k} = k_0 + k_1 2^1 + k_2 2^2 + \ldots + k_{n-1} 2^{n-1}$$

or

$$Q = (\ldots(((Pk_0) + 2Pk_1) + 4Pk_2)\ldots + 2^{n-1}Pk_{n-1})$$

Using the affine method, we compute the x and y co-ordinates at each step in the computation. This requires the calculation of an inverse at each step in the squaring process; this will be the most time consuming part of the computation as each inverse generation requires several multi-plications in the underlying field.

Computations Using Projective Co-ordinates

Another method of computing points on the curve involves projective coordinates. The projec-tive equation for the non-super singular elliptic curve is written as:

$$z\, y^2 + z\, x\, y = x^3 + z\, x^2 a + b\, z^3$$

We set $Q = (x_1, y_1, z_1)$, $P = (x_2, y_2, 1)$ and define $P + Q = (x_3{}'', y_3{}'', z_3{}'')$.

For the case $P \neq Q$ (addition operation), let

$$x_3{}' = z_1^2 \{ z_1 (b + y_2^2 + x_2(y_2 + x_2^2) + x_2^2 a)$$

$$+ (x_2 y_1 + x_1(y_2 + x_2^2)) \} + z_1 (x_1^2 x_2)\ .$$

Then

$$x_3{}'' = (x_1 + x_2 z_1) x_3{}'$$

$$y_3{}'' = (x_1 + x_2 z_1)[(y_1 + y_2 z_1) + y_1(x_1 + x_2 z_1)]$$

$$+ [(y_1 + y_2 z_1) + (x_1 + x_2 z_1)]$$

$$z_3{}'' = z_1 (x_1 + x_2 z_1)^3$$

For the case $P = Q$ (doubling operation), let

$$x_3{}' = x_1^4 + z_1^4 b$$

Then

$$x_3{}'' = (x_1 z_1) x_3{}'$$

$$y_3{}'' = (x_1^2 + y_1 z_1) x_3{}' + (x_1^4 + x_3{}') r_1 z_1$$

$$z_3{}'' = (x_1 z_1)^3$$

The advantage of using this method is that only one inverse operation is required at the end of the calculation to divide out the z coordinate.

A Hardware Implementation

In our previous investigations, it was found that a very fast and efficient multiplier architecture could be developed for fields of characteristic two which had optimal normal basis representa-tions [8], [9]. This led to the development of a single chip public key processor ([3]) which performed computations in the field $GF(2^{593})$. We used this device in conjunction with a Motorola M56000 DSP to implement an elliptic curve cryptosystem. As reported in [10], this system achieved a throughput of about 5 Kbps..

Our experience with this system showed that the bottleneck in computation was the I/O section of the VLSI device. Thus, any processor designed specifically for an elliptic curve system would have to have a very fast I/O structure.

With the experience gained we set out to construct a simple, fast arithmetic processor which can be used to perform elliptic curve computations in conjunction with a control processor. Our objective was to build a device which could be fabricated in a custom gate array and yet be quite secure. Our choice was to implement an optimal normal basis multiplier for $GF(2^{155})$. The gate array device uses three registers to implement the multiplier structure and interconnection, a controller to implement the elementary operations (such as shifts, XOR's, multiplies, etc.) as well as incorporating a very fast 32 bit wide I/O structure. The device was fabricated using a 1.5 micron HCMOS gate array with a clock speed of 40 MHz and required less than 12,000 gates. For the primary operations, the speed can be calculated as:

OPERATION	SIZE	CLOCK CYCLES
Multiplication	155 bit blocks	156
Calculation of Inverse	24 multiplications	approx. 3800
I/O	5 - 32 bit transfers per read/write to registers @ 2 clock cycles per transfer	10
Addition (XOR) and elementary register operation	155 bit parallel operation	2

To realize the elliptic curve system, a module was constructed with a 40 Mhz. Motorola M68030 as the control processor. The main reasons for this choice was the 32 bit internal bus structure and the availability of fast I/O between the 68030 and the $GF(2^{155})$ coprocessor.

Throughput Calculations

If we consider point multiplication by an integer with Hamming weight 30, this will require about 154 point doublings and 29 point multiplications. Using projective coordinates for a non-super singular curve, doubling requires 6 multiplies and point addition requires 13 multiplies. At the end of the computation, a single inverse operation followed by 2 multiplies must be performed to return to affine coordinates. Allowing for I/O overhead in the doubling and multiplication routines, the device will be able to perform at least 145 integer multiplications per second. For use in an encryption system, each of X and Y coordinates can be used so 310 bits can be sent per point calculation for a throughput of approximately 50 Kbps..

In the above calculations, we note that a significant portion of the time is spent in doing the point doublings[2]. In elliptic curve systems, the same base point P can be used repeatedly. If this is the case, then all of the squares can be precomputed which will increase the throughput by a factor of 4 to approximately 200 Kbps.. The storage requirements for the point squarings is less than 6 Kbytes, a relatively small amount.

We have also investigated the area required to implement the elliptic curve coprocessor in a Smart Card wafer. Using new techniques, our estimates show that the registers necessary to implement the elliptic curve system will require less than 1 sq.mil. (or < 4% of the area available on the Smart Card). Design and construction of a Elliptic Curve Smart Card is currently underway and we believe that this will be the first full implementation of a fast, efficient and compact public key system on a card.

Summary

In this paper, we have described the deign and implementation of a fast processor for performing non-super singular elliptic curves over a base field of $GF(2^{155})$. This device is capable of realizing encryption speeds of up to 50 Kbps when clocked at 40 MHz. In addition to its high speed, it is exceptionally simple in implementation requiring less than 12,000 gates to fabricate. This device will have future applications in such areas as smart cards where compactness, high speed and high levels of security are desirable.

References

1. Diffie, W., M. Hellman, "New Directions in cryptography", IEEE Trans. on Info. Theory, vol. IT-22, pp.644-654, Nov. 1976.

2. Brickell, E., "A survey of hardware implementations of RSA", Proceedings of CRYPTO'89, Springer Verlag, pp. 368-370., Aug. 1989.

3. CA34C168 Data Encryption Processor Data Sheet, Newbridge Microsystems, Kanata, Ontario, Canada.

4. CY1024 Data Sheet, Cylink Corp., Sunnyvale, California, USA.

---

[2]As an aside, one of the main advantages of using normal basis representation for fields of characteristic 2 in discrete exponentiation is that squaring is effectively free, that is, a simple cyclic shift. Unfortunately, this is not the case in elliptic curve point doubling as many operations are required.

5. Koblitz, N., "Elliptic curve cryptosystems" Mathematics of computation - 48, 1987 pp.203-209.

6. Miller, V., "Use of elliptic curves in cryptography", Proceedings of CRYPTO'85, Springer Verlag, Aug. 1985, pp.417-426.

7. Menezes, A., S. Vanstone, T. Okamoto, "Reducing elliptic curve logarithms to logarithms in a finite field", STOC 1991, ACM Press, pp. 80-89, May 1991.

8. Agnew, G., T. Beth, R. Mullin, S. Vanstone, "Arithmetic operations in $GF(2^n)$, to appear, Journal of Cryptology.

9. Agnew, G., R. Mullin, S. Vanstone, "An implementation of a fast public key cryptosystem", to appear, Journal of Cryptology.

10. Agnew, G., R. Mullin, S. Vanstone, "A fast elliptic curve cryptosystem", Lecture Notes in Computer Science #434, Proceedings of Eurocrypt'89, Springer Verlag, Apr. 1989, pp. 706-708.

11. Agnew, G., R. Mullin, S. Vanstone, "An implementation of elliptic curve cryptosystems over $F_{2^{155}}$", submitted to IEEE Journal on Selected Areas in Communications.

A MONTGOMERY-SUITABLE FIAT-SHAMIR-LIKE AUTHENTICATION SCHEME

DAVID NACCACHE

PHILIPS TRT, SMART-CARDS & SYSTEMS, BP1 95,

5 AVE. REAUMUR, BP 21, F 92352, LE PLESSIS ROBINSON CEDEX, FRANCE.

EMAIL : NACCACH1@FRPRS0CS.SNADS.PHILIPS.NL

ABSTRACT

Montgomery's algorithm [2] is a process for computing $A B 2^{-|n|}$ modulo n in $O(\text{Log}(n))$ memory space.

Here we construct a Fiat-Shamir-like authentication scheme [1] suitable for Montgomery environnments without introducing any overhead in the number of modular multiplications requested for the execution of the normal protocol.

A very recent result [3] establishes (in a constructive way) that $A B 2^{-|n|} \mod n$ can be computed with the same complexity (timewise and hardwarewise) as A B (not mod n).

This theoretical reduction of the problem of modular multiplication, recently applied to the design of today's fastest hardware modular multiplier, is very important since it implies that the protocol presented hereafter can be executed in the same time as a Fiat-Shamir where all modular multiplications are replaced by standard multiplications.

The fact that no constants are to be precalculated beforehand and the small amount of RAM requested for software implementation of the new protocol makes it highly convenient for smart-card applications.

INTRODUCTION

All along this paper, $|n|$ denotes the length of n (in bits) and $\|n\|$ the Hamming weight of n.

Montgomery's algorithm for modular multiplication [2] is a process for computing $A B 2^{-|n|}$ modulo n in $O(\text{Log}(n))$ memory space.

The method only assumes that n is odd. No other restrictions are imposed on the modulus.

R.A. Rueppel (Ed.): Advances in Cryptology - EUROCRYPT '92, LNCS 658, pp. 488-491, 1993.

Very much simplified, this algorithm works as follows :

Let X[i] denote X's i^{th} bit (with X[0] as LSB) and $K = 4^{|n|}$ mod n.

Kernel Algorithm(A, B)
{
c = 0

For i = 0 to |n|-1

| If A[i] == 1 then c = c + B
| If c[i] == 1 then c = c + n
| c = c/2

If c ≥ n then c = c - n
return(c)
}

It can be shown [1] that $c = A B 2^{-|n|}$ mod n and that :

$$Kernel(K, Kernel(A, B)) = A B \text{ mod } n$$

A more comprehensive and complete approach is presented by Arazi in [3].

THE NEW FIAT-SHAMIR-LIKE PROTOCOL.

[3] formally shows how to reduce the complexity of the computation Kernel(A, B) into that of A B.

From this remark and the last section it clearly appears that if it is possible to transform the Fiat-Shamir scheme in such a way that the parasites ($2^{-|n|}$) will not disturb the protocol (using only one Kernel operation for each modular multiplication), then it would be possible to perform the Fiat-Shamir in about the half of the time requested with a full Montgomery multiplier.

Moreover, we do not require the precalculation or usage of the constant K.

This new approach to the problem is somewhat new since by opposition to the classical process of designing computationnal tools for comfortable execution of number-theoretic protocols, here we transform a cryptographic scheme in order to meet a given computationnal limitation.

Actually, after a proper modification of the relation between the public and the secret keys, we will see that not only the Montgomery parasites don't disturb, they even help ! (As said by Abraham Lincoln : "The best way to destroy your enemies is to transform them into friends").

From now on we will denote by D the parasite factor $2^{-|n|}$ mod n and assume the availibility of a half Montgomery multiplier (that is a Kernel procedure performing only the operation A B D mod n).

THE NEW PROTOCOL

Redefine the Fiat-Shamir [1] public v_j-s by : $D^3 v_j s_j^2 = 1$ mod n

And assume that the v_j's are already known to the verifier (for instance by Fiat and Shamir's F(ID,j) method).

Step 1

The prover computes $Z = R^2 D$ mod n = Kernel(R, R) and sends it to the verifier.

Step 2

The verifier sends the random binary vector e.

Step 3

The prover computes and sends $y = r \prod_{e_i=1} s_i D^{|e|}$ mod n.

This value is easily computed by :

$y = \text{Kernel}(s_{i_1}, \text{Kernel}(s_{i_2}, \ldots \text{Kernal}(s_{i_{|e|}}, r))\ldots)$

Here the i_j-s denote the $|e|$ indices selected by vector e.

Step 4

The verifier computes (similar way to that of the previous step) :

$A = \text{Kernel}(v_{i_1}, \text{Kernel}(v_{i_2}, \ldots \text{Kernal}(v_{i_{|e|}}, \text{Kernel}(y, y)))\ldots) =$

$y^2 D \prod_{e_i=1} v_i D^{|e|-1} D$ mod n $= r^2 \prod_{e_i=1} v_i s_i^2 D^{2|e|} D^{|e|} D$ mod n $=$

$$r^2 \prod_{e_i=1} v_j \ s_j^2 \ D^3 |e| \ D \bmod n = r^2 \ D \bmod n.$$ And tests if $A \overset{?}{=} Z$.

Similarly, this can be applied to the digital signature as well.

CONCLUSION AND IMPLEMENTATION DETAILS.

With a 68HC05 running at 3.5 MHz the Kernel operation (for 512 bit numbers) was implemented in less than 135 ms, RAM usage is less than 70 bytes. A special Kernel-Squaring version runs at 85 ms but requires a double RAM space.

In [3] it is shown that "It is possible to compute ABD mod n in |n|+1 clock cycles. That is, a modular multiplication is performed with the same complexity (timewise and hardwarewise) as that of a standard multiplication operation", it is thus possible to execute a Fiat-Shamir identity check (and signature, with similar modifications of the scheme) in hardware and time equivalent to that required for the execution of the protocol without modular reductions.

The same strategy of modifying the relationship between public and secret keys in order to meet or cancel the effect of parasite constants introduced by modular reduction tools can be applied to a big variety of number-theoretic authentication and signature protocols.

REFERENCES

[1] Fiat Feige and Shamir, "Zero-Knowledge Proofs of Identity", J. Cryptology, vol 1, pp. 77-94.

[2] Montgomery, "Modular Multiplication without Trial Division", Mathematics of Computation, vol 44, pp 519-521.

[3] Arazi, "Modular Multiplication is Equivalent in Complexity to a Standard Multiplication", Fortress U&T Internal Report (1992) available from Fortress U&T Information Safeguards, P.O. Box 1350, Beer-Sheva, IL-84110, Israel.

Author Index